Biomedical Fluid Dynamics

Flow and Form

Biomedical Fluid Dynamics

Flow and Form

Troy Shinbrot

OXFORD
UNIVERSITY PRESS

Great Clarendon Street, Oxford, OX2 6DP,
United Kingdom

Oxford University Press is a department of the University of Oxford.
It furthers the University's objective of excellence in research, scholarship,
and education by publishing worldwide. Oxford is a registered trade mark of
Oxford University Press in the UK and in certain other countries

First Edition published in 2019

Impression:1

Published in the United States of America by Oxford University Press
198 Madison Avenue, New York, NY 10016, United States of America

British Library Cataloguing in Publication Data
Data available

Library of Congress Control Number: 2018951252

ISBN 978–0–19–881258–6 (hbk)
ISBN 978–0–19–881259–3 (pbk.)

DOI: 10.1093/oso/9780198812586.001.0001

Printed in Great Britain by
Bell & Bain Ltd., Glasgow

Alfred North Whitehead asserted that there are two kinds of people: the muddle headed like himself, and the simple minded like his friend Bertrand Russell. This book is dedicated to the simple minded: those of us who believe in kindness, beauty, and truth over the alternatives.

Contents

Introduction

This book presents material from a one semester graduate course focusing on fluid flow for biomedical engineers. Speaking as a generality, the backgrounds of biomedical engineering students are bimodal: either students have a biosciences background, and are strong in chemistry, biology, and laboratory skills, or they have an engineering background, and are strong in analytic and numerical methods. Several books have attempted to bridge this gap; to my mind perhaps the best is E.N. Lightfoot's *Transport Phenomena and Living Systems*, still an excellent book 40 years on.

In the present book, I have attempted to draw from Lightfoot and other sources to teach both analytic skills needed by biologists and biological examples of importance to engineers. In doing so, it is likely that the book may also be of interest to advanced undergraduates in other fields who may want a grounding in fluid dynamics and pattern formation. I have begun with the fundamentals, and advanced readers may want to skim sections and pedagogical definitions (e.g. for $\vec{\nabla}$) that they are already familiar with. The organization follows an approach of sequential embellishment, starting with simple examples and building complications as needs arise, focusing largely on the paradigmatic problem of blood flow and culminating in ideas dealing with biological patterns and structures. Thus I initially describe equations of motion for simple fluids in a rigid pipe. Next, I note that the vasculature consists of flexible tubes, so I derive equations applicable to flexible membranes and describe how these can be coupled to flow equations for simple fluids. I then consider pulsatile flow, then entrance effects and flow through tubes with bends, and then turn to discussions of complex fluid behaviors. The hope is that this systematic approach will provide tools for understanding a broad range of problems.

Two caveats. First, the topic of fluid transport is full of surprising, amusing, and stimulating effects that I mention in each chapter. My hope is that the reader will experiment on her own with some of these effects. In so doing, she may well uncover new science, for many parts of this field are still only marginally understood. Students who do not enjoy such distractions will surely dislike this book, and I recommend that they search out other books that will be more to their liking: there are plenty to be found.

Second, I have included brief historical vignettes throughout this book. The field of fluid dynamics is rich with history that is delightful, bittersweet, and often totally unexpected—like the report that René Descartes refused to accept William Harvey's evidence that the heart mechanically pumped blood, preferring to believe that the heart heated the blood, which caused it to move by expanding. Or the eminent fluid dynamicist who engaged in an acrimonious public dispute over details of very detailed observations—although it was widely known that he was clinically blind. Or the brilliant scientist who won a prestigious prize for a crucial proof—which turned out to be wrong, and so the original proof and all its copies were destroyed. Or the revolutionary mathematician who arguably wrote the single theorem most influential to all of modern physics, but because she was a woman she was not allowed to attend university, and for many years taught low-level courses to male students without pay.

I hope the reader will forgive me these indulgences. We in the twenty-first century can so easily turn to the computer to provide a solution (that we trust, sometimes to our peril, to be right) that we forget the extraordinary sparks of insight that scientists in centuries past had to rely on to provide us with the foundational understanding and analytic tools that we now depend on. This book is dedicated to those scientists.

1 Fundamental Equations

1.1 A Motivational Allegory

On November 9, 1975, the SS Edmund Fitzgerald, called the "pride of the American fleet," left Superior, Wisconsin bound for the Sault St. Marie docks after nearly 20 years of service. She was carrying 26,000 tons of iron ore, was 222 meters long, and had a crew of 29, including a captain with 44 years' experience, Ernest McSorley. The drama of this ship's final voyage has been captured in poetry and song.[1]

By 7 pm on November 9, the Fitzgerald and her companion ship, the SS Arthur M. Anderson, reported gale winds (65–75 km/h) from the northeast. By 2 am on November 10, the winds shifted from the northwest and grew to 90–100 km/hr. At 3:30 pm, Captain McSorley reported that the Fitzgerald was taking on water and had lost two hatch covers. At 7 pm, after a full day and night without sleep or food, Captain McSorley reported that the ship was holding its own, despite some list and damage. At 7:10 pm, the lights of the Fitzgerald vanished, and the ship disappeared from radar, later found to have plunged over 500 feet to the bottom of Lake Superior. There were no survivors.

Why did this ship sink, while its companion ship, 5 years older, survived? Answering this question hinges on an understanding of the complex dynamics of fluid flows, and perhaps surprisingly has important lessons for the understanding of biological fluids. It is not at all obvious that there is any connection between the sinking of a 200 meter long steam ship carrying iron ore and flow of blood in an artery. A central theme of this book—and a crowning achievement of the last two centuries of fluids research—is to show that apparently different systems, sharing little in scale, scope, or appearance, in fact share the same features, can be described by the same equations, and exhibit the same phenomenology.

To understand why, let us focus on two factors important to the SS Fitzgerald, "fetch" and "yield stress." Fetch is a nautical term that refers to the distance over which winds blow over water without obstruction, and yield stress refers to the tendency of some materials to support forces up to a critical load, after which they will "yield," or slip. The relevance of fetch is illustrated in Fig. 1.1(a), where we indicate the wreck of the SS Fitzgerald: a location with a fetch of over 250 km from the upwind northwest shore of Lake Superior. Because of this long fetch in the presence of high winds, as depicted in Fig. 1.1(b) a small wave rising up out of the water will catch the wind like a sail, drawing more water up and steepening as it grows. This leads also to the phenomenon of "rogue waves,"[2] that can rise tens of meters from trough to peak.

Biomedical Fluid Dynamics: Flow and Form. Troy Shinbrot.

DOI: 10.1093/oso/9780198812586.001.0001

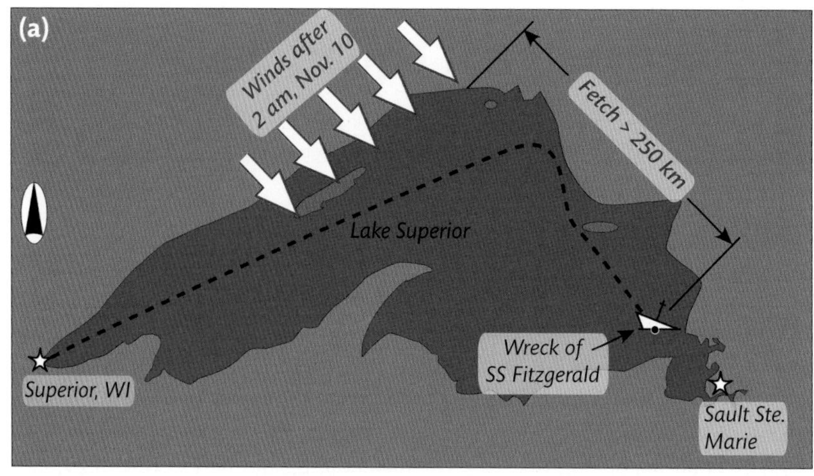

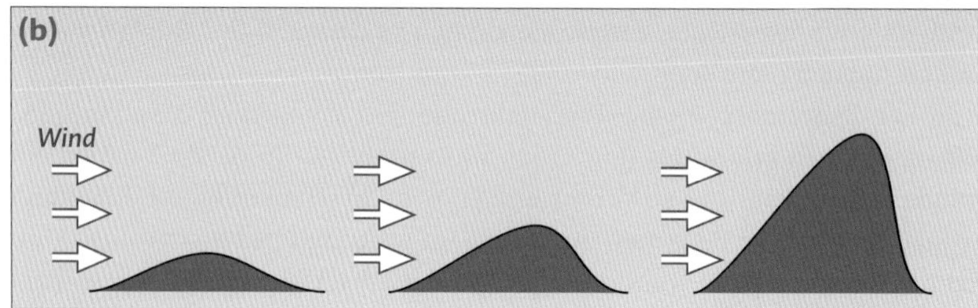

Fig. 1.1 Effects of wind and "fetch" (the distance of open water that the wind acts upon) on wave growth. (a) Lake Superior, showing course of the SS Fitzgerald (dashed line). As indicated, the fetch was over 250 km where the Fitzgerald sank, leading to very large waves (waves over 11 meters were reported). (b) Depiction of wind blowing on a small wave causing it to draw up more water and steepen as it travels.

So the first thing working against the SS Fitzgerald was that after 2 am, November 10, the Fitzgerald had passed the apex of Lake Superior, and the fetch grew with every kilometer that it traveled—and so the Fitzgerald was threatened by ever-larger waves as time wore on. At this stage, we already know enough to analyze when this problem of waves growing with fetch will take place. The cartoon of the mechanism shown in Fig. 1.1(b) implies that waves will only grow so long as the force due to the wind that lifts water from the lake exceeds the force due to gravity that draws water back down again. So we expect that waves will grow only so long as:

$$\frac{F_{wind}}{F_{gravity}} > 1. \qquad [1.1]$$

We can argue about how to calculate the two forces, how the forces change as the size of the wave grows, and so forth, but these are details: the point that we emphasize here is that by balancing the two influences known to drive the phenomenon, we can determine when it will grow (when the ratio in Eq. [1.1] > 1) and when it will shrink (when the ratio < 1). Notice that we can make this determination without knowing very much at all about important details—oceanography, weather, ship design, etc. We just need to correctly identify the dominant effects involved. We will see that this is a common theme: by appropriately defining a ratio between competing effects, we can identify qualitatively distinct behaviors that the effects will produce.

The second effect that worked against the Fitzgerald was more subtle: after all, the Anderson—a smaller and older ship—made it safely to port. The Fitzgerald differed from the Anderson in that its hatch covers—which sealed the cargo from water—had been found to be leaking prior to the trip, and two of these were reported lost before the ship sank. As a result, the Fitzgerald steadily took on water during its fateful journey, and it is believed that the iron ore became suspended, or "fluidized," in the accumulated water. This caused it to suddenly shift to the bow and drive the Fitzgerald nose-first to the ocean floor. There was no time for a warning to the crew, an SOS, or release of lifeboats before the bulkheads broke under the weight of 13 million pounds of ore, causing the Fitzgerald to split apart and sink to the bottom of Lake Superior.

As with the case of fetch, this second effect is amenable to simple attack by analyzing the central influences at work. Here, the bed of ore suspended in water can be expected to shift when particles of ore take longer to settle than the characteristic time of agitation of the bed—that is, of rocking of the Fitzgerald due to the ocean's waves. This occurs when:

$$\frac{T_{settling}}{T_{agitation}} > 1. \quad [1.2]$$

We will learn in this book how the settling time can be evaluated, but as before, whatever calculus is used to compute this time, the same balance between settling and agitation times governs behaviors at any scale. When the ratio in Eq. [1.2] is small, particles will rapidly settle so as to rest on one another – at which point they can collectively support loads like a single solid body. On the other hand, when Eq. [1.2] is large, the particles will always be separated by the liquid, and so will flow past one another with the fluid. At smaller scales this transition between solid-like and liquid-like behavior is critically important for the mixing of pharmaceuticals, the formation of blood clots, and processing in so-called "fluidized beds" that are used in the manufacture of everything from polymers to antibiotics.

The use of dimensionless ratios such as Eqs. [1.1] and [1.2] is a deceptively important and broad-reaching tool, as it permits us to deduce effects independent on scale or details of application. A commonplace example can be found in children's antibiotics such as amoxicillin: as every parent knows, this is dispensed for young children in a thick, sweet-flavored pink suspension. What parents may not know is that suspensions such as this are designed so that when vigorously shaken, the contents will mix in the bottle, but when shaking is stopped, the liquid will gel to prevent the active drug from separating out of the suspension. The ability to perform these useful tricks depends on an understanding of relative forces and timescales as in Eqs. [1.1] and [1.2].

Example Effects of fluidization

Other examples of fluidization are shown in Fig. 1.2, including three very different behaviors. In Fig. 1.2(a), we show that hollow spheres (Q-CEL®) flow down an inclined surface and settle more slowly in air than the mean speed of flow. So $T_{settling}/T_{agitation} > 1$,

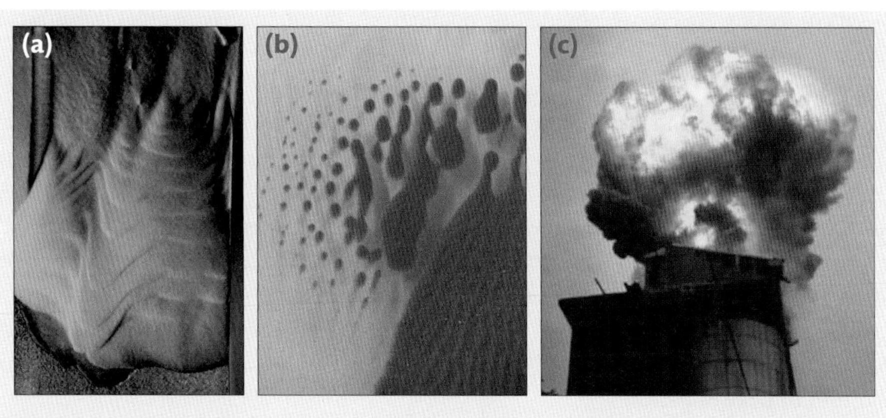

Fig. 1.2 Effects of fluidization of particles. (a) Hollow glass spheres flow down a slope: because they are hollow, they settle slowly in air and so they remain fluid-like for long periods. (b) Droplets formed by glass particles flowing down an acrylic surface: here the droplets are produced by fine particles flowing on a vibrated surface. (c) Small dust explosion in Norway. (Photo of a vented wheat grain dust explosion in a 500 m^3 silo in Norway. Courtesy of Professor Rolf K. Eckhoff, University of Bergen, Norway.) Common powders such as flour can contain more explosive energy than TNT: all that is needed to release this energy is a spark combined with enough air surrounding each powder particle to permit combustion. As a result, dozens of workers are killed every year by explosions in dusty mines, factories, and storage buildings.

and consequently these particles behave in a liquid-like manner, including flowing freely, sloshing and forming multiple wavefronts as shown. In Fig. 1.2(b), we show a different example, in which colored glass particles flow down an inclined acrylic plate: when the plate is vibrated more rapidly than particles can settle onto the plate, a partial vacuum is created under the particle bed, drawing air in along its sides. This produces an effective surface tension that leads to the formation of the droplets shown, provided again that $T_{settling}/T_{agitation} > 1$. Finally, in Fig. 1.2(c), we show a comparatively small industrial dust explosion: these events occur regularly during processing of essentially any powdered material, leading to explosions that can be felt many tens of kilometers away—once more provided that the combustible powder is surrounded by oxidizing air in a fluidized state, that is, $T_{settling}/T_{agitation} > 1$.

With the goal in mind of learning how to analyze such problems, we turn to defining equations that fluids (either liquids or gases) obey. Semantically, the Engineering literature terms these equations "balances" of mass or of momentum, while the mathematics and physics literature describes the equations as following from "conservation" laws. Semantic differences such as this invariably lead to confusion—as for instance when an engineer refers to "balancing momentum," which a physicist will find inexplicable, or when a mathematician refers easily to a derivative as "nabla," which is anachronistically obtained from the shape of a Hebrew harp as described in Greek (!) and is meaningless to an engineer. In the derivations following, we try to remain agnostic with respect to this semantic conflict, but point conflicts out as we go along so that the reader will become acquainted with semantics on both sides of the divide.

1.2 **Continuity Equation**

The first thing that we can say about the behavior of a fluid is that if we put ten grams of it into a box and take eight grams out, we will have added two grams to the box. To make this concrete, let's consider a small box of sides Δx, Δy, and Δz, as shown in Fig. 1.3. We express this "continuity equation," as mathematicians would term it, or "mass balance" as engineers would say, in heuristic terms as:

$$\left\{\begin{array}{l} rate\ of \\ increase \\ in\ mass \end{array}\right\} = \left\{\begin{array}{l} rate\ of \\ mass\ in \end{array}\right\} - \left\{\begin{array}{l} rate\ of \\ mass\ out \end{array}\right\} \qquad [1.3]$$

Mathematically, we write each of these terms as:

$$\begin{aligned} \left\{volume\cdot\frac{\partial\rho}{\partial t}\right\} = & \ \{area_x\cdot\rho V_x|_x\} - \{area_x\cdot\rho V_x|_{x+\Delta x}\} \\ & + \{area_y\cdot\rho V_y|_y\} - \{area_y\cdot\rho V_y|_{y+\Delta y}\} \\ & + \{area_z\cdot\rho V_z|_z\} - \{area_z\cdot\rho V_z|_{z+\Delta z}\} \end{aligned} \qquad [1.4]$$

where the density of the fluid is ρ, the volume is $\Delta x\Delta y\Delta z$, we conventionally denote the velocity in the x-direction, V_x, evaluated at the point $x + \Delta x$ as $V_x|_{x+\Delta x}$, and $area_x = \Delta y\Delta z$, etc. Dividing Eq. [1.4] by the volume produces

$$\left\{\frac{\partial\rho}{\partial t}\right\} = \rho\left\{\frac{V_x|_x - V_x|_{x+\Delta x}}{\Delta x}\right\} + \rho\left\{\frac{V_y|_y - V_y|_{y+\Delta y}}{\Delta y}\right\} + \rho\left\{\frac{V_z|_z - V_z|_{z+\Delta z}}{\Delta z}\right\}, \qquad [1.5]$$

or as the volume becomes infinitesimal, $\Delta x\Delta y\Delta z \to 0$, we obtain the "continuity equation:"

$$\begin{aligned} \frac{\partial\rho}{\partial t} &= -\rho\left\{\frac{\partial V_x}{\partial x} + \frac{\partial V_y}{\partial y} + \frac{\partial V_z}{\partial z}\right\} \\ &= -\rho\,\vec{\nabla}\cdot\vec{V}. \end{aligned} \qquad [1.6]$$

We make two brief remarks to avoid possible confusion. First the minus sign appears simply because derivatives are conventionally defined using the rightmost minus the leftmost position

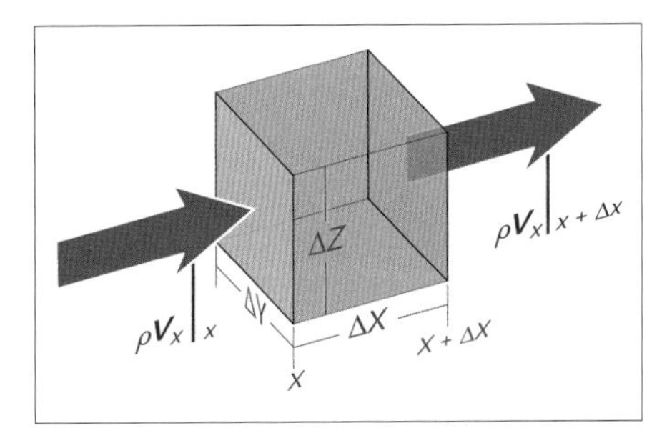

Fig. 1.3 Flow of material into and out of a small unit volume, $\Delta x\Delta y\Delta z$, indicated in blue arrows in the x-direction. Similar flows, not shown, are typically present in the y- and z-directions as well.

$V_x|_{x+\Delta x} - V_x|_x$, while Eq. [1.5] uses opposite signs. Second, the final term uses the "nabla," $\vec{\nabla} \equiv \hat{x}\frac{\partial}{\partial x} + \hat{y}\frac{\partial}{\partial y} + \hat{z}\frac{\partial}{\partial z}$, notation referred to earlier, so that the "divergence" of a velocity, $\vec{\nabla} \cdot \vec{V}$, can be seen to explicitly define the rate at which material is expelled from (if $\vec{\nabla} \cdot \vec{V}$ is positive) or absorbed into (if $\vec{\nabla} \cdot \vec{V}$ is negative) a tiny volume. For most problems of biomedical interest, the fluid is incompressible, so that the density is constant and we write:

$$\vec{\nabla} \cdot \vec{V} = 0 \qquad [1.7]$$

Example Continuity and counter-rotating flow

The continuity equation is in one sense a very basic and straightforward constraint on fluid flow, yet on the other it implies important behaviors at a number of levels, many of which will be discussed in later chapters of this book. To give a flavor for these behaviors, consider flow in a hurricane, such as Irene in 2011, shown in Fig. 1.4. It is well known that Coriolis forces cause hurricanes in the northern hemisphere to rotate counterclockwise. Yet any satellite video of an active hurricane in the northern hemisphere will invariably show both counterclockwise rotation near the eye of the hurricane, and *clockwise* rotation near its outer edges. These are termed "cyclonic" and "anticyclonic" flows respectively, and without an understanding of the continuity equation, their coexistence could seem to be counterintuitive.

Fig. 1.4 Hurricane Irene. Satellite videos show that hurricanes in the northern hemisphere appear to rotate counterclockwise near the eye, but rotate clockwise further out. This can be explained by analyzing the effect of continuity on Coriolis forces as described in the text (Source: NOAA).

In view of the continuity equation, however, the matter is easily explained. The counterclockwise Coriolis force is caused by the Earth's rotation combined with *inward* flow of fluid toward the low pressure region near the hurricane's eye. But continuity implies that the inward flow has to go somewhere, and indeed it does: the inward flow occurs at low elevations, and produces upwelling in the eye and *outflow* at higher elevations. This outflow—again combined with Earth's rotation—produces a Coriolis force in the opposite direction to that caused by inflow. This is what inevitably results in opposing rotational flows: counterclockwise near the eye where inflow dominates, and clockwise further out where outflow dominates. We will see other important applications of the continuity equation (see, for example, effects of stability of bronchial branching discussed in Chapter 5); hopefully this example illustrates that the equation's implications are not as trivial as they may seem at first glance.

1.3 **Navier-Stokes Equation**

Now that we know how to calculate what goes on in small volumes, we can calculate how a small volume will move in response to forces at work on it—for example, for fluid traveling through a pipe in response to an applied pressure, or for fluid flow around a particle falling in gravity.

Responses to forces are governed by Newton's equations of motion, and it is worthwhile to recapitulate what these are so that we are clear about precisely what mechanism causes a volume of fluid to move in response to a prescribed force. As is well known, Newton defined three laws of motion, which can be expressed as follows.

(1) An object in motion will continue in motion, and an object at rest will stay at rest.

(2) The rate of change in momentum of an object is given by the sum of forces on it.

(3) Every action has an equal and opposite reaction.

Before we ask how these laws apply to fluid motion, let's take a moment to appreciate how remarkable the laws are—and how remarkable Newton's insight was—for *none of them is actually observed* in day to day life. When we roll a ball along the ground, we never observe it to continue rolling as Law 1 requires. When we push on most bodies—a couch, a table, a car—they don't move at all to start with, and then begin to move abruptly and only after an effort that is often disproportionate to the small change in force exerted. Both of which are in apparent contradiction to Law 2. And when we run forward, we never see that the earth moves backward—which Law 3 demands.

Thus it took profound insight at the time for Newton to unveil the reality that these are the essential laws by which the universe operates, and to recognize that the observations that we make seem to differ because other, in a sense less essential, effects overcome the essential laws under practical circumstances. Thus friction, air drag, and the mass of the Earth (effects also studied by Newton) dominate in many practical experiments. Newton's great contribution was to provide a simple and comprehensible mechanism by which bodies move, from which one can obtain practically useful results *despite the fact* that these simple mechanisms fail, by themselves, to describe what is seen.

This is important, because throughout this book we too will make use of simple models that themselves do not completely describe problems of interest. For instance, we will shortly study steady flow of a simple fluid in a smooth rigid pipe as a first model for blood flow. Yet realistic blood flow is not steady, it is pulsatile; blood is not simple, it is a very complicated fluid, and veins and arteries are neither smooth nor rigid. Nevertheless, we will see that through this approach we can obtain results from simple models that we will later embellish to include effects of pulsatility, elasticity of vessel walls, and so on. This situation has been well described by the statistician, GEP Box, who wrote that "essentially, all models are wrong, but some are useful." In this spirit, we begin by making use of the simplest model for fluid response to applied forces, recognizing its limitations at the outset.

We begin with Newton's second law of motion, which is most relevant for fluid motion. We write the law as the rate of change of momentum, $m\vec{V}$, in terms of the sum of all of the forces, $\vec{F}$, acting on a small volume of fluid with mass m:

$$\frac{D(m\vec{V})}{Dt} = \sum \vec{F}. \qquad [1.8]$$

We will have more to say about the forces in a moment, but first let's focus on the peculiar way that we wrote the time derivative: D/Dt. This intentionally differs from the partial derivative, $\partial/\partial t$, or the total derivative, d/dt, and the difference is crucial to understanding how fluids behave and why. This new derivative is termed the "substantive," or "material," derivative, and is meant to define how forces on a volume of fluid behave *both* as a result of changes over time *and* as a result of motion of the volume.

To understand why we have invented this new derivative, consider a simpler problem: suppose that you want to sample the concentration of fish in a stream. There are three ways of doing this. First, you could stand on a bridge overlooking the stream and ask how many fish you can count per unit time. From this, you could plot the concentration, C, versus time, and obtain the rate of change of C at a fixed point in space (on the bridge). This would be a time derivative with spatial coordinates held constant, that is, the partial derivative, $\partial C/\partial t|_{bridge}$.

Second, you could take note of the fact that the fish are moving with the stream, and that the previous measurements were taken at a fixed location with respect to land, but at different locations with respect to the water. That is, the measurements taken initially were obtained for fish in the water that, by the end of the experiment, would by that time be far downstream of the bridge. So if fish concentrations changed at different locations within the stream, $\partial C/\partial t|_{bridge}$ might not be what you want to measure. So you could alternatively measure concentration by counting fish from the prow of a motorboat traveling through the stream with velocity, $\vec{V}_{motorboat}$. In this case, we take explicit recognition of the fact that the concentration, C, depends on x,y,z as well as t, that is, $C = C(x,y,z,t)$, so that by the chain rule:

$$\left.\frac{dC}{dt}\right|_{motorboat} = \left.\frac{\partial C}{\partial t}\right|_{motorboat} + \left.\frac{dx}{dt}\frac{\partial C}{\partial x}\right|_{motorboat} + \left.\frac{dy}{dt}\frac{\partial C}{\partial y}\right|_{motorboat} + \left.\frac{dz}{dt}\frac{\partial C}{\partial z}\right|_{motorboat}. \qquad [1.9]$$

Since by definition, $\vec{V}_{motorboat} = \left(\hat{x}\frac{dx}{dt} + \hat{y}\frac{dy}{dt} + \hat{z}\frac{dz}{dt}\right)$, Eq. [1.9] is the same as:

$$\left.\frac{dC}{dt}\right|_{motorboat} = \left.\frac{\partial C}{\partial t}\right|_{motorboat} + \left.\vec{V}\cdot\vec{\nabla}C\right|_{motorboat,} \quad [1.10]$$

where the nabla shorthand raises its head again in the definition of the "gradient" of C: $\vec{\nabla}C \equiv \hat{x}\frac{dC}{dx} + \hat{y}\frac{dC}{dy} + \hat{z}\frac{dC}{dz}$, which is dotted into $\vec{V}$ to produce Eq. [1.10].

So, all is well and good: now we know what happens to a derivative for a prescribed velocity, $V_{motorboat}$. But what if instead we want to know how a derivative behaves if the velocity isn't predetermined, from a motorboat, but is that of a volume of fluid itself? In that case, to track a concentration, we simply obtain the behavior that would be seen if the motorboat drifted at exactly the speed of the fluid itself—a third way of determining the rate of change of concentration:

$$\left.\frac{DC}{Dt}\right|_{fluid} = \left.\frac{\partial C}{\partial t}\right|_{fluid} + \left.\vec{V}\cdot\vec{\nabla}\,C\right|_{fluid.} \quad [1.11]$$

Similarly, to track a vector—like the velocity, $\vec{V}$—we find

$$\frac{D\vec{V}}{Dt} = \frac{\partial\vec{V}}{\partial t} + \vec{V}\cdot\vec{\nabla}\vec{V}, \quad [1.12]$$

where the specification $|_{fluid}$ is understood, and where we write the derivative using capitals to emphasize that it is obtained as a volume of fluid moves.

Physically, the meaning of this substantive derivative is that it provides information about how a volume behaves *both* as time passes *and* as the volume moves from one location to another. Thus influences on the volume might change because time passes, or it might change because the volume has moved to a different location, where for example different forces may be at play. The final term in Eq. [1.12] can be expanded, and in Cartesian coordinates (x,y,z) it becomes the following:

$$\begin{aligned}\vec{V}\cdot\vec{\nabla}\vec{V} = &\left(V_x\frac{\partial V_x}{\partial x} + V_y\frac{\partial V_x}{\partial y} + V_z\frac{\partial V_x}{\partial z}\right)\hat{x}\\ &+\left(V_x\frac{\partial V_y}{\partial x} + V_y\frac{\partial V_y}{\partial y} + V_z\frac{\partial V_y}{\partial z}\right)\hat{y}\,.\\ &+\left(V_x\frac{\partial V_z}{\partial x} + V_y\frac{\partial V_z}{\partial y} + V_z\frac{\partial V_z}{\partial z}\right)\hat{z}\end{aligned} \quad [1.13]$$

In other coordinate systems, $\vec{V}\cdot\vec{\nabla}\vec{V}$ can become rather baroque. We don't present the alternative forms of this expression in this book; if needed these can be looked up.

Example Chaotic flow

A useful terminology that underscores the distinction between a fixed frame of reference and a frame that moves with a fluid involves what are termed "Eulerian" and "Lagrangian" descriptions of fluid flow. The Eulerian description concerns measurements made at a fixed time and place—as in counting fish on a bridge. By comparison, the Lagrangian description involves following trajectories of points—as in tracking the trajectories of the fish in space as a function of time.

This distinction, and its relevance to fluid flow, can be made clear by considering a "sine-flow," which oscillates between two perpendicular flows, as shown in Fig. 1.5. At any location and instant in time, the velocity is completely defined for integer n and amplitude, A:

$$\begin{aligned} &when\ 2n < t < 2n+1, \begin{cases} V_x = A \cdot sin(2\pi y) \\ V_y = 0 \end{cases} \\ &when\ 2n+1 < t < 2n, \begin{cases} V_x = 0 \\ V_y = A \cdot sin(2\pi x) \end{cases} \end{aligned} \qquad [1.14]$$

What this velocity does is to alternate between moving any volume in the x-direction along a sinusoidal profile as shown in Fig. 1.5(a) and then moving the volume in the y-direction, as shown in Fig. 1.5(b). The Eulerian velocity here is perfectly prescribed: at any location and instant in time, the motion of every infinitesimal volume is exactly known. On the other hand, the Lagrangian trajectories produced by this Eulerian velocity field are much more complex, and can in fact rapidly become unpredictable. This is shown in Figs. 1.5(c) and (d).

Fig. 1.5(c) shows that for small amplitude, A, in Eq. [1.14], most of the flow is "regular," meaning that trajectories of marker particles travel in prescribed closed orbits, but a small region of trajectories is "chaotic," meaning that markers—all of which begin on a well-defined circle—rapidly become scattered throughout the region. So two points arbitrarily close together separate rapidly in time—exponentially rapidly, it can be shown.

As the amplitude grows, the chaotic regions grow as well, and the regular regions shrink. This continues until the entire domain is chaotic. Remarkably though, at still larger amplitudes regular regions re-emerge: this behavior is what we term "generic," meaning that it occurs in almost all systems. We will return to this point later when we discuss aspects of mixing, for as we will see, systems like Fig. 1.5(d) scatter initially localized points more rapidly over much of the available volume (the chaotic region) than systems like Fig. 1.5(c). As a consequence, chaotic systems are well suited for problems involving reaction or bacterial migration that depend on good mixing—for example, between two chemicals or between food and bacteria.

This example is straightforward to program, which we leave as an exercise to the reader (below), and the example provides us with two important lessons. First, there is a decided difference between Eulerian descriptions (Figs. 1.5(a) and (b)), in which trajectories are perfectly prescribed in time and space, and the Lagrangian description (Figs. 1.5(c) and (d)), in which the resulting trajectories can be very complex and are not easily deduced from the underlying Eulerian velocities. The Lagrangian description is relevant to understanding transport and mixing, and it is for this reason that the substantive derivative—that tracks moving fluid volumes—is of importance.

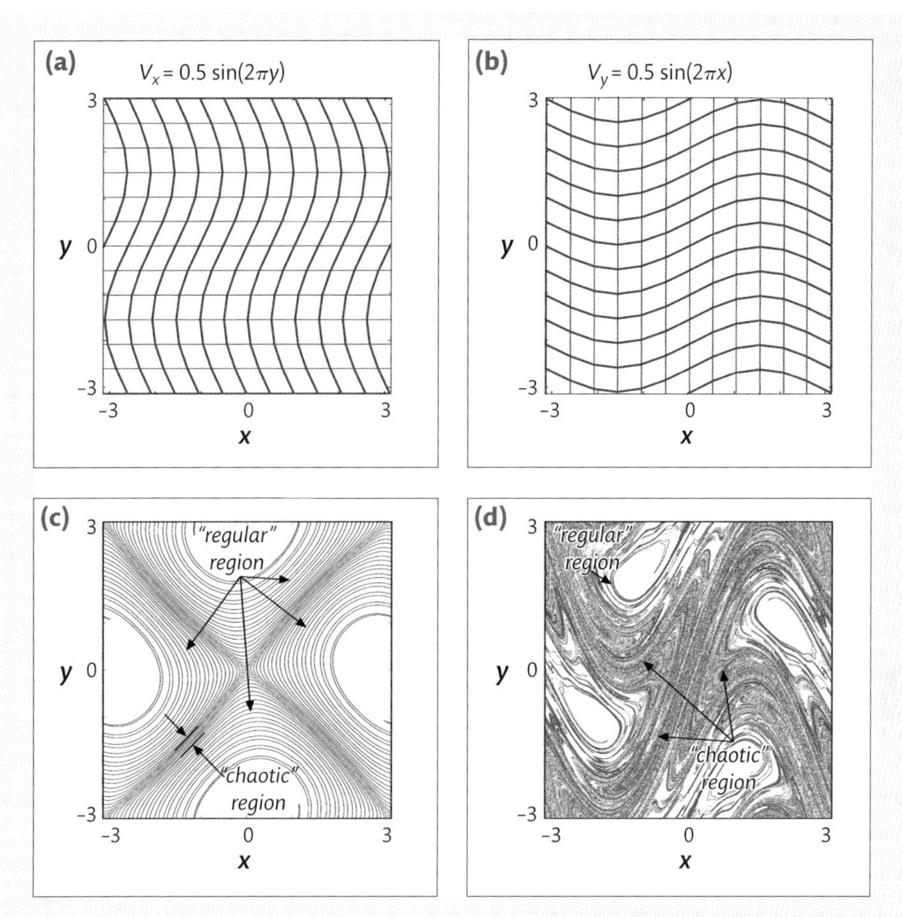

Fig. 1.5 Simulation displaying the difference between "Eulerian" and "Lagrangian" frames of reference for the "sine flow." This flow consists of alternating horizontal and vertical velocities (Eq. [1.14]) as shown in Eulerian descriptions (a) and (b), but produces complex trajectories as shown in Lagrangian descriptions (c) and (d). Panel (a) shows horizontal transport of an initially rectangular grid, and (b) shows vertical transport of the same initial grid. Panel (c) shows transport of a million points initially in a circle surrounding the origin after 100 iterations of combined horizontal and vertical velocities, for amplitude A = 0.5 in Eq. [1.14]. Panel (d) shows transport of the same initial points after only 10 iterations with amplitude A = 2.5.

Second, as we have mentioned, the rapidity of mixing and transport depends on the amplitude, A, of the *nonlinear* sinusoidal term. This is a central issue that we will return to. For now, we merely state for the student to cogitate over the fact that the complex behaviors shown in Fig. 1.5 are a direct result of the fact that the sine function in Eq. [1.14] is not linear: algebraically, it does not take the form of a line, $f(x) = a{\cdot}x + b$ for constant a and b, and geometrically it deforms points along curved trajectories as shown in Figs. 1.5(a) and (b). The term, $\vec{V}\cdot\vec{\nabla}\vec{V}$, in Eq. [1.12] similarly is not linear ($\vec{V}$ appears twice, so for example doubling V will quadruple this term), and we will see that this simple observation has profound consequences for fluid flow.

EXERCISE 1.1

Program Eq. [1.14] to deform a circle of initial points so that you reproduce Figs. 1.5(c) and (d). Show that as the amplitude, A, is increased, the chaotic region grows until it fills the entire available region. Show also that as A is increased further, new regular regions appear. Note that you will need to write your program so that the computational domain is periodic on $(-\pi, \pi)$, meaning that when a point moves to the left of $x = -\pi$ (e.g. at $x = -\pi - 0.1$) it reappears to the left of $x = +\pi$, (at $x = +\pi - 0.1$), and similarly for y. This makes use of the periodicity of the sine function, and can easily be performed using the "*mod*" function in whatever programming language you choose. If you do not do this, you will find that your data points wander away from the domain $-\pi < x < \pi$, $-\pi < y < \pi$. If you use Matlab for this and other exercises, you may find some of the tricks and tools in the Appendix to be helpful.

ADVANCED EXERCISE 1.2

In Fig. 1.6, we show successive enlargements of regions within the sine flow, Eq. [1.14], including numerous chains of regular "islands" surrounding each larger island. It can be proven that this island-around-island structure is generic (again, meaning present in almost all) to nonlinear flows such as this, and that there is a one-to-one correspondence between these islands and the rational and irrational numbers.

The way that this works is that every island chain has a "periodicity:" a number of iterations of the sine flow until a trajectory returns to that same island. In the central panel of Fig. 1.6, we identify a set of period-6 islands: a point precisely in the center of any of these islands will return to the same location after 6 iterations of Eq. [1.14]. Surrounding this central point are infinitely many closed, concentric orbits like the ones shown in Fig. 1.5(c). If any point on

Fig. 1.6 Sine flow, Eq. [1.14], for A = 1.5. Note in successive enlargements that numerous "island chains" are present, each of which corresponds to a central trajectory that returns in an integer number of applications of Eq. [1.14]. In the central panel, we identify a period-6 island chain, and in the rightmost panel we show a period-5 chain that surrounds the period-6 points. This island-within-island structure is present throughout the chaotic sea, as well as surrounding regular islands. Each island corresponds to an integer number (its periodicity), and between each chain is a trajectory corresponding to an irrational number.

these concentric orbits returned exactly to the same point, it would do so forever, so the orbit could never fill the space between one point on the orbit and its neighbor. This means that there must be gaps between points on the orbit: this is true of any integer periodicity. The only way to make a truly closed orbit is to forbid the orbit—or any multiple repetition of the orbit—to ever return precisely to the same point. Numbers with the property that neither they themselves nor any integer multiple of them is an integer are irrational. Thus island chains correspond to integers, or more generally to rational numbers, and closed orbits within the chains correspond to irrational numbers.

Using this knowledge, and the fact that between every pair of irrational numbers lies a rational one, we can conclude that the apparently closed orbits in Figs. 1.5 and 1.6 must in fact consist of an infinitely dense set of irrational, closed, orbits and rational island chains. Use your program for the sine flow to identify some of these, and write a new program that permits you to place additional points (e.g. with a mouse click) near a location of interest so that you can zoom in on your map and convince yourself that these closed orbits and island chains are indeed dense. It can also be proven that within the chaotic sea there are rational and irrational orbits (though these may be extremely small and the periodicity may be very large), so as an added, and more difficult, exercise, zoom in to search for some of the small islands within the chaotic sea. You will need to use very large numbers of points to do this and to either possess great patience or a clever strategy to automatically identify such islands.

Armed with an understanding of why the substantive derivative is of importance for practical problems, we return to Eq. [1.8], which we now rewrite as:

$$m\left(\frac{\partial\vec{V}}{\partial t}+\vec{V}\cdot\vec{\nabla}\vec{V}\right)=\sum\vec{F}. \qquad [1.15]$$

Dividing by the unit volume, $\Delta x\Delta y\Delta z$, gives

$$\rho\left(\frac{\partial\vec{V}}{\partial t}+\vec{V}\cdot\vec{\nabla}\vec{V}\right)=\sum\frac{\vec{F}}{\Delta x\Delta y\Delta z}, \qquad [1.16]$$

where ρ is the density, as before. The careful reader will notice that we have factored out the mass of the fluid volume from the time derivative: doing so assumes that the mass of a unit volume—the density—doesn't change in time. This is correct for incompressible fluids, but for compressible flows, the derivatives on left hand side of Eq. [1.15] would have to be expanded in terms of derivatives of m as well as of $\vec{V}$.

The right hand side of Eq. [1.16] should include all forces that can act on a volume of fluid: due to pressure, to viscosity, and any additional forces present – for example gravity or another external effect. We write each of these forces here, and will derive the forces next:

$$\rho\left(\frac{\partial\vec{V}}{\partial t}+\vec{V}\cdot\vec{\nabla}\vec{V}\right)=-\vec{\nabla}P+\mu\nabla^2\vec{V}+\vec{F}_{ext}, \qquad [1.17]$$

where P is the pressure, μ is the fluid viscosity, and $\vec{F}_{ext}$ is the sum of all external forces per unit volume, and where again we have made use of the nabla shorthand (more on this below). $\vec{\nabla}P$ has previously been defined (beneath Eq. [1.10]), and for completeness, $\nabla^2\vec{V}$ is termed the "vector Laplacian," and is explicitly given in Cartesian coordinates by:

$$\nabla^2\vec{V} = \hat{x}\nabla^2 V_x + \hat{y}\nabla^2 V_y + \hat{z}\nabla^2 V_z, \quad [1.18]$$

where $\nabla^2\vec{V}_i$ is the "scalar Laplacian,"

$$\nabla^2 V_i = \frac{\partial^2 V_i}{\partial x^2} + \frac{\partial^2 V_i}{\partial y^2} + \frac{\partial^2 V_i}{\partial z^2}. \quad [1.19]$$

Both terms are named for French mathematician and astronomer, Pierre-Simon Laplace (1749–1847), who made contributions to statistics, differential equations, fluid mechanics, celestial mechanics, and even presaged the possibility of black holes. Eq. [1.17] is thus a straightforward expression of Newton's second law of motion and is termed the "Navier–Stokes equation," after French civil engineer Claude-Louis Navier (1875–1936) and British mathematician, George Gabriel Stokes (1817–1903). We derive next the forms of the pressure and viscous terms so that we are clear on their meaning and limitations.

1.3.1 **Pressure**

Let us again consider forces on two representative surfaces bounding a unit volume, as shown in Fig. 1.7. Here we imagine that there may be a difference in forces because, say, the volume is being compressed from left to right, or the fluid is escaping to the right. Regardless of the cause, the force on the left surface shown in blue is $F_x = P|_x \Delta y \Delta z$, and on the right the force is $F_{x+\Delta x} = P|_{x+\Delta x} \Delta y \Delta z$, where P is the fluid pressure, defined to be the compressive force per unit area. So the difference in force from left to right is $\Delta F = (P|_x - P|_{x+\Delta x})\Delta y \Delta z$, and that difference divided by volume, $\Delta x \Delta y \Delta z$ (recall all forces in Eq. [1.17] are per unit volume), is

$$\frac{Force}{volume} = \frac{P|_x - P|_{x+\Delta x}}{\Delta x}, \quad [1.20]$$

and as $\Delta x \to 0$, this goes to

$$\frac{Force}{volume} = -\frac{\partial P}{\partial x}. \quad [1.21]$$

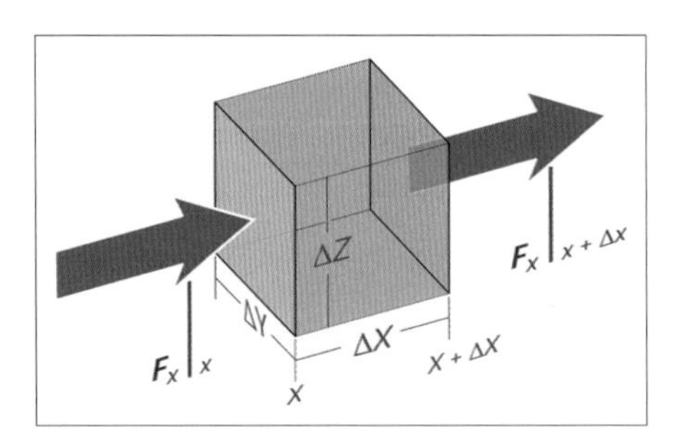

Fig. 1.7 Forces acting perpendicular to x-facing surfaces of a unit volume of fluid.

As before (Eq. [1.6]), a minus sign appears due to the convention of defining derivatives. Eq. [1.21] provides the force on a unit area in the x-direction; extending this result to all three dimensions produces the pressure term shown in Navier–Stokes:

$$\frac{Force}{volume} = -\vec{\nabla}P. \quad [1.22]$$

Review of static pressure

It is worthwhile at this point to briefly review the meaning of pressure in a few simple situations so as to make clear how the pressure term in Eq. [1.22] contributes to fluid behavior. In particular, we emphasize that the pressure that appears in Navier–Stokes equation refers solely to pressure *differences* that lead to fluid motion, and not to the average value that leads to other effects also attributable to pressure.

Average, or "static," pressures, on the other hand, are perhaps more commonly studied, in elementary problems such as a hydraulic jack. A jack operates by sustaining a fixed pressure, P, which on the one hand equals the force applied to the jack handle, F_1, divided by a small area, A_1, of one piston, and on the other hand equals the weight of a car, $M_{car}{\cdot}g$, divided by a much larger area, A_2, of a second piston. Equating these terms gives $F_1/A_1 = M_{car}{\cdot}g/A_2$ and by choosing A_1 and A_2, the force applied to the jack handle can be made as small as one wants—for example, if $A_1 = 1\ \text{cm}^2$ and $A_2 = 100\ \text{cm}^2$, a 1000 kg car can be lifted using a force of $F_1 = 100$ N (about the weight of six textbooks). Similarly a parlor trick involves striking a cork in the mouth of a milk jug: it is easily shown that a 5 cm diameter cork tapped into a 25 cm diameter jug with a small force of 10 N will generate 1000 N at the bottom of the jug—which is more than enough force to tear the bottom off of the jug.

These elementary behaviors are based on Pascal's law, which states that the pressure in a static fluid is the same everywhere. Though correct, this is seldom useful for problems of biomedical interest, which are typically not static, and Pascal's law is not relevant at all to the pressure gradient that appears in Eq. [1.22]. More typical of the problems that we will consider in this book involve pressure drops across a pipe, or differences in pressure between the bottom and the top of a falling object. In both of these problems, the force exerted is due to a *difference* in pressure, and would be the same at any average pressure.

EXERCISE 1.3

Although static pressure problems are of limited biomedical importance, some devices are either truly static or are sufficiently slow that the pressure can be treated as "quasi-static," meaning that the pressure changes much more slowly than other variables. High pressure liquid chromatography (HPLC) is such an example; flow in some microfluidic devices is another.

In the present exercise, we challenge the student's understanding of pressure with a couple of additional parlor tricks, sketched in Fig. 1.8. In Fig. 1.8(a), we show water being poured into a funnel that is connected to an empty coiled tube. As the reader can confirm, the water will overflow the funnel rather than flow through the tube. If the coils are removed and the tube is straightened, however, water will flow freely. Explain. Show in a diagram what the pressures are at relevant locations in the tube.

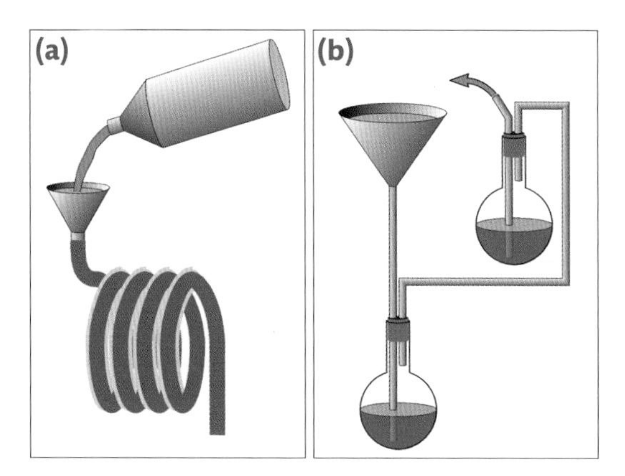

Fig. 1.8 Parlor tricks showing (a) water being poured into a coiled tube, and (b) a self-generating fountain of water (blue arrow). The reader is asked to explain why water does not flow out of the tube in panel (a), and what generates the fountain in panel (b).

In Fig. 1.8(b), we show another parlor trick, in which water flows out of the fountain indicated by the blue arrow. Here we ask the reader to define what forces produce the fountain and to explain whether or not this is a perpetual motion machine, and why.

1.3.2 **Viscosity**

The second term on the right-hand side of Navier–Stokes equation (Eq. [1.17]) is intended to model viscous forces, which we have asserted should be $\mu\nabla^2\vec{V}$. It is not at all obvious that a viscous fluid will produce this particular force per unit volume, and some care is called for in understanding why—and under what conditions—the viscous force takes this form. Indeed, we will see later that many biological fluids obey different forms than this one.

Before we derive the effect of viscosity, let us understand what, mechanistically, viscosity is. Blood flows through a vein under the influence of a pressure difference generated by contractions of the heart. If the heart stops, the flow will stop as well—this occurs due to forces exerted on the blood by the walls of the vein. These forces are what we term viscosity. To clearly distinguish such viscous forces from pressure or other forces, let's consider the mechanism that slows fluid flow through a tube.

In Fig. 1.9(a) we depict flow through a tube, driven by higher pressure, P_{in}, at its inlet, and lower pressure, P_{out}, at its outlet. As we will derive shortly, velocity is highest at the center of the tube and lowest at the tube walls, producing a velocity profile that we sketch as small arrows. If we enlarge a region of flow near the lower wall as shown in Fig. 1.9(b), we see that in the frame of a volume element, fluid appears to be moving backward below and forward above the element. This produces what we term a "shear:" formally a change in velocity one direction (horizontal, here) at different locations along an orthogonal direction (vertical). We call the quantitative movement incurred due to such a change the shear "strain:"

$$Strain = \frac{\Delta x}{\Delta y}, \qquad [1.23]$$

where, as shown in Fig. 1.9(b), Δy is the height of a small volume and $\Delta x = (V_{top} - V_{bottom}) \cdot \Delta t$ is the distance moved horizontally in some time Δt. Similarly we term the "strain rate:"

$$Strain\ Rate = \frac{V_{top} - V_{bottom}}{\Delta y}. \qquad [1.24]$$

Finally then, viscosity is the tendency of a fluid to resist changes in shear strain. We note that this differs both from forces due to pressure, which are associated with compression (along the axis exhibiting changes in velocity) rather than shear (perpendicular to that axis), and from forces due to friction. Frictional forces act between solids, rather than liquids, typically depend on normal forces (related to mean pressure), and crucially do not to lowest order change with speed of relative motion. Eq. [1.24], however, depends explicitly on the relative speeds, V_{top} and V_{bottom}. This distinction between viscosity and friction is important, and we take a moment to understand the mechanism by which viscosity comes about.

To do so, let's consider shear at the molecular level, depicted in Fig. 1.9(d), where we sketch two imaginary layers of fluid molecules, the upper layer moving faster than the layer beneath. This speed difference will cause particles in the two layers to collide, and these collisions will cost the fluid energy—resulting ultimately in heating of the fluid. Slow speed differences will cause fewer collisions per unit time—and so will cost less energy in that unit time—than fast speed differences. Through this reasoning, we argue that the rate of energy dissipation in a fluid (viscosity) depends on the rate of shear strain: how fast neighboring layers are moving with respect to one another. Indeed, this is what we observe in nature: it costs more energy per unit time to swim rapidly than slowly, and fast flow through a tube consumes more power than slow flow. The viscosity itself also depends on the rate of collisions between fluid

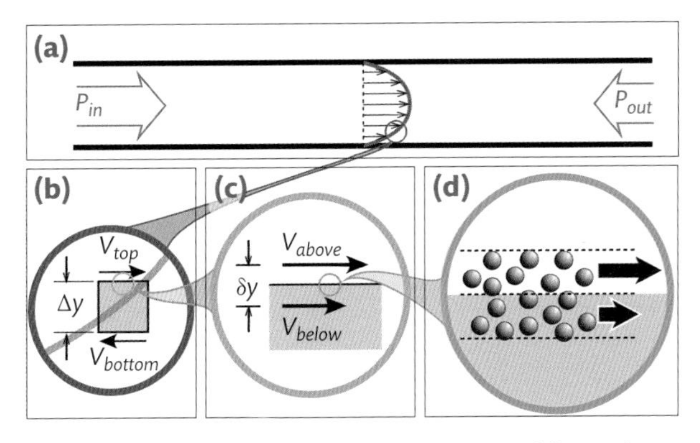

Fig. 1.9 Forces on a volume of fluid within a tube. (a) Schematic showing the difference between pressure at the inlet and that at the outlet of the tube, which results in velocities indicated by thin arrows that are fastest at the tube center and slowest at its walls. (b) Enlarged view of volume element, indicating that the velocity on the top face of the element is faster than that at the bottom—or in the frame of the volume element, the velocity is forward on its top and backward on its bottom. (c) Further enlarged view of top of volume element, showing shear at surface of element, and (d) microscopic cartoon suggesting layers of molecules moving at different mean speeds.

molecules—that is, the temperature—but that is a story for a later chapter: for now, we focus on the fact that the rate of shear strain (Eq. [1.24]) is the relevant quantity that resists flow in a fluid.

So how does this resistance to flow translate into a force per unit volume in Navier–Stokes equation? Let's return to the unit volume sketched in Fig. 1.9(b). If we knew the forces acting on the upper and lower surfaces, we could readily calculate the force/volume desired. What we actually know, however, is not the force but the *displacement* per unit time: the strain rate, Eq. [1.24]. Something has to be done to obtain the force on the surfaces from the strain rate there. In principle, the cartoon of Fig. 1.9(d) indicates that we should expect these quantities to be related, but in practice it is not clear how to obtain forces from molecular collisions in a practicable way.

We could image many possible relations between force on a surface and strain rate, and indeed we will see that different fluids exhibit different such relations. The simplest possible relation was proposed by Newton himself, namely that the force on a surface is proportional to the strain rate:

$$\frac{Force}{\Delta x \Delta z} = \mu \cdot Strain\ Rate, \qquad [1.25]$$

where we have inserted the area, $\Delta x \Delta z$, of the top or bottom surface of a volume element. The force per unit area in general is termed the "stress," and the constant, μ, that determines the magnitude of the stress is termed the viscosity. We reiterate that this is only the simplest possible viscosity relation, and Newton was in a sense fortunate that many fluids, notably water, obey this relation to a good approximation. Correspondingly, we call fluids that obey Eq. [1.25] "Newtonian," to distinguish them from other fluids (most biological fluids being among them), which are "non-Newtonian" and which obey different relations that we will discuss later.

To minimize confusion between values around a volume element (Fig. 1.9(b)) and values around its surface (Fig. 1.9(c)), we denote a small vertical separation at a surface δy, and velocities above and below the surface V_{above} and V_{below}, so that at the top surface, Eq. [1.25] implies

$$\begin{aligned}\frac{Force_{top}}{\Delta x \Delta z} &= \mu \cdot \frac{V_{above} - V_{below}}{\delta y}, \\ &= \mu \cdot \frac{\delta V_{top}}{\delta y}\end{aligned} \qquad [1.26]$$

and similarly at the bottom surface. The difference between forces at top and bottom surfaces is then:

$$\frac{Force_{top} - Force_{bottom}}{\Delta x \Delta z} = \mu \cdot \frac{\delta V_{top}}{\delta y} - \mu \cdot \frac{\delta V_{bottom}}{\delta y}, \qquad [1.27]$$

so the differences produced across a box of height Δy is

$$\frac{Force_{top} - Force_{bottom}}{\Delta x \Delta y \Delta z} = \frac{\mu \cdot \frac{\delta V_{top}}{\delta y} - \mu \cdot \frac{\delta V_{bottom}}{\delta y}}{\Delta y}. \qquad [1.28]$$

This goes to the following as the size of the box goes to zero:

$$\frac{Force}{volume} = \mu \cdot \frac{\partial^2 V}{\partial y^2}. \quad [1.29]$$

This viscous term is due to displacements only in the y-direction; more generally, the force per unit volume goes as $\mu \nabla^2 \vec{V}$, as advertised in Eq. [1.17].

1.3.3 **Nondimensionalization**

Let's recap. We have seen that conservation of mass leads to the continuity equation, Eq. [1.7], and that Newton's second law of motion—combined with Newton's law of viscosity, Eq. [1.26]—leads to the Navier–Stokes equation, Eq. [1.18]. We will occupy much of this book with applications of this equation, so let us spend a moment understanding it. First, we note that the equation appears in the literature in several different forms. We present two of these forms here; for other (especially tensorial) forms, the reader may want to consult *Transport Phenomena*, by RB Bird, WE Stewart, and EN Lightfoot.

The first of the forms that we introduce is really a formality, but presenting it here will be helpful later on. If we divide Eq. [1.18] by the density, we obtain

$$\frac{\partial \vec{V}}{\partial t} + \vec{V} \cdot \vec{\nabla} \vec{V} = -\frac{1}{\rho} \vec{\nabla} P + \nu \nabla^2 \vec{V} + \frac{\vec{F}_{ext}}{\rho}. \quad [1.30]$$

In this equation a new form of viscosity appears: $\nu = \mu/\rho$, which is called the "kinematic viscosity". This terminology causes confusion, which is exacerbated by the practice of calling μ the "dynamic viscosity," or sometimes the "absolute viscosity." There is nothing kinematic, dynamic, or absolute about either μ or ν: these are just conventional expressions for respectively the viscosity of Newton's law, Eq. [1.25], and that viscosity divided by the fluid density, ρ.

The second form of Navier–Stokes that we will discuss is more significant: this involves "nondimensionalizing" the equation. At the beginning of this chapter, we referred to the merits of constructing dimensionless numbers that provide measures of the relative strengths of competing influences. By taking the dimensions—mass, time, and length—out of Navier–Stokes equation, we will see that an important dimensionless number can readily be found.

We can remove the dimensions from Eq. [1.30] simply by dividing the equation through by all masses, times and lengths that appear. Dividing Eq. [1.17] by the density has removed all mass terms, and by inspection, the resulting Eq. [1.30] has units of velocity/time, or equivalently velocity^2/length, so we will just multiply each term by a "characteristic" length and divide by a "characteristic" velocity squared. What length or velocity is "characteristic" may seem somewhat vague at this point: if one is studying swimming of a fish, is the relevant scale the length of the fish? Its width? Height, perhaps? Is the relevant speed its body speed with respect to the water or the speed of its flippers with respect to its body? We will discuss examples later in this book (see especially the Dimensional Analysis section of Chapter 2), and we will see that the identification of what scales one uses depends on what question one seeks to answer. To evaluate the drag encountered by a gliding fish, likely the mean speed and length of the fish will be most relevant, whereas to evaluate the force exerted by its flippers, probably the flipper speed and height would be more pertinent. This vagueness can be important to resolve, and again we will touch on this point later. For now, we will assume that there exists a well-defined velocity, $\langle V\rangle$, and length, $\langle L\rangle$, for a problem of interest.

Dividing Eq. [1.30] by $\langle V\rangle^2/\langle L\rangle$ produces the following results for each of the terms, from left to right, in the Navier–Stokes equation:

$$\begin{aligned}
&\left[\frac{\langle L\rangle}{\langle V\rangle^2}\right]\frac{\partial\vec{V}}{\partial t}=\frac{\partial(\vec{V}/\langle V\rangle)}{\partial(t\langle V\rangle/\langle L\rangle)}=\frac{\partial(\tilde{V})}{\partial(\tilde{t})}\\
&\left[\frac{\langle L\rangle}{\langle V\rangle^2}\right]\vec{V}\cdot\vec{\nabla}\vec{V}=(\vec{V}/\langle V\rangle)\cdot(L\vec{\nabla})(\vec{V}/\langle V\rangle)=\tilde{V}\cdot\tilde{\nabla}\tilde{V}\\
&\left[\frac{\langle L\rangle}{\langle V\rangle^2}\right]\frac{1}{\rho}\vec{\nabla}P=(\langle L\rangle\vec{\nabla})\frac{P}{\rho\langle V\rangle^2}=\tilde{\nabla}\tilde{P}\\
&\left[\frac{\langle L\rangle}{\langle V\rangle^2}\right]\nu\nabla^2\vec{V}=\frac{\nu}{\langle V\rangle\langle L\rangle}(\langle L\rangle\nabla)^2(\vec{V}/\langle V\rangle)=\frac{\nu}{\langle V\rangle\langle L\rangle}\tilde{\nabla}^2\tilde{V},
\end{aligned}\qquad [1.31]$$

where the dimensionless variables have tildes, and are given by:

$$\tilde{V}=\vec{V}/\langle V\rangle$$

$$\tilde{t}=t\langle V\rangle/\langle L\rangle$$

$$\tilde{\nabla}=\langle L\rangle\nabla$$

$$\tilde{P}=P/\rho\langle V\rangle^2.$$

Incorporating Eq. [1.31] into Eq. [1.30], we obtain the nondimensionalized Navier–Stokes equation:

$$\frac{\partial\tilde{V}}{\partial\tilde{t}}+\tilde{V}\cdot\tilde{\nabla}\tilde{V}=-\tilde{\nabla}\tilde{P}+\left[\frac{\nu}{\langle V\rangle\langle L\rangle}\right]\tilde{\nabla}^2\tilde{V}+\tilde{F}_{ext}. \qquad [1.32]$$

All parts of this equation are dimensionless, including the term in square brackets. Crucially, this implies that *all experiments that are performed with the same value of that term will yield exactly the same solution* to Navier–Stokes equation. An experiment with viscosity and characteristic velocity of order one and length a kilometer will give exactly the same result as an experiment with velocity and length of order one and viscosity of a thousandth. Likewise at a given viscosity, the detailed velocity of fluid surrounding a body is guaranteed to be the same provided the product of $\langle V\rangle\cdot\langle L\rangle$ is fixed.

This remarkable property of experiments to produce identical results provided that a dimensionless number such as $\nu/(\langle V\rangle\cdot\langle L\rangle)$ is held constant is termed "dynamical similarity."

For the Navier–Stokes equation, the relevant dimensionless number is called the Reynolds number, $Re \equiv \bar{V} \cdot \bar{L}/\nu$, after Osborne Reynolds (1842–1912), an inventive Irish scientist with writings ranging from effects of oil droplets on calming stormy waters,[3] to speculations that the luminiferous ether could be a granular fluid.[4]

The principle of dynamical similarity is exceedingly useful. For example, before meticulously fabricating a 10 μm wide channel to perform a microfluidic mixing experiment using water, one could perform a test using a 10 mm wide channel using glycerol (kinematic viscosity about a thousand times that of water). Both experiments are guaranteed to yield precisely the same velocity field, and the 10 mm wide channel is certain to be easier to build and instrument. Similarly, there exists no air tunnel even half as large as a commercial jumbo jet—so no such airplane is ever tested on the ground before it is flown. Instead, scale models, a fraction as large as the full size plane, are tested with wind speeds correspondingly larger than those to be used in operation so as to hold the Reynolds number constant. In this way, velocity fields obtained from Navier–Stokes equation can be made identical in an unfeasibly large experiment and in a more moderate sized and higher speed one.

We caution that while this is a fact, the Reynolds number is only one of several dimensionless quantities that govern fluid flow. So to test a full-scale airplane flying at 1000 km/h by using a tenth scale model would require speeds of 10,000 km/h, which is about eight times the speed of sound. At that speed, the Mach number, $Ma \equiv \bar{V}/V_{sound}$, which characterizes shock formation, is vitally important, and both Re and Ma must be matched to produce true dynamical similarity. In a later chapter we will discuss methods of obtaining all relevant dimensionless numbers for a given problem. For now, we content ourselves with an appreciation for the power of matching dimensionless numbers that we began the chapter describing.

EXERCISE 1.4

Apply and interpret the method of nondimensionalization described above to two common equations.

(1) Consider Hooke's law for the motion of a spring: $m\ddot{x} = -kx$, where $\ddot{x} \equiv d^2x/dt^2$. What dimensionless number do you obtain, and what two competing *physical* effects (not merely mathematical expressions) does the number describe?

(2) Consider the diffusion equation (introduced in a later chapter): $\partial C/\partial t = D \cdot \nabla^2 C$, where C is the concentration of some substance and D is its diffusivity. Again, what dimensionless number do you obtain, and what two competing physical effects does it describe? Notice that the diffusion equation and parts of Navier–Stokes equation are similar: what does this tell you about a relation between how a substance diffuses and how viscosity operates?

1.3.4 **Limiting Behaviors**

As with other dimensionless numbers, the Reynolds number is a ratio between two competing quantities, here the "inertia," $\langle V \rangle \cdot \langle L \rangle$, and the viscosity, ν. At its heart, for $Re \gg 1$, a large enough body of fluid is moving fast enough that viscosity can't immediately slow it down, and correspondingly we call such flows "inertial." At the other extreme, for $Re \ll 1$, viscosity dominates over inertia and the motion of the fluid is slaved to the forces being immediately applied: when the forces stop, the fluid motion stops essentially instantaneously. We call such flows "creeping."

Let us briefly show, then, what the Navier–Stokes equation becomes at each of these extremes. This will be important, for Navier–Stokes equation taken in its entirety is enormously difficult—most would say impossible—to solve, and the only way to solve, or even cope with, it is to find conditions under which one or more of its terms can be gotten rid of. Examining limiting behaviors at very low or very high *Re* is one way of getting rid of bothersome terms. The reader should view this process precisely as we have described it: the assumptions that we make throughout this book are not made because they are especially meritorious or because we are wise to make them, rather we typically make these assumptions because this is the only way to make any headway into understanding this august equation. With this in mind, the limiting low and high *Re* forms of Navier–Stokes equation are as follows.

$Re \ll 1$

At small *Re*, the Laplacian term, $\tilde{\nabla}^2\tilde{V}/Re$, in Eq. 1.32 becomes much larger than the nonlinear term, $\tilde{V}\cdot\tilde{\nabla}\tilde{V}$, and so we neglect the latter, leaving only linear terms:

$$\frac{\partial \tilde{V}}{\partial \tilde{t}} = -\tilde{\nabla}\tilde{P} + \left[\frac{\nu}{\bar{V}\bar{L}}\right]\tilde{\nabla}^2\tilde{V} + \tilde{F}_{ext}. \qquad [1.33]$$

This is known as the "Stokes equation," which governs low speed, small scale, and high viscosity flows that characterize many biomedical problems including microbial transport and blood flow in smaller veins and arteries. Often we will consider problems in which the flow is steady (so that $\partial\tilde{V}/\partial\tilde{t} = 0$) and external forces such as gravity are small or absent, in which case we will write the more manageable equation:

$$\tilde{\nabla}^2\tilde{V} = Re\cdot\tilde{\nabla}\tilde{P}. \qquad [1.34]$$

$Re \gg 1$

At the other extreme where speeds and lengths are large and viscosities are low, the Laplacian term can be neglected, leaving:

$$\frac{\partial \tilde{V}}{\partial \tilde{t}} + \tilde{V}\cdot\tilde{\nabla}\tilde{V} = -\tilde{\nabla}\tilde{P} + \tilde{F}_{ext}. \qquad [1.35]$$

This is termed the "Euler equation," after Leonhard Euler (1707–1783), the prolific Swiss mathematician after whom the number *e*, the base of the natural log, is named, and of whom Laplace said, "Read Euler, read Euler: he is the master of us all." We will study Euler's equation later in this book. We begin with the Stokes equation, Eq. [1.34], which is linear and so is comparatively straightforward to deal with.

1.4 Application: Poiseuille Flow in a Rigid Tube

We now know enough to begin to study applications. The first of these is Poiseuille (pronounced pwahzay, or pwahzuwee, depending on taste) flow: steady, slow flow through a tube, first introduced in Fig. 1.9(a) and shown again in Fig. 1.10. Flow through tubes is ubiquitous in biology: there are for example over 100,000 km of blood vessels in the human body, which circulate 9500 liters of blood daily.

To solve the Stokes equation (Eq. [1.34]) for this flow, let us first write the equation in its simplest imaginable form, and then try to work out how to find the detailed solution from this. If the problem were 2D as in Fig. 1.10, we could simplify Eq. [1.34] to become:

$$\frac{\partial^2 \widetilde{V}_x}{\partial y^2} = constant, \qquad [1.36]$$

where we have taken the liberty of assuming that the velocity is only in the x-direction and depends only on y, in the coordinate system of Fig. 1.10(a). Let's not worry for the moment about what the constant above is or whether a 2D form is reasonable—for now let's just ask what the solution of this simplest form could be.

It has been said that the canonical method for solving differential equations is to stare at them until a solution occurs to you, and if we apply this wisdom, sooner or later it may occur that the second derivative of a function like the following will fit the bill:

$$\widetilde{V}_x(y) = c_0 + c_1 y + c_2 y^2, \qquad [1.37]$$

for constants c_0, c_1, and c_2.

If we differentiate Eq. [1.37] twice, we find that $2c_2 = constant$, and c_0 and c_1 can be anything at all. To pin down c_0 and c_1, we need more information, which comes in two forms. First, we're neglecting gravity, so there is no difference between flow in the top and bottom halves of the flow. This isn't always the case (e.g. we'll consider curved tubes later on), but let's assume top–bottom symmetry. Additionally, life is simplest if we put the origin at the center of the tube, as shown in Fig. 1.10(a). This isn't at all necessary (see Exercise 1.5), but it makes the algebra

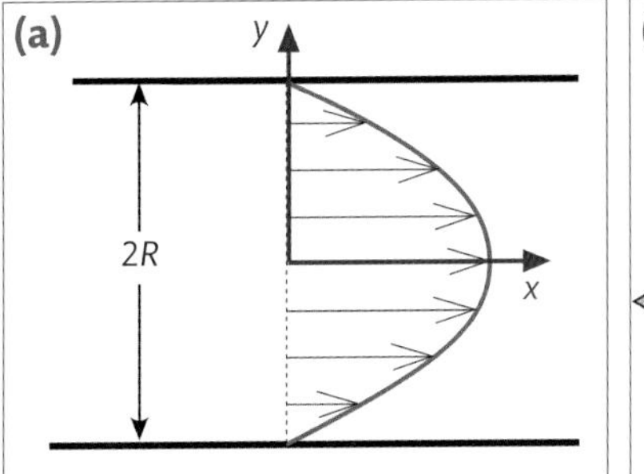

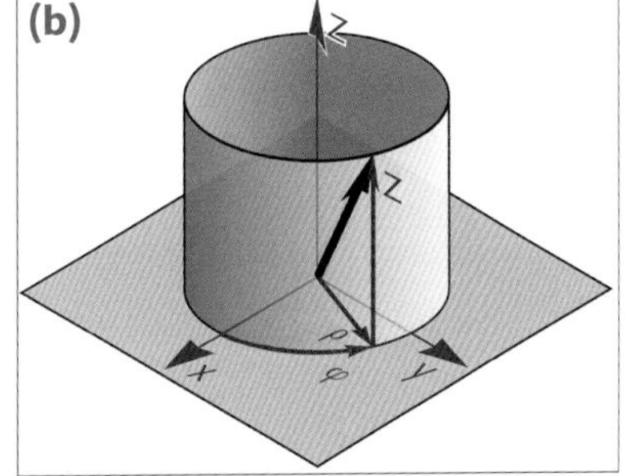

Fig. 1.10 Flow through a tube of radius R in two and three dimensions. (a) Coordinate system used for two-dimensional (2D) flow: shear is in the vertical, y, direction, and flow, V_x, occurs in the horizontal, x, direction. (b) Coordinate system for three-dimensional (3D) cylindrical flow. In keeping with convention, flow, V_z, is now in the vertical, z, direction, and any vector, for example, the one shown in black, is defined by coordinates (ρ, φ, z), shown in red.

simple, for we can see that c_1 must then be zero: otherwise the velocity would be different above the centerline (positive y) and below (negative). This leaves us with c_0 to sort out. But we know something else: we know that the fluid touches the top and bottom boundaries of the tube. There is a large body of literature on what happens to flows near boundaries, and we will have more to say about some of these behaviors later. For now, in the continuing spirit of keeping things simple, we consider a "no-slip" condition: $\tilde{V}_x(y_{boundary}) = 0$, which by fortune is very nearly exact for all but exotic surfaces (e.g. the "superhydrophobic" surface of the lotus leaf). No slip implies $V_x = 0$ at the tube walls (i.e. at $y = \pm R$) so $c_0 = -c_2 R^2$, and

$$\tilde{V}_x(y) = \tilde{V}_{max} \cdot \left(1 - \frac{y^2}{R^2}\right). \qquad [1.38]$$

If we substitute in the value, $constant = Re\, \tilde{\nabla}\tilde{P}$, from Eq. [1.34], we get the tidy form above using the abbreviation. $\tilde{V}_{max} = [Re \cdot \tilde{\nabla}\tilde{P}] \cdot R^2/2$. Eq. [1.38] tells us that the velocity is parabolic in shape, as sketched in Fig. 1.10(a), and it is easily confirmed from Eq. [1.38] that the velocity is zero at $y = \pm R$, and reaches a maximum $\tilde{V}_{max}$ at the center of the tube, $y = 0$.

Eq. [1.38] is the general result for low speed, steady flow through a channel. All that remains are embellishments to do with such things as extending the solution to three dimensions, including effects of curvature or disturbances at the entrance of a channel, and so forth. In the remainder of this section, we derive 3D results, and discuss other effects in subsequent chapters.

1.4.1 **3D Poiseuille Flow in a Cylindrical Tube**

In this section, we extend the 2D result that we have derived to three dimensions. We will find that the arithmetic is more cumbersome in three dimensions, but the essential idea is the same: we will write Stokes equation, we will stare at it until a solution occurs to us, and we apply symmetry and boundary conditions to solve for any remaining constants. We will see that the solution is quadratic, like Eq. [1.38], and although the making of this quadratic solution is a bit more complicated in three than in two dimensions, the underlying reason that the velocity is quadratic will be as before: that Stokes equation has two derivatives, and differentiating a quadratic function produces a constant. We'll see in other examples that things aren't always this simple, but they are for this problem.

So let's begin. We have solved Stokes equation (Eq. [1.35]) in dimensionless form; this time we will do so in dimensional form. This is only for completeness so that both forms are presented; the solutions are almost indistinguishable saving for tilde's and substituting $1/\mu$ for *Re*. In dimensional form, Stokes equation is:

$$\nabla^2 \vec{V} = \frac{\vec{\nabla} P}{\mu} \qquad [1.39]$$

As mentioned, this becomes a bit more cumbersome in three dimensions, but the complications will soon be dealt with. In cylindrical coordinates (because we're solving for flow inside a tube, or cylinder), Eq. [1.39] becomes

$$\hat{\rho}[\nabla^2 V]_\rho + \hat{\varphi}[\nabla^2 V]_\varphi + \hat{z}[\nabla^2 V]_z = \frac{\vec{\nabla} P}{\mu}, \qquad [1.40]$$

where the terms in each coordinate direction are

$$[\nabla^2 V]_\rho = \frac{\partial}{\partial \rho}\left[\frac{1}{\rho}\frac{\partial}{\partial \rho}(\rho \cdot V_\rho)\right] + \frac{1}{\rho^2}\frac{\partial^2 V_\rho}{\partial \varphi^2} + \frac{\partial^2 V_\rho}{\partial z^2} - \frac{2}{\rho^2}\frac{\partial V_\varphi}{\partial \varphi},$$

$$[\nabla^2 V]_\varphi = \frac{\partial}{\partial \rho}\left[\frac{1}{\rho}\frac{\partial}{\partial \rho}(\rho \cdot V_\varphi)\right] + \frac{1}{\rho^2}\frac{\partial^2 V_\varphi}{\partial \varphi^2} + \frac{\partial^2 V_\varphi}{\partial z^2} + \frac{2}{\rho^2}\frac{\partial V_\rho}{\partial \varphi},$$

$$[\nabla^2 V]_z = \frac{\partial}{\partial \rho}\left[\frac{1}{\rho}\frac{\partial}{\partial \rho}(\rho \cdot V_z)\right] + \frac{1}{\rho^2}\frac{\partial^2 V_z}{\partial \varphi^2} + \frac{\partial^2 V_z}{\partial z^2}.$$

Now we can deal with the complications. The first thing we do is to use symmetry. As in two dimensions, we assume that flow is only in the "streamwise" direction, so we keep terms with V_z, but we dispose of all velocities in the "spanwise" directions, that is, we assume that there is no flow outward against the walls of the tube (so that $V_\rho = 0$) and we assume that there is no spiraling flow (so $V_\varphi = 0$). These terms are colored red in Eq. [1.40], and will be neglected.

Also, by the same token that we assume no velocity in the φ direction, we assume that the velocity that remains, V_z, doesn't *change* in this direction. So the derivative in blue also vanishes. Finally, we note that since the fluid doesn't move in either ρ or φ, the continuity equation implies that V_z cannot change along the z-direction. If, for example, the velocity increased as z increased, then either mass would accumulate downstream, or mass would have to leave in some other direction—both of which we assume don't occur. So the magenta term vanishes as well.

This is good: by using symmetry and continuity, we've polished off everything in Eq. [1.40] except:

$$\frac{\partial}{\partial \rho}\left[\frac{1}{\rho}\frac{\partial}{\partial \rho}(\rho \cdot V_z)\right] = \frac{\vec{\nabla} P}{\mu}, \qquad [1.41]$$

and we expand the left hand side of this equation to obtain

$$\frac{1}{\rho}\frac{\partial V_z}{\partial \rho} + \frac{\partial^2 V_z}{\partial \rho^2} = \frac{\vec{\nabla} P}{\mu}. \qquad [1.42]$$

Having solved this physical problem in two dimensions, we guess that the solution could be a quadratic, $V_z(\rho) = c_0 + c_1\rho + c_2\rho^2$, and substituting this guess into Eq. [1.42], we get

$$\frac{c_1}{\rho} + 4c_2 = \frac{\vec{\nabla} P}{\mu}. \qquad [1.43]$$

The only way this can be true for all choices of ρ is if $c_1 = 0$, leaving us with the solution $V_z(\rho) = c_0 + \vec{\nabla} P \rho^2/4\mu$. As before, we use the no-slip boundary condition, $V_z(R) = 0$, which gives us c_0, so finally:

$$V_z(\rho) = V_{max} \cdot \left(1 - \frac{\rho^2}{R^2}\right), \qquad [1.44]$$

where $V_{max} = \vec{\nabla} P \cdot R^2/4\mu$.

We can now calculate something useful: suppose we want to know how much pressure drop is needed to sustain a required rate of flow of blood through a vein of known radius. We're still considering only a limited problem (slow steady flow assuming that the fluid is simple and the vein is smooth and rigid), but still we'll find that we can obtain a useful result.

To calculate the rate of flow of a fluid through a tube, we imagine that a small portion of cross-sectional area, Δa, within the fluid is swept forward by the flow a distance $\Delta z = V_z \cdot \Delta t$ in a time Δt. The volume of fluid transported in time Δt is thus $\Delta z \cdot \Delta a = V_z \cdot \Delta t \cdot \Delta a$. Of course V_z depends on radius, so different locations in the cross-section will sweep out different volumes in this time, but we can calculate all of the quantities needed. In Fig. 1.11(a), we show a typical cross-sectional area element being swept forward in time under the action of an axial flow, and in Fig. 1.11(b) we enlarge the resulting volume element. Since V_z changes with ρ, we must integrate over all such elements, or explicitly we must calculate

$$\begin{aligned} \frac{volume}{\Delta t} &= \iint V_z \cdot da \\ &= \int_0^{2\pi}\int_0^{R} V_z \rho d\varphi d\rho \end{aligned} \qquad [1.45]$$

We pause here to note a point that can easily be overlooked using cylindrical coordinates, namely that, as shown in Fig. 1.11(b), an infinitesimal area is the change in radius, $d\rho$, times the azimuthal distance, which is the radius *times* the change in angle: $\rho \cdot d\varphi$. So integrals in cylindrical coordinates must include an extra factor of ρ, which we highlight in blue: without this factor, the surface $\rho d\varphi d\rho$ would have units of length rather than area.

By convention, the volumetric flow rate is denoted Q, and Eq.[1.45] becomes

$$\begin{aligned} Q &= 2\pi \int_0^R V_{max} \left(1 - \frac{\rho^2}{R^2}\right) \rho d\rho \\ &= \frac{\pi \nabla P R^2}{2\mu} \left[\frac{\rho^2}{2} - \frac{\rho^4}{4R^2}\right]_0^R . \\ &= \frac{\pi \nabla P R^4}{8\mu} \end{aligned} \qquad [1.46]$$

This gives the classic, and remarkable, result that to maintain a constant volumetric rate of flow, the pressure gradient, ∇P, must grow with the *fourth* power of the radius. So if the radius of a blood vessel drops by half, the pressure across the vessel must grow by a factor of $2^4 = \mathbf{16}$ to maintain the same flow rate. If the radius drops by another half, the pressure must grow a total of **256** fold!

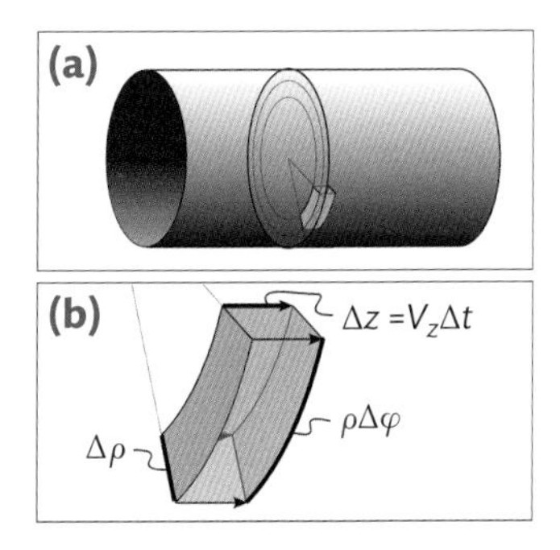

Fig. 1.11 Geometry for calculation of volumetric flow rate through a tube. (a) Schematic of tube; (b) enlargement of small volume element of cross-sectional area $\Delta a = \Delta\rho{\cdot}\rho\Delta\varphi$ that is transported a distance $\Delta z = V_z\Delta t$ in a time step Δt.

This basic fact puts atherosclerosis in a stark light, and explains why we should avoid greasy chips and similar junk food: blood pressure must grow dangerously high to maintain flow through narrowed blood vessels—similarly for constrictions elsewhere, such as in the lungs, or kidney, or as all men discover if they live long enough, through a hypertrophied prostate.

Example Poiseuille viscometer

We note that Eq. [1.46] can be easily inverted to find an expression for the viscosity:

$$\mu = \frac{\pi \nabla P R^4}{8Q}, \qquad [1.47]$$

so by measuring the pressure drop, radius and flow rate of a fluid through a tube, we can easily calculate the viscosity of the fluid. This is termed "Poiseuille viscometry," or sometimes "capillary viscometry," and is the first of several methods of measuring viscosity that we will touch on in this book.

A final remark to presage discussions to come: we mentioned that we have so far neglected realistic features including elasticity of tubes, non-Newtonian fluid behaviors and so forth. Each of these features produces unexpected and intriguing effects, and indeed one of the things that has attracted so much research in fluid dynamics is that almost any departure from idealized flow produces novel behaviors. As just one example, the two geometries that we have considered so far—in a cylinder and between parallel plates—lack edges and corners. To solve for flow in a rectangular duct with corners, one has to resort to a "Fourier series:" infinite sums of sines and cosines. This is algebraically unpleasant, but introduces no new phenomenology: the flow is as you would expect, slowest in the corners and smoothly growing to a maximum at the center of the duct.

As soon as sharp bends appear, however, all bets are off: mathematically we will show that gradients "diverge" (i.e. the velocity formally changes infinitely rapidly at a convex edge), but more interestingly, new physical effects appear. Downstream of a single bend, shown to the left of Fig. 1.12(b), a subtle swirling "secondary" flow appears—and in fact persists as the Reynolds

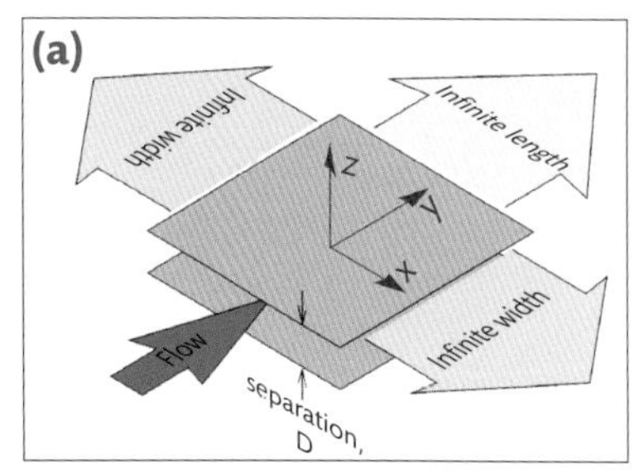

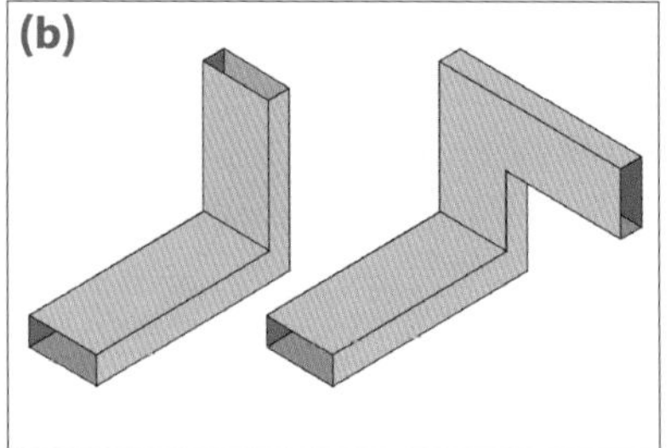

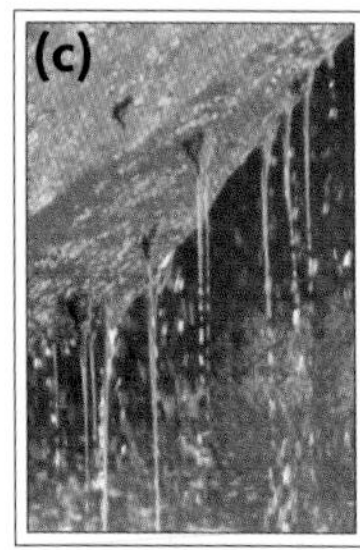

Fig. 1.12 (a) Parallel plate geometry for Exercise 1.5. (b) Embellishments to geometry: a single bend produces swirling secondary flow that concentrates bacteria into tendrils, and out-of-plane bends generate chaotic mixing (discussed also in Chapter 5). (c) Naturally occurring biofilm tendrils, here at the bottom of a rock.

number becomes vanishingly small. This swirling flow traps small cells and particles, and has been shown to lead to concentration of bacteria into biofilm tendrils.[5] Biofilms have long been known to be present in nature, ranging from plaque on teeth to algae on rocks (Fig. 1.12(c)), however only recently have mechanisms for the concentration of microbes in seemingly benign flows been discovered. Even more surprisingly, if out-of-plane bends are present, complex flows—complete with regular and chaotic regions illustrated in Fig. 1.6—spontaneously appear, and these flows can be manipulated to produce efficient mixing in microfluidic channels (see Chapter 5).[6] These two effects are very typical of fluid dynamics: wherever one looks, there are surprises.

EXERCISE 1.5

Repeat the calculations performed above for flow between two parallel plates separated by a distance, D, and infinite in the spanwise and streamwise directions, as shown in Fig. 1.12(a). In particular using the coordinate system shown:

(1) Solve for the velocity in the y-direction as a function of height, in the z-direction, *with the origin chosen to be on one of the plates* rather than at the symmetry plane. Determine how the velocity changes with height.

(2) Solve for the flow rate per unit width. How does the pressure required to maintain a constant flow rate change with the separation between the plates, D? Given this information, if you were to choose between a microfluidic flow device that was prone to fouling of its surfaces, would you choose a tubular or a parallel plate design?

REFERENCES

1. Lightfoot, G. and Balent, A. (1976) *The wreck of the Edmund Fitzgerald.* Heath Levy Music Company.

2. Chabchoub, A., Hoffmann, N., Onorato, M. and Akhmediev, N. (2012) Super rogue waves: observation of a higher-order breather in water waves. *Physical Review X, 2*(1), 011015.

3. Reynolds, O. (1875) On the action of rain to calm the sea. *Proceedings of the Manchester Literary and Philosophical Society, 12–15*, 72–4.

4. Reynolds, O. (1885) On the dilatency of media composed of rigid particles in contact. *Philosophical Magazine, Series 5, 20*, 469–81.

5. Rusconi, R., Lecuyer, S., Guglielmini, L. and Stone, H.A. (2010) Laminar flow around corners triggers the formation of biofilm streamers. *Journal of the Royal Society Interface, 7*(50), 1293–9.

6. Aref, H. (2002) The development of chaotic advection. *Physics of Fluids, 14*, 1315–25; see also Stone, H.A., Stroock, A.D. and Ajdari, A. (2004) Engineering flows in small devices: microfluidics toward a lab-on-a-chip. *Annual Review of Fluid Mechanics, 36*, 381–411.

2 Elastic Surfaces

We have begun to characterize how fluids behave in the simplest of circumstances. As we have mentioned, however, circumstances are often not simple, and an important complication that historically led to intense, even acrimonious, controversy was the nature of fluid surfaces, as are present in bubbles, unbounded interfaces, and boundaries such as blood vessels and cell membranes. This controversy was largely led by a dispute between the respected Belgian physicist, Joseph Plateau (1801–1883), and comparative upstart, Carlo Marangoni (1840–1925), a high school teacher. Plateau cited Descartes in support of his contention[1] that the surface of water was no different from its bulk, and so must possess a viscosity (which he termed viscidity), but not an elasticity, while Marangoni asserted[2] that the surface was in fact elastic. Their disagreement centered on detailed observations of a needle floating on the surface of water, an argument made somewhat more curious by the fact that Plateau was well known to have been completely blind for 20 years at the time that he claimed to be making these observations.

Under the test of time, it has been determined that the water surface is in fact elastic, and we now know that water has one of the highest surface tensions of any liquid by virtue of the fact that H_2O can make two hydrogen bonds per molecule, and so water actually forms a crystalline lattice at free surfaces.

2.1 Young-Laplace Equation

Notwithstanding this dispute, equations for surface elasticity had already been obtained[3] by Laplace, whom we mentioned earlier, and Thomas Young (1773–1829) who is perhaps better known for the wave theory of light and for Young's modulus of solids. The behavior of elastic membranes might be most easily understood by considering the simplest elastic problem, a mass–spring, as sketched in Fig. 2.1(a). That problem is well known to obey Hooke's law, $F = -k \cdot x$, from which we can surmise that the material property that defines a spring is $k = -F/x$. This turns out to be a general relation for elastic materials, and for a surface this leads to the definition of a surface tension constant as force per unit length. The constant itself is commonly written as either γ or σ; in this text we will use σ:

$$\sigma = F/L \qquad [2.1]$$

Biomedical Fluid Dynamics: Flow and Form. Troy Shinbrot.

DOI: 10.1093/oso/9780198812586.001.0001

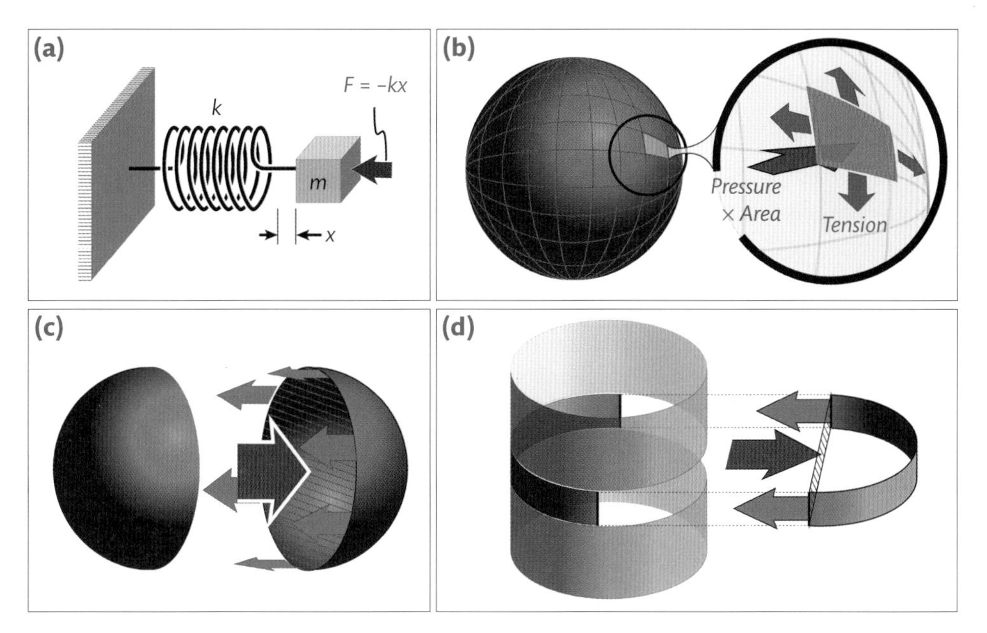

Fig. 2.1 (a) Mass-spring, producing elastic force $F = -k \cdot x$ for spring constant, k. (b) Spherical bubble, with enlargement of area element showing balance between outward pressure and inward components of tension. (c) Analysis of forces tending to pull hemispheres apart (red) competing with those holding them together (blue). The outward force acts on the cross-sectional area, πR^2 (red cross-hatching), and the inward forces act on the circumference of the sphere, $2\pi R$. (d) Similar analysis of segments of a cylindrical pipe, where the inward force acts on two line segments of length Δz, and the outward force acts on the area $2R\Delta z$ (again red cross-hatched).

This same relation is often derived by calculating the work needed to incrementally enlarge a surface element: there is nothing wrong with that approach: however it is arrived at, our starting point will be that the surface tension is given by a force divided by a length.

So let's start with this relation and ask what it tells us about forces on elastic membranes. We begin with the special case of a spherical bubble, as shown in Fig. 2.1(b). If we enlarge a small area element, we see that the outward force due to a pressure difference between the inside and the outside of the bubble, pressure × area, has to be balanced by the inward, radial, component of the tensions surrounding the element. In principle, we could integrate the radial components of each tension element and set these equal to the integrated pressure over the sphere: this would give us a relation between the magnitude of tension required to maintain equilibrium and the opposing pressure.

A more elegant way of reaching the same endpoint is due to Lev Landau (1908–1968), iconic and controversial Soviet physicist who received the Nobel Prize for modeling superfluidity, but who also made major contributions to everything from astrophysics to statistical mechanics. Landau's approach was to cut the bubble in half, as sketched in Fig. 2.1(c), and to define conditions on the elastic force holding the halves together. The elastic forces (blue arrows) act on the circumference of the sphere, $2\pi R$, for a sphere of radius R, so the total force holding the hemispheres together must be:

$$F_{in} = 2\pi R \cdot \sigma \tag{2.2}$$

Equilibrium demands that this force exactly equals the rightward force due to pressure (red arrow in Fig. 2.1(c)), which acts on the cross sectional area of the bubble, πR^2:

$$F_{out} = \pi R^2 \cdot P, \tag{2.3}$$

where we write P as shorthand for the pressure drop across the bubble. Setting F_{in} and F_{out} equal gives the surface tension required to hold a spherical bubble in equilibrium:

$$P_{sphere} = 2\frac{\sigma}{R_{sphere}}. \tag{2.4}$$

The reader may notice that this relation is geometry dependent, since for a sphere, tension holds the surface in along two directions against the influence of pressure (cf. Fig. 2.1(b)), whereas on a tube most tension is applied around the perimeter of the tube, with little or no tension along its axis. We can apply the same analysis to a half-ring of a cylindrical tube (exploded view in Fig. 2.1(d)), which tells us that the inward force on a half-ring of radius R and width Δz is $F_{in} = 2\Delta z \cdot \sigma$. This must be balanced against an outward force of $F_{out} = 2R\Delta z \cdot P$, so that:

$$P_{cylinder} = \frac{\sigma}{R_{cylinder}}. \tag{2.5}$$

Equations [2.4] and [2.5] can be generalized to an arbitrary surface with two orthogonal radii of curvature, R_1 and R_2:

$$P = \sigma \cdot \left(\frac{1}{R_1} + \frac{1}{R_2}\right). \tag{2.6}$$

This is the "Young–Laplace equation," which leads to a number of medically important results. To begin with, we can tell by inspection that as R_1 and R_2 decrease, the pressure must increase—and in fact diverges as the radii go to zero. This simple conclusion is in agreement with the common experience of anyone who has tried to blow up a child's balloon: large balloons are very easily inflated, but small balloons can be impossible to inflate without a pump.

This same effect is also important for expansion of alveoli (air sacks in the lungs) of newborns: alveoli are only about 200 μm across, and because they are so small, they require high pressure differences to inflate. At the same time, alveolar surfaces are necessarily thin and delicate, for they must permit intimate contact between circulating blood on one side and air in the lung on the other. Nature copes with these contradictory needs by providing us with a surfactant that reduces the surface tension at the alveolar surface by about a factor of 15—and by geometrical good fortune, the concentration of surfactant grows as the size of the alveoli shrinks, so the surfactant does its job best when it is needed most. Premature infants, however, often have not yet developed this surfactant layer, and for this reason clinical measures are taken to promote the secretion of surfactants into the lungs of infants (such as providing the mother with glucocorticoids that have this side effect)—or failing this, surfactants are directly administered through breathing tubes.

At the other end of life, healthy adults have on the order of 300 million alveoli, which provide for effective oxygenation by occupying the surface area of two tennis courts. Patients with chronic obstructive pulmonary disease (COPD), however, have dramatically fewer functioning alveoli, and these are often enlarged and merged into large sacs. Making matters worse for COPD patients, the tissue that remains is typically scarred and has diminished elasticity. The Young–Laplace equation tells us that decreasing the surface tension, σ, and increasing the radii, R, both diminish the pressure drop, P. Reducing numbers of alveoli only worsens this effect, and as a result, a COPD sufferer often cannot generate enough air pressure to blow out a candle.

Example Asymmetric lung inflation

The Young–Laplace equation is simple enough, but applying it to lung inflation is surprisingly complicated, as can be illustrated by a simple demonstration. Eq. [2.4] tells us that a small sphere is more difficult to inflate than a large one. Consequently, if two identical balloons are connected to a Y-junction as sketched in Fig. 2.2(a), only one will inflate. This occurs because Eq. [2.6] implies that whichever balloon—either by chance or by initially holding it closed—has smaller radius will have larger pressure drop. This will cause the *other* balloon to inflate more, which will increase the pressure differential, making the larger balloon even easier to inflate.

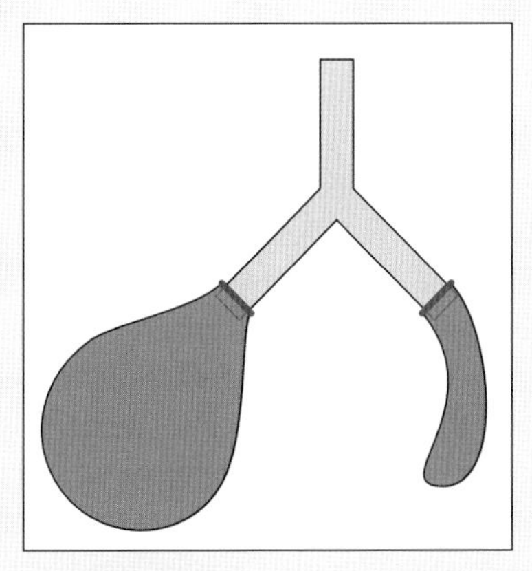

Fig. 2.2 Identical balloons attached to tube with symmetric Y-junction and inflated. Invariably, one balloon or the other will inflate, as can be deduced from the Young–Laplace equation.

EXERCISE 2.1

This simple effect raises an obvious question: since even identical balloons inflate asymmetrically, shouldn't lungs also suffer from this problem? All that the chest and diaphragm do during inflation is to maintain a pressure difference between the mouth and the interior of the body: isn't this the same as blowing into the Y-tube of Fig. 2.2? So why doesn't one lung inflate at the expense of the other? Moreover, if there is a defect in one limb of a branch, doesn't the Young–Laplace equation imply that the weaker limb will tend to enlarge all the more, compounding the defect?

Even without a defect, each lung has millions of branches, and as early as the fifteenth century Michelangelo performed dissections of cadavers (against objections by the church) and found throughout the body that asymmetry is the rule rather than the exception. So shouldn't intrinsic asymmetry in the lungs exacerbate the problem?

Perhaps part of the answer is that the pleural space between the lung and the chest lining is filled with fluid, so the lungs are physically pulled into an inflated state as the chest expands. But then what happens during pneumothorax? Pneumothorax can result from a puncture wound to the chest that allows air into the space between the lung and the chest cavity. The immediate treatment for pneumothorax has two parts: first, the puncture is left open during exhalation—this causes air to be expelled from the puncture as the chest cavity contracts. Fair enough. But second, the puncture is held closed (e.g. with pressure from an open palm) during inhalation—this causes air to be drawn into the lung through the mouth, after which the process is repeated until all the air is expelled from the chest cavity. This seems logical, but why does air enter the damaged lung when the puncture is closed: isn't this precisely the situation shown in Fig. 2.2 in which one balloon is more inflated than the other?

These questions don't have simple answers, and the reader may want to consult some of the literature on this topic (e.g. Mauroy et al.[4]).

We turn next to examining consequences of the factor of two difference between pressure on a sphere, Eq. [2.4], and pressure on a cylinder, Eq. [2.5]. This difference implies that a sphere can sustain twice the pressure difference that a cylinder of the same radius can. It is for this reason that a poodle tail in children's balloon figure can be made: as sketched in Fig. 2.3(a), the small radius of a tube can co-exist with the larger radius of a sphere at the same pressure. We note that in a real balloon, the actual sphere has more than twice the radius of the uninflated tube: this is due to the fact that real materials do not behave ideally, an issue that we have mentioned previously and that we will return to later in this book. Nevertheless, the fact that two different radii are stably present can be deduced as a straightforward consequence of differences between the Young–Laplace equations for a sphere and for a cylinder.

By similar reasoning, we can conclude that a spherical bleb on a tube (as in Fig. 2.3(b), representing for example an aneurism or an intestinal diverticulum) will tend to shrink (because the inward tensile force on the spherical bleb exceeds that on the cylindrical tube) *so long as* $R_{sphere} < 2{\cdot}R_{tube}$. This assumes that the material surface tension, σ, is the same in both spherical and tubular regions: again this is not fully realistic, for blebs tend to form due to a defect in the tubular wall. Nevertheless it is instructive to carry through an analysis of this idealized problem, which provides some generally applicable lessons.

To this end, let's analyze the forces on a small bleb on an elastic tube, shown in cross section in Fig. 2.3(c), with the goal of determining whether the bleb is stable, or if it will grow until it pops. The bleb is pushed outward by the pressure difference multiplied by the bleb's area, and as shown in the enlarged vector diagram in Fig. 2.3(c), the bleb is held inward by a force given by twice the radial component of tension at either end of the bleb, that is, $2T \cdot \sin(\vartheta)$. The bleb

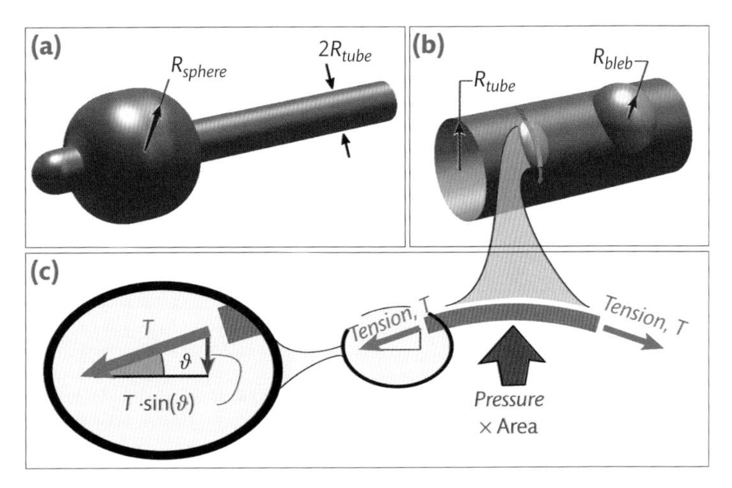

Fig. 2.3 (a) Depiction of a "poodle tail" on a tubular balloon, maintained in equilibrium because of the difference between spherical and cylindrical solutions for the Young-Laplace equation. (b) Small and large blebs on an elastic tube. (c) Free body diagram of central ribbon of bleb showing balance between outward force due to pressure (red) and inward component of tensions (blue), along with an enlargement identifying the inward component on one side of the ribbon, $T \cdot \sin(\vartheta)$.

is weakest at its center, so in Fig. 2.3(b) we sketch an area consisting of a ribbon traveling through the bleb center: if this area grows, the bleb will enlarge, and if the area shrinks, so will the bleb.

The area of a ribbon subtending an angle 2ϑ is $R_{bleb} \cdot 2\vartheta \cdot \Delta L$, where ΔL is the axial length of the ribbon, so the outward force caused by fluid pressure is

$$F_{out} = P \cdot R_{bleb} \cdot 2\vartheta \cdot \Delta L. \quad [2.7]$$

The tensile force, on the other hand, is the surface tension times the length:

$$F_{in} = 2\sigma \cdot \sin(\vartheta) \cdot \Delta L, \quad [2.8]$$

and at equilibrium $F_{in} = F_{out}$. In this case, and in many others throughout this book, equilibrium does not always exist, and when it does exist it isn't always stable. The emergence and disappearance of stable and unstable states is a common theme in biomedical fluid dynamics, and it is worthwhile to spend a moment understanding this, comparatively simple, example.

To analyze the problem of bleb growth, in Fig. 2.4 we plot Eqs. [2.7] and [2.8] for respectively the outward and inward forces per unit length in a few representative cases. In Fig. 2.4(a), we show the case $P \cdot R_{bleb} < \sigma$. There are two equilibrium points in this case, one at $\vartheta = 0$, and a second, circled in the figure, at a larger angle that we'll call ϑ_c. Here $\vartheta_c \approx 75°$, but it could be any angle depending on choices of P, R_{bleb}, and σ. As indicated in Fig. 2.4(a), at $\vartheta < \vartheta_c$ the force due to restraining tension is larger than that due to expanding pressure, so the bleb would contract. On the other hand, at $\vartheta > \vartheta_c$, the forces would be reversed so the bleb would grow—and as it does so, R_{bleb} would grow as well, increasing the difference between inward and outward forces.

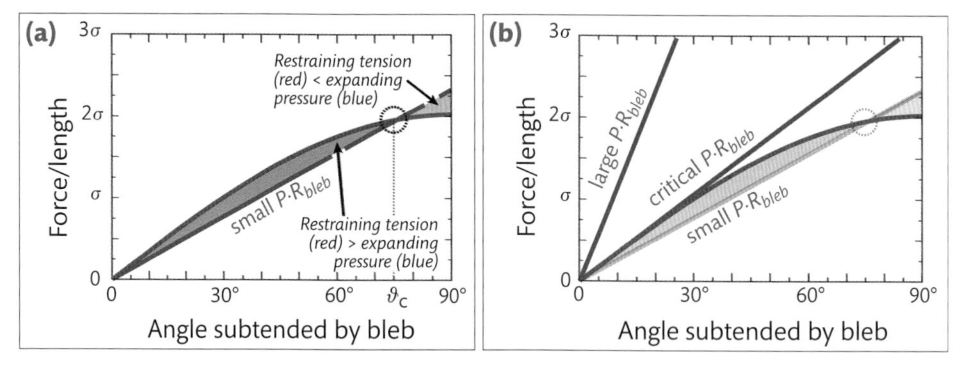

Fig. 2.4 (a) Restraining force (red, from Eq. [2.8]) and expanding force (blue, from Eq. [2.7]) on ribbon at center of bleb, both per unit length. To the left of ϑ_c (circled), the restraining force is larger than the expanding force, so the bleb will shrink; to the right, the bleb will grow without limit, that is, until it pops. (b) For larger pressures or bleb radii (or equivalently for lower bleb surface tension), the expanding force will always exceed the tension, so the bleb will always grow until it pops.

We call ϑ_c an "unstable fixed point:" "fixed" because at precisely $\vartheta = \vartheta_c$ there is no force on the center of the bleb so it would not change, and "unstable" because any tiny change from $\vartheta = \vartheta_c$ would increase with time. This system has a second fixed point, which is stable, at $\vartheta = 0$: at $\vartheta > 0$, the tension would reduce the angle, drawing the bleb closer to $\vartheta = 0$, while for $\vartheta < 0$ (corresponding to a dimple, rather than a bump), the tension would be outward with magnitude greater than the pressure difference—again reducing the angle ϑ. If we either increase the pressure, P, or the size of the bleb, R_{bleb} (or equivalently decrease the surface tension, σ, beyond a critical point at which $P \cdot R_{bleb}/\sigma = 1$), we would obtain the situation shown in Fig. 2.4(b), where the outward force exceeds the inward force for all angles. In this case, a bleb of any size would grow until it pops.

The upshot of this analysis is that if we know the pressure drop across a membrane and the surface tension at a defect location, we can easily calculate the size of bleb that can be sustained without popping. A bleb larger than this will need to be stabilized, for example, by surgery; a smaller bleb will not need intervention provided that the defect doesn't grow. Moreover, this analysis immediately tells us that the largest blood vessels, which can accommodate large radii of blebs, are more prone to catastrophic aneurism growth, as shown for $\vartheta > \vartheta_c$ in Fig. 2.4(a). Defects on smaller vessels can grow (as occurs in the formation of varicose veins), but tend to become stabilized on their own, as shown for $\vartheta < \vartheta_c$ in Fig. 2.4(a).

EXERCISE 2.2

The problem of bleb growth as we have modeled it is idealized in several respects. To consider just one idealization, we have neglected the inward pressure associated with tissues that may surround a tube—for example, a blood vessel. Rewrite Eq. [2.8] to include such an inward pressure: for simplicity, assume that there is a constant inward force, independent of ϑ. This implies that the blood vessel will have a fixed surface area and no (or negative) tension until a minimum force is applied, meaning that the vessel would collapse under the external pressure until the force inflates it. You can achieve this by including either a negative constant in Eq. [2.7] or a positive constant in Eq. [2.8]: the result will be the same. Replot Fig. 2.4, and describe what happens as $P \cdot R_{bleb}$ is increased from a small value. How does this behavior differ from the behavior plotted in Fig. 2.4? The presence of a minimum value that must be overcome before something occurs (e.g. growth of a bleb) is often termed an activation energy.

ADVANCED EXERCISE 2.3

Set P and σ to be constants of your choosing (1 is convenient, but anything reasonable will work). Increase the values of R_{bleb} starting from zero, and plot ϑ for all fixed points as a function of R_{bleb}, using solid lines if the fixed point is stable, and dashed lines if it is unstable. You should get a plot that looks like Fig. 2.5. The phenomenon by which a system exhibits a qualitative change in behavior—for example, from having three fixed points to only one—is termed a "bifurcation." This particular bifurcation is further called a "pitchfork bifurcation" because it looks like a pitchfork pointing to the left. For further details, the reader may want to consult texts on the subject, for example, Strogatz.[5]

Repeat the process for your new model, now associated with a finite activation energy. The two "bifurcation plots" will look very different. Refer to the literature and name this second bifurcation. Finally, note that it is a bit unsatisfactory that Figs. 2.4 and 2.5 involve axes with unknown units. Reproduce both plots using dimensionless variables by calculating the ratio between inward and outward forces rather than their dimensional (but not known) values.

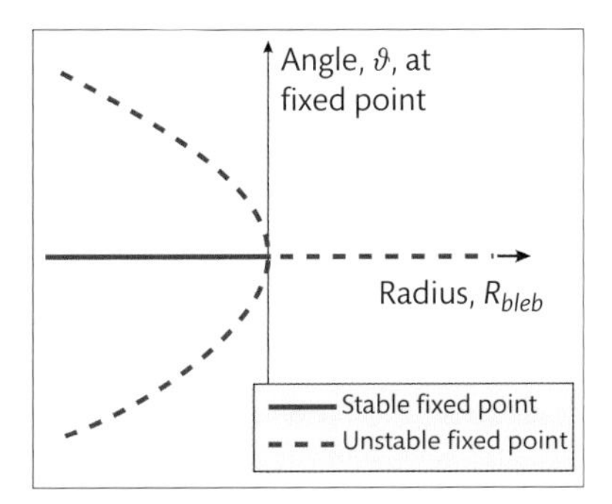

Fig. 2.5 Bifurcation plot for Eqs. [2.7] and [2.8]. The negative angle fixed point occurs when $F_{out} < 0$ and isn't physical.

EXERCISE 2.4

A simple kitchen experiment that illustrates some of the ideas that we have been discussing can be performed by adjusting a faucet to produce a thin stream of water, as sketched in Fig. 2.6. If the reader puts her finger beneath the stream and raises it up toward the faucet, something interesting and unexpected will be observed in the water stream. The assignments here are to perform this experiment, (1) sketch the interesting phenomenon, (2) develop a hypothesis to explain the interesting effect, and (3) propose an experiment to test this

Fig. 2.6 Kitchen experiment. Adjust a faucet so that a *thin stream* of water emerges. Put a finger in the stream and raise it toward the faucet. When the finger is 2 cm or less from the faucet, observe what happens to the stream *between the finger and the faucet.* Something interesting will occur; if you do not see anything interesting, repeat the experiment until you do. Provide a hypothesis to explain the interesting thing that you see. *Advanced:* plot the shape of the water stream that you see in three dimensions, for example, using Matlab's "surf" or "surfl" functions.

hypothesis. As a more advanced extension, (4) do the test and report whether the hypothesis is confirmed or refuted and (5) plot the interesting effect in three dimensions.

2.2 **Plateau-Rayleigh Instability**

A milestone in the analytic understanding of surface tension effects was placed by Joseph Plateau and Lord Rayleigh (1842–1919). We mentioned Plateau earlier in the chapter: he largely studied the curvature of soap films, and interestingly also developed the first prototype of a moving picture. Rayleigh was famous for "Rayleigh scattering," which makes the sky blue and for which he received the Nobel Prize, as well as for studies of heat transfer, after which the dimensionless Rayleigh number is named. Through a uniquely British tradition, Rayleigh is not the man's name at all: he was born John Strutt, but is known as Lord Rayleigh after succeeding his father, the 2nd Baron of Rayleigh, a feudal title with origins in the thirteenth century. The use of peerage titles occasionally causes confusions in citations, such as for Lord Kelvin, whose birthname was William Thomson and whose elder brother, James Thomson, studied beautiful patterns that form when ink is dropped into water—but never carried the Kelvin title.

The Plateau-Rayleigh effect causes water droplets to form from a tubular stream. Superficially, the reason that droplets form is that surface tension tends to contract an enclosed volume to the smallest possible area—a state described by a sphere. The detailed formation mechanism and size selection of droplets, however, is not evident from this simple picture. To clarify these details, we consider an elastic tube of water, and ask what happens to a small initial disturbance to the surface as shown in Fig. 2.7(a):

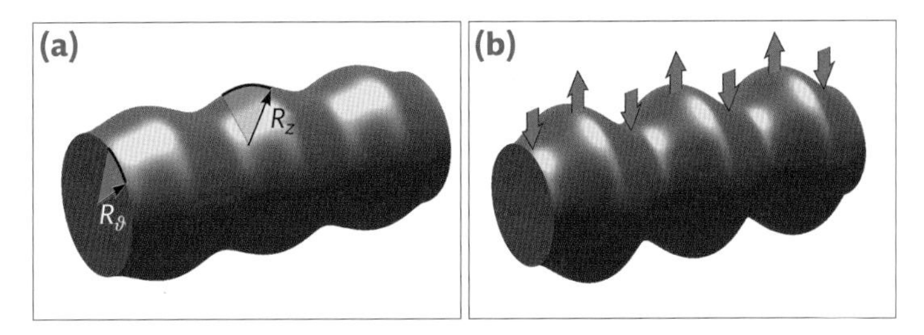

Fig. 2.7 Schematic of a sinusoidal perturbation to the surface of a liquid tube. (a) Azimuthal radius of curvature, R_ϑ, and axial radius, R_z, associated with perturbation. (b) Resulting growth of perturbation due to pressure differentials along tube.

$$R_\vartheta = R_0 + \epsilon_k \cdot \sin(kz), \qquad [2.9]$$

where k is the wavenumber ($k = 2\pi/wavelength$) and ϵ_k is the amplitude of the perturbation at that k.

In this problem, there is no applied pressure, and the stress produced within the tube is due to the fact that surface tension squeezes the tube. For the unperturbed tube, this stress is

$$S_{tube} = \frac{\sigma}{R_0}, \qquad [2.10]$$

whereas from Eq. [2.6], the stress in the presence of an initial disturbance as in Fig. 2.7(a) is

$$S_{perturbed} = \sigma\left(\frac{1}{R_\vartheta} + \frac{1}{R_z}\right). \qquad [2.11]$$

Conveniently, a theorem of analytic geometry provides that the radius of curvature in direction z for a surface defined by a function *f(z)* is given by:

$$\frac{1}{R_z} = \frac{d^2 f(z)}{dz^2}\left[1 + \left(\frac{df(z)}{dz}\right)^2\right]^{-3/2.} \qquad [2.12]$$

In our case, *f(z)* is given by Eq. [2.9], and to leading order in ϵ_k, Eq. [2.12] gives

$$\frac{1}{R_z} = -\epsilon_k k^2 \cdot \sin(kz), \qquad [2.13]$$

so the additional stress produced by the perturbation, again to leading order in ϵ_k, is

$$\begin{aligned}\delta S &= S_{perturbed} - S_{tube} \\ &= \sigma\left[\frac{1}{R_0 + \epsilon_k \cdot \sin(kz)} - \epsilon_k k^2 \sin(kz) - \frac{1}{R_0}\right] \\ &= \sigma \cdot \epsilon_k\left[\frac{1}{R_0^2} - k^2\right]\sin(kz).\end{aligned} \qquad [2.14]$$

Noting the minus sign, this says that the inward stress is reduced where the tube bulges out, and is increased where it indents in. This stress is strictly due to forces imparted by the surface tension from an equilibrium calculation, so the lower stress at the bulges makes the bulges grow.

This analysis implies that sinusoidal surface perturbations will grow as indicated in Fig. 2.7(b), until ultimately isolated droplets form. Recall, however, that we have made use several times of ϵ_k being small, so this result only applies to the initial growth of sinusoidal perturbations. Considerable research on what happens after the initial growth of a small perturbation has been carried out since Plateau and Rayleigh's original work, and we highlight two aspects of that research.

First, Eq. [2.14] shows that there is a pressure gradient that drives water from troughs to peaks of a sinusoidal perturbation. In reality, we know from experience that water droplets tend toward a particular size, while our analysis depends on wavenumber k, but gives no indication of how large or small k is. The size of droplets can be approximated by asking a somewhat different question, namely how does the speed of growth of perturbations depend on k, and at what k do sinusoidal perturbations grow fastest? That is, all wavenumbers appear to grow from our lowest order analysis, but in fact through detailed analysis of the rate of growth it has been shown that one particular wavenumber grows fastest: $k_{fastest} = 0.697/R_0$, which works out to be a wavelength of about 9 times the tube radius, R_0. The tube is pinched off at that wavelength before any other wavelength disturbance can grow large enough to pinch off the tube, and so this sets the size of droplets that result. The presence of a fastest growing mode—which is the one that nature chooses—is a particular example of a quite general result. The same thing occurs when a horn is blown or a violin is bowed: an infinite number of frequencies are possible, but the fastest growing one dominates.

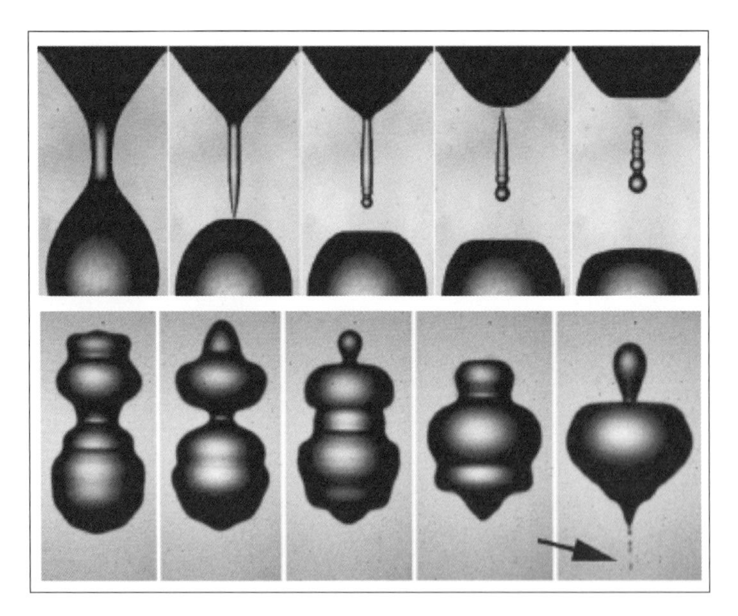

Fig. 2.8 Examples of complex patterns that form when water droplets pinch. Top row: pinch off of water from a 5 mm nozzle; bottom row: sequence of elaborate satellite patterns, culminating in the ejection of a high speed jet of microdroplets only 30 μm wide (arrow).[6] (From Thoroddsen, S.T., Etoh, T.G. and Takehara, K. (2007) Microjetting from wave focusing on oscillating drops. *Physics of Fluids, 19*(5). By permission of AIP Publishing LLC.)

Second, the actual dynamics of droplet formation are typical of the surprises that we mentioned characterize fluid behaviors. We can see this in the intricate patterns of Fig. 2.8. These patterns form spontaneously as a droplet falls from a faucet, but they appear and disappear too rapidly for the human eye to capture. Nevertheless, such transient effects are important for the function of inkjet printers, for the fabrication of novel materials such as controlled diameter microspheres, and for microfluidic technologies that make use of droplet breakup to produce encapsulated droplets.

2.3 **Dimensional Analysis**

A fundamental aspect of scientific analysis that we have referred to previously concerns the definition of dimensionless numbers, also called dimensionless groups, that define a behavior of interest. This topic has led historically to some misunderstanding, even exasperation, and we flesh out some of the underlying issues here.

The significance of performing "dimensional analysis" was formally put on record by Rayleigh, who wrote a somewhat cutting letter to *Nature* in 1915 on what he termed the principle of "similitude," which we have referred to (see beginning of Chapter 1) as dynamical similarity. In retrospect, it is almost amusing how oddly the matter was treated historically, but at the time, it must have provoked great and justifiable fury. Some appreciation for Rayleigh's attitude can be surmised from his opening paragraph,[7]

> I have often been impressed by the scanty attention paid even by original workers . . . to the great principle of similitude. It happens not infrequently that results in the form of "laws" are put forward as novelties on the basis of elaborate experiments, which might have been predicted *a priori* after a few minutes' consideration.

Rayleigh went on to describe several specific examples of exactly how dimensionless analysis could be performed to obtain dimensionless groups that correctly predict the terminal velocity of a ball falling through air, the strength of bridges, the period of a tuning fork, and other quantities.

At this point, anyone fluent in the English language would recognize that Rayleigh, an eminent scientist who had received the Nobel prize 11 years previous, was irritated by widespread ignorance of a basic issue that he felt was deserved due consideration. Remarkably, his letter provoked an objection 4 months later by a largely unknown aeronautical engineer named M. Riabouchinsky, who wrote a half column, 8 line objection to a technical point in Rayleigh's letter. We can only imagine Rayleigh's frustration. His predictable terse and sarcastic reply was supplemented by an important clarification after four further months by Edgar Buckingham (1867–1940), who formally defined what is now known as the "Buckingham pi theorem:" a second general method for obtaining dimensionless groups.[8] In this section, we describe both Rayleigh's and Buckingham's approaches to obtain dimensionless groups.

2.3.1 **Rayleigh's Approach**

Rayleigh's method can be summarized as follows. Suppose that a given problem has dimensions of mass, M, length, L, and time, T, and depends on quantities A, B, C, D, etc. These quantities can be anything, and need not be either unique or independent. For example, to model a tuning fork, A could be the oscillation period (of dimension T), B could be its length (L), C its density ($\rho \sim M/L^3$), and D its Young's modulus ($Y \sim M/(LT^2)$). Rayleigh would write these quantities as an equation:

$$\begin{aligned} const. &= A^a B^b C^c D^d \\ &= T^a L^b \left(\frac{M}{L^3}\right)^c \left(\frac{M}{LT^2}\right)^d \cdots, \end{aligned} \qquad [2.15]$$

where a, b, c, d, etc. are numbers (not necessarily integers). Rayleigh reasoned that if the product really is a constant, then it shouldn't depend on the units chosen for mass, length, or time, and so must have no dimensions. Hence he would collect terms in like powers of M, L, and T:

$$const. = T^{a-2d} L^{b-3c-d} M^{c+d}, \qquad [2.16]$$

from which he would obtain equations for the powers a, b, c, d, etc. Here:

$$\begin{aligned} 0 &= a - 2d \\ 0 &= b - 3c - d, \\ 0 &= c + d \end{aligned} \qquad [2.17]$$

so that $d = -c$, $b = 2c$, and $a = -2c$. So Eq. [2.15] becomes

$$const. = T^{-2c} L^{2c} \rho^c Y^{-c}, \qquad [2.18]$$

or in more familiar terms, the period of a tuning fork can be concluded to be:

$$T = C \cdot L \sqrt{\frac{\rho}{Y}}, \qquad [2.19]$$

where C is a new constant ($C = \sqrt{1/const.^c}$). The square root in Eq. [2.19] has dimensions time ÷ length, and is in fact the inverse propagation speed of sound in the tuning fork.

Thus we obtain a standard result that is usually derived from more sophisticated differential equations, here using little more than the knowledge that there has to be some dimensionless way of grouping the quantities that define a problem (Eq. [2.15]).

2.3.2 **Buckingham's Approach**

Buckingham's method is equivalent, but is a little more general. Buckingham proposed the following thought experiment. Suppose we already *knew* the equations governing a system. Buckingham would ask, "how many dimensionless constants would there be?" If there were k known equations in n quantities, then if $k = n$, the problem would be exactly solvable, but if $k < n$, there would be $n - k$ constants (or freedoms) to be prescribed. And, significantly, it is up to us—the scientists studying the problem—to decide which constants we want to know about.

We'll look at examples shortly, but Buckingham's idea was that if we wanted to know how the inertia of a body moving through a fluid competes with the fluid's viscosity, we would look for a dimensionless group that measures that competition (which we know from Chapter 1 is the Reynolds number). If, on the other hand we wanted to know how kinetic and potential energies of a fluid stream play off of one another, we might define a different number (which we will see is termed the Froude number). Or, we could be interested in the buoyancy (the Rayleigh number) or shock waves (the Mach number) associated with a moving body. All of these numbers (and more) are obtainable, and the choice of which to look for depends on what problem is of interest—as we will see next.

2.3.3 Comparison between Rayleigh's and Buckingham's Approach

Let's take a concrete example: suppose we want to know how a solid body falling through a fluid under gravity behaves (we will analyze this problem in greater detail in Chapter 6). A partial list of quantities that we expect this problem to depend on could include the following:

F_D: drag force
λ: size of the object (say its diameter)
μ: viscosity of the fluid
ρ: density (e.g. the difference in density between the object and the fluid)
V: speed of fall
g: gravity

Rayleigh:

To make things simple, let's solve for F_D: this is essentially the same as Eq. [2.15], but puts the constant on the right hand side and has the very slight advantage that when we're done, we'll have an expression for the drag. We could instead use the form shown in Eq. [2.15] and then calculate F_D, but the present expression saves one step at the end. So we start with:

$$F_D = const \cdot \lambda^a \rho^b \mu^c V^d g^e. \qquad [2.20]$$

Performing dimensional analysis as in Eq. [2.16], we obtain:

$$\frac{ML}{T^2} = const \cdot L^a \left(\frac{M}{L^3}\right)^b \left(\frac{M}{LT}\right)^c \left(\frac{L}{T}\right)^d \left(\frac{L}{T^2}\right)^e, \qquad [2.21]$$

or:

$$M^1 L^1 T^{-2} = const \cdot M^{b+c} L^{a-3b-c+d+e} T^{-c-d-2e}, \qquad [2.22]$$

so:

$$\begin{aligned} 1 &= b + c \\ 1 &= a - 3b - c + d + e \, . \\ -2 &= -c - d - 2e \end{aligned} \qquad [2.23]$$

This gives us three equations in six unknowns, so as Buckingham made explicit we must choose which three quantities we want to know about and which three we will leave unknown. For the sake of this exercise, let's solve for a, b, and d, and leave F_D, c, and e unknown. Eq. [2.23] can be solved to produce

$$\begin{aligned} a &= 2 - c + e \\ b &= 1 - c \\ d &= 2 - c - 2e \end{aligned} \, , \qquad [2.24]$$

so that Eq. [2.20] gives us:

$$F_D = const \cdot \lambda^{2-c+e}\rho^{1-c}\mu^c V^{2-c-2e}g^e. \quad [2.25]$$

Finally, these terms can be grouped to isolate dimensionless ratios, for example:

$$F_D = const \cdot \left(\frac{V^2}{\lambda g}\right)^{-e}\left(\frac{V\lambda\rho}{\mu}\right)^{-c}\rho\lambda^2V^2. \quad [2.26]$$

This equation gives the drag force on a falling body. It is correct for any e or c, so the quantities in parentheses must be dimensionless; likewise we can divide both sides by $\rho\lambda^2V^2$ to produce a third dimensionless group, in all providing three dimensionless numbers, as follows:

$\left(\frac{V^2}{\lambda g}\right)$.........“Froude” number

$\left(\frac{V\lambda\rho}{\mu}\right)$.........“Reynolds” number [2.27]

$\left(\frac{F_D}{\rho\lambda^2V^2}\right)$.........“Drag coefficient”

We have seen the Reynolds number previously, and we will derive the other numbers later in this book. Even without detailed derivations, however, we can deduce unexpectedly useful knowledge. For example the drag coefficient equation suggests, all other things being equal, that the drag force on a moving object can be expected to grow with its area (λ^2) and speed squared. Neither conclusion is obvious on its face, but we will see later that, details aside, both conclusions are essentially correct.

Buckingham:

Rayleigh’s approach is entirely rigorous, but the reader may have noticed that its derivations are a little haphazard: for example we chose to solve for *a*, *b*, and *d* more or less arbitrarily with no physical guidance for what this choice might imply, and likewise we grouped terms at the end without a clear or deliberate intention. Buckingham’s approach is more methodical in these respects. As before, we identify the same six quantities, $F_D, \lambda, \mu, \rho, V$, and g that the problem is believed to depend on, and we anticipate that there will be three dimensional quantities, *M*, *L*, and *T*, involved. Buckingham would then conclude that there must be $6 - 3 = 3$ dimensionless groups to be obtained.

Just as in Rayleigh’s method, this means that three things will be solved for (previously *a*, *b*, and *d*) and other things will be left unknown (previously *c*, *e*, and F_D). In the Buckingham method, however, the things that will be solved for will be deliberately chosen based on what, physically, we want to know about. We will solve for these things in terms of dimensionless groups that Buckingham denoted π_1, π_2, π_3. Because of his choice of the letter π to define the dimensionless groups, his method is called the “Buckingham π” method. Additionally, the things that are left unknown are termed “repeating variables” in his method, but this is purely semantic.

The algorithm for the Buckingham method begins by *choosing* the three π quantities that we want to know about. Suppose, for example, that we want to know what competes against the drag force, F_D, the viscosity, μ, and gravity, g. Then we would choose these to explicitly appear linearly in each of the three π quantities, as follows:

$$\begin{aligned}\pi_1 &= F_D \cdot \rho^{a_1} V^{b_1} \lambda^{c_1} \\ \pi_2 &= \mu \cdot \rho^{a_2} V^{b_2} \lambda^{c_2} \\ \pi_3 &= g \cdot \rho^{a_3} V^{b_3} \lambda^{c_3}.\end{aligned} \quad [2.28]$$

Now our job is to find $a_1 \ldots c_3$ that make π_1, π_2, π_3 dimensionless. This is much more methodical and systematic than before: when we are done, we will know that each of F_D, μ, and g will appear respectively in one dimensionless group. In the Rayleigh method, by comparison, we knew that F_D would appear in one group, and we could maybe have figured out how the choice of a, b, and d would affect what remained, but much was left to chance.

So, let's set about to find $a_1 \ldots c_3$. We do this by making use of the fact that the π s are dimensionless, so that:

$$\pi_1 = M^0 L^0 T^0 = \left(\frac{ML}{T^2}\right)\left(\frac{M}{L^3}\right)^{a_1}\left(\frac{L}{T}\right)^{b_1}(L)^{c_1}, \quad [2.29]$$

which we manipulate to set dimensions equal, giving us three equations in a_1, b_1, and c_1:

$$\begin{aligned}M &: 0 = a_1 + 1 \\ L &: 0 = 1 - 3a_1 + b_1 + c_1 \\ T &: 0 = -2 - b_1.\end{aligned} \quad [2.30]$$

These can be solved to give $a_1 = -1, b_1 = -2, c_1 = -2$, so that:

$$\begin{aligned}\pi_1 &= \frac{F_D}{\rho\lambda^2 V^2} \\ \pi_2 &= \frac{\mu}{V\lambda\rho} \\ \pi_3 &= \frac{g\lambda}{V^2}.\end{aligned} \quad [2.31]$$

Thus Buckingham's approach gives the same result as Rayleigh's approach in this case, but the utility of the former should be obvious: the method lends itself naturally to providing explicit control over what quantities are studied. Moreover if we included an additional quantity—for example, the speed of sound—or removed one—for example, gravity—the method would readily and directly let us establish what the effect would be.

EXERCISE 2.5

We have seen that the Reynolds number appears as the relevant dimensionless group in derivation of flow through a smooth and rigid pipe. An important problem that arises in many applications is the analysis of the onset of turbulence in pipes—we haven't looked at all at turbulence yet, but it turns out that the surface roughness of the pipe plays an important role in its onset. Consider then flow in a rough-walled pipe, where there are five relevant parameters: the pressure drop Δp, the pipe length, λ, the fluid viscosity, μ, its density, ρ, and the pipe roughness, ϵ (defined for example to be a mean distance from peaks to troughs of surface asperities).

For this problem, answer several questions. (1) How many πs will there be? (2) Before you perform a dimensional analysis, what does your intuition tell you about what the roughness will compete against (i.e. guess a dimensionless group involving ϵ)? (3) Use the Buckingham π approach to solve for however many dimensionless groups can be obtained. Include ϵ and λ as two of these. (4) Comment on the physical meaning: what do roughness and length compete with in pipe flow? Why should the pipe length matter (we will answer this question in a later chapter)?

EXERCISE 2.6

We have seen that droplets can form from a tube of water subjected to sinusoidal surface perturbations. Consider water flowing from a faucet (freely, without an interruption as in Exercise 2.4). This differs from the problems studied previously: in the present problem, the water is accelerated downward due to gravity, and so the water stream must travel faster lower down. To conserve mass, this implies that the stream must either become infinitely thin or break up into separating drops. Perform both a Rayleigh and a Buckingham dimensional analysis to determine at what distance from the faucet droplets can be expected. Use any parameters that you feel are relevant. Comment on the physical meaning of your result, especially compare your result with the droplet formation problem associated with the Plateau-Rayleigh instability: are the two processes related?

Example Measurement of red blood cell membrane tension

We have seen that the Young–Laplace equation can be used to study forces acting on tubular membranes and the dynamics that result. This equation has also been used as an assay for cell membranes. Red blood cells (RBCs), in particular, have many unique properties that the equation can be used to probe, including dynamical transitions between tumbling and so-called "tank-treading," as well as complex shape transitions. Indeed, the entire structure of these cells is unusual: they have no nucleus, they have a truss-like cytoskeletal network, and their membrane is more rugged than one would anticipate.

Importantly, RBCs withstand extreme stresses in passage through the heart, and squeeze in and out of narrow capillaries about once a minute. The cells are 8 µm at their largest diameter, yet deform to fit into capillaries as small as 3 or 4 µm. With a half-life of about 40 days, each cell undergoes deformations tens of thousands of times in its lifetime. Yet RBCs sustain this abuse with membranes that are only 7–10 nm thick.

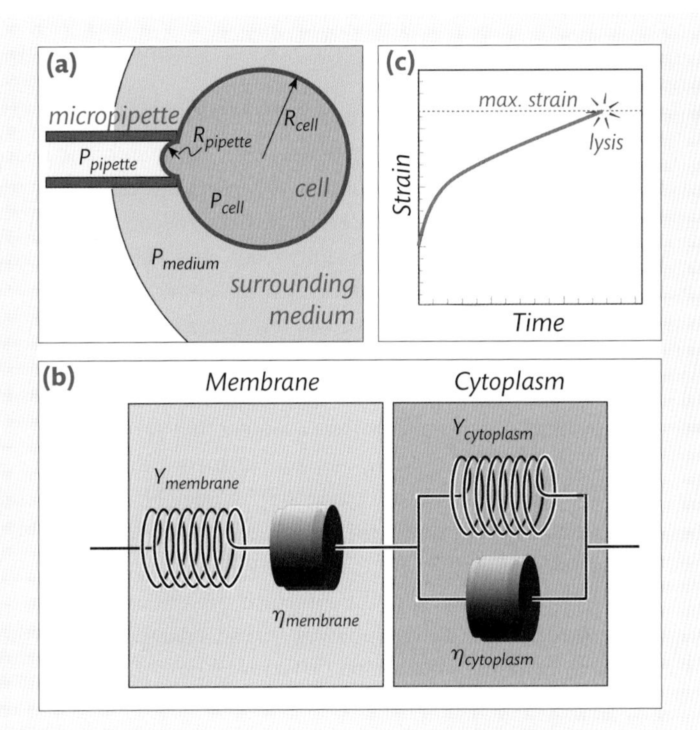

Fig. 2.9 (a) Schematic of cell membrane being drawn into pipette. (b) Compartment model consisting of a membrane compartment (blue) and a cytoplasm compartment (pink), each modeled with a spring and damper. (c) Response of this model to sudden stress, assuming that cell lysis occurs once the deformation exceeds a maximum value.

One method that scientists have employed to probe the RBCs impressive mechanical properties has been to perform the simple experiment shown in Fig. 2.9(a). As sketched in the figure, a pipette draws part of a RBC inside, and the pressures and radii of curvature indicated are then measured. These measurements are then input into the Young–Laplace equation for a sphere to infer the surface tension involved in deforming the cell.

Explicitly, the Young–Laplace equation applied to the pressure drop between the cell and its surrounding medium gives:

$$P_{cell} - P_{medium} = \frac{2\sigma}{R_{cell}}, \qquad [2.32]$$

while the pressure drop between the cell and the pipette obeys:

$$P_{cell} - P_{pipette} = \frac{2\sigma}{R_{pipette}}. \qquad [2.33]$$

We can eliminate P_{cell} from Eqs. [2.32] and [2.33] to obtain:

$$\sigma = \frac{P_{medium} - P_{pipette}}{2\left(\frac{1}{R_{pipette}} - \frac{1}{R_{cell}}\right)}. \qquad [2.34]$$

Everything here can be measured, so in principle σ can be determined directly to calculate the surface tension of the RBC membrane.

This has been done, but in practice this approach suffers from two significant shortcomings. First, it is observed experimentally that when a cell membrane is drawn into a small pipette, it will lyse (break) after some time—yet the Young–Laplace equation is time independent and says nothing about this observed behavior. Second, the cell cytoplasm is known to play a significant role in supporting loads imposed on the cell, and the cytoplasm isn't included in the Young–Laplace equation either.

To address these shortcomings, a second way of probing RBC properties has been constructed.[9] This involves a so-called "compartment" model: one that encapsulates complicated behaviors into separate modules containing the essential dynamics of interest. Very often, these compartment models treat a system that varies over both space and time in terms of a set of simplified modules, or compartments, that only vary over time: semantically this is also termed a "lumped parameter" model. However the semantic pie is sliced, a simplified model for the mechanics of RBCs is sketched in Fig. 2.9(b) based on work that appears in Buckingham.[8]

In this model, both the membrane and the cytoplasm are considered to have an elastic component characterized by the Young's modulus, Y, and a damping component defined by a term, η. In recognition of the observation that a membrane will deform continuously under stress, the membrane is provided with a viscous damper in *series*, as shown in Fig. 2.9(b). This approach is widely used to describe materials that deform continuously (called "creep"), and is referred to as a "Maxwell" model, after James Clerk Maxwell, who is discussed in Chapters 7 and 10. The complete model in Fig. 2.9(b) has an analytic solution, provided in Buckingham,[8] to a sudden imposition of stress as would occur with sudden initiation of suction from a pipette:

$$Strain = Stress \cdot \left[\frac{1}{Y_{membrane}} + \frac{1}{\eta_{membrane}} t + \frac{1}{Y_{cytoplasm}} - \frac{1}{\eta_{cytoplasm}} e^{-Y_{cytoplasm} t / \eta_{cytoplasm}} \right]. \qquad [2.35]$$

In Fig. 2.9(c) we plot this solution for arbitrary choices of constants. In actual experiments, the constants have been fit to data, resulting in estimates:

$$Y_{membrane} \approx Y_{cytoplasm} \approx 10^7 - 10^8 \ dyne/cm^2$$
$$\eta_{membrane} \approx 10^9 - 10^{11} \ Poise$$
$$\eta_{cytoplasm} \approx 10^8 - 10^9 \ Poise.$$

These Young's moduli are close to that of collagen, and are in agreement with more recent and comprehensive measurements and computations based on a detailed understanding of the RBC spectrin cytoskeleton. This, then, isn't a surprise. The viscosities, on the other hand, are surprising: these are extremely high—about the same as tar at room temperature, and it is believed that the resulting very strong damping plays an important role in dissipating the stresses associated with the RBCs continual and energetic deformations.

EXERCISE 2.7

The behavior of elastic surfaces can be remarkably ornate. The reader may want to have a look at the journal The *Gallery of Fluid Motion*, which presents winners of an annual American Physical Society competition, for examples of some of these extraordinarily beautiful structures. We provide two exercises here involving kitchen-sink experiments that the reader can easily perform to get a taste for some of these effects. A first example is shown in Fig. 2.10: here we show patterns that appear spontaneously when a drop of food coloring is carefully placed on the surface of a shallow (about 4 mm) dish of clean water.

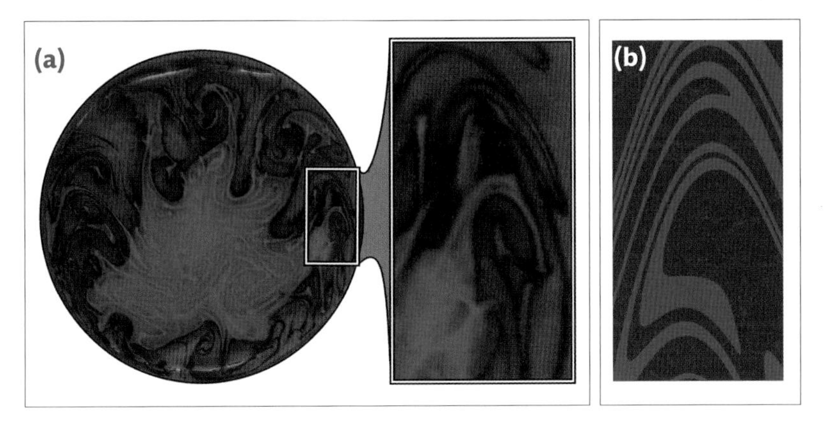

Fig. 2.10 (a) False colored snapshot of food coloring (blue) dropped into water (red). Enlargement shows stretching and folding produced by oscillations caused by periodic surface tension variations. (Source: Carlos E. Caicedo-Carvajal.) (b) Comparable stretching and folding patterns produced in a portion of the sine flow described in the previous chapter as an example of chaotic mixing produced by periodic flow.

The food coloring reduces the water's surface tension, and causes periodic oscillations on the surface that lead to the pattern shown in Fig. 2.10(a). This snapshot is false colored; notwithstanding this embellishment, a wide variety of intricate and surprising patterns can be produced in this simple experiment. In Fig. 2.10(b), we show a comparable pattern taken—again false colored—from the sine flow described in Chapter 1 as an example of a chaotic pattern that arises due to periodic forcing of a simple flow.

For this exercise, repeat the experiment described and document the patterns that you see: there are several effects to be observed, of which the chaotic mixing shown is only one.

EXERCISE 2.8

A second remarkable effect involving surface tension that can be observed in the kitchen is shown in Fig. 2.11. Here, clean water is poured into a cup containing floating tea leaves. The leaves reduce the surface tension of the water, so that there is greater tension upstream, in the clean water, than downstream, in the cup. This draws leaves up into the pot as shown in the figure. This works with cold or hot water and with a variety of floating materials.

Fig. 2.11 Preparation of mate tea, showing tea leaves climbing up a waterfall into a pot of water. Enlargement to the right shows tea leaves that have made their way from the cup into the pot, up a waterfall about 1 cm in height. (Source: Sebastian Bianchini, Alejandro Lage, and Ernesto Altshuler.)

Reproduce the effect and explain why the tea travels in the patterns indicated by arrows in the enlargement to the right of Fig. 2.11, up the sides of the flow and down the center. Carefully observe the flow patterns in the cup as well, and explain what makes the patterns that you see occur.

REFERENCES

1. Plateau, J. (1880) On the superficial viscidity of liquids. *Science, 1*, 298–302 (translated by C. Lanza).
2. Marangoni, C. (1878) Difesa della teoria dell'elasticita superficiale dei liquidi. Plasticá superficiale. *Nuovo Cimento (ser 3), 3*, 50–68, 97–115, and 193–211.
3. Young T. (1805) An essay on the cohesion of fluids. *Philosophical Transactions of the Royal Society, 95*, 65–87.
4. Mauroy, B., Filoche, M., Andrade, J.S., Jr. and Sapoval, B. (2003) Interplay between geometry and flow distribution in an airway tree. *Physical Review Letters, 90*, 148101; Mauroy, B., Filoche, M., Weibel, E.R. and Sapoval, B. (2004) An optimal bronchial tree may be dangerous. *Nature, 427*, 633–6.
5. Strogatz, S. (1994) *Nonlinear Dynamics and Chaos with Applications to Physics, Biology, Chemistry and Engineering*. Reading, MA: Perseus Books.
6. Thoroddsen, S.T., Etoh, T.G., and Takehara, K. (2007) Microjetting from wave focusing on oscillating drops. *Physics of Fluids, 19*, 052101.
7. Rayleigh, Lord J. (1915) The principle of similitude. *Nature, 95*, 66–8.
8. Buckingham, E. (1915) The principle of similitude. *Nature, 96*, 396–7.
9. Rand, R.P. (1964) Mechanical properties of the red cell membrane. *Biophysics Journal, 4*, 115–32 and 303–16.

3 Flow through Elastic Tubes

Now that we have some basic equations describing how both fluids and elastic surfaces behave, we can put the information together to understand how a fluid, such as blood, flows in an elastic tube, such as a vein or artery, as sketched in Fig. 3.1. We describe here how the two different effects, flow and elasticity, interact.

3.1 Flow in Elastic-Walled Tubes

This will be something of an involved derivation, as it combines all of the elements of analysis that we have introduced up to this point, and it is worth outlining the elements of the derivation before setting to work. The plan of attack will be in five parts.

(1) First, we will use the continuity equation to determine how flow of the fluid alone produces a change in radial velocity.

(2) Next, we will use the Young–Laplace equation to evaluate how the elastic membrane alone produces a change in pressure due to a change in radial velocity.

(3) We will combine the results of steps 1 and 2 to relate the pressure to the fluid velocity.

(4) Then we will use an approximate treatment to model how the fluid velocity responds to changes in pressure.

(5) Finally, we will solve the resulting equation for both pressure and velocity.

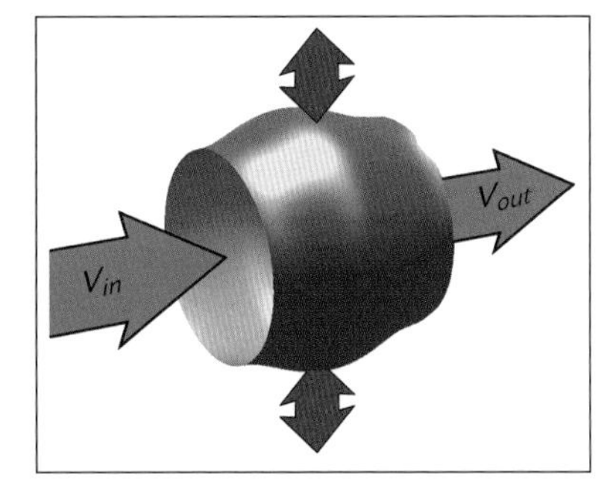

Fig. 3.1 Fluid flow through an elastic tube. Because the tube is elastic it can expand or contract (red arrows) to accommodate a difference in rates of flow into and out of (blue arrows) a region of the tube.

Biomedical Fluid Dynamics: Flow and Form. Troy Shinbrot.

DOI: 10.1093/oso/9780198812586.001.0001

This derivation will provide us both with an answer to the specific question of how surface elasticity affects flow and with tools for the analysis of problems in which different systems are coupled together: a ubiquitous situation in biological systems.

3.1.1 **Continuity Equation**

We begin by assuming that our fluid is incompressible and (as in the Poiseuille flow derivation of Chapter 1) that there is no flow in the azimuthal, φ, direction. In this case, there is only flow in the radial and axial directions, V_r and V_z, and the continuity equation tells us that:

$$\begin{aligned} 0 &= \vec{\nabla}\cdot\vec{V} \\ &= \frac{1}{r}\frac{\partial}{\partial r}(r\cdot V_r) + \frac{\partial V_z}{\partial z}. \end{aligned} \quad [3.1]$$

If we integrate this equation across a cross section of the tube, we get:

$$\begin{aligned} 0 &= \int_0^{2\pi}\int_0^R \frac{1}{r}\frac{\partial}{\partial r}(r\cdot V_r) r\, dr\, d\varphi + \int_0^{2\pi}\int_0^R \frac{\partial V_z}{\partial z} r\, dr\, d\varphi \\ &= 2\pi R V_r|_R + \frac{\partial}{\partial z}\int_0^R (2\pi r) V_z dr. \end{aligned} \quad [3.2]$$

We divide this equation by the area of the cross section, πR^2, to give:

$$\frac{2}{R}V_r|_R = -\frac{\partial \langle V_z \rangle}{\partial z}, \quad [3.3]$$

where $\langle V_z \rangle$ is the axial velocity, V_z, averaged over the cross sectional area of the tube. Eq. [3.3] tells us that the tube moves outward at the wall in response to changes in average velocity through the tube: a fact that makes sense, as suggested by Fig. 3.1.

3.1.2 **Young-Laplace Equation**

The next thing we want to know is how the elastic surface of the tube responds to the radial velocity in Eq. [3.3]. To solve for this, we start again from Newton's second law, applied here to a unit area on the surface, which heuristically is just:

$$\frac{mass\cdot acceleration}{area} = pressure\, drop - \frac{elastic\, force}{area}, \quad [3.4]$$

or mathematically:

$$\rho h \frac{\partial^2 (r - R)}{\partial t^2} = P - P_o - \frac{\sigma}{R}, \quad [3.5]$$

where ρ is the density of the elastic surface, h is its thickness, r is the radius of the surface, and R is a nominal radius of the unperturbed surface. Also, P is the pressure inside the tube, P_o is the pressure outside, and σ is the surface tension, as described in Chapter 2, Eq. [2.5]. In Chapter 2, we defined the surface tension, σ, as a material property, but we remarked that realistically it may not be a simple constant. For an elastic tube, such as a rubber band or a blood vessel, we can approximate how σ depends on deformation (which after all is what we are studying here) as follows.

Suppose we stretch a simple rubber band. We expect the force to follow Hooke's law (as described in Chapter 2), $F = -k \cdot x$, where x is the deformation of the band. In this equation, Hooke's constant k goes as $Y{\cdot}h$, where Y is Young's modulus for the rubber, and h is the band's thickness. In our problem, the deformation $x = 2\pi(r - R)$, and the unit circumferential, or "hoop," length of the tube is $2\pi R$, so the force per unit length, σ, should obey

$$\sigma = \frac{F}{L} = \frac{Yh \cdot (r - R)}{R}, \qquad [3.6]$$

so Eq. [3.5] becomes:

$$\rho h \frac{\partial^2 (r - R)}{\partial t^2} = P - P_o - \frac{Yh \cdot (r - R)}{R^2}. \qquad [3.7]$$

We want to make our problem as simple as possible (we will see that it becomes complicated quickly enough as it is), so let's ask if there is anything that we can reasonably get rid of. The pressure drop and the surface tension are central to our entire problem, which leaves us only with the left hand side of Eq. [3.7] to potentially discard. Let us then ask how large the left side of the equation is compared with the other terms. The pressure drop is externally applied (from the heart in the problem of blood flow) and could be of any value, so let's compare the terms in the displacement, $(r - R)$, on the left and right sides of Eq. [3.7]. If we perform dimensional analysis on these terms, we find, after cancelling factors in common, that to determine how large the left side of Eq. [3.7] is, we need to compare $1/T$ on the left side with $\sqrt{Y/\rho R^2}$ on the right. The T here is a characteristic time over which acceleration takes place, and for the problem of blood flow, the T is provided by forcing by the heart, which occurs on the order of once a second. This is to be compared with the term on the right side, which necessarily also has dimensions of inverse time, and represents the natural frequency of oscillation of the wall of the blood vessel. This can be calculated by inserting typical values for Young's modulus, thickness and radius of a blood vessel, yielding a frequency on the order of 1000 times a second. So the right side of Eq. [3.7] is around a thousand times larger than the left side, so we take the liberty of neglecting the latter.

The physical meaning of what we have done is to observe that arterial walls respond much more rapidly than the rate at which they are forced, so the wall motion is tightly slaved to the driving. Looked at another way, the arterial wall is much lighter and more responsive than the blood it contains, so the inertia of the wall is tiny compared with that of the blood. Disregarding acceleration terms (the left side of Eqs. [3.4] and [3.7]) is therefore termed an "inertia-less approximation." This makes Eq. [3.7] become:

$$P = P_o + \frac{Yh \cdot (r - R)}{R^2}. \qquad [3.8]$$

We differentiate this to obtain the rate of change of pressure:

$$\frac{\partial P}{\partial t} = \frac{Yh}{R^2}\frac{\partial r}{\partial t} = \frac{Yh}{R^2} V_r|_R \ . \qquad [3.9]$$

3.1.3 Combining Continuity and Young-Laplace results

This is encouraging, since we now have two equations, Eqs. [3.3] and [3.9], containing $V_r|_R$ that we can combine, producing:

$$\frac{\partial P}{\partial t} = -\frac{Yh}{2R}\frac{\partial \langle V_z \rangle}{\partial z}. \qquad [3.10]$$

Example Balloon farts and whistling

Before we move on, let's take a moment to understand our latest result. Because of the minus sign, Eq. [3.10] tells us that the pressure will *increase* in time when the downstream velocity *decreases* with z. Consider an example: suppose we hold the mouth of an inflated balloon between pinched fingers, as sketched in Fig. 3.2(a). Because the balloon is constricted by the fingers, to conserve mass the air must travel faster in the constriction than away from it. This means that the air has to speed up going into the constriction and slow down coming out of it. That is, in Fig. 3.2(a), $\partial\langle V_z\rangle/\partial z > 0$ to the left of the fingers, and $\partial\langle V_z\rangle/\partial z < 0$ to their right—and so from Eq. [3.10], $\partial P/\partial t > 0$ to the right of the fingers. So as air leaves the balloon, the pressure will increase in time, pushing the elastic walls out.

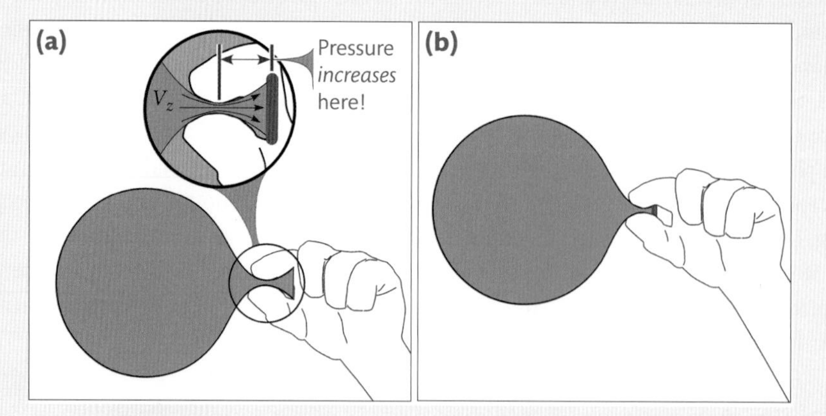

Fig. 3.2 Air release from balloon produces (a) a farting sound according to Eq. [3.10] if slack at balloon mouth is long enough to produce fast pressure growth, or (b) no sound if mouth is closer to fingers.

The elastic walls will do what elastic does: spring back. This produces the farting sound that we are all acquainted with when air is let out of a balloon as the balloon walls expand outward and recoil back inward.

Notice that Eq. [3.10] tells us that the sound results from pressure growth in the extended elastic end of the balloon—so if we move our fingers to cinch up on the end as shown in Fig. 3.2(b), the farting must stop. To some degree one can argue that there is no farting at that point because there is no free balloon left, but we raise two points. First, the reader can confirm that even when there is slack to the left of the fingers (which occurs after a balloon has been inflated for a time), no noise will be produced.

Second, another example of the same thing can be produced by whistling: it is the free end (the lips) that produces the whistling sound. Any reader who can whistle can confirm that by pulling her lips inward: the whistling sound will stop—and any reader who cannot whistle now knows the cause. The reader can demonstrate this either by pushing her lips outward—so that their elasticity enables vibrations that result in a whistle—or by sucking air in with lips fixed—so that the elasticity on the *inside* of the lips can produce a whistle. One way or the other, the cause of vibrations with elastic membranes is the growth in pressure defined by Eq. [3.10] that occurs in an *enlarging* elastic passageway where $\partial \langle V_z \rangle / \partial z < 0$.

3.1.4 Dependence of Fluid Velocity on Pressure: Darcy's Law

So far, we haven't used Navier–Stokes equation at all: we have only used mass conservation and elasticity relations. At some point, surely the momentum of the fluid must matter. What we want now is some knowledge of how $\langle V_z \rangle$ behaves. For the Poiseuille problem, we assumed that there was no variation in velocity in the z-direction. We can't use the Poiseuille result in an elastic tube, however, because we have shown using continuity that $\langle V_z \rangle$ changes with z.

What we ought to do is to repeat the Poiseuille derivation for the present situation. That is, we should start with the full Navier–Stokes equation, remove what terms we can, solve for V_z, and average over the cross section of the tube. This is, as we say, what we *ought* to do, but it isn't what we *will* do. The reasons for this are several. First, the flow we are looking at is intrinsically time dependent, so we can't remove the $\partial \vec{V} / \partial t$ term from Navier–Stokes. This alone makes the problem much harder to solve, but additionally even if we could obtain a solution, we don't know where the boundaries of the flow are, and even if we did know this, the boundaries would themselves be moving, and our whole problem is that we don't know how the boundaries move in response to the flow. All in all, we can't do what we would like.

On the other hand, we don't really need the entire solution to Navier–Stokes' equation, what we need is the z-velocity averaged over the cross section of the tube, and this we can say something about. To do so, let's take a brief diversion and consider averaged flow of something else that is at least as complicated as fluids: the flow of electrons. For that problem, we have all at some point memorized Ohm's law, which tells us that the rate of flow of electrons (the current, $\hat{I}$) is related to the voltage, $\hat{V}$, and the resistance, $\hat{R}$:

$$\hat{V} = \hat{I} \cdot \hat{R} \quad [3.11]$$

where we have put hats on the variables so as not to confuse $\hat{V}$ with velocity or $\hat{R}$ with radius. The resistance for that problem ought in principle to be a terribly complicated function of metal lattice defects, skin depth effects, temperature fluctuations, and so forth, but in fact to a high degree of accuracy, the average voltage drives the average current in proportion to the average resistance. The fluid equivalent to Ohm's law is Darcy's law, after Henry Darcy (1803–1858), a French civil engineer who was interested in resistance to flow in city water supplies. His equation for flow in the z direction reads:

$$\frac{\partial P}{\partial z} = \langle V_z \rangle \cdot \hat{R}, \quad [3.12]$$

where the voltage that drives flow of electrons in Ohm's law is replaced with the pressure gradient that drives flow of fluid, and the current of electrons is replaced with the mean speed of water. As is the case for Ohm's law, $\hat{R}$ is an unknown empirical constant that defines how much the system resists flow.

We note that for our problem, the resistance $\hat{R}$ depends both on fluid viscosity and on the resistance to flow that deforming the surface causes. This is an important point that we will return to, but for the moment we just remark to presage discussions to come that the viscosity will unavoidably cause a loss of energy, but the elastic surface can in principle either take energy away from or return it to the fluid as the surface oscillates.

We can now plug Eq. [3.12] into Eq. [3.10] to obtain:

$$\frac{\partial P}{\partial t} = -\frac{Yh}{2R\hat{R}} \frac{\partial^2 P}{\partial z^2}, \quad [3.13]$$

an equation only in the pressure within the tube, P. If we can solve this for P, we can make use of Eq. [3.12] to find $\langle V_z \rangle$, and Eq. [3.9] to find $V_r|_R$.

3.1.5 **Solution at Last!**

We can now set about to solve Eq. [3.13]. Doing so will require some knowledge of complex analysis: if the reader is acquainted with this, she can continue from here; if not, she should refer to the section dealing with this topic at the end of this chapter.

We begin by performing "separation of variables," an old standard for reducing a partial differential equation to a set of ordinary differential equations. Separation of variables works like this: let's assume that the solution to Eq. [3.13] can be written as a product:

$$P(z, t) = Z(z) \cdot T(t), \quad [3.14]$$

where $Z(z)$ is a function of z alone, and $T(z)$ is a different function of t alone. There is no guarantee that the solution will look like this, for instance if the solution were $\sin(z \cdot t)$ or even just $x + V \cdot t$, we would be out of luck. In this case, the trick works, and plugging Eq. [3.14] into Eq. [3.13] gives:

$$Z(z)\frac{\partial T(t)}{\partial t} = -\frac{Yh}{2R\hat{R}}T(t)\frac{\partial^2 Z(z)}{\partial z^2}, \tag{3.15}$$

which we divide by $Z(z) \cdot T(t)$ to get:

$$\frac{1}{T(t)}\frac{\partial T(t)}{\partial t} = -\frac{Yh}{2R\hat{R}}\frac{1}{Z(z)}\frac{\partial^2 Z(z)}{\partial z^2}. \tag{3.16}$$

The way the separation of variables game is played at this point is to observe that the left side of the equation only depends on t, but not on z, and the right side depends on z but not on t. The only way something can be neither dependent on z nor dependent on t is if it is a constant, which we'll call c. So Eq. [3.16] is the same as the two equations:

$$\frac{1}{T(t)}\frac{\partial T(t)}{\partial t} = c \tag{3.17}$$

and:

$$-\frac{Yh}{2R\hat{R}}\frac{1}{Z(z)}\frac{\partial^2 Z(z)}{\partial z^2} = c. \tag{3.18}$$

So we have separated the partial differential equation, Eq. [3.13], into two ordinary differential equations, Eqs. [3.17] and [3.18], which we will solve next. Let's start with Eq. [3.17], which is the same as:

$$\frac{\partial T(t)}{\partial t} = c \cdot T(t). \tag{3.19}$$

The reader may recognize that this equation has the solution:

$$T(t) = T_0 \cdot \exp(c \cdot t). \tag{3.20}$$

If c is negative, the behavior is stable, so any disturbance will eventually decay to zero. Stable motions are nice for the mathematician in that they can be predicted for all time, however this is often unrealistic and is always boring. If c is positive, however, the solution is unstable, and can lead to much more interesting behaviors.

Example Double pendulum

An example of such an interesting behavior can be found in the double pendulum, shown in Fig. 3.3. In Fig. 3.3(a), we show the double pendulum, which appears in every dynamics text as the prototypical example of a coupled oscillator. Each pendulum by itself has a simple solution that, for small angles ϑ_1 and ϑ_2, is perfectly periodic in time. Coupling the two pendula together produces slightly more complicated behavior (discussed later in this book) consisting of a sum of two periodic motions, but remains completely predictable for all time.

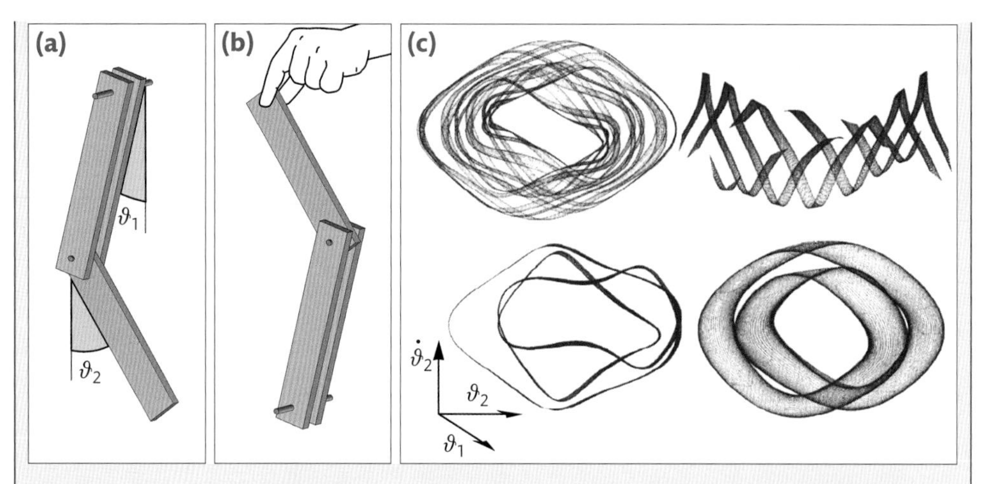

Fig. 3.3 Double pendulum. (a) Pendulum in stable configuration: here the pendulum oscillates periodically and reproducibly. (b) Pendulum in unstable configuration: here the pendulum executes complex motions. (c) A few of the motions executed by the pendulum, here plotted against ϑ_1,ϑ_2 and $\dot{\vartheta}_2 = \partial\vartheta_2/\partial t$: these motions are manifested in the physical pendulum by an infinite number of periodic states that the pendula sample over time. The presence of infinite number of coexisting periodic states is a characteristic of chaotic motion (see, for example, Alligood et al.[1]).

On the other hand, if the angles are permitted to become large, as shown in Fig. 3.3(b), the pendula can both become unstable, resulting in a rich variety of complex motions. This problem is not solvable in the senses that (1) one cannot write down an exact analytic solution and (2) the actual motion seen will diverge from any approximate solution exponentially rapidly in time. So the motion is not predictable for long times, despite the facts that this is a very simple physical system consisting of only two pendula, and that the equations of motion are very well known. A small sampling of the infinite number of possible states is shown in Fig. 3.3(c). As an advanced exercise, the student may want to obtain the equations of motion (available on the web or in the literature[2]), program a computational simulation, for example, using Runge–Kutta integration discussed in the Appendix, and confirm that two initial states arbitrarily close to one another diverge on average exponentially in time.

We so far have only the time part, Eq. [3.17] of the full solution, so let's work out the space part from Eq. [3.18]:

$$\frac{\partial^2 Z(z)}{\partial z^2} = -c\frac{2R\hat{R}}{Yh}Z(z). \quad [3.21]$$

If the solution is stable, then $c < 0$, and this has an exponential solution as well:

$$Z(z) = Z_0 \exp\left(\pm\sqrt{-2R\hat{R}c/Yh}\cdot z\right), \quad [3.22]$$

where we note that the minus sign under the square root cancels with the negative c to produce a positive value. Combining this with Eq. [3.20] makes the pressure go as:

$$P(z,t) = T(t)Z(z) = T_0 Z_0 \exp\left(c \cdot t \pm \sqrt{-2R\hat{R}c/Yh \cdot z}\right). \quad [3.23]$$

This, then, is the stable and boring solution: any disturbance in the pressure (and again, equivalently $\langle V_z \rangle$ and $V_r|_R$ from Eqs. [3.12] and [3.9], respectively) decays exponentially rapidly in time at any fixed location (because of the $\exp(c \cdot t)$ term), and propagates as it does so with speed $U_z = \mp c / \sqrt{-2R\hat{R}c/Yh}$ (because the pressure is constant for $z = z_0 \mp U_z \cdot t$).

The more interesting solution appears when $c > 0$. We'll discuss the physical meaning of this change in sign in a moment, but for now we remark that with the help of complex analysis (again, the reader can refer to the final section in this chapter for a review), we recognize that we don't have to make any change to the solution (Eq. [3.23]): if c is positive, then the argument of the square root is negative, so the pressure can either be left alone, or can be rewritten as:

$$P(z,t) = T_0 Z_0 \exp(c \cdot t \pm i\sqrt{2R\hat{R}c/Yh \cdot z}). \quad [3.24]$$

This makes explicit that the spatial term is imaginary, and so is oscillatory.

But wait, this is interesting: at precisely the same point (c becoming positive) that the time term becomes unstable, the space term becomes oscillatory. What does this mean physically? The answer is that the pressure grows in time as the fluid moves downstream, and as it does so, the membrane produces oscillations in space. Of course oscillations can't grow without limit: physically the radius can't become negative, but this limit isn't included anywhere in our derivation. In a real elastic tube, these solutions would lead either to oscillations that periodically expand and collapse downstream (the farting balloon solution) or, if the membrane is part of the fluid, to pinch off of droplets (see Fig. 3.8).

This interpretation is satisfying, but something else isn't quite right. Let's imagine that we are studying a tube that is fixed at one end: perhaps we are interested in pumping of blood from the heart, or in the exit of water from a rigid faucet. If the mean speed downstream is U_z, then as time passes, fluid travels downstream, so we can see what happens at time t by looking at position $U_z \cdot t$. We aren't trying to be precise here: doing so would involve far too many complications and would distract from the point, which is that the pressure downstream at position z should go like:

$$P(z) \sim \exp\left(c \cdot z/U_z \pm i\sqrt{2R\hat{R}c/Yh \cdot z}\right). \quad [3.25]$$

For $c > 0$, the real term implies that the pressure—and so the expansion of the tube—grows downstream, while the imaginary term dictates that the pressure must oscillate as well. So the tube must take on a shape something like the sketch shown in Fig. 3.4(a) (cf. Fig. 3.8 (for $c < 0$).

However in reality, in yet another example of unexpectedly rich fluid behaviors, elastic tubes do more than either simply exponentially converge (for $c < 0$) or exponentially grow and oscillate (for $c > 0$). Some of these behaviors are shown in the snapshots of Figs. 3.4(b) and (c).

So, apparently elastic tubes do more complex things than our solutions (Eqs. [3.23] and [3.25]) indicate. Where are the solutions for these complexities: do we need to add something

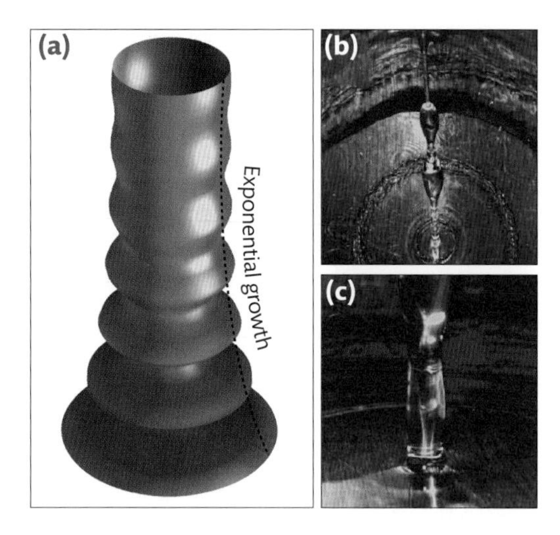

Fig. 3.4 (a) Exponentially growing oscillatory tube solution. (b) False colored snapshot of water being released from a kitchen faucet onto a stainless steel sink, showing strong oscillations. (c) The same situation as in panel (b) at a slightly higher flow rate, showing decaying oscillations. (Source: Carlos E. Caicedo-Carvejal.)

to our model or did we overlook something in our derivation? The answer is that we overlooked something which, now that we have adjusted to the idea of complex-valued solutions like Eq. [3.25], is really quite simple. Back when we separated variables to produce Eqs. [3.17] and [3.18], we wrote the constant, c, and without a second's notice we took it to be real valued. But there is no reason that c cannot be complex, right? That is, we could let $c = a + ib$, where a and b are real.

If we do this, we find that Eqs. [3.23] (stable) and [3.25] (unstable) can both be written as the following:

$$\begin{aligned} P(z,t) &= T_0 Z_0 \exp\Big((a+ib)\cdot t \;\pm\; \sqrt{-2R\hat{R}(a+ib)/Yh\cdot z}\Big) \\ &= T_0 Z_0 \exp\big[(at \mp \sqrt{r-a}\cdot\hat{z}) \;+\; i(bt \pm \sqrt{r+a}\cdot\hat{z})\big], \end{aligned} \quad [3.26]$$

where we have made use of a result derived later in this chapter that $\sqrt{a+ib} = \big(\sqrt{r+a} + i\cdot\sqrt{r-a}\big)/\sqrt{2}$, and where we introduce the abbreviations $r = \sqrt{a^2+b^2}$ and $\hat{z} = \sqrt{2R\hat{R}/Yh}\cdot z$. Eq. [3.26] has a distinct real, exponential part, $\exp\big[(at \mp \sqrt{r-a}\cdot\hat{z})\big]$, and an imaginary, oscillatory part, $\exp[i(bt \pm \sqrt{r+a}\cdot\hat{z})]$. Moreover, this problem has two limiting cases. For purely real c, that is, $b = 0$, we recover Eq. [3.24]:

$$P(z,t) = T_0 Z_0 \exp[a(t \pm i\hat{z})], \quad [3.27]$$

while for purely imaginary c, that is, $a = 0$, we get:

$$P(z,t) = T_0 Z_0 \exp[b(\mp\hat{z} + i(t \pm \hat{z}))]. \quad [3.28]$$

This new set of solutions represents waves traveling either up- or downstream (depending on the sign of $\hat{z}$) with speed $\sqrt{Yh/2R\hat{R}}$, that can grow or diminish downstream, as in Fig. 3.4.

Now that we have a general solution for flow through an elastic tube, we can adapt it to fit our particular problem, namely vascular flow driven by pulses from the heart. In that case, Eq. [3.23] provides that the pressure goes as e^{ct}, so it is natural to let c be imaginary to accommodate periodic pulses in time. This is the solution (Eq. [3.28]) where b is the frequency of driving from the heart. We can obtain the velocities as well by using Eqs. [3.9] and [3.12]:

$$\begin{aligned} V_r|_R &= \frac{R^2}{Yh}\frac{\partial P}{\partial t} \\ &= i\frac{bR^2 T_0 Z_0}{Yh}\exp\left[b(\mp\hat{z}+i(t\pm\hat{z}))\right]. \end{aligned} \quad [3.29]$$

$$\begin{aligned} \langle V_z \rangle &= \frac{1}{\hat{R}}\frac{\partial P}{\partial z} \\ &= \mp(1-i)b\sqrt{\frac{2R}{Yh\hat{R}}}\cdot T_0 Z_0 \exp\left[b(\mp\hat{z}+i(t\pm\hat{z}))\right]. \end{aligned} \quad [3.30]$$

The attentive reader may be disturbed to notice that these velocities are explicitly complex-valued. One can't measure a complex velocity or pressure, so what does this mean? The answer is that anything we measure is real, so, for example, if we were to measure $V_r|_R$, we would obtain the real value of Eq. [3.29]. Since this equation contains i as a pre-factor, and since the exponent goes as a cosine plus i times a sine, the *real* value will go as −sine, or explicitly:

$$Real(V_r|_R) = -C\cdot\exp(\mp b\hat{z})\cdot\sin(t\pm\hat{z}), \quad [3.31]$$

where we have collected all of the constants into C. This raises two related points. First, if this is what we can measure, what are the imaginary parts of the solution for? And second, notice that the radial velocity, $V_r|_R$, has a pre-factor of i, but the pressure, $P(z,t)$, doesn't. So the real part of $P(z,t)$ must depend on cosine, while the real part of $V_r|_R$ depends on sine. This observation resolves both points: evidently $P(z,t)$ and $V_r|_R$ are out of phase by $\pi/2$.

Physically this occurs because the pressure pushes the wall out hardest when its velocity is traveling inward fastest. Mathematically, the presence of imaginary terms encodes for the phase of terms that affect the dynamics, but aren't apparent for a given measurement at a specific instant. So if we only measured $P(z,t)$ at one instant in time and knew nothing else, we wouldn't know about the imaginary terms and we wouldn't be able to conclude anything about how $P(z,t)$ would change over time. If, however, we measured the derivatives of $P(z,t)$, we would be able to predict how $P(z,t)$ would evolve over time. And what would differentiating do? It would bring down the imaginary terms, hidden in the exponent, as pre-factors.

The same thing could be obtained without using complex notation, by separately keeping track of the pressure and its derivatives. Every term in every equation would then have both a magnitude (which we can measure) and a phase (which we cannot). Complex notation permits us to keep both pieces of information in one compact form—and does away with three separate functions (exponential, sine, and cosine) in favor of only one function (exponential) with complex arguments. We will make use of this stratagem repeatedly, and to foretell what will come, we will find that $c = a + ib$ can be defined as a complex viscosity: a will give the rate of damping (which we normally term viscosity) and b will give the rate of oscillations (which we associate with elasticity). In the present problem, the elasticity comes from something outside the fluid—the elastic tube—but in other problems, elasticity can be intrinsic to the fluid itself. We call such fluids "viscoelastic."

Example Viscoelasticity

Viscoelasticity is exceedingly common in biological fluids, ranging from blood to mucus. In blood, elasticity arises from red blood cells: it shouldn't be a surprise that blood, which contains about five billion red blood cells per cc, is elastic, since we showed that each red blood cell exhibits elasticity. Viscoelasticity in mucus is produced by mucin, a complex

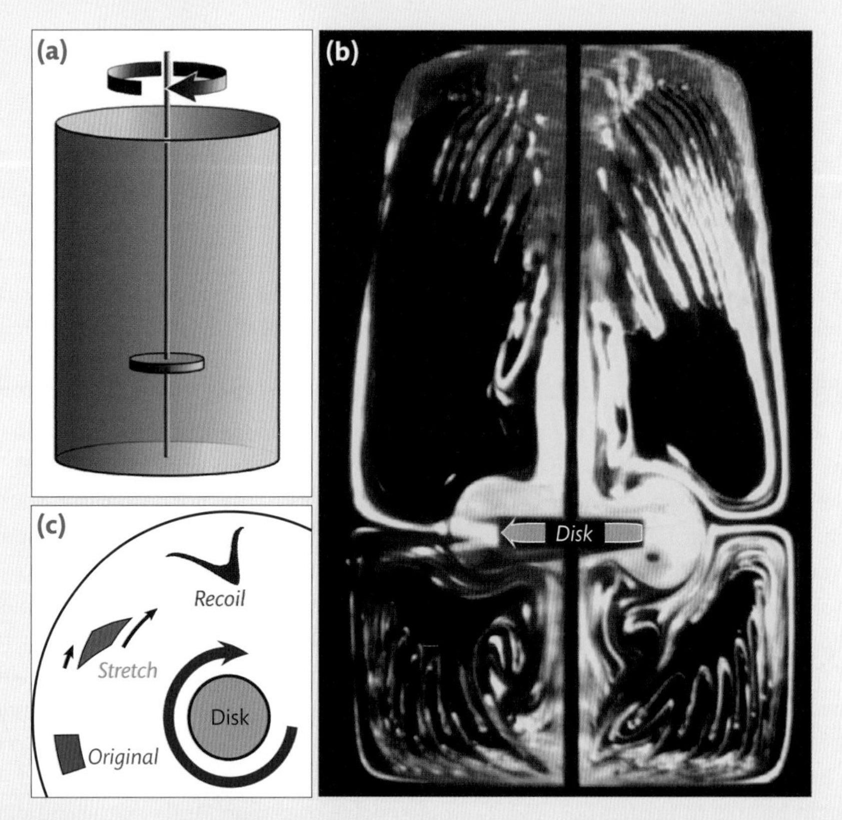

Fig. 3.5 Example of viscoelastic fluid flow. (a) Geometry of a tank stirred steadily with a single symmetric disk. (b) Flow in the tank is made visible by injecting fluorescent dye and illuminating with a vertical sheet of laser light. The fluid is 1% carboxymethyl cellulose in water, which is elastic as well as being viscous, and so periodically recoils after being stretched to form the folds shown. (c) Schematic of the recoil, indicating that the steadily spinning disk stretches a region of fluid (blue→green), which then recoils to produce a fold (green→red). In a Newtonian fluid, no such folds would be present. (Source: Mario Alvarez, Paulo Arratia, and Fernando Muzzio.)

protein whose structure is characterized numerous coiled branches, each literally acting as a spring. We will discuss effects of blood and mucus rheology (characteristics of response to deformation) later. Here we illustrate one of many remarkable viscoelastic properties in a material almost as common as blood or mucus: carboxymethyl cellulose (CMC). CMC is a widely used water-soluble thickener used in foods, pastes, paints, skin creams, and eye drops: this is the material that makes fast food shakes thick and keeps pearls suspended in specialty shampoos. At very low concentrations, CMC can dramatically alter the flow of water—as an example, in Fig. 3.5 we show an experiment in which a 1% solution of CMC is dissolved in water. As described in the figure caption, the CMC causes the fluid to recoil elastically, producing the nested folds shown in Fig. 3.5(b).

The reader can confirm that common fluids are elastic at home by stirring commercial canned tomato soup: when the stirring is abruptly stopped, a visible recoil can be observed that causes the soup to briefly but visibly turn in the direction opposite to the direction stirred. In this case, it is stretchy tomato pulp, rather than an artificial polymer or cell membrane, that makes the fluid elastic. Viscoelasticity is a surprising and fun topic, and we will discuss examples throughout this book.

EXERCISE 3.1

Consider Eq. [3.30] for flow through an elastic tube. What must you change to make the tube stiffer? In the limit of infinite stiffness, this should approach the Poiseuille limit, right? Show algebraically whether the equation does this or not. If it doesn't approach the Poiseuille limit, identify where the derivation failed. Additionally, name several physiological pathologies of elastic tubes. Suppose that you wanted to stabilize one of these oscillatory pathologies: what parameters could be changed, and how would these changes affect stability? What, arithmetically, would you want to do in order to make the fluid-elastic tube system stable?

3.2 **Aside: A Brief Encounter with Complex Analysis**

The underlying idea behind complex analysis is very simple; however, its implications are remarkably profound. We devote only limited space to the topic here, however the reader is encouraged to investigate the topic further: she is unlikely to be disappointed.

For our purposes, let's consider a complex number, $z = x + iy$, where $i = \sqrt{-1}$. We can plot this point in Cartesian coordinates (named anachronistically for René Descartes (1596–1650)), as shown in Fig. 3.6(a). We are familiar with the concept of polar coordinates, so that we could instead define the same point as:

$$z = r\cos(\vartheta) + i \cdot r \cdot \sin(\vartheta). \qquad [3.32]$$

A singularly productive contribution from complex analysis is Euler's formula:

$$e^{i\vartheta} = \cos(\vartheta) + i\sin(\vartheta). \tag{3.33}$$

This can be obtained in a number of ways, for example from the Taylor series for the exponential:

$$\begin{aligned} e^{i\vartheta} &= 1 + i\vartheta + \frac{(i\vartheta)^2}{2!} + \frac{(i\vartheta)^3}{3!} + \frac{(i\vartheta)^4}{4!} + \frac{(i\vartheta)^5}{5!} + \\ &= \left[1 - \frac{\vartheta^2}{2!} + \frac{\vartheta^4}{4!} + \ldots\right] + i \cdot \left[\vartheta - \frac{\vartheta^3}{3!} + \frac{\vartheta^5}{5!} + \ldots\right], \end{aligned} \tag{3.34}$$

where we have grouped the even terms in the first bracket and the odd terms in the second. The reader can confirm that these brackets contain the Taylor expansions for cosine and sine respectively.

Euler's formula tells us two things that will be central to the study of flows discussed in this book. First, since an imaginary exponent can be resolved into sines and cosines, Euler's formula tells us that imaginary exponents necessarily represent oscillatory behavior. This will be a recurring theme throughout this book: wherever we see imaginary numbers we can expect that oscillations are involved. Real exponents, on the other hand, code for stability if they are negative or instability if they are positive.

Second, less central but nonetheless useful, Euler's formula tells us that Eq. [3.32] can be rewritten so that the two expressions are exactly equivalent: the first describing a point in Cartesian coordinates and the second in polar coordinates:

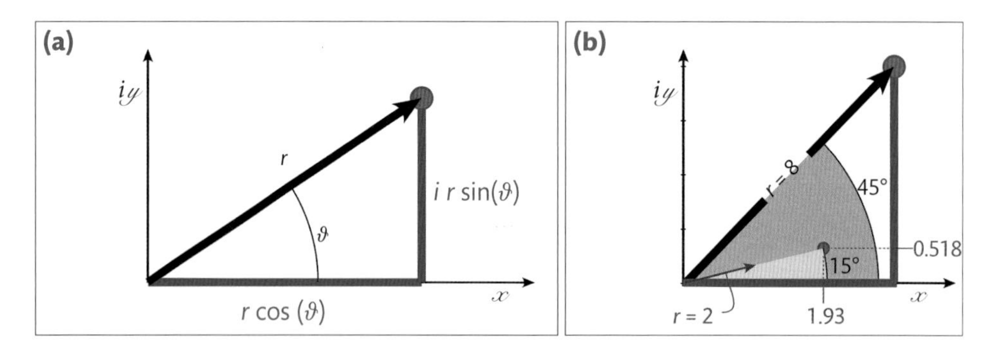

Fig. 3.6 Geometric interpretation of Euler's formula. (a) The red circled point can be described either by $z = x + iy$ or by $z = r \cdot e^{i\vartheta}$. (b) A tiny taste of difficult calculations that complex analysis makes simple is the geometric solution for the cube root of a line of length $r = 8$, with angle $\vartheta = 45°$. Eq. [3.36] tells us that the cube root must have length $\sqrt[3]{8} = 2$, and angle $45/3 = 15°$. This is clearly a contrived example: we will see physically relevant cases throughout this book.

$$z = x + iy. \tag{3.35}$$

$$z = r \cdot e^{i\vartheta}. \tag{3.36}$$

This may appear to the reader to be what the eminent mathematician David Ruelle has described as "dividing an incomprehensible problem into two incomprehensible parts," but in fact it is a very useful modification. To gain a glimpse into its usefulness, suppose we need to know $\sqrt{x + iy}$ (which indeed we do—see Eq. [3.26] earlier in this chapter). Solving for this square root in Cartesian coordinates would be extraordinarily cumbersome: if x or y was small, we could expand in a Taylor series, but nobody in their right mind would begin such an enterprise without at least looking for an alternative. Eq. [3.36] provides that alternative, for we can see right away that:

$$\begin{aligned} \sqrt{r\,e^{i\vartheta}} &= \sqrt{r} \cdot e^{i\vartheta/2} \\ &= \sqrt{r} \cdot \left[\cos(\vartheta/2) + i \cdot \sin(\vartheta/2)\right]. \end{aligned} \tag{3.37}$$

Moreover, this has a simple geometric interpretation that we will make use of later in this book, namely that to find the nth root of a complex number, we take the nth root of the radius and multiply it by the exponential of the angle divided by n. This approach is shown schematically in Fig. 3.6(b); algebraically this works out to be:

$$\begin{aligned} \sqrt[3]{8\,e^{i(45^\circ)}} &= \sqrt[3]{8} \cdot e^{i(15^\circ)} \\ &= 2 \cdot \left[\cos(15^\circ) + i \cdot \sin(15^\circ)\right] \cdot \\ &= 1.93 + i \cdot 0.518 \end{aligned} \tag{3.38}$$

Example Flows from complex analysis

The problem leading to Eq. [3.38] is, of course, an artificial example, but the same ideas can be applied in more practical problems. For example, later in this book we will see that there is a class of problems known as "potential flows" that can be rapidly solved using complex analysis. One of these that relates to our artificial example is flow into, or over, a wedge. For instance, flow within a wedge of 45° open angle can be solved exactly, as shown in Fig. 3.7(a), and once this flow is obtained, flow within a 15° wedge consists merely of taking the cube root of the prior solution—shown in Fig. 3.7(b). Moreover, flows outside of wedges are equally easily obtained, as shown in Figs. 3.7(c) and (d) for 90° and 90/3 = 30° wedges. And more complicated flows such as flow past the airfoil shown in Fig. 3.7(e) can be obtained using the same mathematical machinery, discussed in Chapter 8. All of this will follow soon from tools related to Euler's formula.

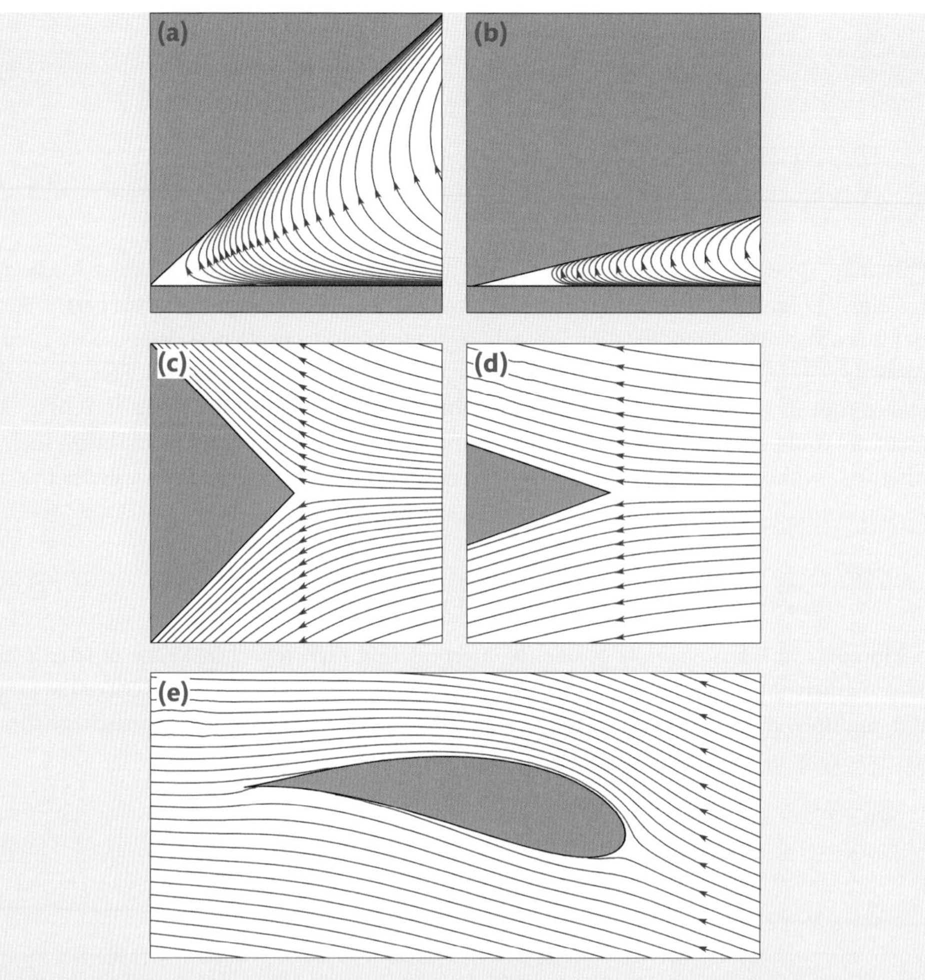

Fig. 3.7 Complex analysis can be used to obtain exact solutions for various flows: we will derive these results in a later chapter. Flow within (a) a 45° and (b) a 15° wedge. Flow outside of (c) a 90° and (d) a 30° wedge. These solutions will be derived later in this book, and we will show that it is just a short step to obtaining flows for more complicated geometries, as in the example of flow past an airfoil, as shown in panel (e).

EXERCISE 3.2

Johann von Neumann (1903–1957), one of the nimblest and most prolific mathematicians who ever lived, once said, "In mathematics you don't understand things. You just get used to them." Complex analysis epitomizes this philosophy—how, after all, do we understand the square root of -1? Despite its literally imaginary nature, $i = \sqrt{-1}$ is a highly productive thing to get used to, and in this spirit, we provide several exercises here.

(1) Solve $\sqrt{3+4i}$ both analytically and using the computer, for example, with Matlab. For the analytic solution, sketch the vector $z = x + iy$ and the vector $\sqrt{x + iy}$.

(2) In Eq. [3.26], we used the more general formula: $\sqrt{a+ib} = (\sqrt{r+a} + i \cdot \sqrt{r-a})/\sqrt{2}$, where $r = \sqrt{a^2+b^2}$. Rewrite $\sqrt{a+ib}$ in polar coordinates, $\vec{r} = r \cdot e^{i\vartheta}$, as shown in Fig. 3.6(a), and use trigonometric half-angle formulas to derive the general formula for the square root.

(3) Take the square root of the solution to the exercise above, and show analytically and by computer that this is the same as $\sqrt[4]{3+4i}$.

(4) Formally, one can perform any operation in complex space that is performable in real space—so, for example, one can solve for $\cos(x + iy)$ or $\ln(x + iy)$. Solve analytically for each of these expressions. Note that the first can be expanded using the sum of angles formula from trigonometry using $x + iy$, and the second is more easily analyzed by writing $z = r \cdot e^{i\vartheta}$—thus both Cartesian and polar notations are useful. Notice also that something strange happens when you take the logarithm: $\vartheta = 0$ and $\vartheta = 2\pi$ are indistinguishable and should give the same answer for $\ln(z)$, but don't. This is referred to as a "branch cut." Branch cuts aren't used in the analysis presented in this book, but they may arise in some fluids problems, and they are worth being aware of.

(5) Later in this book, we will make use of contour plots to solve for fluid behaviors (see, for example, the discussion of streamlines in Chapter 6). Just like cosines or logarithms, evaluating contours is formally straightforward for complex numbers. Using Matlab or a similar language, plot a contour of the imaginary part of $(x + iy)^{3/2}$. You will find that this looks very much like Fig. 3.7(c). Use Euler's formula to explain why this contour is 3-fold symmetric. What function would you use to plot Fig. 3.7(a)? Confirm your prediction.

(6) Use what you know of complex analysis to choose real and imaginary parts of Eq. [3.25] that look like droplets forming and falling from a faucet at an instant in time. Note the term, "look like:" Eq. [3.25] defines the pressure, while in this exercise we are asking for the surface shape—assume that the surface shape obeys the same functional form as the pressure.

Plot the result in three dimensions: your plot should look something like Fig. 3.8. Using Matlab, you may find that modifying the function "cylinder" will help. Notice that the droplets appear to pinch off in this plot—this can be accomplished in Matlab using a line of code like *r = r./(r > 0)*. Explain what this line does and why the plot of the result produces the appearance of droplet pinch off.

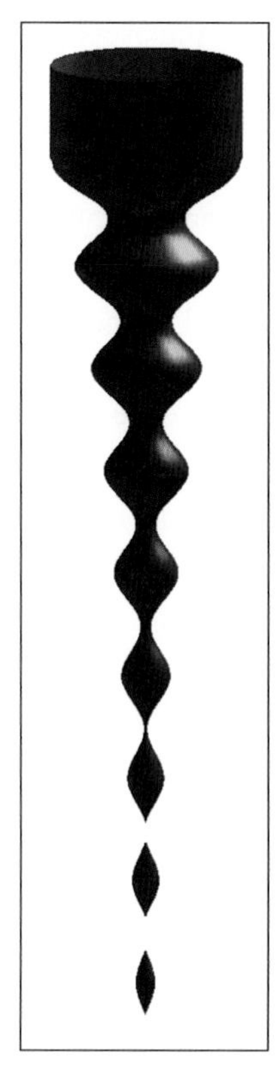

Fig. 3.8 3D plot of surface resembling formation and pinch off of droplets from a faucet.

REFERENCES

1. Alligood, K.T., Sauer, T. and Yorke, J.A. (1996) *Chaos, an Introduction to Dynamical Systems*. New York: Springer.
2. Shinbrot, T., Grebogi, C., Wisdom, J. and Yorke, J.A. (1992) Chaos in a double pendulum. *American Journal of Physics*, *60*, 491–9.

4 Pulsatile Flows

4.1 Introduction

In the last chapter, we mentioned almost in passing that blood flow is driven by periodic pulsations from the heart. In the presence of periodic forcing, the time dependence that we conveniently dropped from Navier–Stokes up to this point has to be reintroduced. We will see that this produces a qualitatively different flow from the Poiseuille solution that we introduced in Chapter 1.

To include time dependence, let's return to the starting point, Navier–Stokes:

$$\rho\left(\frac{\partial \vec{V}}{\partial t}+\vec{V}\cdot\vec{\nabla}\vec{V}\right)=-\vec{\nabla}P+\mu\nabla^2\vec{V}+\vec{F}_{ext}. \quad [4.1]$$

As we did in the Poiseuille derivation, we'll consider first the low Reynolds number case so that $\vec{V}\cdot\vec{\nabla}\vec{V}$is small, and we'll neglect external forces, $\vec{F}_{ext}$. Also as before, we'll consider a smooth cylindrical pipe in which flow is only axial, so that $V_r=V_\varphi=0$. Again the continuity equation tells us that consequently $\partial V_r/\partial r=\partial V_\varphi/\partial\varphi=0$, and we'll additionally assume that $\partial P/\partial r=\partial P/\partial\varphi=0$. Once we've made these assumptions, what we're left with is:

$$\rho\frac{\partial \vec{V}_z}{\partial t}-\frac{\mu}{r}\frac{\partial}{\partial r}\left(r\frac{\partial V_z}{\partial r}\right)=-\frac{\partial P}{\partial z}. \quad [4.2]$$

We've rearranged the order of the terms so that the left side of the equation depends only on r and t, while the right side depends only on z and t. Similar to separation of variables, this implies that both sides are the same function of time:

$$\rho\frac{\partial V_z}{\partial t}-\frac{\mu}{r}\frac{\partial}{\partial r}\left(r\frac{\partial V_z}{\partial r}\right)=f(t) \quad [4.3]$$

$$-\frac{\partial P}{\partial z}=f(t). \quad [4.4]$$

We're interested in pulsatile flow, so let's try the simplest possibility, $f(t)=Ae^{i\omega t}$, for a single frequency, ω. We note that Eq. [4.2] is a linear equation, so if $f_1(t)=Ae^{i\omega_1 t}$ is a solution and $f_2(t)=Ae^{i\omega_2 t}$ is also a solution, then $f_1(t)+f_2(t)$must be a solution. This implies that any sum of sines and cosines will also be a solution, and Fourier's theorem states that any sufficiently smooth and non-pathological function will also be a solution. Thus by solving

Biomedical Fluid Dynamics: Flow and Form. Troy Shinbrot.

DOI: 10.1093/oso/9780198812586.001.0001

for $f(t) = Ae^{i\omega t}$, we aren't claiming that the heartbeat is a simple sine wave, we are opening a door to almost any shape driving function.

Eq. [4.4] is then uninteresting: it just says that the pressure gradient is constant (let's call it A) in the downstream direction and goes as $e^{i\omega t}$ in time. What else could it be for a simple fluid in a rigid pipe, after all? Eq. [4.3] is a bit more interesting. It is easiest to solve this equation without constants like ρ, μ, and ω floating around, so we'll nondimensionalize to consolidate these pesky terms:

$$\alpha^2 \frac{\partial U}{\partial \tau} - \frac{1}{\eta}\left(\frac{\partial}{\partial \eta}\eta\frac{\partial U}{\partial \eta}\right) = Ae^{i\tau} \quad [4.5]$$

where we've used the dimensionless variables:

$$U = \frac{V_z}{R\omega}$$
$$\tau = \omega t \quad [4.6]$$
$$\eta = \frac{r}{R},$$

and where R is the radius of the tube. These constants are commonly combined into the new dimensionless group, the Womersley number:

$$\alpha = R\sqrt{\frac{\omega\rho}{\mu}}, \quad [4.7]$$

after British mathematician John Womersley (1907–1958). The boundary conditions are now no slip at the wall: $U = 0$ at $\eta = 1$, and speed at the center of the tube: $U = U_{max}$ at $\eta = 0$. Let's try separation of variables using:

$$U = u(\eta)Ae^{i\tau}. \quad [4.8]$$

This makes Eq. [4.5] into:

$$i\alpha^2 Ae^{i\tau}u - \frac{1}{\eta}\left(\frac{\partial u}{\partial \eta} + \eta\frac{\partial^2 u}{\partial \eta^2}\right)Ae^{i\tau} = Ae^{i\tau}. \quad [4.9]$$

The $Ae^{i\tau}$ terms cancel, leaving:

$$\frac{\partial^2 u}{\partial \eta^2} + \frac{1}{\eta}\frac{\partial u}{\partial \eta} - i\alpha^2 u + 1 = 0. \quad [4.10]$$

4.2 **Aside: Who's Afraid of Bessel Functions?**

Dang: Eq. [4.10] is a differential equation that we can stare at as long as we want without a solution that we have seen before occurring. This happens from time to time, and typically the person who first finds such an equation gets to name it after himself. In this case, Daniel Bernoulli (1700–1782) discovered the equation while studying oscillations of a hanging chain. The

Bernoullis were a Dutch family; James, Nicholas, and several Johns were accomplished mathematicians, but Daniel outshone them all. This equation, however, was not named after any of the Bernoullis: a different equation that we will mention later achieved that status. Instead, Eq. [4.10] was in a class of equations named for Friedrich Bessel (1784–1846), a German astronomer who described its solutions in more general form.

So when a solution to an equation doesn't occur to us, what can we do? The answer is that we invent a new function that solves the equation. No, really. Eq. [4.10] is a second order equation, so two functions are really needed, and these are called the Bessel function of the first kind, J_a, and the Bessel function of the second kind, Y_a. The subscripts, a, can be any number (usually an integer) and play the role that the frequency plays for sines and cosines (see Fig. 4.1).

Bessel functions have been embellished and go under various guises under other names: Weber functions, Neumann functions, Hankel functions, Macdonald functions, modified Bessel functions, hyperbolic Bessel functions, Ricatti–Bessel functions, and on and on. These, however, are merely names to frighten small children with. Despite this caution, all scientists when confronted with Bessel functions are prone to an unfortunate yet unavoidable feeling of intimidation and foreboding.

Bessel functions are really quite simple. They are solutions to Eq. [4.10], which tends to arise in cylindrical coordinates. Without naming its solutions, we wouldn't be able to talk about what waves look like in cylindrical tubes or tanks. If we had grown up swimming through

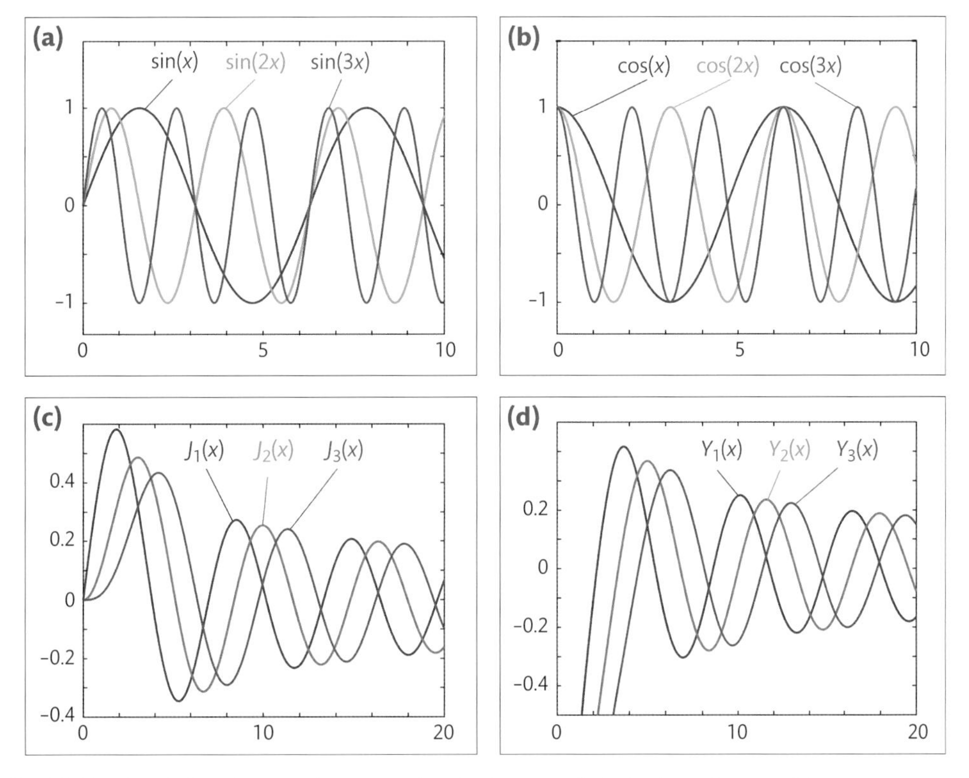

Fig. 4.1 Functions in Cartesian space (above) and in cylindrical space (below). (a) Sine functions, (b) cosine functions, (c) Bessel functions of the first kind, and (d) Bessel functions of the second kind.

cylindrical pipes or tapping on cylindrical drumheads, we'd be used to describing waves as Bessel functions. Instead, we have grown up jumping ropes and waving strings for cats to chase after, so we are used to sines and cosines, which are merely solutions to differential equations in Cartesian coordinates rather than in cylindrical coordinates.

To illustrate this point, we show sines and cosines for several frequencies in Figs. 4.1(a) and (b), above Bessel functions of the first and second type for several a values in Figs. 4.1(c) and (d). Computer programs being immune to intimidation, will just as happily work with Bessel functions as with sines and cosines, and we should do so too.

4.2.1 **Technicalities**

A few minor technicalities need to be invoked to put our solution in the standard form for Bessel equations. Bessel equations are usually written like this:

$$\frac{\partial^2 v}{\partial x^2} + \frac{1}{x}\frac{\partial v}{\partial x} + \left(1 - \frac{a^2}{x^2}\right)v = 0, \qquad [4.11]$$

where a is called the "order" of the resulting Bessel function. Eq. [4.10] has no $1/x^2$ term, so a must be zero in our case, that is, our solutions are "zeroth order" Bessel functions. Also, Eq. [4.10] has a $-i\alpha^2 v$ term in place of the $+v$ term above. This can be finessed by the substitution $x \to i^{3/2}\alpha\eta$, which turns Eq. [4.11] into:

$$\frac{\partial^2 v}{\partial \eta^2} + \frac{1}{\eta}\frac{\partial v}{\partial \eta} - i\alpha^2 v = 0. \qquad [4.12]$$

Finally, we can substitute $v \to u - 1/i\alpha^2$to get Eq. [4.10]. Incorporating these modifications along with the boundary conditions (no slip at the cylindrical walls of radius R) gives a final solution:

$$u = \frac{u_{max}}{i\alpha^2}\left[1 - \frac{J_0(i^{3/2}\alpha\eta)}{J_0(i^{3/2}a)}\right], \qquad [4.13]$$

which the reader can confirm both solves Eq. [4.10] and matches the boundary conditions. Recall that this is a function of $\eta = r/R$ and the Reynolds-like parameter $\alpha^2 = R^2\omega\rho/\mu$, and the solution oscillates in time as $e^{i\omega t}$. In Fig. 4.2 we plot the velocity profile in cross section at several values of α.

Fig. 4.2 shows two things. First, at small α (i.e. in small vessels and low heart rates, ω), the velocity profile is much like that of Poiseuille flow. This shouldn't be surprising since we have assumed that the velocity is independent of azimuthal angle, φ: this implies that the flow must be fully symmetric, and so can only go as even powers of the argument $i^{3/2}\alpha\eta$. For very small α, the second power of the argument will be much larger than the fourth or higher powers, so the velocity has to be quadratic in the radius, η.

Second, as α grows, the flow near the walls increases in speed, and ultimately flow becomes *faster* near the walls—though recall flow is zero *at* the walls. This is exactly what one would want to remove debris from the walls. So whereas in Chapter 1 we learned why we shouldn't eat junk food (because the blood pressure must grow dramatically as the lumen, or inner radius, of blood vessels shrink), now we learn why we should exercise (because this boosts α,

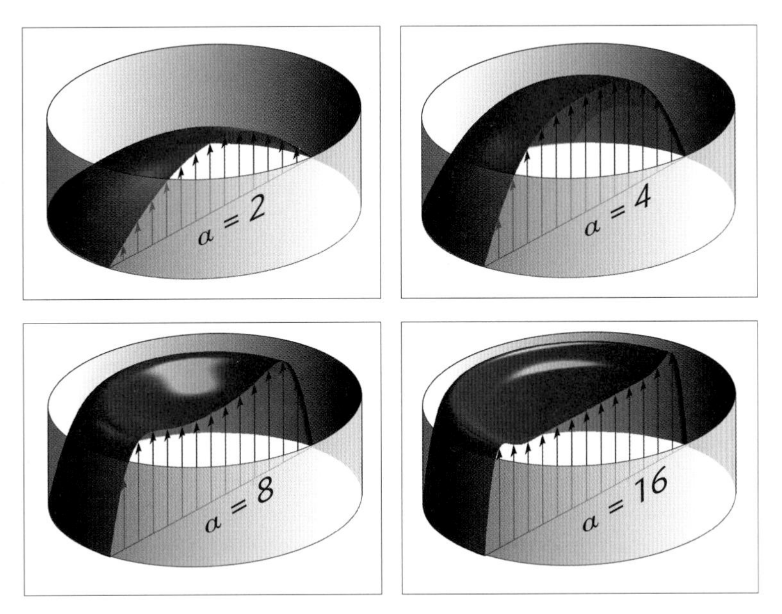

Fig. 4.2 Cutaway views of velocity profiles for pulsatile flow through rigid tube at various values of $\alpha = R\sqrt{\omega\rho/\mu}$. Notice that fluid speed is highest near the walls at large (α). The value of (α) at rest ranges from essentially zero in capillaries to over 15 in larger blood vessels such as the ascending aorta or the main pulmonary artery, and can be 20 or higher during exercise.

which keeps the vessels clear of debris). It should go without saying that neither lesson is complete, for junk food and exercise do many other things than respectively occlude or clean veins and arteries, but this mechanical view is a genuine part of understanding the effects of healthy living.

4.3 **Nonlinear Effects**

Lest we become self-satisfied with the tools that we have learned so far, let's touch base with some data from actual blood vessels. The study of blood flow, termed hemodynamics, has been intensively pursued for a very long time, dating perhaps to Hippocrates, who proposed that a balance between four essential fluids, "humors" (of which blood was the first) was essential to good health. This led to an unfortunate enthusiasm for bloodletting as a curative therapy, but also generated a wealth of surprising results. Among these are two that we will focus on at the moment in our discussion of pulsatile flow.

4.3.1 **Shocks and Lead Edge Steepening**

First, we mentioned that the actual pressure wave from the heart isn't the simple sinusoidal shape used in our analysis. The shape in a healthy subject near the heart is shown in Fig. 4.3 (left), and consists of two well-understood features. First, the forward "compressive" wave from the left ventricular power stroke generates a rapid rise in pressure, which diminishes

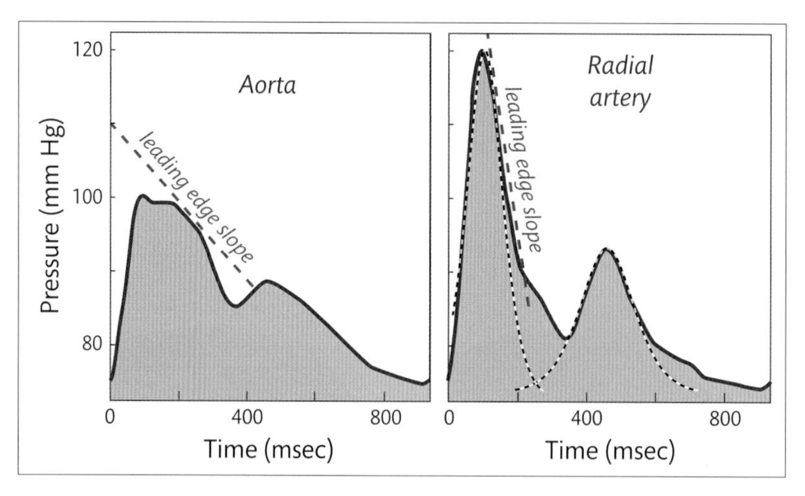

Fig. 4.3 Comparison between pressure waves at aorta and downstream at radial artery from a healthy young subject. Note that the wave amplitude grows and the leading edge steepens (red dashed lines). Black and white dashed lines are sech^2 solutions from Eq. [4.24]. (Data from Mahmud and Feely.[1])

about a third of a second after initiation. Then a second forward "expansion" wave appears, caused by rebound following rapid relaxation of the ventricle. This much is well understood.

Downstream of the heart, something new happens. As shown in Fig. 4.3 (right), by the time that blood reaches the radial artery, the amplitude of the pressure wave has grown, and the leading edge of both peaks have steepened considerably. This is important both for facilitating pulsatile transport of blood and as a clue to something going on in the workings of the vasculature that our models so far have left out. This phenomenon is termed "lead edge steepening," and leads to the formation of "shocks," which we will have more to say about shortly.

Before we do so, we can do a quick calculation to identify what the physical source of this behavior could be. Shocks occur when a signal is pushed into a region faster than it can be carried away—so when a jet plane travels faster than the speed of sound, the nose of the plane pushes a pressure pulse ahead of it faster than the wave can be transmitted away by sound waves. This leads to a variety of effects beyond the scope of this book; nevertheless lead edge steepening is a signature of a shock, and it is worth understanding how shocks occur.

Everything that we need to know for our purposes can be understood by considering a traffic jam. If I am traveling at 100 km/hr and the car ahead of me slows to 99 km/hr, I can easily match its speed *if there are a few car lengths between us*. If I am following too close, however, when the car ahead of me slows by 1 km/hr, I will need to slow by 2 km/hr to prevent a collision, and the car behind me will have to slow by 4 km/hr to keep from hitting me—and so forth. This produces a wave, termed a shock, that travels backward through traffic in response to a tiny 1 km/hr difference in speed. We have all experienced traffic stoppages for no apparent reason: a common cause of this is driving at too high a speed with too high a density of cars: the high speed pushes energy in to a region faster than it can be carried away. This tells us that if the mean speed or the density of cars is reduced, shocks (traffic jams) will not occur. This conclusion may be counterintuitive, but it is exactly the case. Some cities therefore restrict traffic flow into major arteries, which perhaps unexpectedly actually increases flow through the arteries. In another example, rope tows are used on beginners' ski slopes, and often the operators will increase the speed of the rope tow in response to an increased volume of skiers. This invariably causes skiers to collide, *reducing* the throughput.

What about in blood flow? The 5 liters of blood in the typical human body are circulated about once a minute, so the rate of flow is about 5000 cc/min. All of this blood travels through the aorta, with a radius of about 1 cm, so the mean speed at which blood is fed into the aorta is $5000/(\pi{\cdot}1^2) = 1600$ cm/min there. By comparison, the speed at which disturbances are carried away is the speed of sound, which is about 250,000 cm/min. So the speed of flow is less than 1% of the speed at which sound carries disturbances away. This means that the lead edge steepening shown in Fig. 4.3 is surely not caused by a traditional shock: the source cannot simply be in the fluid response itself, and must lie elsewhere.

4.3.2 **Nonlinear Material Response**

Based in part on this simple calculation, careful measurements have been made of the elastic properties of the vasculature. Recall from Chapter 2 that we mentioned that even a simple latex balloon is not just a Hookean spring: here we find that the vasculature is likewise more complex than our simple models would suggest. This shouldn't be surprising: an artery is composed of multiple layers of connective, laminar, and muscular tissues, and blood pressure itself is actively regulated from the medulla, which receives data from a network of baroreceptors.

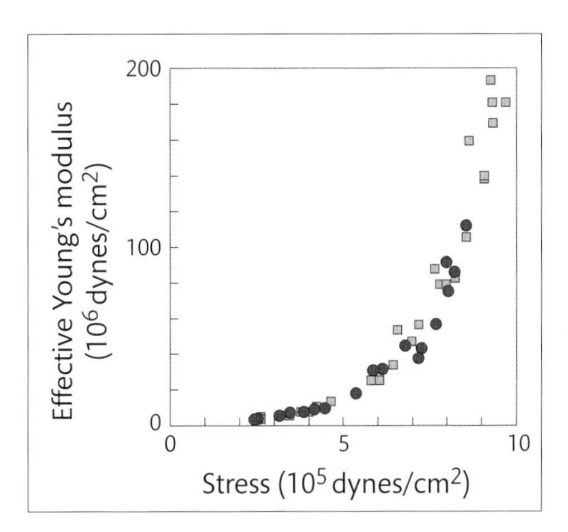

Fig. 4.4 Young's modulus measured from two typical experiments in the vena cava of dogs as a function of tensile stress on vein. Young's modulus is determined by measuring the speed of a wave produced by a piston inserted downstream of two pressure sensors; stress is changed by occluding the vein upstream of the measurement points, causing the vein to enlarge. Stress is evaluated by measuring the inner diameter of the vein using a gauge inserted for this purpose. (Data from Yates.[2])

To better understand vascular mechanics, let's turn to data that have been taken of the Young's modulus itself.[2] In Fig. 4.4, we show data obtained from the vena cava of an anesthetized dog. Here the vein was occluded by a balloon that was pressurized from a canula, and pressure pulses were artificially produced by a piston system provided upstream from another canula fed through the common iliac vein (which supplies the vena cava). Measurements of the speeds of these artificial pulses were taken using pressure transducers, and the Young's modulus, Y, was inferred from Eq. [3.29] of Chapter 3, which relates the speeds of pressure waves to Y. The longer the balloon is left in place, the greater the stress imposed by pressure on the walls of the vena cava: the stress is inferred by measuring the internal radius of the vein. In this way, the Young's modulus of the vein can be measured as a function of applied stress.

Data similar to those shown in Fig. 4.4 have been obtained under a wide variety of conditions. The universal feature that all of these data share is that the Young's modulus is neither constant nor a simple linear function of applied stress. We mentioned in Chapter 1 that we would be considering conditions (low Reynolds number) under which the nonlinear term in Navier–Stokes equation could be neglected. Now we find that even if we make this approximation, blood flow through elastic tubes will be nonlinear because the elastic modulus itself varies nonlinear during deformation.

4.4 What is to be Done?

A strategy is needed to cope with the intrinsic nonlinearity present in this problem. We have seen in prior exercises (e.g. the sine flow) that nonlinearities are not a difficulty that can be addressed by simply finding a different solution: in the presence of nonlinearities, solutions can be, and often are, either not attainable or not unique. This is something that we observe every day of our lives: if we take a snapshot of coffee being poured into a cup, we never see the same thing at every instant, rather we see the coffee surface constantly changing in a complicated way. So any solution that we might obtain for the coffee must also constantly change.

Moreover, the tools that we have used so far to obtain solutions aren't even applicable. As an example, we solved Eq. [4.5] for simple sinusoidal driving, $A{\cdot}e^{i\omega t}$, and we asserted that for linear equations we could superimpose multiple sine waves to produce a more complicated driving function. This does not work for nonlinear equations: indeed, if we could solve for pulsatile driving at one frequency, ω_1, there is no assurance that driving at another frequency, ω_2 would also be a solution, and certainly if $A{\cdot}e^{i\omega_1 t}$ and $B{\cdot}e^{i\omega_2 t}$ each yielded solutions there is no reason to believe that $A{\cdot}e^{i\omega_1 t} + B{\cdot}e^{i\omega_2 t}$ might lead to a solution. In such a case, we say that "superposition" breaks down: we cannot superimpose two solutions to get a third solution.

4.4.1 Mode Coupling

This brings us to a central issue that characterizes nonlinear equations, "mode coupling." Mode coupling can be most easily illustrated by going back to the nonlinear term in

Navier–Stokes, $\vec{V}\cdot\nabla\vec{V}$. To keep things simple, let's imagine that we are interested in a problem only in one dimension, and let's consider a solution that looks like $\vec{V}= V_0\sin(kx)\hat{x}$—perhaps there is a simple sinusoidal disturbance in one direction to an otherwise stationary tank of water. In this simple example, the nonlinear term becomes:

$$\begin{aligned}\vec{V}\cdot\nabla\vec{V} &= \vec{V}\cdot\frac{\partial\vec{V}}{\partial x} \\ &= V_0{}^2k\cdot sin(kx)cos(kx),\end{aligned} \qquad [4.14]$$

and we may recall the trigonometric identity $\sin(2kx) = 2\sin(kx)\cos(kx)$, so that:

$$\vec{V}\cdot\nabla\vec{V}=\frac{V_0{}^2k}{2}\cdot\sin(2kx). \qquad [4.15]$$

Uh oh: if we start with a simple sine wave with wavenumber k, we end up with a solution that depends on a sine wave with twice k. And that solution must produce also 4 times and 8 times that number, and so on. This is "mode coupling:" the $\sin(kx)$"mode" is coupled to the $\sin(2kx)$, and higher modes. Mode coupling means that barring a miraculous cancellation somewhere, we can't have a simple sinusoidal solution to (most) nonlinear equations.

So we can't have a simple sinusoidal solution to typical nonlinear equations, and if we did obtain a solution it probably wouldn't be unique and its components would couple to other waves. Additionally, superposition is what allows two waves to collide and pass through one another unaltered, and since superposition doesn't hold, if a solution ran into anything at all, it couldn't continue and ought to inevitably be destroyed. This is where the understanding of nonlinear equations languished for many years.

Example: Jupiter's Great Red Spot

Yet mysteriously, stable and long-lasting solutions to insanely nonlinear equations are often observed in nature. Let's look at a couple of examples. The first example is the Great Red Spot (GRS) on Jupiter, shown in Fig. 4.5. The GRS was first reported by Robert Hooke (1635–1703). In addition to producing Hooke's law and studying everything from gravitation to architecture, Hooke built one of the first reflecting telescopes, which in 1664 he trained on Jupiter and observed the GRS. It is known that Jupiter is entirely fluid: it has no solid surface, and the Reynolds number in the vicinity of the GRS is, well, astronomical. The nonlinear term in Navier–Stokes becomes important around Re ~ 1, and depending on geometry, turbulence sets in at around Re ~ 1000. Near the GRS, the Reynolds number is at least 100,000.

In Fig. 4.5 we show the GRS alongside a snapshot of the earth for scale; the reader can clearly see strong turbulent eddies near the spot that one would expect would tear the spot apart. Yet the GRS has persisted for well over 300 years. There are many different effects

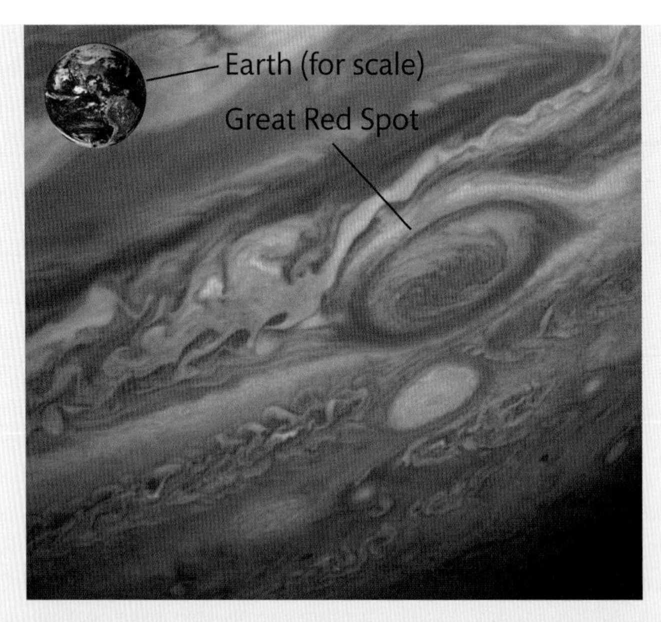

Fig. 4.5 Great Red Spot on Jupiter, with snapshot of Earth included for scale. Note the strong eddies surrounding the spot, confirming calculations that the Reynolds number is very large, at least 100,000. (Figures courtesy NASA/JPL-Caltech.)

at work here beyond simple fluid dynamics as defined by Navier–Stokes: the spot is significantly influenced by strong temperature gradients (Jupiter produces more heat than it absorbs), and there are strong Coriolis forces, to name two effects. Nevertheless, there is no doubt that this structure is associated with nonlinear fluid dynamics that follow from Navier–Stokes equation alone. This has been shown computationally as well as experimentally: notably, an experiment in a meter-sized tank of water modified to mimic the relevant fluid conditions (recall the principle of dynamical similarity discussed in Chapter 1) reproduced a stable vortex under highly nonlinear conditions.[3] Other spots are also visible in Fig. 4.5, and similar long-lasting structures are seen on other planets.

4.4.2 **Solitons: the Korteweg–de Vries Equation**

So the GRS is both pretty and pretty strange. What does this have to do with pulsatile flows in the vasculature? To answer that question, let's turn to a second historical example that has been directly connected to our present problem.

In 1834, a naval engineer named John Scott Russell (1808–1882), studying designs for canal boats,[4] was riding a horse along the Union Canal in the outskirts of Edinburgh, Scotland. At that time, goods were transported along canals in barges pulled by horses on towpaths lining the canal. The horses stopped, but Russell noticed that the wave of water that had been pushed ahead of the barge continued on. He followed the wave and found that it traveled along the canal with no loss in energy or change in height for over a mile before he lost sight of it.

Waves of this type are now called "solitary waves," and are observed in numerous different systems, ranging from tidal bores that surfers flock to (e.g. the famous "Silver Dragon" on the Chinese Qiantang River) to proposed use in fiber optic cables. And indeed, to solitary pulsatile waves in elastic tubes.[5] It can be shown that these waves appear only in nonlinear systems—which also exhibit more exotic phenomena such as large "rogue waves."[6] We now know that in shallow canals, the nonlinearity is associated with a pinching off of the troughs of the waves near the bottom of a shallow canal: this prevents the water held in the peak from leaking away to dissipate outward. In fiber optic cables, the nonlinearity is due to optical media in which the index of refraction depends on the intensity of light, and in elastic tubes the nonlinearity is due to the nonlinear Young's modulus described in Fig. 4.4.

These mechanisms were not understood at all by Russell or his contemporaries. Russell's descriptions were investigated by many scientists, including Rayleigh. A breakthrough was obtained by pair of Dutch mathematicians, Diederik Korteweg (1848–1941) and his doctoral student Gustav de Vries (1866–1934). De Vries' dissertation described a special kind of solitary wave in water that is now termed a "soliton," which has the additional—and quite astonishing—property that *despite the abject and demonstrable failure of superposition*, two solitons can pass through one another without damage to either. We will see an example of this effect shortly; first let's examine the equation that Korteweg and de Vries produced to describe nonlinear waves in shallow water. Their derivation, incidentally, isn't rigorous, and is a good example of an *ad hoc* model that describes the phenomenon of interest well without being completely accurate.

The Korteweg–de Vries (or KdV) equation can be written as:

$$\frac{\partial V}{\partial t} + \frac{\partial^3 V}{\partial x^3} - 6V \cdot \frac{\partial V}{\partial x} = 0, \qquad [4.16]$$

where the factor of 6 isn't necessary, but leads to a convenient algebraic simplification. To understand this equation, let's break it apart and first neglect the final term, leaving just the first two terms:

$$\frac{\partial V}{\partial t} = -\frac{\partial^3 V}{\partial x^3}. \qquad [4.17]$$

So the change of velocity in time, $\partial V/\partial t$, depends on the third derivative of V in space, $\partial^3 V/\partial x^3$.

EXERCISE 4.1

Eq. [4.17] is sometimes referred to as being "diffusive." We will discuss diffusion equations later in this book. This particular equation differs from diffusion equations in that it has a third derivative, which is a bit of a curious beast. In mechanics, the third derivative with respect to time is termed the "jerk," and is a measure of the lack of smoothness of motion: spastic motions have large jerk, and smooth motions have small jerk. Amusingly, the fourth derivative is called the "jounce" or "snap," and higher derivatives are so scarcely used that no dour mathematician has so far objected to their playful terms "crackle" and "pop." In our case, Eq. [4.17], the third derivative also corresponds to lack of smoothness, here in space, and since the change in velocity with time depends on the third derivative in space, small spatial fluctuations become amplified in time. This produces serious difficulties in performing a straightforward numerical solution to the equation. For an excellent and very readable description of a stable numerical approach, the reader is referred to Trefethen[7] (see also summary of Fourier series in Chapter 12). For our purposes, to see the numerical difficulty and to get a foothold into solving equations numerically, the reader can perform the following exercise.

Confirm that simulating the time evolution of V from Eq. [4.17] of even a simple function, for example, V = cos(x), becomes numerically unstable almost immediately. A numerical solution can be obtained by rewriting Eq. [4.17] so that the infinitesimal time and space steps are replaced with finite steps—this is termed "finite differencing" (described in the Appendix)—and evolve the velocity after a time Δt to become:

$$V(t+\Delta t) = V(t) + \frac{\partial^3 V(t)}{\partial x^3}\Delta t. \qquad [4.18]$$

The third derivative can be calculated in any of several ways. We discuss technicalities of this step later; for now, the reader can use the formula:

$$\frac{\partial^3 V_i}{\partial x^3} \approx \frac{V_{i+2} - 2V_{i+1} + 2V_{i-1} - V_{i-2}}{2\Delta x^3}. \qquad [4.19]$$

where V is broken into a large number (say 10,000) of spatial grid points identified by the subscript, i. That is, V_1 is the velocity at the leftmost point being considered, V_2 is the velocity a distance 1/10,000 to the right, and so on. This exercise will confirm that a smooth function, V, rapidly becomes very noisy under Eq. [4.17]. Analyze what the third derivative does that produces this noise and explain. Try smoothing the result (e.g. use Matlab's "smooth" function): explain what this does and explain why it doesn't completely resolve the problem. What else could you do to prevent this numerical problem?

The preceding exercise shows that the second term in Eq. [4.16] produces noise that spreads smooth disturbances out. This is what we might expect, but is not what is seen either in the turbulent atmosphere of Jupiter or on the Union Canal. What then does the last term in Eq. [4.16] do? If this time we consider just the first and last term, we have:

$$\frac{\partial V}{\partial t} = 6V\cdot\frac{\partial V}{\partial x}. \qquad [4.20]$$

To understand this equation, let's compare it with a simpler equation:

$$\frac{\partial V}{\partial t} = A\cdot\frac{\partial V}{\partial x}. \qquad [4.21]$$

It is not hard to see that A has units of distance/time, and indeed Eq. [4.21] is a form of wave equation that translates a disturbance uniformly at speed A, as shown in Fig. 4.6(a). So what about Eq. [4.20], where $A = 6V$? Evidently, at least in the short term, this will translate a

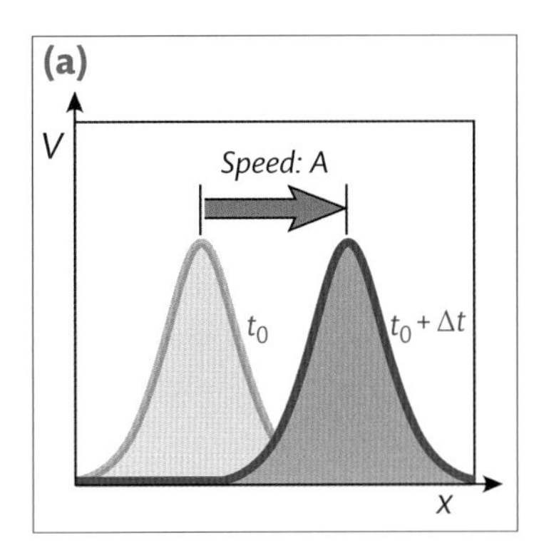

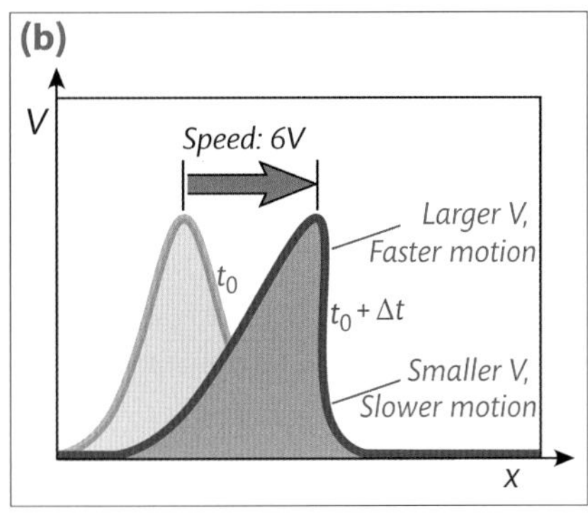

Fig. 4.6 Wave equations [4.20] and [4.21]. (a) A disturbance at time t_0 (light blue) is translated forward in time after time increment Δt (dark blue) under the linear wave equation, Eq. [4.20]. (b) The same disturbance is translated a distance that grows with the magnitude *V(x)* under the nonlinear equation, Eq. [4.21]. Notice that because points with larger *V* are translated forward faster than points with smaller *V*, the leading edge of the disturbance becomes steeper, as we saw in Fig. 4.3.

disturbance, *V(x)* forward in time with speed $6V$—meaning that a large disturbance will be translated faster than a small one. This also causes the top of a disturbance (with larger *V*) to be translated forward faster than the bottom of the same disturbance. As shown in Fig. 4.6(b), this produces lead edge steepening.

Apparently the KdV equation (Eq. [4.16]) has two parts: a third derivative that spreads out a disturbance (Eq. [4.17]), and a nonlinear term that sharpens the leading edge of the disturbance (Eq. [4.20]). By a bit of algebraic magic that we have done nothing to deserve, these two effects exactly cancel out. This cancellation is simple to see in the KdV equation, but similar results are found in other nonlinear equations as well.

For the KdV equation, we can seek a solution that travels at constant speed, *c*. This is to say that *V* depends only on $\chi = x - ct$. Plugging this ansatz into Eq. [4.16], we obtain:

$$-c\frac{\partial V}{\partial \chi} + \frac{\partial^3 V}{\partial \chi^3} - 6 \cdot V \cdot \frac{\partial V}{\partial \chi} = 0. \qquad [4.22]$$

We referred in Chapter 3 to the idea that the standard method for solving partial differential equations is to find a trick to reduce them to ordinary differential equations: we see an example here. We can integrate with respect to χ to obtain:

$$\frac{\partial^2 V}{\partial \chi^2} = C + cV + 3 \cdot V^2, \qquad [4.23]$$

where *C* is an integration constant. The reader can confirm that this equation has solution:

$$V = -\frac{c}{2} sech^2\left(\frac{\sqrt{c}}{2}\chi\right), \qquad [4.24]$$

where $sech(X) = 2/(e^X + e^{-X})$ is the hyperbolic secant (pronounced "seech," to rhyme with "beach").

EXERCISE 4.2

Insert Eq. [4.24] into the KdV equation, and confirm that the third derivative and the non-linear terms exactly cancel. This simplifies Eq. [4.16] to become $\partial V/\partial t = 0$. Either perform this calculation by hand, or use a symbolic programming language (Maple, Mathematica, Matlab, Mupad, etc.). Eq. [4.24] is a single solitary wave that travels to the right with speed, c. We mentioned earlier that the KdV equation also admits solitons: *multiple* solitary waves that travel through one another—despite the nonlinear term that would seem to forbid superposition.

As a second exercise, confirm that the two soliton solution following also is a solution to Eq. [4.16], and plot the evolution of this solution over time, from $t = -1$ to $t = +1$:

$$V = 12\frac{3 + 4\cdot\cosh(2x - 8t) + \cosh(4x - 64t)}{[3\cdot\cosh(x - 28t) + \cosh(3x - 36t)]^2}, \qquad [4.25]$$

Confirm that the larger soliton travels faster than the smaller (as anticipated from Eq. [4.21]), and that the larger overtakes and passes through the smaller leaving both intact. Notice that superposition manifestly does *not* hold: when the two solitons overlap, the total amplitude actually *decreases*. For comparison, plot two colliding sine waves that are subject to superposition, and show that the amplitude increases when the waves overlap.

ADVANCED EXERCISE 4.3

We do not include the derivation here; however, it has been shown that the KdV equation can be used to model pulsatile flow in an elastic tube.[5] A final property of solitary wave equations is that these equations spontaneously separate an arbitrary input into two sets of functions:

(1) solutions like Eqs. [4.24] and [4.25], which travel forward coherently and without change;

(2) noise, which does not.

This is germane to blood flow because there is no reason for the pressure wave from the heart (Fig. 4.3(a)) to be a solution to the relevant nonlinear wave equation, yet its evolution approaches *sech*^{2} waves (dashed lines in Fig. 4.3(b)). Numerically confirming this separation of waves into solutions and noise is left as an exercise for the advanced reader: the evolution of an arbitrary initial state under the KdV equation can be simulated using so-called "spectral" methods, with code provided in Chabchoub et al.[6]

Example Vortex rings

We have mentioned solitary waves in the context of Jupiter's Great Red Spot and J.S. Russell's bow wave. Solitary waves are seen in a variety of other contexts as well, including optical solitons used to transmit signals long distances without degradation, solitary "barchan" dunes that travel intact hundreds of kilometers across deserts, even a weapon has been designed to transmit a coherent pressure pulse in air strong enough to knock over a (admittedly rickety) building.

Also in the first half of the twentieth century, tobacco companies made use of a type of solitary wave by advertising a competition for their users to blow three smoke rings inside of one another. Two smoke rings are easily produced, and travel in tandem, repeatedly leapfrogging one another, but a third ring is more difficult to incorporate – hence the competition. Smoke rings are a type of "vortex ring," many of which appear in nature, and we show several examples in Fig. 4.7. In Fig. 4.7(a) we show a 200 m diameter vortex ring in air produced by Mt. Etna during a small eruption; in Fig. 4.7(b) we show multiple rings produced by a falling droplet of dye in water; and in Fig. 4.7(c) we show a bubble ring emitted by a beluga whale. Belugas, dolphins, and other marine mammals produce these rings, mostly in play, but some whales also deploy bubbles in a shrinking net to corral fish into a "bait ball." Finally, in Fig. 4.7(d), we show a different kind of coherent ring, here produced by a school of barracuda in the Bismarck Sea. This final case involves active swimmers and so plainly differs from the previous examples of passive fluid flow; in another sense, it involves material objects that necessarily obey continuity and flow equations and that conspire to produce a coherent vortical structure.

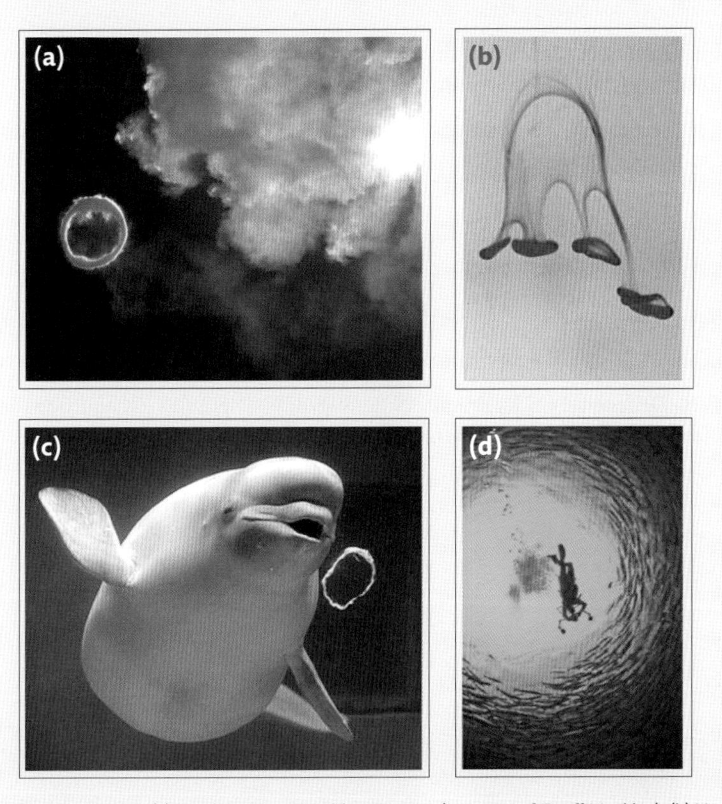

Fig. 4.7 Solitary vortex rings. (a) Smoke ring emitted by Mt. Etna (courtesy of Geoff Mackley). (b) Vortex rings from falling ink droplet (courtesy of Lasse Johansson). (c) Bubble rings produced by beluga whale (KAZUHIRO NOGI/AFP/Getty Images). (d) School of barracuda (Christopher Swann/Science Photo Library).

For other photogenic examples of complex fluid structures, the reader may want to visit the "Gallery of Fluid Motion," mentioned previously in Chapter 2, and for examples of cooperative animal behaviors including circulating schools and fountain effects in fish, bird murmurations, and ant rafts, the reader may want to look at the books by Sumpter or Camazine et al.[8,9]

REFERENCES

1. Mahmud, A. and Feely, J. (2003) Spurious systolic hypertension of youth: fit young men with elastic arteries. *American Journal of Hypertension, 16*, 229–32.

2. Yates, W.G. (1969) *Experimental studies of the variations in the mechanical properties of the canine abdominal vena cava.* PhD Dissertation, Stanford University, USA.

3. Sommeria, J., Meyers, S.D. and Swinney, H.L. (1988) Laboratory simulation of Jupiter's Great Red Spot. *Nature, 331*, 689–93.

4. A captivating history of Russell and his contemporaries appears in Ferreiro, L.D. and Pollara, A. (2017) Clippers, yachts, and the false promise of the wave line. *Physics Today, 70*, 52–8.

5. Demiray, H. (2001) Solitary waves in fluid-filled elastic tubes: weakly dispersive case. *International Journal of Engineering Science, 39*, 439–51; also Korteweg, D. J. (1878) *Over voortplantingssnelheid van golven in elastische buizen.* Van Doesburgh, Holland.

6. Chabchoub, A., Hoffmann, N., Onorato, M. and Akhmediev, N. (2012) Super rogue waves: observation of a higher-order breather in water waves. *Physical Review X, 2*, 011015.

7. Trefethen, L.N. (2010) *Spectral Methods in MATLAB.* Philadelphia, PA: SIAM.

8. Sumpter, D.J.T. (2010) *Collective Animal Behavior.* Princeton, NJ: Princeton University Press.

9. Camazine, S., Deneubourg, J.L., Franks, N.R., Sneyd, J., Theraulaz, G. and Bonabeau, E. (2003) *Self-organization in Biological Systems.* Princeton, NJ: Princeton University Press.

5 Entrances, Branches, and Bends

We have seen how fluids behave in simple tubes under ideal circumstances. Biological systems are seldom ideal, however, and we now begin examining effects of some of these non-idealities. In this chapter, we consider what happens at the inlet to a simple tube; then we describe what occurs when tubes branch or bend. The importance of these issues can be made clear by noting that the longest straight arterial segment is the aorta, between the mesenteric and renal arteries. In this segment, the aspect ratio (AR = length $\div$ diameter) is only about $AR = 20$; more commonly blood vessels have aspect ratio around $AR = 5$. Between these straight segments are significant curved (e.g. the circle of Willis), and branched regions. So let's first ask whether aspect ratios of 5–20 are sufficiently long to justify the use of solutions that we have obtained so far, which intrinsically neglect entrance or curvature effects.

5.1 Entrance Effects

In Fig. 5.1, we sketch what we expect to see for low Reynolds number flow in a two-dimensional (2D) channel. To the left in Fig. 5.1(a), we assume that perfectly uniform flow feeds into a channel, and at a large distance downstream, we assume that flow is fully developed. Near the entrance to the channel, however, we expect flow near the boundaries to be slowed by viscosity (lavender in Fig. 5.1), and flow at the center of the channel to travel essentially un-influenced by boundary effects (blue). We term flow at the center "plug" flow because all of the fluid travels at once like a solid plug. Separating the boundary and the plug flow regions is some boundary layer thickness function, $y = \delta(x)$, shown in red in the figure. We define the entrance length, L_e, to be the distance at which the plug flow region vanishes, at which point flow is assumed to be fully developed and to be defined by the Poiseuille formula derived in Chapter 1.

Our problem, then, is to obtain L_e from what we know about fluid flow and continuity. We will solve this problem by matching its two parts—that is, we will solve for flow near the boundary, we will then solve for flow near the center, and finally we will match the two solutions so that the velocity smoothly transitions from one solution to the other at $y = \delta(x)$. Matching is common in fluids problems, and the approach will be used again in Exercise 5.2 and elsewhere in this book.

To obtain the two matching solutions, we note that a small parcel of fluid only responds to its local pressure and shear strain, and so has no information about what may be going on far

Biomedical Fluid Dynamics: Flow and Form. Troy Shinbrot.

DOI: 10.1093/oso/9780198812586.001.0001

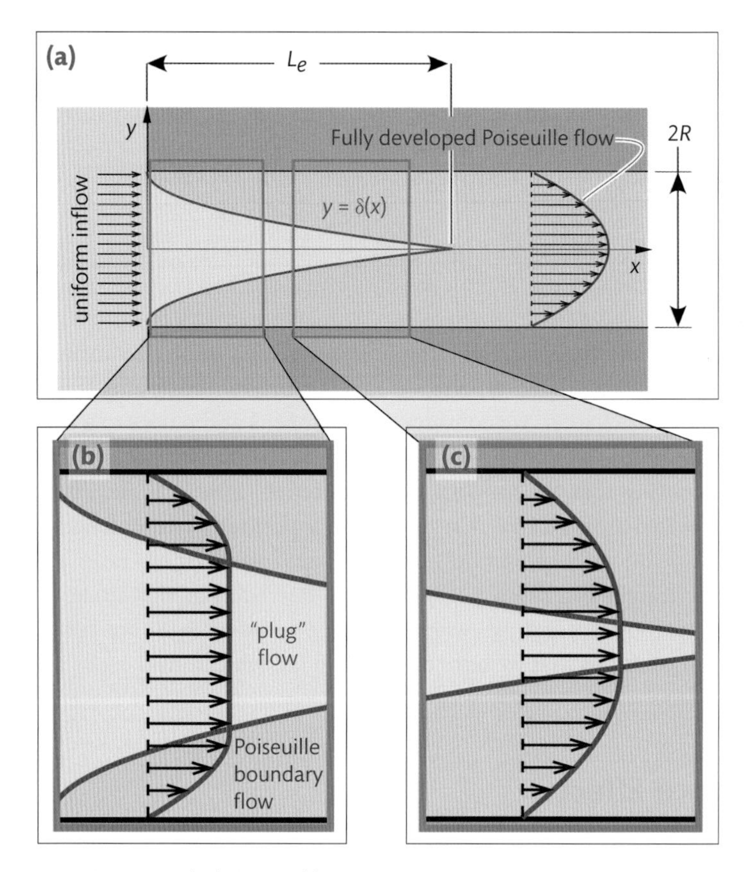

Fig. 5.1 Inlet to a two-dimensional (2D) channel. (a) Schematic of a 2D channel with uniform flow at entrance. Beyond an entrance length, L_e, the flow is fully developed (lavender), and for small Reynolds number, we will take that to be the Poiseuille solution obtained in Chapter 1. At the center of the channel and close to the entrance, the fluid (blue) has not yet had a chance to be influenced by the walls, so flow is still uniform. Between flow near the walls and flow near the center is a boundary, $y = \delta(x)$ (red). (b) Highlight of the flow profile, showing the uniform "plug" flow at the center and Poiseuille boundary flow near the walls. (c) Similar highlight further downstream: note that to conserve volume, flow must be faster at the center downstream than upstream (as derived in the text).

away. So if we knew $\delta(x)$, we could solve for flow in either of the two regions entirely independently of what the fluid is doing in the other region.

An elegant approach to solving for flow in both regions was described by Landau and Lifshitz.[1] We discussed Landau in Chapter 2; Evgeny Lifshitz (1915–1985) was his student and friend, known among other things for his wry assertion that "mathematical rigor . . . in theoretical physics often amounts to self-deception." Their approach follows this philosophy: it is clever and produces an answer that has proven over time to be correct, but in several places would fail in the eyes of a strict mathematician to be completely rigorous. Their approach is as follows.

If we assume that flow is steady and neglect external forces, the Navier–Stokes equation can be written:

$$\vec{V}\cdot\vec{\nabla}\vec{V} - \nu\nabla^2\vec{V} = -\frac{1}{\rho}\vec{\nabla}P, \tag{5.1}$$

and if we neglect flow in the y- and z-directions, this becomes:

$$V_x \frac{\partial V_x}{\partial x} - \nu \left(\frac{\partial^2 V_x}{\partial x^2} + \frac{\partial^2 V_x}{\partial y^2} \right) = -\frac{1}{\rho} \frac{\partial P}{\partial x}. \quad [5.2]$$

We note that to obtain Eq. [5.2], we have implicitly assumed that $\partial P/\partial y = 0$, because otherwise the pressure gradient would produce flow in the y-direction. Additionally we assume that viscous resistance by the side walls produces much more shear than elongation, so we neglect the elongational term, $\partial^2 V_x/\partial x^2$ in comparison with the shear term, $\partial^2 V_x/\partial y^2$:

$$V_x \frac{\partial V_x}{\partial x} - \nu \frac{\partial^2 V_x}{\partial y^2} = -\frac{1}{\rho} \frac{\partial P}{\partial x}. \quad [5.3]$$

We can solve this equation separately in the central, plug flow, region and in the boundary, Poiseuille flow, region. In the central region, we have assumed that flow isn't influenced by viscosity, so we neglect the viscous term and write:

$$V_{plug} \frac{\partial V_{plug}}{\partial x} = -\frac{1}{\rho} \frac{\partial P}{\partial x}. \quad [5.4]$$

Now we employ a first trick: Eq. [5.3] gives the pressure gradient in the x-direction. If we assume that the pressure is in equilibrium, then we can use the same expression for $\partial P/\partial x$ *in the boundary region*. Using the pressure gradient obtained in one region across both regions is in one sense a mathematical sleight of hand, but on the other hand if this were not true, there would be a pressure difference—and so a flow—in the y-direction, which we have assumed does not occur. Accepting this argument, Eq. [5.3] becomes:

$$V_x \frac{\partial V_x}{\partial x} - \nu \frac{\partial^2 V_x}{\partial y^2} = V_{plug} \frac{\partial V_{plug}}{\partial x}. \quad [5.5]$$

At this point, we employ a second, much more svelte, trick by nondimensionalizing Eq. [5.5] in a special way:

$$\tilde{V}_x = \frac{\vec{V}_x}{\langle V \rangle}, \qquad \tilde{x} = \frac{x}{L},$$
$$\tilde{V}_y = \frac{\vec{V}_y}{\langle V \rangle} \sqrt{Re}, \quad \tilde{y} = \frac{y}{L} \sqrt{Re}. \quad [5.6]$$

The square roots are very important for a reason that will not be obvious to those of us with less remarkable minds than Landau. If we insert Eqs. [5.6] into Eq. [5.5] we obtain:

$$\tilde{V}_x \frac{\partial \tilde{V}_x}{\partial \tilde{x}} - \frac{\partial^2 \tilde{V}_x}{\partial \tilde{y}\ 2} = \tilde{V}_{plug} \frac{\partial \tilde{V}_{plug}}{\partial \tilde{x}}. \quad [5.7]$$

Landau noticed that Eq. [5.7] lacks any scale—that is, it does not contain any constant that depends on the size of the system, L. Consequently, as Landau points out, the solution to Eq. [5.7] cannot contain any information about scale, and so if we plot the solution, it must look identical at every scale. This means that the elusive function $\widetilde{y} = \delta(\widetilde{x})$ which defines the change from plug to boundary flow must also look identical at every scale.

So how can we construct a function relating $\widetilde{x}$ to $\widetilde{y}$ that will be scale independent? Rewriting $\widetilde{x}$ and $\widetilde{y}$ from Eqs. [5.6] we find:

$$\widetilde{x} = \frac{x}{L} \quad and \quad \widetilde{y} = y\sqrt{\frac{\langle V \rangle}{\nu L}}, \tag{5.8}$$

which can be combined without the scale L only if:

$$\widetilde{y} \sim \sqrt{\widetilde{x}}. \tag{5.9}$$

In this case, L cancels on left and right sides, leaving:

$$y\sqrt{\frac{\langle V \rangle}{\nu}} \sim \sqrt{x}. \tag{5.10}$$

This defines a family of curves:

$$y = \phi\sqrt{x}, \tag{5.11}$$

where:

$$\phi = c\sqrt{\frac{\nu}{\langle V \rangle}} \tag{5.12}$$

is a scale-free constant.

There is no other scale-free function $y = f(x)$ than Eq. [5.11], and the flow illustrated in Fig. 5.1(a) depends on only one function $y = \delta(x)$ defining the transition between boundary and plug flow. Consequently Eq. [5.11] must be the function $\delta(x)$. As sketched in Fig. 5.2, for large viscosity or small velocity, ϕ will be large and so the entrance length will be short, and contrariwise for small viscosity or large velocity.

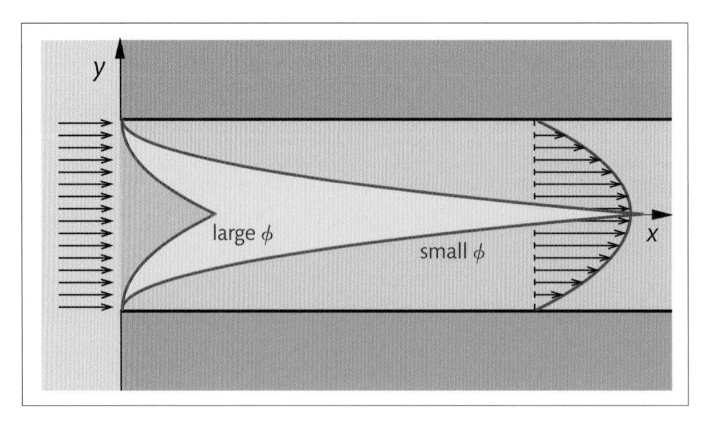

Fig. 5.2 Scale-free function $\delta(x)$ shown for large and small ϕ (defined in Eq. [5.12]). As sketched here, the entrance length, L_e, at which fully developed Poiseuille flow appears is short for small ϕ and long for large ϕ.

The entrance length, L_e, is defined to be where $y = \delta(x) = R$, or $\phi\sqrt{L_e} = R$, that is, at:

$$L_e = \frac{R^2\langle V\rangle}{c^2\nu}, \qquad [5.13]$$

or since $Re = R\langle V\rangle/\nu$:

$$L_e = \frac{1}{c^2}R{\cdot}Re. \qquad [5.14]$$

So the entrance length grows with both the channel radius and the flow Reynolds number. More detailed analysis[2] provides c^2:

$$L_e = 0.208{\cdot}R{\cdot}Re. \qquad [5.15]$$

We can now compare the expected aspect ratio, $AR = L_e/2R = 0.104\ Re$, at which fully developed Poiseuille flow can be expected for typical blood vessels. A typical speed for blood flow in a 0.1 cm diameter vessel is 10 cm/s, and the kinematic viscosity, ν, of whole blood is about 0.1 Stokes. This is the cgs unit for ν, abbreviated "St;" for μ the cgs unit is the Poise, abbreviated "P." These typical values produce $Re \sim 10$, so the aspect ratio must be over 1 to produce fully developed flow. Evidently, near the entrance of a typical blood vessel, flow will be significantly boundary dependent, however most typical vessels, with aspect ratios over $AR = 5$, will obey the equations derived in earlier chapters.

Now that we know $\delta(x)$, we can easily obtain the full velocity field in boundary and plug regions. In the boundary region, flow is given approximately by the Stokes equation subject to boundary conditions that are no-slip at the walls and reach a maximum, V_{plug}, at $y = \delta(x)$. We have already solved for this: this is simply Poiseuille flow:

$$V_x(y) = V_{plug}\left(1 - \frac{y^2}{\delta(x)^2}\right). \qquad [5.16]$$

When we derived this equation earlier (Eq. [1.37] in Chapter 1), V_{plug} was the maximum velocity at the center of the channel, V_{max}, and $\delta(x)$ was a constant channel radius, but otherwise the analysis and logic of the derivation was identical to the present problem.

However, if we carefully compare Chapter 1's derivation with the geometry shown in Fig. 5.1, we find that the origin used in Chapter 1 actually lies in the present case on the curve $y = \delta(x)$. Having an origin that moves along a curve is awkward, to put it mildly, so to avoid this situation we fix the origin at only one location by replacing y with $y - \delta(x)$. This way, when y shown in Eq. [5.16] is $\delta(x)$, y in the "fixed origin" frame is identically zero. This changes Eq. [5.16] to:

$$\begin{aligned} V_x(y) &= V_{plug}\left\{1 - \frac{[y - \delta(x)]^2}{\delta(x)^2}\right\} \\ &= V_{plug}\left[\frac{2y}{\delta(x)} - \frac{y^2}{\delta(x)^2}\right]. \end{aligned} \qquad [5.17]$$

Finally, in the plug region flow is constant in the y-direction, however because flow is slowed near the boundaries, in order to conserve mass the speed in the plug region must increase with distance, x, that is: $V_{plug} = V_{plug}(x)$. We can solve for V_{plug} by demanding that the flow into the channel equals the flow in the plug region plus the flows in the surrounding boundary regions. Expressed over the bottom half of the channel (the top half is identical), this is:

$$R \cdot V_{inflow} = V_{plug}\left(R - \delta(x)\right) + \int_0^{\delta(x)} V_x(y)dy. \qquad [5.18]$$

Using Eq. [5.17], this becomes:

$$\begin{aligned} R \cdot V_{inflow} &= V_{plug}[R - \delta(x)] + V_{plug}\int_0^{\delta(x)} \left[\frac{2y}{\delta(x)} - \frac{y^2}{\delta(x)^2}\right] dy \\ &= V_{plug}\left\{R - \delta(x) + \left[\delta(x) - \frac{\delta(x)}{3}\right]\right\} \qquad [5.19] \\ &= V_{plug}\left[R - \frac{\delta(x)}{3}\right], \end{aligned}$$

so:

$$V_{plug} = \frac{3RV_{inflow}}{3R - \delta(x)}. \qquad [5.20]$$

This tells us that the velocity in the plug flow region grows from V_{inflow} at the channel entrance to $^3/_2 V_{inflow}$ at L_e.

EXERCISE 5.1

A second type of entrance effect concerns what happens to an impulse, or bolus, of material injected into a long channel. This problem was first studied by G.I. Taylor (1886–1975), an icon in fluid mechanics who worked in everything from turbulence and rotating fluid flow to single-photon diffraction and the effect of electric fields on fluids. We will discuss rotating flow in a later chapter, and Taylor's work on "electro-spinning" of fine tendrils of polymers will likely interest the biomechanics-oriented reader, for electrospinning is used to produce artificial tendons and ligaments in the laboratory.

Taylor's idea was that a bolus of fluid, shown as a blue rectangle in Fig. 5.3, would be spread out, or dispersed, by Poiseuille flow, as shown in consecutive parabolic trajectories in the figure. The effect of "Taylor dispersion" on concentration is shown in the corresponding consecutive plots in red. For this exercise, (1) reproduce Fig. 5.3 and explain why the leading edge of the bolus of fluid carries the highest concentration, and (2) repeat this exercise for the entry flow defined by Eqs. [5.11], [5.12], [5.17], and [5.20]. Discuss any differences between the fully developed Poiseuille result shown in Fig. 5.3 and transport subject to the entry flow.

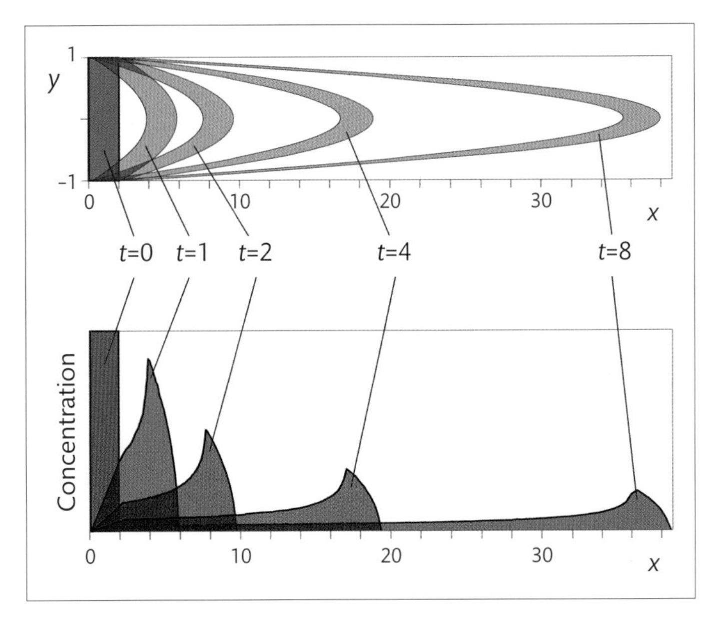

Fig. 5.3 "Taylor dispersion" of a bolus of fluid. Top panel shows an initial rectangular blue region of fluid that is advected (jargon for "moved") forward in time under the influence of Poiseuille flow, and lower panel shows the resulting concentration as a function of downstream distance.

EXERCISE 5.2

Matching of different solutions—for example, one solution near a pipe boundary and another closer to its center—is a general and widely applicable technique. We include a second, simpler, exercise to reinforce the idea. In this problem, we consider the problem of flow of two different viscosity fluids. There are many examples of this problem: we will see in a later chapter that blood separates into a central flow predominantly occupied by red blood cells, encased in a surrounding flow containing smaller materials such as platelets and proteins, through a process referred to as shear induced migration (discussed further in Chapter 10). In another example, Venezuelan and Canadian crude oils have very high viscosities, and one method of transporting these oils through pipes is to infuse small amounts of low viscosity lubricants, which migrate to the pipe walls, allowing the heavy crudes to flow nearly as a plug in the center of the pipes.

The geometry for the problem that you will solve in this exercise is depicted in Fig. 5.4(a). Your task is to match solutions above and below a centerline separating two fluids with different viscosities, μ_{top} and μ_{bottom}. To help with the exercise, assume that flow is at low Reynolds number and note that within either top or bottom half of the channel, flow will therefore be Poiseuille. So assume quadratic forms like $V_{top} = a_{top} + b_{top}{\cdot}y + c_{top}{\cdot}y^2$ and $V_{bottom} = a_{bottom} + b_{bottom}{\cdot}y + c_{bottom}{\cdot}y^2$. Assume also no-slip conditions at top and bottom boundaries of the channel: that gives you two constraints on the six constants a_{top}, a_{bottom}, etc. Match the top and bottom velocities at the interface, $y = 0$: that's a third constraint: only three

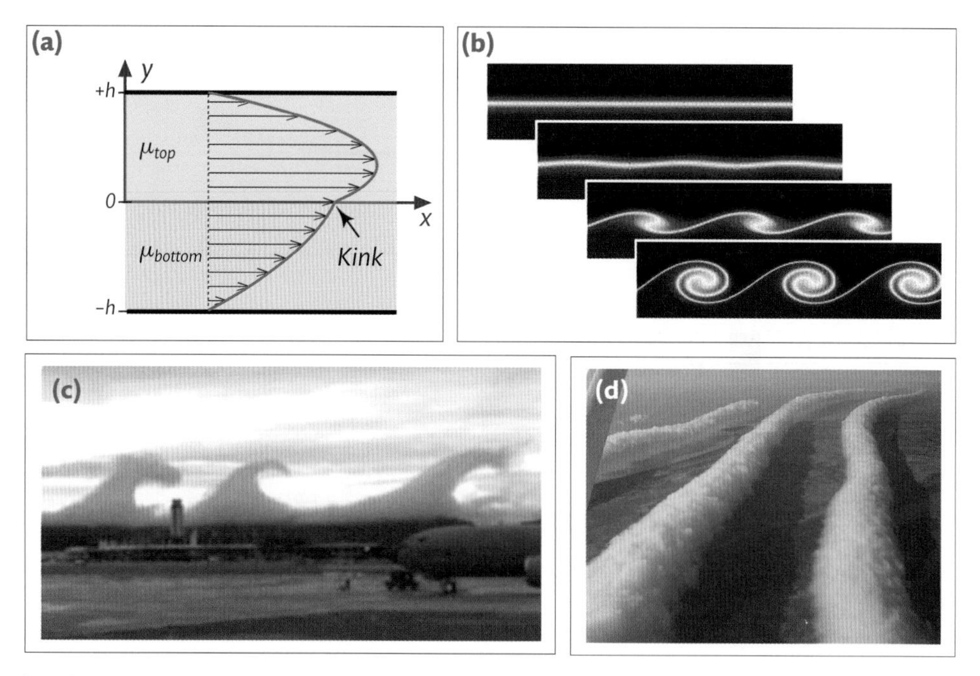

Fig. 5.4 Flow differentials across an interface. (a) Geometry for this exercise, in which two different viscosity fluids are driven through a 2D channel. Notice that the stress is matched at the interface: this produces a kink, or jump, in the derivative, $\partial V_x/\partial y$, which isn't present in other common matching problems. (b) Kelvin–Helmholtz instability causes rolls to develop when velocity gradient across interface exceeds a critical value. This figure shows computations by Changhong Hu (Kyushu University) at four equally spaced time steps. (c) Example of roll clouds that form when high speed air flows over low-lying clouds, here in Alabama (© ABC 33/40 in Birmingham, Alabama). (d) Similarly, "morning-glory" roll clouds can develop, here in Australia (Mick Petroff/CC BY-SA 3.0).

more to go. Then match the pressures at $y = 0$ for a fourth constraint, and finally match the stresses at the interface.

This last condition, matching stresses across an interface, appears in some problems and isn't difficult to implement. Recall that Newton's law of viscosity says that the stress is the viscosity times the strain rate, so set this equal above and below $y = 0$. Notice that this will force $\mu \cdot \partial V/\partial y$ to be equal across the interface, which is different from requiring $\partial V/\partial y$ to match, and as a consequence the velocity profile will have a noticeable kink, or discontinuity, in its derivative as shown in Fig. 5.4(a).

Once you have applied all 5 of these matching conditions, you will be left with one unknown. This is as it should be: after all, in the physical problem, the pressure gradient—and so the flow rate—has been undetermined. To complete the problem, define the velocity at $y = 0$ to be a free parameter, V_o, and you're done. Plot the velocity profile that results at a few different values of the relative viscosity, $\mu_r = \mu_{bottom}/\mu_{top}$. Logically enough, you'll find that as μ_r grows, the kink shown in Fig. 5.4(a) will become more extreme.

Discontinuities across an interface can result in many curious effects—see, for example, an article by Dan Joseph about "Rollers" that develop using unusual fluids.[3] One of the best known examples of dynamical behaviors across an interface appears when a high speed flow meets a low speed one: this results in Kelvin–Helmholtz rolls, as shown in Fig. 5.4(b). The jump in shear stress across the interface causes fluid to form characteristic rolls, for example in roll clouds shown Figs. 5.4(c) and (d). Rolls can be formed in simple experiments as well, for example in water or water–cornstarch mixtures flowing down an inclined surface.

5.2 **Branching Effects**

Biological vessels ranging from blood vessels to bronchi to neuronal trees are typically too complicated to admit analytic solutions. Nevertheless, some lessons can be derived from the principles of mass and momentum conservation, and we discuss a few of these lessons here.

First, let's examine what happens when a flow rapidly expands as sketched in Fig. 5.5(a), where flow is from left to right. Such a flow can be driven by a negative pressure gradient, $\partial P/\partial x$ (meaning that the pressure is larger to the left as shown in the plot beneath the sketch of the expanding channel. If the expansion angle, ϑ, shown is large enough, the pressure gradient $\partial P/\partial x$ can become positive, as indicated in Fig. 5.5(b): we saw the same thing in the outlet of a balloon in Fig. 3.2 of Chapter 3. Since flow in a channel is driven by pressure gradients, this will cause flow to reverse. If the pressure gradient is negative (pink in Fig. 5.5(b)) to the left of a "separation point" and positive (blue) to its right, then the gradient—and so the flow—must be zero at the separation point itself. This produces recirculating regions as shown in an expanding channel in Fig. 5.5(c) and in other geometries such as those shown Fig. 5.5(d). Recirculating flows are important in biological situations, for they can lead to plaque formation, coagulation, and other pathologies in blood vessels, and to biofilm development in other systems.[4]

Another important aspect of branching concerns differential cross sections of branching channels. Consider for example the bronchial tree. The trachea is a single channel, which branches once into the two lungs, and branches another 16 times to produce over 130,000 bronchioles, each of which branches several more times until terminating at about 250 million alveoli in each lung.

As shown in Fig. 5.6(a), the bronchial airways reduce in diameter by a consistent ratio of about $2^{-1/3}$, or about 80%, with each generation. It may seem odd that this doesn't conserve area—that is, if we simply ask what ratio of diameters is needed to maintain a constant average flow speed $\langle V \rangle$, we find that at a first airway of $Area_1$, the volumetric flow rate Q_1 will be:

$$Q_1 = \langle V \rangle \cdot Area_1, \qquad [5.21]$$

while after a single branch, the flow rate will become:

$$Q_2 = 2\langle V \rangle \cdot Area_2. \qquad [5.22]$$

Flow rates, $Q_{1,2}$, must be conserved, and to additionally conserve flow speed, we might expect $Area_2 = ½ \cdot Area_1$, as shown in Fig. 5.6(b), or $h = D_2/D_1 = 1/\sqrt{2}$, where (because everything by this point in the twenty-first century has acquired a name) h is termed the "homothety

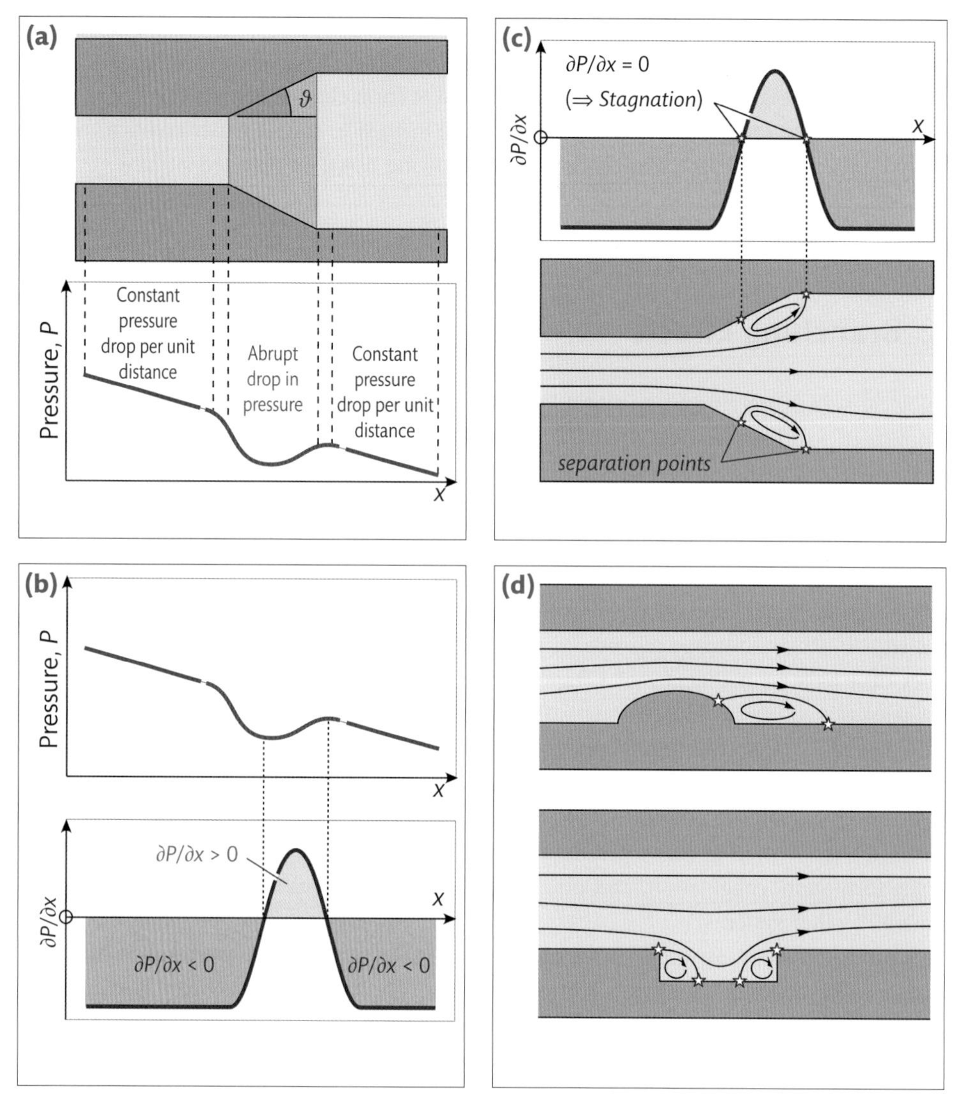

Fig. 5.5 Flow separation. (a) Expanding channel plotted above the pressure driving flow from left to right. As the expansion angle, ϑ, grows, the resulting pressure drop also increases. (b) Comparison between pressure and pressure gradient, showing that at sufficiently large ϑ, a region of positive pressure gradient can appear, meaning that flow will be pushed from right to left. (c) Comparison between pressure gradient and velocity profiles. Notice that where the gradient, $\partial P \partial x$, is zero (dotted lines), the flow will stagnate since there will be no net force on a fluid parcel in that region. This produces separation points, indicated, surrounding a "separation bubble" of recirculating flow. (d) Other geometries exhibiting separation points.

ratio." This would reduce vessel diameters by a factor of about 71% per generation as shown in the red line of Fig. 5.6(a). Yet as we have said, in the bronchi diameters decrease by a factor of 80% per generation, and as is also shown in Fig. 5.6(a), in the acini (respiratory vessels close to the terminal alveoli), diameters decrease by close to 90% every generation.

We can analyze the effect of multiple branches in more detail if we assume that every branch is "self-similar," meaning that every generation is an exact proportional copy of

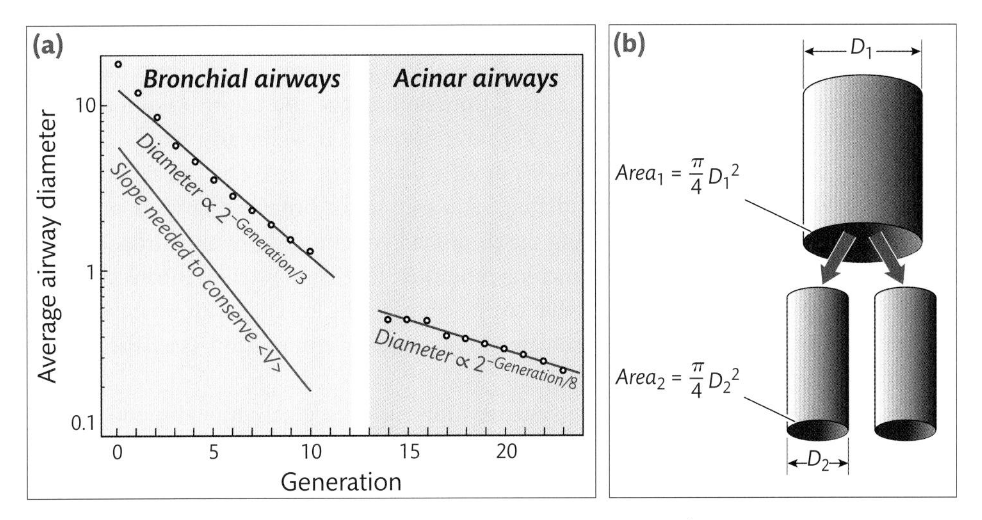

Fig. 5.6 (a) Scaling of airway diameters with generation of branching, after Weibel.[5] The data show that each generation of bronchi has diameter $2^{-1/3}$ ~ 0.8 times smaller than the last, and each generation of acini has diameter $2^{-1/8}$ ~ 0.9 times smaller. By comparison, to conserve both volumetric flow rate and mean flow speed, about 70% reduction per generation would be needed (red line). (b) Schematic of the cross-sectional areas due to a single branch from a tube of diameter D_1 to two tubes, each of diameter D_2.

the previous one. In this case at generation g there are 2^g branches, and if we assume that both diameters and lengths of vessel segments are proportional at each branch, then the volume after branching is reduced by a factor of h in each dimension, hence by h^3 in volume. The total volume occupied by the bronchial tree after N generations is then:

$$Vol_N = Vol_0\left[1 + \sum_{g=1}^{N} 2^g h^{3g}\right]. \quad [5.23]$$

This says that the total volume grows, rather rapidly, with homothety ratio, h, so there is a cost to keeping airway sizes large at each branch.

On the other hand, the pressure drop needed to drive a flow of rate Q through any segment of a branched tree can be approximated using the Poiseuille formula (Chapter 1):

$$\Delta P = \frac{8\mu}{\pi}\frac{QL}{R^4}, \quad [5.24]$$

where L is the length of the segment, R is its radius, and μ is the fluid viscosity. Assuming self-similarity again, L/R is constant, so the pressure drop is proportional to $1/R^3$, and decreases with $1/h^3$ every generation due to this Poiseuille term, and by another ½ because two tubes have less resistance to flow than one, so that:

$$\Delta P_N = c\cdot Q\left[1 + \sum_{g=1}^{N} \frac{1}{2^g h^{3g}}\right], \quad [5.25]$$

where c is an easily derived constant.

Thus the total volume of a bronchial tree goes as $(2h^3)^g$, while the pressure drop needed to supply the tree goes as $1/(2h^3)^g$. To optimize the problem one wants to both keep the volume and the pressure drop low, which constrains h from both above and below. Additionally, it is not difficult to see that if $h = h_c = (½)^{1/3}$, Vol_N and ΔP_N both grow linearly with N, otherwise either Vol_N grows as a power law (if $h > h_c$) or ΔP_N does (if $h < h_c$). The critical homothety ratio, $h_c = 0.79$, is very close to the branching ratio seen in the bronchial airways, and so the bronchi can be thought of as minimizing the combined volume and pressure drop.

On the other hand, by setting the branching ratio close to h_c, the system is intrinsically close to a dangerous state:[6] Eq. [5.25] tells us that constrictions in the lower airway, which can occur in asthma, will on average lower h and so demand pressures that must grow as a large power of h to maintain airflow.

There are many biological branching systems, bronchial, vascular, lymphatic, and so forth, that transport fluids, and there is no unique solution to this balance between minimizing volume, pressure drop, and sensitivity to pathological conditions. Indeed as shown in Fig. 5.6, even for a single system, two very different solutions are obtained, one for the larger bronchial scale and a second for the smaller acinar scale. In all branching systems, volume and flow are ultimately constrained by the simple geometric relations, Eqs. [5.23] and [5.25].

5.3 **Bending Effects**

A final important effect was reported in 1910 by John Eustice, an engineering professor who discovered that water flows more slowly through curved than straight pipes. Eustice found experimentally that the reduction in flow depends on the pipe curvature according to:

$$\frac{\Delta Q}{Q} \propto \frac{1}{R}, \qquad [5.26]$$

where Q is the flow rate and R is the radius of curvature of the pipe. This may seem strange: how, after all, does fluid know that it is traveling in a curved pipe?

To answer this question, we note that there are two flows in a curved pipe, as shown in Fig. 5.7: a primary flow transporting fluid lengthwise along the pipe, and a secondary flow recirculating fluid in the transverse direction. The secondary flow appears because fluid entrained in the primary flow will tend to move in a straight line (Newton's first law), and so in a curved pipe a parcel of fluid will be pushed outward through centrifugal action. As sketched in Fig. 5.7(b), this outward force acts on all fluid in the pipe. However, fluid near the pipe walls feels an opposing drag due to viscosity, and this force differential sets up a circulatory secondary flow as shown in Fig. 5.7(c) for a circular pipe, and in Fig. 5.7(d) for a square pipe.

We can use dimensional analysis as described in Chapter 2 to construct a dimensionless group that characterizes the strength of this secondary flow. Alternatively, given this picture of secondary flow being driven by centrifugal force opposed by viscous damping, we can more directly construct a dimensionless group. The centrifugal acceleration acting on a fluid parcel moving with primary velocity V along a circular trajectory of radius R goes as:

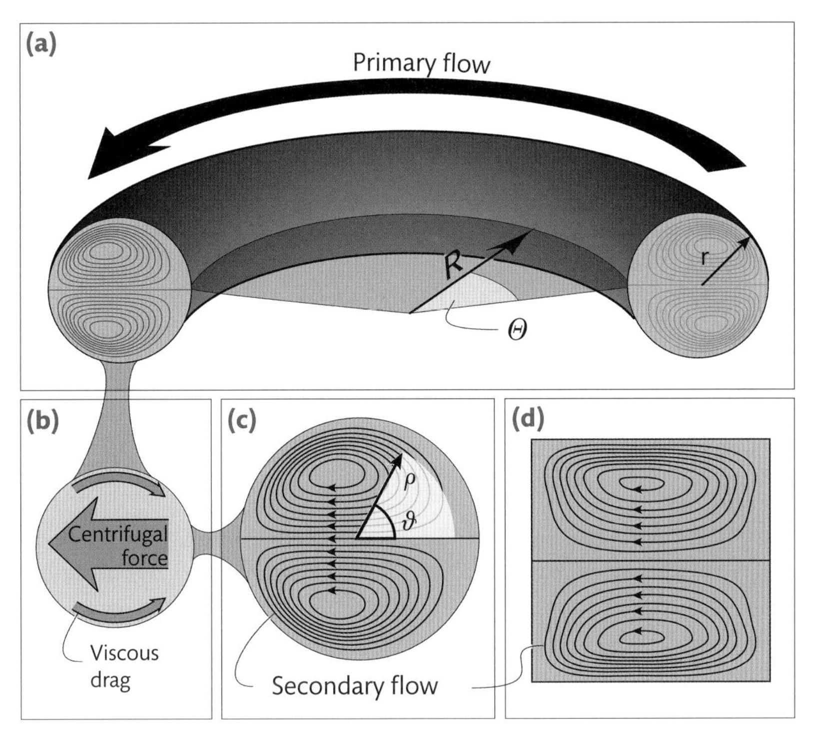

Fig. 5.7 "Dean's flow" in a curved tube. (a) Curved tube of radius *r* and curvature *R* with primary flow indicated. (b) Cross-sectional view of centrifugal force pushing fluid outward is resisted by viscous drag against the walls. (c) This force differential drives secondary recirculating flows as shown (here $De = 606$, after McConalogue and Srivastava[8]). (d) Similar flow is seen in other shaped tubes (here $De = 100$, after Zhang et al.[7]).

$$a_{centrif} = \frac{V^2}{R}. \tag{5.27}$$

To estimate the viscous drag that opposes centrifugation, we note that kinematic viscosity, ν, has dimensions area ÷ time: this is evident from examination of Navier–Stokes, Eq. [5.1]. From Fig. 5.7, the area over which viscosity acts goes as r^2, and the quantity produced from ν and r with the dimension of acceleration is:

$$a_{visc} = \frac{\nu^2}{r^3}. \tag{5.28}$$

The natural dimensionless group is thus $a_{centrif}/a_{visc}$. For historical reasons, the dimensionless group most commonly used is the "Dean number," *De*, which is the square root of this ratio; likewise historically *De* is expressed in terms of diameters rather than radii, which introduces a factor of two, as follows:

$$De = \frac{2Vr}{\nu}\sqrt{\frac{r}{R}}. \tag{5.29}$$

De is also often written in terms of Reynolds number: $De = 2Re\sqrt{r/R}$. The Dean number is named for William Dean (1896–1973), who produced in two papers in 1927 and 1928 an analytic expression for the solution to curved cylindrical pipe flow. Much later, in 1959, Dean developed an expression for curved square pipe flow as well. Navier–Stokes equation for cylindrical pipe flow in the "toroidal" coordinates $(\rho, \vartheta, \Theta)$ shown in Fig. 5.7(a) and 5.7(c) is:

$$U\frac{\partial U}{\partial \rho} + \frac{V}{\rho}\frac{\partial U}{\partial \vartheta} - \frac{V^2}{\rho} - \frac{W^2\cos(\vartheta)}{R} = \frac{1}{\delta}\frac{\partial P}{\partial \rho} - \frac{\nu}{\rho}\frac{\partial}{\partial \vartheta}\left(\frac{\partial V}{\partial \rho} + \frac{V}{\rho} - \frac{1}{\rho}\frac{\partial U}{\partial \vartheta}\right) \qquad [5.30]$$

$$U\frac{\partial V}{\partial \rho} + \frac{V}{\rho}\frac{\partial V}{\partial \vartheta} + \frac{UV}{\rho} + \frac{W^2\sin(\vartheta)}{R} = -\frac{1}{\delta\cdot\rho}\frac{\partial P}{\partial \vartheta} + \nu\frac{\partial}{\partial \rho}\left(\frac{\partial V}{\partial \rho} + \frac{V}{\rho} - \frac{1}{\rho}\frac{\partial U}{\partial \vartheta}\right) \qquad [5.31]$$

$$U\frac{\partial W}{\partial \rho} + \frac{V}{\rho}\frac{\partial W}{\partial \vartheta} = -\frac{1}{\delta\cdot R}\frac{\partial P}{\partial \Theta} + \nu\left(\frac{\partial^2 W}{\partial \rho^2} + \frac{1}{\rho}\frac{\partial W}{\partial \rho} + \frac{1}{\rho^2}\frac{\partial^2 W}{\partial \vartheta^2}\right). \qquad [5.32]$$

Here U is the velocity in the radial, ρ, direction, V is the velocity in the "toroidal," ϑ, direction; W is the velocity in "azimuthal,"Θ, direction, and ν and δ are the fluid kinematic viscosity and density, respectively. For completeness, the continuity equation in these coordinates is:

$$\frac{\partial U}{\partial \rho} + \frac{U}{\rho} + \frac{1}{\rho}\frac{\partial V}{\partial \vartheta} = 0. \qquad [5.33]$$

Some of the terms in Eqs. [5.30]–[5.32] have recognizable origins in the Navier–Stokes equation. For example, there are nonlinear terms like $U\cdot\partial U/\partial\rho$, there are pressure derivatives, and there is a viscosity multiplied by a set of second derivatives. The important centrifugal forces are shown in blue, and logically enough depend on the primary velocity squared, W^2, divided by the radius R. On the whole, however, most would agree that this is not a readily understandable set of equations.

By comparison, most of the equations that we have confronted thus far are comprehensible. The Poiseuille formula, for example, can be understood by recognizing that the expression whose second derivative is constant is a quadratic. Bessel's equation can be understood as resulting from the same second derivative in cylindrical coordinates. And so forth. Eqs. [5.30]–[5.33], however, are not as comprehensible, and it is impressive that Dean obtained an analytic solution.

Dean's solution for flow in a curved cylindrical pipe, which we do not express here, turns out to be accurate provided $De < 34$ (of all things!) and $\sqrt{r/R} \ll 1$. Solutions can be further extended up to $De \sim 600$ (shown in Fig. 5.7(c)) using Fourier analysis,[8] and in more complicated situations using numerical simulations.[9] All of these approaches are difficult to implement and, while it is in a way commendable to make the effort, as a practical matter these solutions have already been laboriously obtained, and little is to be learned by repeating the exercise here. On the other hand, the problem of flow in a curved pipe is an excellent archetype to introduce other methods that are both easily understood and broadly applicable. We do this next.

5.4 **Model Building**

As we have seen the problem of secondary recirculating flow in a curved cylindrical pipe has been solved. Suppose we want to solve for a similar flow in a rectangular vessel. Let's begin with the simplest case, that of a single symmetric recirculation loop, as shown in Fig. 5.8(a). Note that we are assuming that we know what the flow looks like, and our job here is to construct an analytic solution that behaves as we expect. This could at its simplest just be a matter of curve-fitting: finding a function that looks like experimental data, but as we will see later in the book, the same approach can be used to construct predictive models in complicated geometries and situations.

Moreover, taking the flow shown in Fig. 5.8(a) as an example, we note that similar flows are seen in many different physical situations—for example, the cross section of a layer of fluid heated from below and cooled from above exhibits square "Rayleigh–Bénard convective cells," illustrated in Fig. 5.8(d). Or in a later chapter we will discuss flow in the annulus between rotating cylinders: this can produce "Taylor–Couette vortices," which again consist of square cells, illustrated in Fig. 5.8(e).

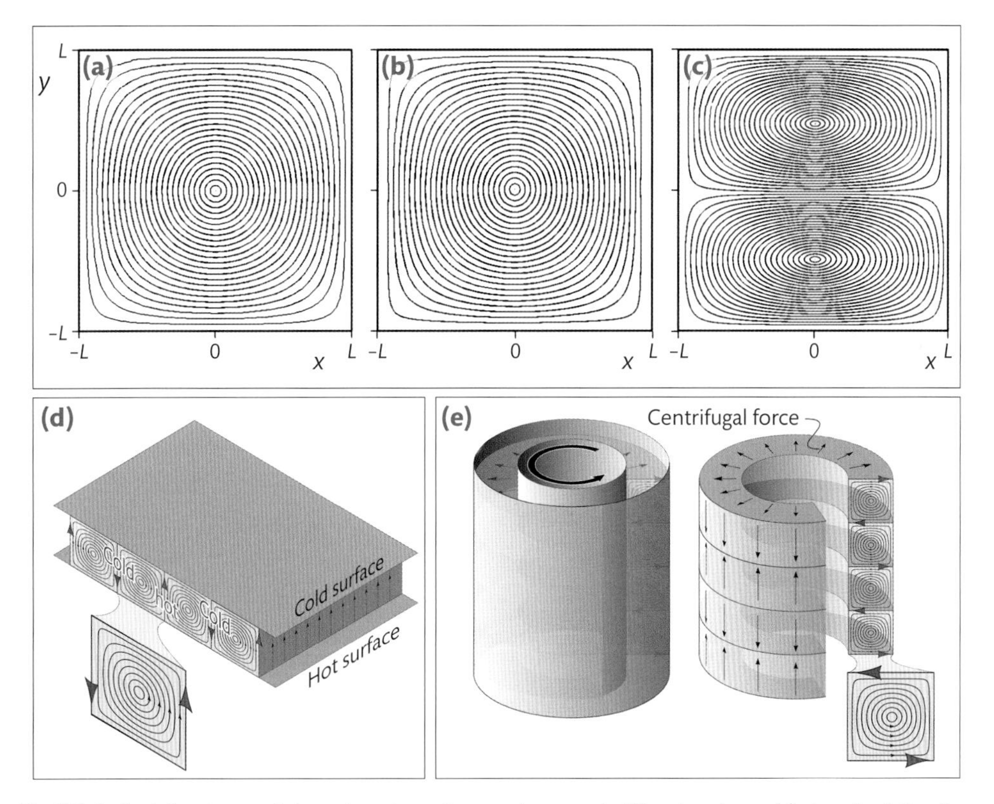

Fig. 5.8 Recirculating transport in rectangular cells seen in several different systems. (a) A recirculating flow produced using sinusoidal functions as described in the text. (b) A recirculating map produced again using sinusoidal functions. (c) A double recirculating map. (d) Square cellular flow produced in a heated "Rayleigh–Bénard" system. (e) Square cellular flow produced in a rotating "Taylor–Couette" system.

These very different problems produce similar flows because the force balance leading to the formation of recirculating flow is mechanically the same in all three problems. In curved pipe flow and in the Taylor–Couette problem, centrifugal force pushes fluid outward; in the Rayleigh–Bénard problem, fluid is pushed upward, here by buoyancy of the heated fluid. The fluid doesn't know or care what the source of the force is, it merely responds by moving up or out in one region, which is necessarily compensated by motion down or in, leading to a recirculating cellular flow that is qualitatively similar in all three problems. Thus a model generated in one problem can be applied to an apparently different system: this is a common situation.

So, if we are to produce a flow like that shown in Fig. 5.8(a), how would we go about the task? One thing that we could do would be to observe that the flow has to be purely vertical at left and right boundaries, and horizontal at top and bottom. So we want the horizontal component of the flow to be zero at $x = -L$ and $x = +L$, and similarly the vertical component must be zero at $y = -L$ and $y = +L$. There is also no flow in either direction at the center, so the velocity must vanish at $x = y = 0$. So let's work with a function that vanishes at 0 and at $\pm L$ — for example, a sine function. Maybe this won't be precisely right—for example, the flow in Fig. 5.7(d) is noticeably asymmetric—but we can always modify the sine function later on to better describe the problem at hand.

Based on this reasoning, let's choose:

$$V_x = f(x)\cdot\sin\left(\frac{\pi y}{L}\right)$$
$$V_y = \sin\left(\frac{\pi x}{L}\right)g(y). \qquad [5.34]$$

As required, V_x clearly vanishes at $y = 0$ and $y = \pm L$, and V_y similarly vanishes at $x = 0$ and $x = \pm L$, leaving $f(x)$ and $g(y)$ to be determined. For an incompressible fluid, we require from the continuity equation that:

$$\nabla\cdot\vec{V} = \frac{\partial V_x}{\partial x} + \frac{\partial V_y}{\partial y}$$
$$= f'(x)\cdot\sin\left(\frac{\pi y}{L}\right) + \sin\left(\frac{\pi x}{L}\right)\cdot g'(y), \qquad [5.35]$$

or:

$$f'(x) = -f_0\sin\left(\frac{\pi x}{L}\right)$$
$$g'(y) = g_0\sin\left(\frac{\pi y}{L}\right), \qquad [5.36]$$

where f_0 and g_0 are positive constants (and the minus sign can be in either one of these two equations). Plainly, letting f and g be cosines (f positive and g negative) will solve this problem: this is what is plotted in Fig. 5.8(a).

EXERCISE 5.3

Reproduce the plot shown in Fig. 5.8(a). For this purpose, use "Euler integration:"

$$x(t + dt) = x(t) + V_x(t)\cdot dt$$
$$y(t + dt) = y(t) + V_y(t)\cdot dt, \qquad [5.37]$$

where $x(t + dt)$ is x at time $t + dt$, etc. You will find that dt must be extremely small (less than 0.0001), otherwise numerical errors will appear. In a later chapter (and in the Appendix), we will discuss numerical methods to correct these errors. It is worthwhile at this point to identify the source of the errors and to leave the reader to cogitate over possible approaches to resolve them.

The mechanism by which numerical error is introduced by Euler integration is illustrated in Fig. 5.9. Fig. 5.9(a) shows the desired result, and Fig. 5.9(b) shows an enlargement of the actual numerical trajectory produced by setting $dt = 0.001$. Starting from the red star, an individual trajectory travels around the recirculation loop and after one complete revolution returns to a point *outside* of the actual loop, rather than landing on the star as it should. This occurs with all trajectories, so the numerical trajectories expand in time, violating the continuity equation. The cause of this expansion of trajectories is that Euler integration, Eq. [5.37], produces a sequence of straight lines, from the initial point, x_0, shown in Fig. 5.9(c) to x_1, x_2, x_3, etc. Since Euler integration produces lines that never exactly conform to a curved true trajectory, Euler trajectories will always tend toward the outside of any curve. Reducing dt will decrease this error, but it will always be present. To eliminate the error, curvature must be incorporated into numerical trajectories—we will see how this is done in a later chapter (and again also in the Appendix).

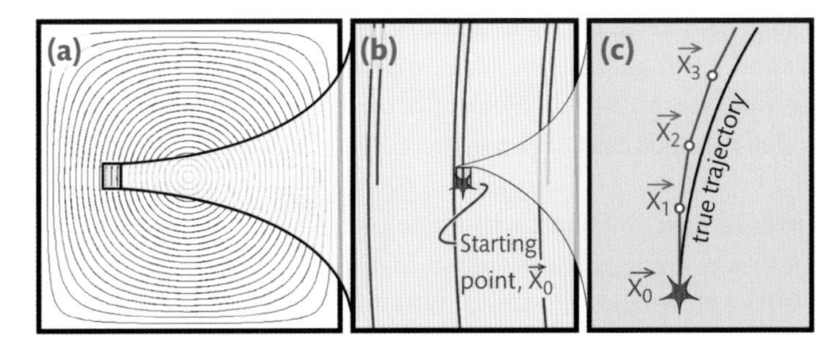

Fig. 5.9 Euler integration error. (a) Desired recirculating trajectories. (b) Enlargement of numerical trajectories for $dt = 0.001$—note that trajectories should return to their starting points, but actually spiral to the outside. (c) Euler integration connections successive points with straight lines: this fails to fit the curvature of the true trajectory. As dt is reduced, this error will diminish, but numerical trajectories will always tend toward the outside of any curve.

The root source of the difficulty in numerical integration—using the Euler approach or more sophisticated methods—is that the differential equation at hand has to be converted into an algebraic equation for a digital computer to deal with. This wasn't always the case, and there is a long and fascinating history of analog computers that are free of this type of error and that predate modern digital computers. The first analog computer was perhaps the Greek "Antikythera mechanism," a hand-cranked machine that was used over 2000 years ago: it is believed that this device was used to predict eclipses and positions of the planets. The interested reader may also want to look into other mechanical analog computers, for example, Charles Babbage's (1791–1871) "difference and analytic engines," which inspired the first record of a list of computer instructions (i.e. a program) written by Lord Byron's daughter, Ada Lovelace (1815–1852). Other mechanical analog computers were developed specifically to solve differential equations, notably the "differential analyzer" developed during WWII by Vannevar Bush (1890–1974). These were superseded by electronic analog computers that could instantaneously solve integral as well as differential equations. One of the first of these was the "Mathematical Analyzer, Numerical Integrator and Computer" (MANIAC), so named by Nicholas Metropolis (1915–1999) to ridicule bureaucratic acronyms. This machine was programmed in the 1950s by Mary Tsingou-Menzel (1928–) in the first-ever molecular dynamics simulation—a simulation that astonishingly led to historic solitary wave solutions (known as Fermi–Pasta–Ulam recurrence).

Notwithstanding these mechanical and electronic developments, it is worthwhile to recognize that up to the 1960s a "computer" was a person—invariably a woman—assigned to perform specified calculations by another person—almost always a man [see the movie *Hidden Figures* (2016)]. For example, NASA's predecessor NACA (National Advisory Committee for Aeronautics) hired hundreds of women as computers to analyze aerodynamic data leading to low drag airfoil designs and the first supersonic planes.

Nowadays, however, analog computers are seldom seen outside of museums, and we are left to approximate derivatives and integrals in terms of algebraic equations such as Eq. [5.37]. An alternative approach that is free of approximations is to produce models from the algebraic equations themselves. There are two approaches here: one can make use of "symplectic integrators" such as the Verlet velocity method, which generate equations similar to Eq. [5.37] but using methods that are guaranteed to conserve area. There is a large literature on these methods, and we will not describe them in detail here, however if the reader has a set of differential equations that must be numerically solved in an area conserving manner, it is worth knowing that methods exist to obtain these. The methods are not always simple and can be inefficient and exhibit other numerical errors; however, they will rigorously conserve area.

We discuss instead a second approach, which circumvents issues of integration accuracy and efficiency by producing algebraic equations that are themselves rigorously area preserving. This approach is part of modern "dynamical systems" theory, and includes many examples of direct relevance to fluid transport. The essential idea is that rather than starting from a differential equation and performing an integration, one can recognize that the outcome of that process will be a set of algebraic equations, and so one can instead study such equations from the start. Thus we consider equations of the form:

$$\begin{aligned} x_{n+1} &= x_n + f(x_n, y_n) \\ y_{n+1} &= y_n + g(x_n, y_n), \end{aligned} \qquad [5.38]$$

where x_{n+1} is x at discrete times $n+1$ (meaning $n = 1, 2, 3\ldots$ rather than continuous values) and f and g are algebraic, rather than differential, functions. Eqs. [5.38] are termed a "mapping" and are distinguished from continuous differential systems, which are termed "flows." For our purposes, we focus on mappings that conserve area, and we will further restrict ourselves to two dimensions, which is easier to analyze. One class of 2D conservative functions is of the form:

$$\begin{aligned} x_{n+1} &= x_n + f(y_n) \\ y_{n+1} &= y_n + g(x_{n+1}). \end{aligned} \qquad [5.39]$$

If the reader examines Eqs. [5.39], she will see that the first equation changes x by something that is a function of y, and the second subsequently changes y by something that is a function of x: this is precisely what the sine flow shown in Chapter 1 does. To understand why this conserves area, imagine a square block of dough that is sheared purely in the x-direction according to some function f of y, as shown in Fig. 5.10(a). If the block is sheared in this way, each horizontal segment is simply moved with respect to its neighbors so that there is no change in overall area. And if the same operation is performed in the y-direction, the combined process—that is, Eqs. [5.39] must also conserve area. Other ways of deforming a material—for example, rotations, uniform translations, compressions in one direction compensated by expansions in another—obviously can also conserve area, as indicated in the examples of Figs. 5.10(b) and (c).

To evaluate whether a particular mapping conserves area, we note that whatever happens on larger scales, an infinitesimal element of area can either (1) stretch and contract along orthogonal axes, or (2) shear, and these operations can either conserve area or not. The area of an element of dimensions Δx_n, Δy_n is obviously $\Delta x_n \cdot \Delta y_n$, so to conserve area after stretching, we require:

$$\Delta x_{n+1} \cdot \Delta y_{n+1} = \Delta x_n \cdot \Delta y_n, \qquad [5.40]$$

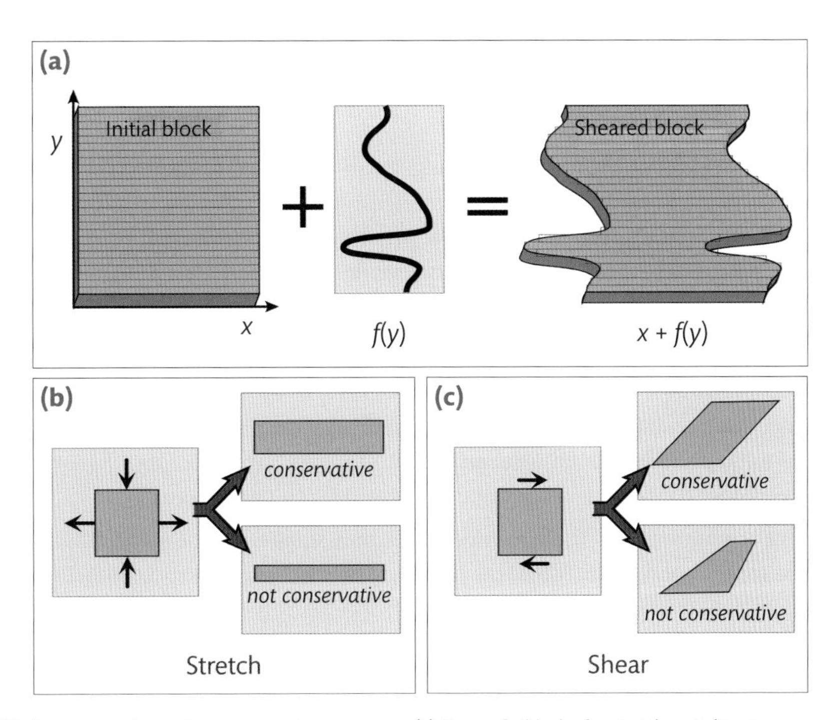

Fig. 5.10 Area conserving and non-conserving mappings. (a) Shear of a block of material according to $x_{n+1} = x_n + f(y_n)$ conserves area because horizontal slices merely slide left or right without change in size. (b) Stretching in one direction ($\Delta x/x$) combined with contraction in another ($\Delta y/y$) can be conservative, but only if the products of rates of stretch in both directions is one ($\Delta x \Delta y / xy = 1$). (c) Similarly shear can be conservative, but only if the product of shear in each direction is zero. The upper right panel shows conservative shear in x depending on y only; the lower right panel shows non-conservative shear depending on both x and y.

or as the area element becomes infinitesimal:

$$\frac{\partial x_{n+1}}{\partial x_n} \cdot \frac{\partial y_{n+1}}{\partial y_n} = 1. \qquad [5.41]$$

Similarly it can be shown that shear is conservative if:

$$\frac{\partial x_{n+1}}{\partial y_n} \cdot \frac{\partial y_{n+1}}{\partial x_n} = 0. \qquad [5.42]$$

Combined, Eqs. [5.41] and [5.42], a mapping is conservative if:

$$|J| = 1, \qquad [5.43]$$

where J is the "Jacobian" of the mapping, defined by:

$$J = \begin{pmatrix} \dfrac{\partial x_{n+1}}{\partial x_n} & \dfrac{\partial x_{n+1}}{\partial y_n} \\ \dfrac{\partial y_{n+1}}{\partial x_n} & \dfrac{\partial y_{n+1}}{\partial y_n} \end{pmatrix}, \qquad [5.44]$$

named after mathematician Carl Jacobi (1804–1851).

EXERCISE 5.4

In this exercise, we will produce a recirculating square mapping, as shown in Fig. 5.8(b), that closely resembles the integrated flow shown in Figs. 5.8(a) and 5.9. This will consist of several parts. Some parts are intended to reinforce the physical and mathematical principles involved in constructing an area preserving mapping, while other parts are intended to develop computational skills. Neither set of tasks is certain to be joyful, but both understanding and skills are important to developing facility with modeling flows, and the reader is encouraged to persist.

(1) Confirm that the determinant of the Jacobian of the mapping in Eq. [5.39] is identically one, as anticipated by the physical argument presented in Fig. 5.10(a). So any functions that we choose for f and g will produce an area preserving mapping.

(2) Produce a mapping that leads to trajectories in a square. By the same reasoning as we applied for the continuous "flow" (as Eqs. [5.34] are called), we expect something like the following may do the trick:

$$\begin{aligned} x_{n+1} &= x_n + k \cdot \sin\left(\frac{\pi y_n}{L}\right) \\ y_{n+1} &= y_n - k \cdot \sin\left(\frac{\pi x_{n+1}}{L}\right) \end{aligned}, \qquad [5.45]$$

Analytically calculate the Jacobian of this mapping and confirm that its determinant is one. Notice that x_{n+1} in the second equation has to be expanded in terms of x_n and y_n in order to calculate the derivatives defined in Eq. [5.44]

(3) Start with a line of points defined by $x = y$ and ranging from 0 to 1.5, and plot how these points evolve under this mapping for $k = 0.1$. Show that this plot looks like Fig. 5.8(b), but rotated by 45°. We have seen that shear defined by $x_{n+1} = x_n + f(y_n)$ conserves area; similarly, rotation defined as follows also conserves area:

$$\begin{aligned} x_{n+1} &= x_n \cos[\vartheta(r_n)] - y_n \cdot \sin[\vartheta(r_n)] \\ y_{n+1} &= x_n \sin[\vartheta(r_n)] + y_n \cdot \cos[\vartheta(r_n)], \end{aligned} \quad [5.46]$$

where $\vartheta(r_n)$ is any function of $r_n = \sqrt{x_n{}^2 + y_n{}^2}$. Rewrite your program that tracks a line of points to first apply the mapping of Eq. [5.45] and then apply the mapping of Eq. [5.46] where ϑ is the constant function $\vartheta = 45°$. Your plot should now look like Fig. 5.8(b).

(4) Plot a small square inside, as shown in green in Fig. 5.11(a), and iterate it forward in time using the mappings of Eqs. [5.45] and [5.46] as shown in blue. Evaluate the areas inside the successive polygons, and confirm that area is indeed conserved. A few notes: *first*, the function "polyarea" will do this for you in Matlab; other languages may use different commands. *Second*, the square will have to enclose the area desired, and if you define the points making up the square in the wrong order, this will not occur and the area will not be conserved. So if you connect top-left to top-right to bottom-right to bottom-left to top-right, you will be successful ("Right" in Fig. 5.11(b)). If you connect top-left to top-right and then bottom-left to bottom-right, you will not ("Wrong" in Fig. 5.11(b)). *Third*, you will need to use enough points—say 400 in all—to smoothly surround the area as the map deforms the square. If you use too few points, you will find that the perimeter of the area will become jagged, and will not conserve area.

(5) Plot the mappings of Eqs. [5.45] and [5.46] for increasing values of k. Explain qualitatively why the behavior changes (refer to Fig. 1.5, Chapter 1). Confirm for a modest value of k

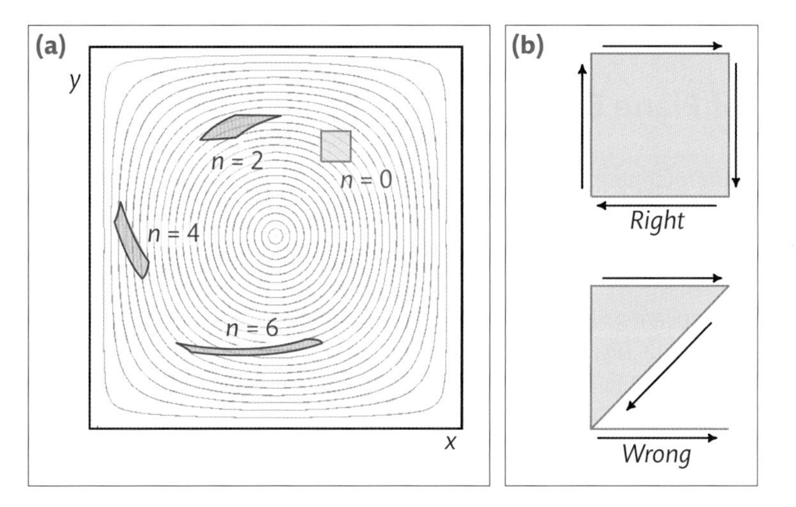

Fig. 5.11 Computational confirmation of area conservation. (a) An initial small square (green) deformed after 2, 4, and 6 iterations of the mappings of Eq's [5.45] and [5.46]. As you will show, the deformed perimeter encloses the same area. (b) To correctly calculate the area enclosed, the perimeter must travel sequentially around the square ("Right"). If the order of the points defining the perimeter is not sequential, the area enclosed may not be what you expect ("Wrong").

(say $k = 1$) that area is still conserved by iterating a square as in part (4) above. Continue to increase k, until the entire domain is chaotic (meaning regular regions have disappeared). Search for still larger k at which regular islands re-emerge: the re-appearance of regular regions is a generic feature of complex flows that is related to the presence of Jupiter's Great Red Spot, mentioned in Chapter 4. For an introduction to a mechanism by which regular behavior appears in chaotic systems, the reader may want to investigate periodic windows in the logistic map.

ADVANCED EXERCISE 5.5

(1) The square mapping in Fig. 5.8(c) contains two counter-rotating recirculation loops. First, produce a flow (similar to Eqs. [5.34]–[5.37]) that contains two loops in $-L< x< L$, $-L< y< L$. This should be straightforward. Then produce a map that does the same thing. *Nota bene*: this does not mean plotting the mappings of Eqs. [5.45] and [5.46] over $-2L< x< 2L$ and then dividing all x-values by 2: this would be a cheat, since dividing by 2 is certainly not conservative. For this exercise, you must produce a conservative mapping that contains two loops in $-L < x < L, -L < y < L$. Hint: contracting by a fixed factor in the x-direction and elongating by the same factor in the y-direction is conservative.

(2) Modify the sine functions in Eq. [5.45] to make the mapping asymmetric as in Fig. 5.7(d). Hint: you are free to choose any functions f and g in Eq. [5.39]. The symmetry of the mapping shown in Fig. 5.11 is produced by the left–right reflection symmetry of the sine function. So if you were to shear the sine to make it lean to right or left, the resulting mapping would become asymmetric.

5.5 Out of Plane Bending

Armed with the mapping-based approach to model area preserving flows, let's return to Dean's problem: flow in a curved pipe. For simplicity, we consider a square pipe, as shown in Fig. 5.7(d). Experiments and simulations have been done of this problem applied to flow in so-called "static" mixers: these mixers have no moving parts, and perform fluid mixing by driving a fluid through repeating segments that deform the flow. Many designs are used to mix fluids in straight pipes with inserts that repeatedly twist, split and recombine flow—commercial brands of epoxy are sold with such pipes to efficiently mix resin and hardener.

In Fig. 5.12, we show two cases of static mixers in curved pipes without inserts. Notice that Fig. 5.12(a) shows a design that deforms the flow back and forth in a plane, while Fig. 5.12(c) shows a design that also deforms the flow out of the plane. Flow in Fig. 5.12(a) has a "separatrix:" a barrier to transport so that dye injected below the separatrix (violet in inset) travels through multiple segments of the mixer without crossing the separatrix. Another way of viewing this system that is widely applicable to other mixing problems appears in Fig. 5.12(b),

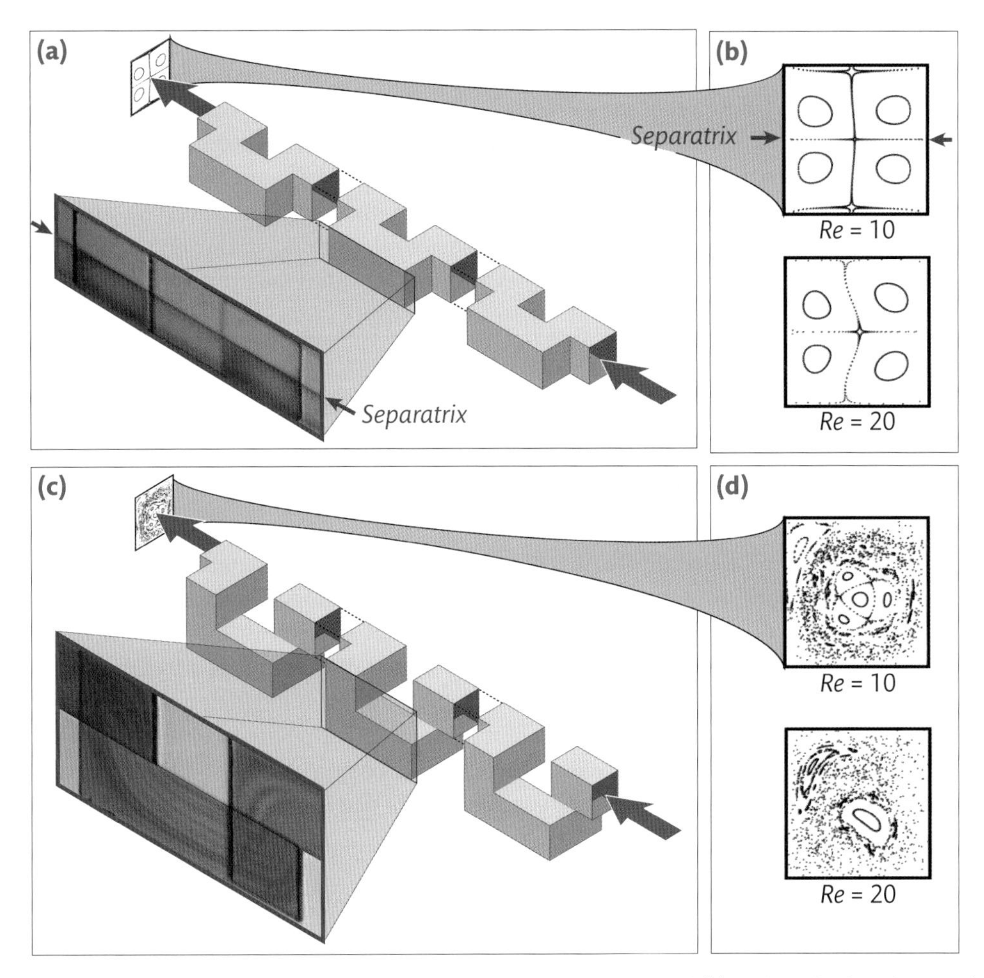

Fig. 5.12 Static mixers that exploit circulation caused by bends in a square channel. (a) Bends confined to a horizontal plane produce predominantly regular flow with very little mixing. So as shown in the "streamwise" inset from a microfluidics experiment, there is negligible mixing between top and bottom of the channel: inset photo shows pH sensitive dye injected below midline after 14 segments of mixer at *Re* = 10. (b) The lack of mixing can also be seen in a computational "spanwise" surface of section, which shows predominantly regular flow and the "separatrix" that prevents transport. (c) Bends in both horizontal and vertical planes: this design, termed a "serpentine" channel, produces chaotic mixing as shown in dye injection the inset after 11 segments at *Re* = 10. (d) Surfaces of section show breakup of separatrixes and chaotic mixing throughout most of the channel. Insets and surfaces of section after Aref.[14]

where we show a "surface of section." Surfaces of section were introduced by Henri Poincaré (1854–1912) to study periodic orbits of planets. Poincaré was perhaps the pre-eminent mathematician at the turn of the twentieth century, and had the unique notoriety of winning a prestigious prize by the King of Sweden, Oscar II, for proving both that the solar system is stable and after finding an error in his earlier proof that it is unstable.[10] In that problem, Poincaré considered a plane through the sun and perpendicular to the orbits of the planets, and imagined recording a point wherever a planet's orbit pierced the plane. In this way, he reduced the planets' three-dimensional orbits to a 2D plane. Equally importantly, he could easily identify periodic orbits, for these traced closed curves in the surface of section, now termed a "Poincaré section."

Similarly in fluid flow problems—especially those with repeating segments such as the mixers shown in Fig. 5.12—periodic orbits are easily identified as closed curves obtained by tracking marker particles placed in the flow. Since these curves are closed, they prevent material on opposite sides from mixing with one another. In Fig. 5.12(b) we identify a separatrix in the surface of section that corresponds to the separatrix shown in the inset to panel (a). Evidently there are two separatrices, dividing flow into four vortical flows.

EXERCISE 5.6

In Fig. 5.7 we showed that there are two recirculating vortices in a curved square channel. Explain why there are four such vortices in Fig. 5.12(b).

ADVANCED EXERCISE 5.7

Assume that flow in the streamwise direction, z, is defined by the Poiseuille formula (i.e. that V_z is parabolic in both spanwise directions, x and y), and that flows in the spanwise directions are defined by four square flows, Eqs. [5.34]–[5.36]. Plot trajectories of points to identify closed concentric tubes. Additionally, solve for the spatial distribution of a large number of points released in a bolus: do the circulatory flows affect the Taylor dispersion results shown in Fig. 5.3? Optionally, plot the deformation of a plane of initial points spanning the channel opening. This plane will rapidly deform beyond your computational ability to track, but should produce an attractive and intricate surface before that occurs.

The channel shown in Fig. 5.12 consists of multiple identical segments, each of which produces the same fourfold vortical spanwise flow. Consequently, no matter how many segments are included, this channel will never mix material across the separatrix boundaries. On the other hand, we saw in the sine flow of Chapter 1 that alternating two different flows can introduce chaotic mixing. Chaos is a broad and deep topic; see, for example, Strogatz.[11] For now we define chaos as a process that crosses flow boundaries (i.e. separatrixes), and produces rapid and efficient mixing free of regular, non-mixing, regions. So if we combine a bend in the horizontal plane with a second bend in the vertical plane, we can expect chaotic mixing to appear. This in fact is the case, as shown in the "serpentine" mixer of Fig. 5.12(c). Now dye introduced below the centerline mixes throughout the channel as shown in the inset, and correspondingly, chaotic trajectories appear in the surface of section shown in Fig. 5.12(d).

EXERCISE 5.8

The absence of chaotic mixing when channels are curved in the plane, and its emergence when the channels are curved out of the plane, is a subtle and important process with applications in manufactured,[12] as well as biological,[9] channels, and its discovery provides us with a valuable teaching moment. Hassan Aref (1950–2011) introduced this and related ideas in a series of now famous papers,[13] summarized in his address on the occasion of receiving a fluid mechanics prize.[14] In this address, Aref outlined his work and discussed the reception of his ideas by authorities in the field. Any graduate student planning to perform original research will encounter aspects of this reception at some point, and in this exercise, the reader is asked to read Aref's address and explain briefly:

(1) Whether you feel that the scientific community is receptive to new ideas.

(2) What the underlying sources of Aref's negative reviews were. Give some thought to the reviewers' scientific background and personal motivations, and differentiate between objective and subjective influences. These are important factors that affect reviews, and it is worthwhile to give these factors due consideration, both now and in the presentation of your future ideas and results.

For further analysis of this topic, the reader may want to consult the book by Kuhn,[15] a seminal and easily read text written by noted philosopher of science, Thomas Kuhn (1922–1996).

REFERENCES

1. Landau, L.D. and Lifshitz, E.M. (1959) *Fluid Mechanics.* London: Pergamon, pp. 145–51.

2. Lightfoot, E.N. (1974) *Transport Phenomena in Living Systems.* New York: John Wiley & Sons, pp.102–4.

3. Joseph, D.D., Nguyen, K. and Beavers, G.S. (1986) Rollers. *Physics of Fluids, 29,* 2771.

4. Rusconi, R., Lecuyer, S., Guglielmini, L. and Stone, H.A. (2010) Laminar flow around corners triggers the formation of biofilm streamers. *Journal of the Royal Society Interface, 7,* 1293–9.

5. Weibel, E.R. (2005) Mandelbrot's fractals and the geometry of life. In: Losa, G.A., Merlini, D., Nonnenmacher, T.F., et al. (eds.) *Fractals in Biology and Medicine.* Basle, Switzerland: Birkhäuser, pp. 2–16

6. Mauroy, B., Filoche, M., Weibel, E.R. and Sapoval, B. (2004) An optimal bronchial tree may be dangerous. *Nature, 427,* 644–6; see also van Ertbruggen, C., Hirsch, C., and Paiva, M. (2005) Anatomically based three-dimensional model of airways to simulate flow and particle transport using computational fluid dynamics. *Journal of Applied Physiology, 98,* 970–80 and Mauroy, B., Filoche, M., AndAndrade, J.S. and Sapoval, B. (2003) Interplay between geometry and flow distribution in an airway tree. *Physical Review Letters, 90,* 148101 and related comments.

7. Zhang, J.S., Zhang, B.Z. and Jü, J.W. (2001) Fluid flow in a rotating curved rectangular duct. *International Journal of Heat and Fluid Flow, 22,* 583–92.

8. McConalogue, D.J. and Srivastava, R.S. (1968) Motion of a fluid in a curved tube. *Proceedings of the Royal Society A, 307,* 37–53.

9. Santamarina, A., Weydahl, E., Siegel, J.M. and Moore, J.E. (1998) Computational analysis of flow in a curved tube model of the coronary arteries: effects of time-varying curvature. *Annals of Biomedical Engineering, 26,* 944–54.

10. A very readable and enlightening description of Poincaré's work appears in Peterson, I. (1993) *Newton's Clock: Chaos in the Solar System*. New York: Macmillan.

11. Strogatz, S. (1994) *Nonlinear Dynamics and Chaos with Applications to Physics, Biology, Chemistry and Engineering* Reading, MA: Perseus Books.

12. Stone, H.A., Stroock, A.D. and Ajdari, A. (2004) Engineering flows in small devices: microfluidics toward a lab-on-a-chip. *Annual Review of Fluid Mechanics, 36*, 381–411.

13. Aref, H. (1984) Stirring by chaotic advection. *Journal of Fluid Mechanics, 143*, 1–21.

14. Aref, H. (2002) The development of chaotic advection. *Physics of Fluids, 14*, 1315–25.

15. Kuhn, T. (1962) *The Structure of Scientific Revolutions*. Chicago, IL: University of Chicago Press.

6 Shearing Flows around Cylinders and Spheres

6.1 Introduction

Last chapter, we saw some examples of secondary flows that arise in simple geometries—for example, in bent pipes and static mixers. These flows can become surprisingly complex, nowhere more so than in "Couette" flow between two concentric rotating cylinders (see Chapter 5, Fig. 5.8). Consider Fig. 6.1(a), where we identify multiple distinct behaviors exhibited in the Couette geometry at a variety of inner and outer cylinder rotation speeds. Details of these flows are topics of entire volumes in their own right, and we will not attempt to reproduce that body of work here. Nevertheless, the Couette system has become a classic problem with many lessons applicable to other problems in fluid dynamics; additionally the geometry has applications in blood filtration[1] and centrifugation, so we will spend a little time understanding some of the simpler features of the flow.

At its heart, the multiple patterns indicated in Fig. 6.1(a) are associated with the need for the fluid to dissipate the energy imparted on fluid trapped between the two cylinders. To get a flavor of how this energy dissipation mechanism functions, let's consider the case illustrated in Fig. 6.1(b), termed "Taylor vortex flow." As we mentioned in Chapter 5, this flow occurs when the inner cylinder rotates sufficiently rapidly to impart centrifugal acceleration on the fluid that is greater than can be damped by viscosity. In this case, some packets of fluid move outward rather than being held in place by viscosity, leading to the production of counter-rotating toroidal flows as indicated in Fig. 6.1(b). This toroidal motion causes neighboring fluid regions to shear against one another, which dissipates energy according to Newton's law of viscosity (Chapter 1).

If still higher stresses are applied, either by increased centrifugation (associated predominantly with the rotation speed of the inner cylinder) or by increased shear (associated with the difference in speeds between the inner and outer cylinders), additional instabilities arise, for example causing the Taylor vortices to undulate up and down, in so-called "wavy vortex flow," shown in Fig. 6.1(c) (and in Fig. 6.2 following).

It is by no means evident what conditions will lead to wavy flow or what produces the many other complex patterns, however we already know enough to identify dimensionless groups that govern these flow transitions. As we have mentioned, the onset of the first instability depends on a balance between centrifugal acceleration and viscous damping. Centrifugal acceleration goes as $\Omega^2 R$, and an acceleration associated with viscosity is ν^2/R^3, where Ω and R are a characteristic angular rotation speed and radius, and ν is the kinematic viscosity of

Biomedical Fluid Dynamics: Flow and Form. Troy Shinbrot.

DOI: 10.1093/oso/9780198812586.001.0001

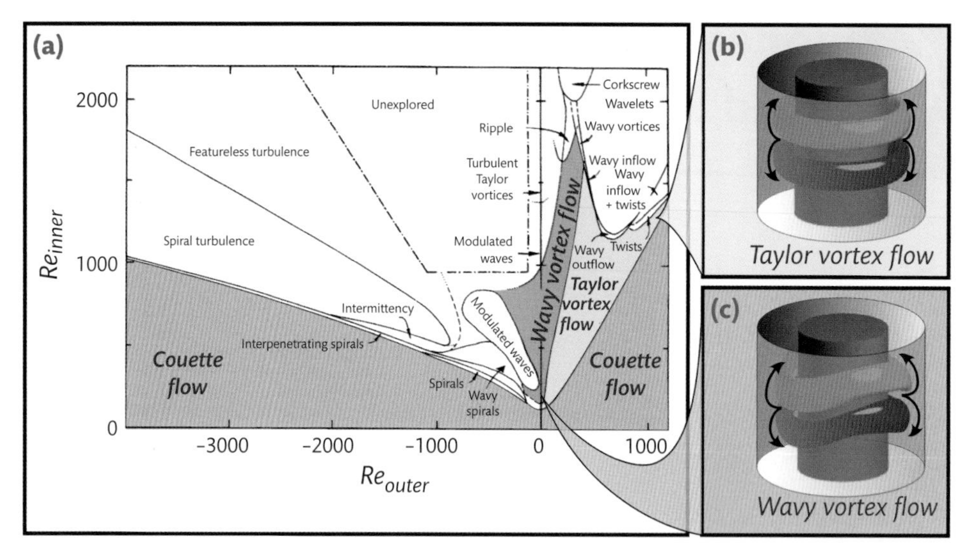

Fig. 6.1 (a) Phase diagram identifying multiple flow behaviors in a Couette cell between rotating cylinders (see also Chapter 5). Simple Couette flow described in text appears when the centrifugal force produced by the inner rotating cylinder isn't strong enough to produce outwelling. Diagram after Andereck et al.[2] (b) If the centrifugal force is strong enough and if both inner and outer cylinders are co-rotating, "Taylor vortex" flow is produced, leading to counter-rotating toroidal flow depicted here (also in Chapter 5, Fig. 6.7(e)). (c) At higher inner rotational speeds, the Taylor vortices become wavy as depicted here, and other, more exotic states listed in panel (a) appear as well. Panels (b) and (c) after Hoffman et al.[3]

the fluid. So we expect Taylor–Couette instabilities to be governed to first order by the "Taylor" number:

$$Ta = \frac{4\Omega^2 R^4}{\nu^2}, \qquad [6.1]$$

where the factor of 4 is a matter of convention. So we expect outward flow to dominate over viscosity—and so for Taylor vortex flow to appear—for large Ta, and viscosity to dominate and suppress outward flow for small Ta. It turns out that the critical Taylor number at which vortices appear is quite high, around 1700, and varies with relative cylinder speeds.

This formulation of the Taylor number doesn't contain any information about shear rate—thus R is some suitable radius (typically the inner) governing centrifugation. To characterize shear, the difference between outer and inner radii, $(R_{outer} - R_{inner})$, must clearly appear somewhere, and so a different formulation of the Taylor number that accounts for shear is:

$$Ta = \frac{\Omega^2 R_{inner}\left(R_{outer} - R_{inner}\right)^3}{\nu^2}. \qquad [6.2]$$

These two expressions for the Taylor number measure different things: the first depends on the average centrifugal acceleration and the second also includes the shear rate. Clearly, though, both of these leave something else out, for the rate of shear depends on the difference between the inner and outer cylinder speeds. For this reason, two different dimensionless groups have been used in Fig. 6.1(a), namely the inner and outer Reynolds numbers:

$$Re_{inner} = \frac{L_{gap} V_{inner}}{\nu}$$
$$Re_{outer} = \frac{L_{gap} V_{outer}}{\nu}, \qquad [6.3]$$

where:

$$
\begin{aligned}
L_{gap} &= R_{outer} - R_{inner} \\
V_{inner} &= \Omega_{inner} R_{inner} \\
V_{outer} &= \Omega_{outer} R_{outer}.
\end{aligned}
\qquad [6.4]
$$

The fact that there are many ways of expressing dimensionless groups in any particular problem should not be surprising: we discussed this issue in Chapter 2 (see the Buckingham pi theorem). What is surprising is that there is quite such a lavish variety of patterns referred to in Fig. 6.1(a), and the interested reader is referred to any of several books describing the problem, or to Taylor's original 1923 paper,[4] available through the online archiving resource, JSTOR (Joint Storage Project).

Example Kitchen Couette experiment

A variety of techniques have been developed to visualize complicated flows. One of the simplest is to seed the fluid with fine, flat particles that align with the flow. This produces alternate light and dark regions: light where the particles are perpendicular to the viewer and so reflect light, and dark where the particles are aligned with the viewer and so absorb light. Originally, fish scales were used for this; nowadays aluminum flakes are used in products such as Kalliroscope™ or Pearl Swirl™. Similar materials are used in other

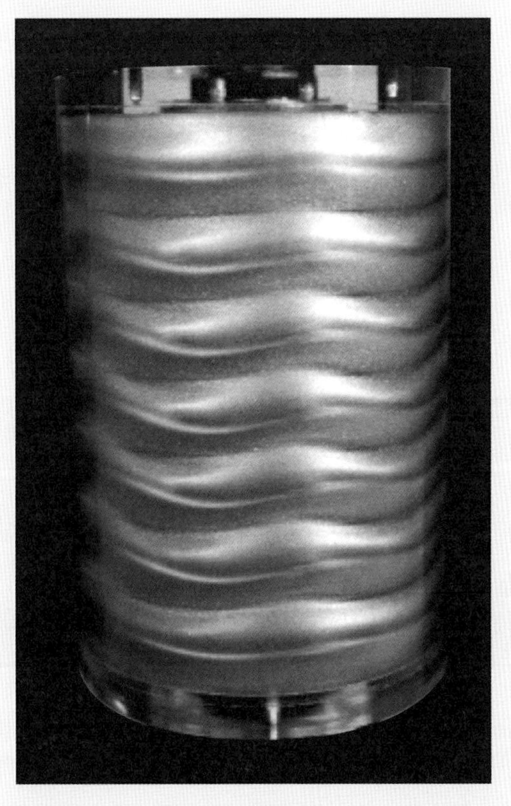

Fig. 6.2 Wavy vortex flow illuminated with rheoscopic fluid. Flow in light regions is parallel to the camera; flow in dark regions is perpendicular (Courtesy of Karl G. Roesner).

commercial products such as shampoos to produce a pearly or iridescent appearance. In Fig. 6.2 we show a snapshot of wavy vortex flow using one of these products. A similar effect can easily be obtained by adding some pearly shampoo to water in a 1 or 2 liter bottle. If the bottle is swirled and then stopped, complex striated structures aligned with the direction of flow will become evident.

For our purposes, we restrict our attention to the simpler—and fully solvable—problem of Couette flow, consisting of purely azimuthal fluid motion, which Fig. 6.1(a) shows appears at small Re_{inner}. This flow has neither radial nor vertical velocity, so Taylor vortices shown in Fig. 6.1(b) do not appear. Flow in this case travels in concentric "streamlines," varying in speed to match the speeds of the inner and outer cylinder walls.

EXERCISE 6.1

As a warm-up problem, let's consider incorporating moving boundaries into the Poiseuille problem that we solved previously. That is, as shown in Fig. 6.3(a), consider a two-dimensional channel with an applied pressure gradient from left to right, and where the lower boundary is stationary, but the upper boundary moves horizontally with speed V_o. To solve this problem, we will assume low Reynolds number so that we can start as in Chapter 1 with the Stokes equation:

$$\nabla^2\vec{V} = \frac{\vec{\nabla}P}{\mu}. \qquad [6.5]$$

Again as in Chapter 1, we will take the simplest possible case, namely that the solution has flow only in the x-direction that varies only in the y-direction, and we will try a solution:

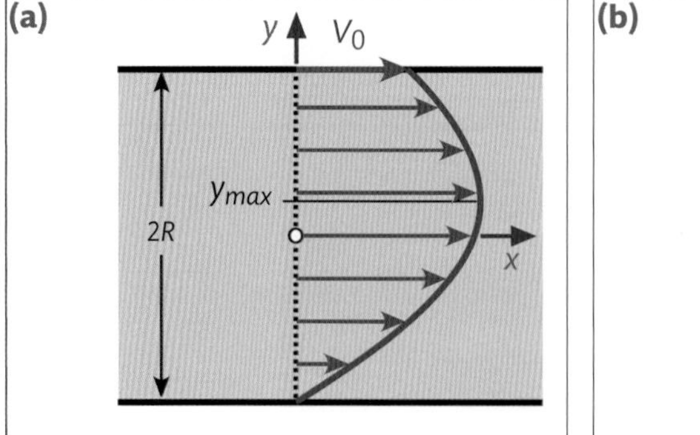

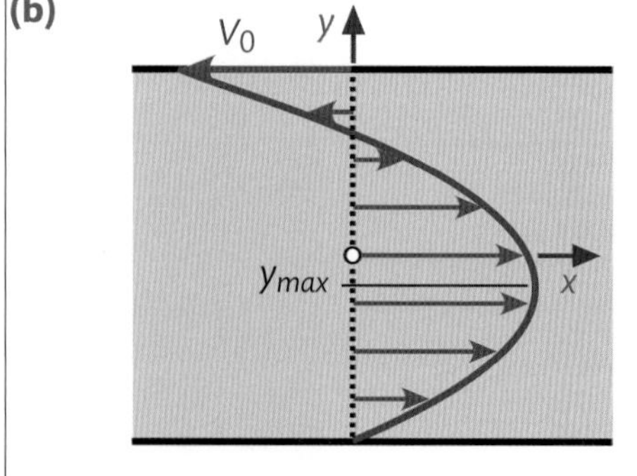

Fig. 6.3 Simple flow with moving boundary. (a) Poiseuille solution with pressure gradient between left and right ends of channel as well as moving upper boundary. (b) The same problem with upper boundary moving to left.

$$V_x(y) = c_0 + c_1 y + c_2 y^2. \quad [6.6]$$

The boundary conditions are now

$$\text{at top:} \quad V_x(+R) = c_0 + c_1 R + c_2 R^2 = V_o, \quad [6.7]$$

$$\text{at bottom:} \quad V_x(-R) = c_0 - c_1 R + c_2 R^2 = 0. \quad [6.8]$$

Additionally, the maximum speed, V_{max}, occurs where:

$$\frac{\partial V_x}{\partial y} = 0. \quad [6.9]$$

or:

$$c_1 + 2c_2 y_{max} = 0. \quad [6.10]$$

We now have three equations, Eq. [6.7], [6.8], and [6.10], in three unknowns, c_0, c_1, and c_2. For this exercise, solve for the unknowns and plot the resulting velocity field for positive V_0 (as in Fig. 6.3(a)) and for negative V_0 (Fig. 6.3(b)). Explain what you would do instead of using Eq. [6.9] if V_0 were large enough that the y_{max} was outside of the flow region.

Now that we know from Exercise 6.1 how to solve for fluid flow in the presence of moving boundaries in a simple context, let's return to a slightly more complicated version of the same problem: Couette flow for small Re_{inner}. As in Exercise 6.1, we begin with Stokes equation, which we now express in cylindrical coordinates as described in Chapter 1 (Eq. [1.41]). As always, we consider the simplest case, in which there is neither radial nor axial flow, so that $V_r = V_z = 0$. Thus we are left only with a velocity in the ϑ-direction, $V_\vartheta(r)$.

We also assume that $V_\vartheta(r)$ depends only on r and not on z or ϑ. This implies that the pressure doesn't change with z or ϑ either, since if it did it would produce a change in velocity with z or ϑ. This leaves a pressure term that can change only with r, that is, $\vec{\nabla} P = \frac{\partial P}{\partial r}\hat{r}$. The resulting Stokes equation is then:

$$\nabla^2 \vec{V} - \frac{\vec{\nabla} P}{\mu} = 0$$

$$\left[-\frac{1}{r^2} V_\vartheta(r) + \frac{1}{r}\frac{\partial V_\vartheta(r)}{\partial r} + \frac{\partial^2 V_\vartheta(r)}{\partial r^2}\right]\hat{\vartheta} + \left[\frac{\left(V_\vartheta(r)\right)^2}{r} - \frac{1}{\mu}\frac{\partial P}{\partial r}\right]\hat{r} = 0. \quad [6.11]$$

Additionally, the inner and outer cylinder boundaries move in the Couette problem, so we impose the boundary conditions from Eq. [6.4]:

$$\begin{aligned} V_\vartheta(R_{inner}) &= \Omega_{inner} R_{inner} \\ V_\vartheta(R_{outer}) &= \Omega_{outer} R_{outer}. \end{aligned} \quad [6.12]$$

We observe that the $\hat{\vartheta}$ and $\hat{r}$ terms in Eq. [6.11] are independent vector components, so for the equation to hold, each must vanish. Considering the $\hat{\vartheta}$ bracket (green in Eq. [6.11]) first, we note that each term has the same power: $1/r^2$. If we stare at such equations, it may occur to us to attempt a power law solution: $V_\vartheta(r) = a \cdot r^b$, where a and b are constants. Why? If we expand the derivatives:

$$\begin{aligned} -\frac{1}{r^2}V_\vartheta(r) + \frac{1}{r}\frac{\partial V_\vartheta(r)}{\partial r} + \frac{\partial^2 V_\vartheta(r)}{\partial r^2} &= -\frac{1}{r^2}a{\cdot}r^b + \frac{1}{r}a{\cdot}b{\cdot}r^{b-1} + a{\cdot}b{\cdot}(b-1){\cdot}r^{b-2} \\ &= \left[-a + a{\cdot}b + a{\cdot}b{\cdot}(b-1)\right]{\cdot}r^{b-2}, \end{aligned} \quad [6.13]$$

we find that *because* each term has the same power, the power law solution produces a common factor of r^{b-2} in each term. Consequently, our differential equation is reduced to an algebraic one.

We pause for a brief aside to remark that Eq. [6.13] illustrates a common theme in solving differential equations: the standard method for solving a partial differential equation is to reduce it to a set of ordinary differential equations (ODEs), and the standard method for solving an ODE is to reduce it to an algebraic equation. There are many of ways of accomplishing this—various transforms, tricks to separate terms (cf. Eq. [6.11]), etc.—but ultimately either a specific function magically solves the equation (as in exponential or Bessel functions) or a stratagem has to be devised to reduce the differential equation to an algebraic one. Viewed in this context, hopefully some of the mystique of differential equations will be reduced: we solve differential equations using the algebra we learned in high school. The secret known only to the cognoscenti is that everything else involved in solving differential equations relies on specialized sleights of hand to get the differential equation into an algebraic form. Or inventing a new name for the solution that we cannot otherwise obtain, as we saw with Bessel functions in Chapter 4.

Returning to Eq. [6.13], the problem we have to solve has been reduced to:

$$\begin{aligned} 0 &= \left[-a + a{\cdot}b + a{\cdot}b{\cdot}(b-1)\right] \\ &= a{\cdot}(b^2 - 1), \end{aligned} \quad [6.14]$$

so apparently $b = \pm 1$, and:

$$V_\vartheta(r) = a_+{\cdot}r^{+1} + a_-{\cdot}r^{-1}. \quad [6.15]$$

Now we can apply the two boundary equations, Eq. [6.12] to solve for the two constants, a_+ and a_-.

EXERCISE 6.2

Solve for a_+ and a_-, and show that they can be expressed as:

$$\begin{aligned} a_+ &= \frac{\Omega_{inner} - \Omega_{outer}}{\Delta} \\ a_- &= \Omega_{inner} + \frac{\Omega_{inner} - \Omega_{outer}}{\Delta{\cdot}R_{inner}^2}, \end{aligned} \quad [6.16]$$

where $\Delta = \dfrac{1}{R_{inner}^2} - \dfrac{1}{R_{outer}^2}$.

Plot several representative cases, as shown in Fig. 6.4—explicitly, as shown in panel (a) start with an initial line or marker points and advect it forward in time to show where the markers arrive sometime later. Notice for a small gap between inner and outer cylinders (Fig. 6.4(c)) that the flow is nearly uniformly shearing: at larger radius the distance that the fluid moves invariably grows. For small or moderate gaps (panels (a) and (b)), however, significant variations in shear appear.

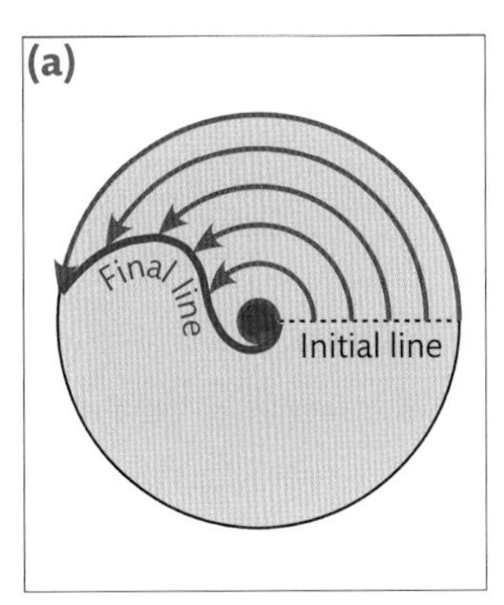

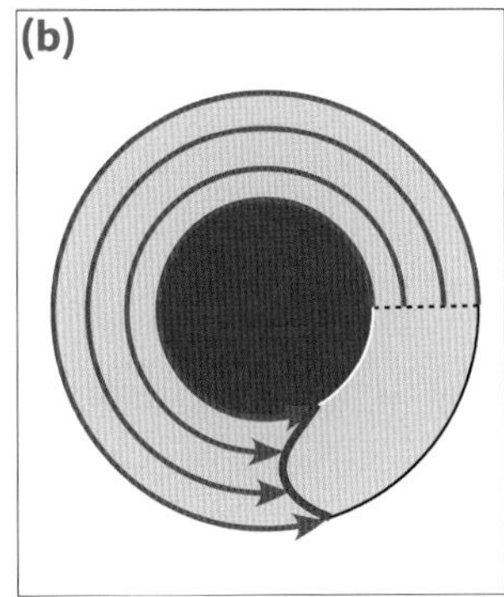

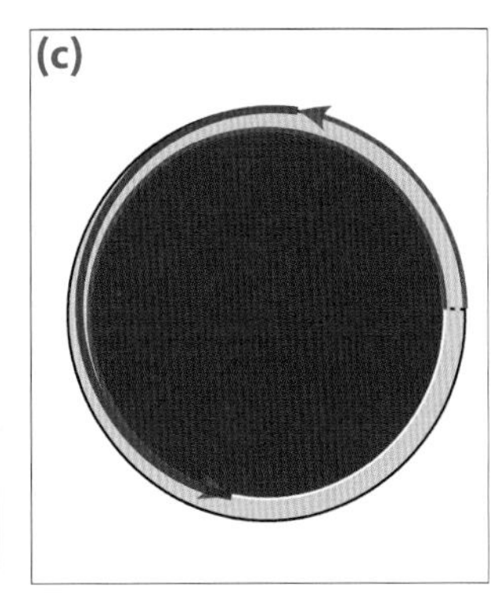

Fig. 6.4 Couette flow between two rotating cylinders (a) with large gap: here, $R_{inner} = 0.1R_{outer}$, (b) with moderate gap: $R_{inner} = 0.5R_{outer}$, and (c) with small gap: $R_{inner} = 0.9R_{outer}$.

Example Couette viscometer

We saw in Chapter 1 that simple tubes can be used to produce Poiseuille viscometers. Couette devices are also commonly used to measure viscosity. For this purpose, the inner or outer cylinder is rotated, and the torque on the other cylinder is measured. Because of the instabilities shown in Fig. 6.1, it is preferable to rotate the outer cylinder: this way the flow will remain "laminar" (meaning it will travel in lamina, or smooth curves, rather than complex trajectories) at all speeds. Assuming the outer cylinder is rotated, the stress on an inner cylinder of height L is simply:

$$\tau_{inner} = \frac{Torque}{2\pi R_{inner}^2 L}. \quad [6.17]$$

If additionally the gap between the inner and outer cylinders is small, we can write that the strain rate is:

$$\begin{aligned}\dot{\gamma}_{inner} &= \frac{\partial V_\vartheta}{\partial r} \\ &= \frac{\Omega_{outer} R_{outer}}{R_{outer} - R_{inner}},\end{aligned} \quad [6.18]$$

and using Newton's law of viscosity (Chapter 1), we obtain:

$$\mu = \frac{\tau_{inner}}{\dot{\gamma}_{inner}} = \frac{Torque \cdot (R_{outer} - R_{inner})}{2\pi R_{inner}^2 L \Omega_{outer} R_{outer}}. \qquad [6.19]$$

So we can obtain the viscosity of a fluid in a Couette viscometer by measuring the torque, angular speed, and geometric parameters of the cylinders. This is only valid for small gaps; there is a literature describing corrections needed for large gaps.[5]

Let's now turn to the $\hat{r}$ bracket (pink in Eq. [6.11]). This gives:

$$\frac{\partial P}{\partial r} = \mu \frac{\left(V_\vartheta(r)\right)^2}{r}. \qquad [6.20]$$

But we have solved for $V_\vartheta(r)$, Eq. [6.15], so we know $\partial P/\partial r$ and can integrate to get P:

$$\begin{aligned} P &= \int \frac{\mu}{r}(a_+ \cdot r^{+1} + a_- \cdot r^{-1})^2 dr \\ &= \mu \cdot \int \left(a_+^2 \cdot r + 2\frac{a_+ a_-}{r} + \frac{a_-^2}{r^3} \right) dr \\ &= \mu \cdot \left(\frac{a_+^2}{2} r^2 + 2a_+ a_- \ln(r) - \frac{a_-^2}{2r^2} \right) + P_o, \end{aligned} \qquad [6.21]$$

where P_o is an integration constant. This final result is revealing. In Exercise 6.2, we found that the magnitude of a_- grows with Ω_{inner}, while the magnitude of a_+ grows with Ω_{outer}. So we can deduce the qualitative behavior, for which two cases emerge. When Ω_{inner} is large, $a_-^2 \gg a_+^2$ — and so pressure will be highest at the smallest radius. If Ω_{inner} is large enough, this will produce a destabilizing radially outward flow, leading to breakdown of the solution in favor of Taylor vortices. At the other extreme, if Ω_{outer} is large, the r^2 and $\ln(r)$ terms will dominate, and pressure will be higher at *larger* radius. This will *stabilize* any tendency to produce outgoing flow, which leads to the enlargement of the stable Couette flow region at large Ω_{outer} (which in turn corresponds to large Re_{outer}), as shown in Fig. 6.1(a).

So although as we have mentioned we will not delve into the rich details Taylor–Couette instabilities shown in Fig. 6.1(a), Eq. [6.21] provides a stepping off point for explaining some of its qualitative features. At small Ω_{outer}, stable Couette flow appears when the centrifugal acceleration—set by Ω_{inner} and so Re_{inner}—is small, and as Ω_{outer} grows, Couette flow becomes increasingly stabilized so that larger Ω_{inner} can be tolerated.

Example Vertical flows

Before leaving this topic, we mention that when we wrote Stokes equation, Eq. [6.11], we explicitly omitted any z dependence. Flows in the z-direction are of interest in their own right, and we mention two examples here: the Boycott effect and colloidal stratification.

Boycott effect: In 1920, the pathologist and conchologist (snail expert), Arthur E. Boycott (1877–1938), wrote a tiny, 24-line article[6] reporting the observation that red blood cells (RBCs) settle more rapidly in inclined tubes than in vertical ones. This effect can be easily understood by considering particles in a fluid-filled vertical tube. As shown on the left of Fig. 6.5(a), in order for the particles to move down (pink arrow), fluid beneath that would be displaced by the particles must move up (blue arrow). Under some conditions (especially fine dense particle beds), the fluid will erode a channel through the particles, but for particles such as blood in a fluid such as plasma, this does not occur, and the fluid needs to slowly filter through the particle bed, slowing the rate of particle settling. This so-called "hindered settling" is the topic of extensive research in the sedimentation and fluidized bed fields (see, for example, Richardson and Zaki[7]). In air-fluidized beds, this can lead to even more extreme behavior: "slugging," in which air escapes from beneath a dense "slug" of particles in periodic bursts, causing the slug to repeatedly lift and crash to the bottom of the container. On the other hand, if the tube is tilted as shown on the right of Fig. 6.5(a), the particles fall to one side of the tube, leaving a natural channel for the fluid on the other side. This allows fluid to leave the bottom of the tube and so speeds the settling of the particles. Modern centrifuges make use of this fact by angling tubes to the horizontal.

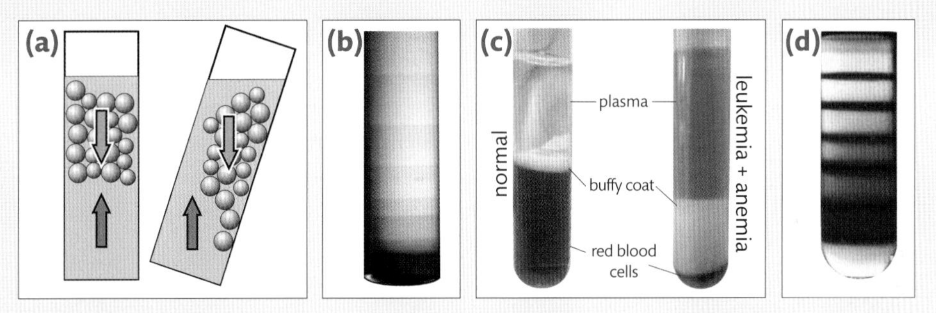

Fig. 6.5 Sedimentation and stratification. (a) In the "Boycott effect," particles in a vertical tube settle more slowly than in an inclined tube, because inclination produces a channel for fluid beneath to pass by the particles. (b) As colloids settle, they can form stratified layers through mechanisms that remain controversial: this image shows an emulsion of decane in water after 100 hours of settling;[8] figure foreshortened vertically. (From Mueth, D.M., Crocker, J.C., Esipov, S.E. and Grier D.G. (1996). Origin of stratification in creaming emulsions. *Physical Review Letters, 77*. By permission of the American Physical Society.) (c) Blood settling can be accelerated using a centrifuge: shown are normal blood and blood from a patient suffering from leukemia and anemia. (From Amrein, P.C., Kumar, J.R., Poulin, R.F., Umlas, J. and Weitzman, S.A. (1982) Comparison of filtration leukapheresis and centrifugation leukapheresis in treatment of lymphosarcoma cell leukemia. *Southern Medical Journal, 75*(8), 969–71. By permission of Wolters Kluwer Health, Inc.) (d) Further "fractionation" of blood can be accomplished, for example, using "Percoll™," a suspension of colloidal particles of multiple distinct densities. Shown is a stratified sample of red blood cells from a patient infected with plasmodium falciparum (malaria), which produces multiple cell shapes and densities. (From Rüssmann, L., Jung, A. and Heidrich, H.G. (1982) *Z. Parasitenkd., 66*, 273. By permission of Springer Berlin Heidelberg.)

Colloidal stratification: This brings us to a second issue: stratification of "colloids," which are microscopic mixtures of a dispersed phase (e.g. droplets or particles) and a continuous phase (e.g. a pure fluid). Examples of colloids include clays, paints, mayonnaise, and gelatin. In 1884, William H. Brewer (1828–1910), an American botanist and soil chemist, discovered that when clay sediments in water, it forms stratified layers. This simple observation has been the subject of study in several labs since that time, resulting in conflicting theories for the phenomenon, including the production of convection cells produced either by minute temperature fluctuations or tiny inclinations in seemingly vertical tubes, differences in sizes or shapes of particles, and "coarsening" (growth by agglomeration) of segregated domains. An example of spontaneous stratification after 100 hours of an emulsion of decane in water is shown in Fig. 6.5(b). Similar, though much more elaborate, patterns appear when settling material reacts—see Chapter 13.

Naturally you do not have to wait a hundred hours to produce sedimentation, you can accelerate the process by centrifugation. In Fig. 6.5(c) we show the results of centrifuging blood: here cells simply separate by density, rather than by a subtle or controversial process as in Fig. 6.5(b). On the left is an example of normal blood, containing nearly half RBCs by volume and a small amount of white blood cells and platelets, called a "buffy coat" because of its appearance. Likewise white blood cells are so termed because of the whiteness of this layer. Fig. 6.5(c) shows both normal blood and blood from a patient with leukemia (hence the thick buffy coat) as well as anemia (hence the small volume of RBCs).

Finally, we remark that additional technologies are available to separate, or "fractionate," materials, for example, cells can be centrifuged in a suspension of colloids of multiple distinct densities: this produces a striated appearance as in Fig. 6.5(d), in which each stripe consists of material of density between steps of density contrast. In Fig. 6.5(d), we show stratified layers of RBCs in "Percoll™," a commercial colloidal suspension: the RBCs are from a patient infected with the malaria bacterium, which causes the cells to exhibit varying degrees of hemoglobin polymerization and so varying densities.

6.2 **Stokes Drag: Flow Past Particles at Low Reynolds Number**

This brings us to an analysis of flow past particles. This has a number of applications, some of which we have already discussed, including flow through fluidized beds and settling of RBCs. Also important is flow through packed beds—which are used in high pressure liquid chromatography as well as in filtration. As we will discuss shortly, particle settling has been used to measure viscosity as well. A final example of flow past particles carries historical as well as practical significance, as described in the following exercises.

EXERCISE 6.3

Cell products ranging from recombinant proteins to vaccines to antibiotics can be manufactured in "roller bottles." As shown in Fig. 6.9(a), these are what they sound like: 1–2 liter horizontal bottles rotated slowly and partially filled with culture medium. The slow rotation provides a low shear environment in which cells, or cells attached to carrier beads, are

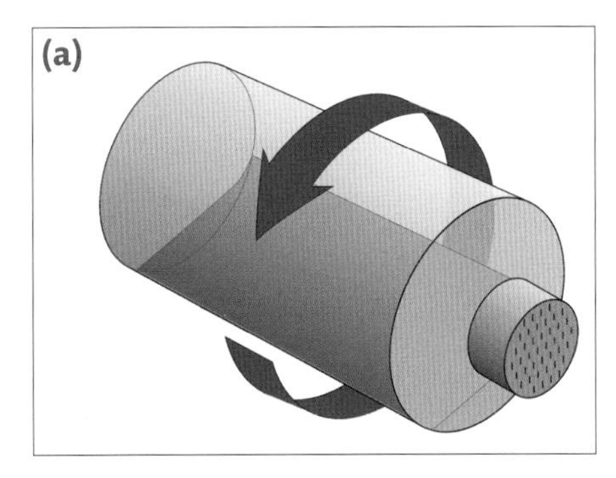

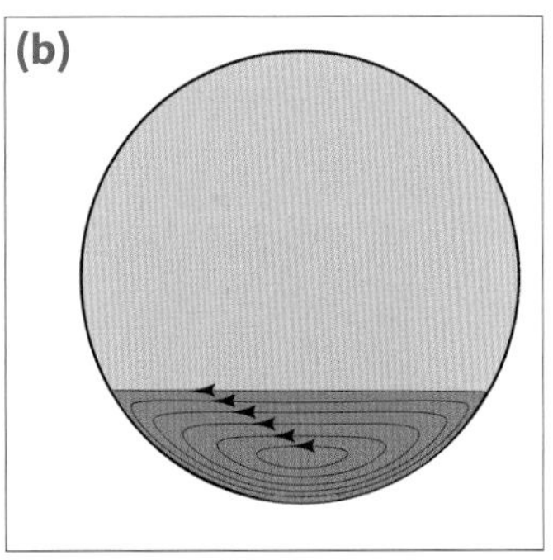

Fig. 6.6 Flow in roller bottle. (a) Geometry of roller bottle used to culture cells. (b) Sketch of flow in cross section of bottle, from Muzzio et al.[9]

exposed to air and nutrients: a process that depends critically on how rapidly particles settle in the bottle.

For this exercise, use what you have learned up to this point to simulate recirculating cells (use any flow, for example recall the discussion of Dean's flow in Chapter 5). The measured flow[9] is sketched in Fig. 6.6(b). Add a constant downward settling velocity, and plot trajectories of marker particles. Assume that particles that cross the bottom boundary of the bottle would attach to the surface, so remove them from the simulation. Finally, show for a finite (and not too large) settling velocity that some particles will remain, recirculating forever and never reaching the surface.

A consequence of this exercise is that for small diameter bottles, either cells themselves or cells attached to carrier beads will always reach the bottom of the bottle – and so will be dragged to the surface. However for larger bottles, cells can become trapped in recirculating loops beneath the surface, and so will effectively drown. For this reason, roller bottles are never larger than about 2 liters: cells in larger bottles are never exposed to air at the surface. Consequently, to scale up production using roller bottles, the bottle size isn't changed; rather, large numbers of 1–2 liter bottles are deployed in racks.

EXERCISE 6.4

Roller bottles were invented by George O. Gey (1899–1970), a biologist better known for discovering the "HeLa" cell line, shown in Fig. 6.7(a). HeLa cells were removed from of a cervical cancer patient named Henrietta Lacks (Fig. 6.7(b)) in January of 1951 by a colleague of Dr. Gey's. These cancer cells were extremely aggressive, and by October of 1951, Ms. Lacks had died at age 31. Dr. Gey cultured the HeLa cells and discovered that they are "immortal," meaning that unlike most mammalian cells, they would reproduce without limit in the lab: descendants of these original cells are still being cultured today. HeLa cells were the first human biological materials to be bought and sold: they readily grow in culture and reproduce rapidly, every 24 hours, and so have become a standard in the laboratory. They were used by Jonas Salk to test the first polio vaccine, and have since been used in studies ranging from AIDS and cancer to gene mapping and toxicity testing. Over 60,000 research papers have been published using HeLa cells, and products with a market value in the billions of dollars have been generated with these cells.

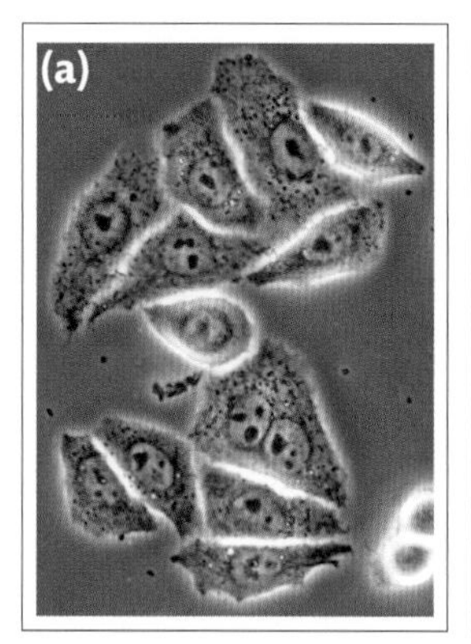

Fig. 6.7 Cell growth. (a) Immortal HeLa cells, continuing to reproduce decades after their discovery. (Source: Nikon MicroscopyU.) (b) Human face of HeLa cells: Henrietta Lacks, after whom the cell line is named (© Lacks Family).

Neither Ms. Lacks nor her family were told of the use of the cells for over two decades, and although the cells were unambiguously Henrietta Lacks', neither she nor her family has ever been paid for their use. Indeed, the family only learned that the cells were being used some 23 years after Ms. Lacks' death, in 1974, when it was discovered that the cells can migrate on airborne dust and so contaminate other cultures. Because of this situation, the Lacks family was contacted for genetic tests to distinguish HeLa from other cells, and thereafter their own investigation revealed the use of the cells. This case, which has been described by Skloot,[10] has parallels in the "MO" cell line (named after John Moore),[11] and more recently in the cases Beleno v. Texas, Higgins v. Texas, and Bearder v. Minnesota.[12]

For this exercise you are asked to consider the human face of biomedical research. It has been argued that countless lives have been saved from research using HeLa cells, and that imposing restrictions on the use, sale, or patent rights of cells or cell products would stifle this research. On the other hand, plaintiffs in several cases have argued that cells are the property of the person from whom they were taken, and patent rights as well as any profits derived from the cells should be shared with that person. This issue continues to evolve: for example, to treat inherited mitochondrial diseases, a donor egg from one woman can be enucleated and replaced with second woman's nucleus. In this case, the baby will have three copies of DNA: a mitochondrial copy from the first woman, as well as (half) nuclear copies from the father and the second woman. So whose baby will this be?[13] Does the mitochondrial mother have any rights?

Provide your opinions on what obligation the biomedical researcher has to the donor of cells or cell products. Do you think that the donor has rights to profits from cell lines? What about control over their use? What about privacy rights: is it an unlawful search and seizure to perform and catalog DNA test on blood samples?[12] Should the donor be permitted to disallow certain lines of research—for example, to produce bioweapons or to provide DNA information that could permit insurance coverage for to be limited for certain people? Or do you feel that the threat of stifling future research outweighs any rights the donor might have? In several cases, knowingly false statements have been given to the donor in informed consent documents;[11] as well, court rulings[12] on the use of donor cells have been violated: would this change your opinion? Finally, comment on whether you feel discussions of ethics belong in a book on biological fluid flow.[14]

After this long preamble, let's return to discussing how particles interact with flow. As always, we start with the simplest case: low Reynolds number flow, and consider spherical or cylindrical particles. We will study high Reynolds number flows and flow past complex shapes in a later chapter.

So let's consider flow past a sphere or a cylinder, as illustrated in Fig. 6.8(a). For reference, and to avoid confusion, Fig. 6.8(b) shows spherical and cylindrical coordinate systems. It is easy to become confused because there isn't uniformity regarding these systems in the literature. For example, for spherical coordinates, the polar angle, ϑ, shown is almost universally the angle to the vertical (shown in blue in Fig. 6.8(b))—except in some computer languages, notably Matlab, in which ϑ is the angle to the horizontal (shown as red broken line in the figure). Likewise the azimuthal angle is invariably ϕ in spherical coordinates, but is written either as ϕ or as ϑ in cylindrical coordinates (and unfortunately corresponds to ϑ in x–y polar coordinates). To deal with this confusion to the extent that we can, in this book we use the conventions shown in Fig. 6.8(b): in spherical coordinates, ϑ is the angle to the vertical, and in both spherical and cylindrical coordinates, the azimuthal angle is ϕ. We note though that the radius, "r," differs: in spherical coordinates, it is the distance from the origin, while in cylindrical coordinates, it is the distance from the z-axis. This is unavoidable: some texts instead use r and ρ to distinguish these quantities; however, this would lead instead to an ambiguity with the density which we also call ρ.

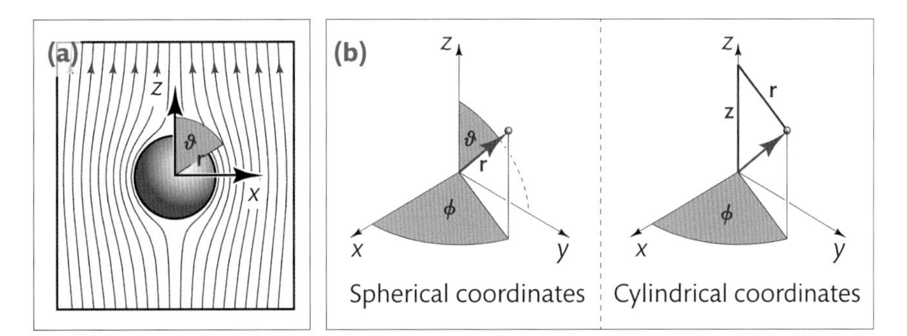

Fig. 6.8 (a) A 2D cross section of flow past a sphere. (b) Spherical and cylindrical coordinate systems—we caution that for spherical coordinates conventionally, ϑ is the angle from the vertical (shown in blue); however, some computing languages (e.g. Matlab) define ϑ to be the angle from the horizontal (red dashed line).

These ambiguities are only semantic; more substantive confusions emerge as soon as we begin to analyze the flow shown in Fig. 6.8(a). For example, the r-component of Stokes' equation in spherical coordinates in the absence of external forces is as follows:

$$\frac{\partial V_r}{\partial t} = -\frac{1}{\rho}\frac{\partial P}{\partial r} + \nu\left[\frac{1}{r^2}\frac{\partial}{\partial r}\left(r^2\frac{\partial V_r}{\partial r}\right) + \frac{1}{r^2\sin^2(\vartheta)}\frac{\partial^2 V_r}{\partial \phi^2} + \frac{1}{r^2\sin(\vartheta)}\frac{\partial}{\partial \vartheta}\left(\sin(\vartheta)\frac{\partial V_r}{\partial \vartheta}\right)\right.$$
$$\left. - 2\frac{V_r + (\partial V_\vartheta/\partial\vartheta) + V_\vartheta\cot(\vartheta)}{r^2} - \frac{2}{r^2\sin(\vartheta)}\frac{\partial V_\phi}{\partial \phi}\right] \qquad [6.22]$$

This is just the r-component: the ϑ and ϕ components are equally abstruse. Remarkably—indeed amazingly—we will see shortly that this rat's nest of equations can be dramatically simplified into a single and (nearly) comprehensible equation. First, let's make use of what we know about our problem to see what can be surmised about the necessary form of the solution.

One thing that we expect of the solution is that very far away, the obstacle shouldn't affect the flow. So if we are considering a fixed obstacle in a uniform vertical flow, that is, $\vec{V} = V_\infty \hat{z}$, then as r becomes large, we expect the "far-field" velocity to become $V_\infty \hat{z}$. So we may want to separate variables (described in Chapter 3), that is, write the velocity as:

$$\vec{V} = V_\infty \cdot V(\vartheta)\cdot V(r)\hat{z}, \qquad [6.23]$$

where $V(\vartheta)$ captures the angular dependence of the velocity far away, and $V(r)$ approaches 1 as r approaches ∞. It's straightforward to show that the far-field velocity can be written in spherical coordinates as:

$$\begin{aligned} V_r(\vartheta) &= V_\infty \cos(\vartheta) \\ V_\vartheta(\vartheta) &= -V_\infty \sin(\vartheta), \end{aligned} \qquad [6.24]$$

so in the spirit of giving dessert first, we should not be surprised to arrive at a solution that looks something like:

$$\begin{aligned} V_r &= V_\infty \cos(\vartheta)\cdot\left[1 - \frac{3R}{2r} + \frac{R^3}{2r^3}\right] \\ V_\vartheta &= -V_\infty \sin(\vartheta)\cdot\left[1 - \frac{3R}{4r} - \frac{R^3}{4r^3}\right], \end{aligned} \qquad [6.25]$$

where R is the radius of the sphere. We'll need some more tools to derive the formulas in square brackets, but before we get too involved with these tools, let's notice that these formulas do approach 1 as r approaches ∞, and Eq's. [6.25] do fit the far-field boundary conditions, Eq. [6.24].

EXERCISE 6.5

Reproduce Fig. 6.8(a) using Euler integration. That is, let $R = 1$ and take a horizontal line of marker points some distance away, say at $z = -5$, and iterate the points forward in time over multiple small time steps. The procedure will be the same as in Chapter 5, Exercise 5.2, however you will have to deal with the spherical coordinate system, either by working entirely in spherical coordinates, or by only transforming to Cartesian coordinates for plotting, or by transforming Eq. [6.25] into Cartesian coordinates. The first two alternatives are equally difficult; the final option is extremely cumbersome, but possible.

However you do this exercise, when you are done *make sure that your trajectories don't converge*: the flow should be area conserving and should look very much like Fig. 6.8(a). If your trajectories do not look like this, it is likely that you may have a dimensional problem—in particular, notice that the angle, ϑ, has no units, while the velocity in the ϑ-direction has dimensions length/time: so at some point, you need to divide the velocity by, or multiply the angle by, the radius. We confronted a similar issue in Eq. [1.46] of Chapter 1, concerning including an extra radius term when integrating in cylindrical coordinates.

Hold onto the code that you write for this exercise: we will compare this early result with plots using better integration techniques, with plots at high Reynolds number and with plots using streamfunctions, which we introduce in the next section.

Before we get into new tools including streamfunctions, let's use the tools that we have already to enjoy our dessert. We have been given the velocity, Eq. [6.25], and we know Stokes equation, Eq. [6.11], lets us calculate the pressure gradient from the velocity:

$$\vec{\nabla} P = \mu \nabla^2 \vec{V}, \qquad [6.26]$$

so inserting the velocity, Eq. [6.25], and carrying out the derivatives, Eq. [6.22], we get:

$$\begin{aligned} \frac{\partial P}{\partial r} &= 3\mu V_\infty \frac{R}{r^3}\cos(\vartheta) \\ \frac{\partial P}{\partial \vartheta} &= 3\mu V_\infty \frac{R}{2r^2}\sin(\vartheta). \end{aligned} \qquad [6.27]$$

We can integrate this to produce:

$$P - P_o = -\frac{3\mu V_\infty}{2}\frac{R}{r^2}\cos(\vartheta). \qquad [6.28]$$

But look: between Eq. [6.25] and Eq. [6.28], we have the shear rate and the pressure—that is, we have everything we need to calculate the total force on the sphere. Let's do so. As illustrated in Fig. 6.9(a) and (b), the pressure vector, ΔP, is normal to the sphere surface, and the shear stress, D_μ, is tangential to its surface. It's then a simple matter of trigonometry to determine that the upward component of ΔP is $-\Delta P\cos(\vartheta)$, and the upward component of D_μ is $D_\mu\sin(\vartheta)$. Of course Eq. [6.28] gives ΔP, and Newton's law of viscosity gives the shear stress, D_μ:

$$D_\mu = \mu \left.\frac{\partial V_\vartheta}{\partial r}\right|_{r=R}, \qquad [6.29]$$

so the total upward stress on an area segment of the sphere at angle ϑ is:

$$\begin{aligned} \sigma(\vartheta) &= D_\mu|_{r=R}\cdot\sin(\vartheta) - \Delta P|_{r=R}\cdot\cos(\vartheta) \\ &= \mu\left.\frac{\partial V_\vartheta}{\partial r}\right|_{r=R}\cdot\sin(\vartheta) - [P - P_o]_{r=R}\cdot\cos(\vartheta). \end{aligned} \qquad [6.30]$$

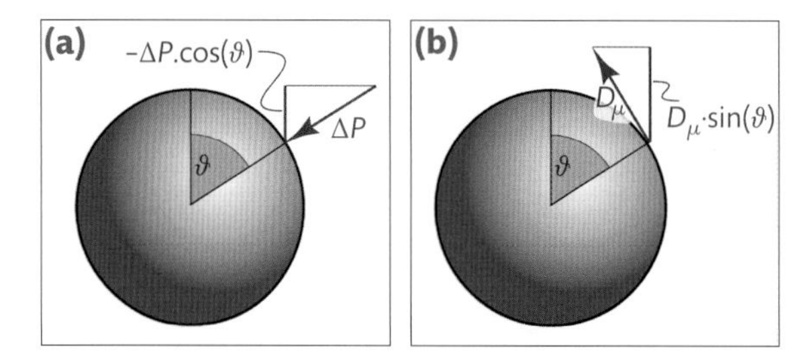

Fig. 6.9 Forces on a sphere in vertical upward flow. (a) Vertical component of pressure, ΔP; (b) vertical component of shear, D_μ.

Now we only have to substitute V_ϑ and P from Eq. [6.25] and [6.28] to get

$$\begin{aligned}\sigma(\vartheta) &= \mu\left[V_\infty\cdot\sin(\vartheta)\cdot\left(\frac{3R}{4R^2}+\frac{3R^3}{4R^4}\right)\right]\cdot\sin(\vartheta)-\left[-\frac{3\mu V_\infty}{2R}\frac{R^2}{R^2}\cos(\vartheta)\right]\cdot\cos(\vartheta)\\ &= \mu\cdot\left[\frac{3V_\infty}{2R}\right]\cdot\sin^2(\vartheta)+\mu\cdot\left[\frac{3V_\infty}{2R}\right]\cdot\cos^2(\vartheta)\\ &= \frac{3\mu V_\infty}{2R}.\end{aligned} \qquad [6.31]$$

Miraculously, the stress is entirely independent of angle: in moving from any location to any other, the decrease in the vertical component of the shear stress always precisely equals the increase in that component of the pressure. So calculating the total force on a sphere in a moving fluid, F_{drag}, is simplicity itself; it's just the total surface area times the stress:

$$\begin{aligned}F_{drag} &= 4\pi R^2\cdot\sigma\\ &= 6\pi\mu RV_\infty.\end{aligned} \qquad [6.32]$$

Notice that F_{drag} depends linearly on μ, R, and V_∞. Probably we could have guessed the μ and the V_∞ dependence, but a reasonable guess might also be that the drag would depend on the cross-sectional area: indeed we will see in a moment that the terminal velocity does depend on the area. Moreover, we will show later that many spheres packed into a tube impede fluid flow according to the pore area (i.e. the total tube area minus the spheres' cross-sectional areas). Nevertheless, the drag force depends linearly on R, and this leads to an important result.

If we consider a sphere falling gravitationally through a fluid, the downward force will go as the sphere's mass, that is, R^3, while the upward drag force will go as R. As illustrated in Fig. 6.10(a), this means that at any prescribed speed and viscosity, there will be a critical radius, R_c, above which gravity will exceed drag, and below which drag will exceed gravity. Consequently, particles larger than R_c will be gravitationally dominated and will settle rapidly, while particles smaller than R_c will be drag dominated and will tend to remain suspended longer.

This same argument applies to particles moving through a fluid with fixed speed: the forward momentum will go as R^3, while the backward drag force tending to change the momentum will only go as R. This argument tells us that larger inhaled particulates will tend

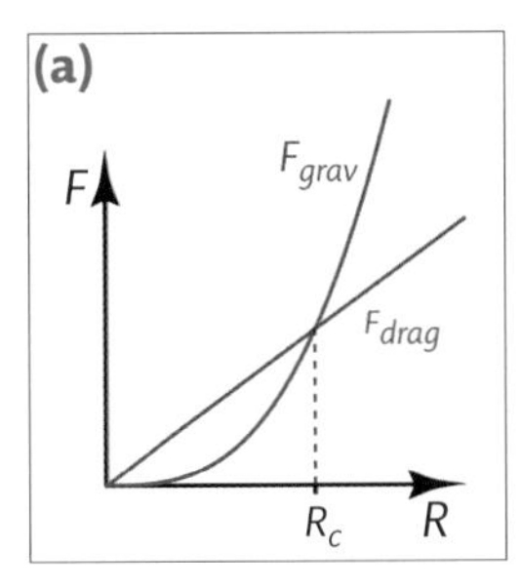

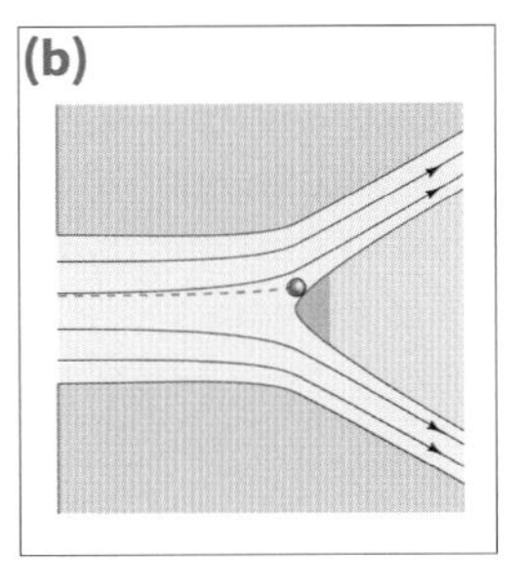

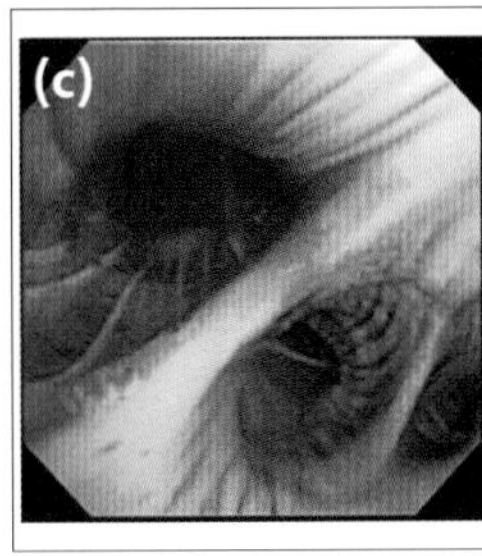

Fig. 6.10 (a) Comparison between drag and gravitational forces on a sphere as a function of radius—note that for $R < R_c$, drag dominates, and for $R > R_c$, gravity dominates. (b) Illustration that larger inertial particles tend to maintain their momentum and so travel in nearly straight lines. This causes larger particles to strike bifurcation points in the lung as indicated. (c) Autofluorescent microscopic imaging identifies dysplastic squamous cells (magenta) at bronchial bifurcation.[15] (© Olympus. www.olympus-europa.com)

to continue traveling in straight lines when flow is diverted, for example by bronchial bifurcations. This leads to an increased incidence in squamous- and adeno-carcinomas at these bifurcations (see Fig. 6.10(b) and (c)). On the other hand, fine particles travel with the airstream and can be carried into the deep lung—for this reason, particulates delivered by medicinal inhalers are designed to be on the order of a micron in diameter. On the other hand, very small particles tend to follow the airstream both in and out, and so can be exhaled without striking the surfaces of the peripheral airways or alveoli. Generating particles that are small enough, but not too small, to be absorbed in the lung poses significant engineering challenges.

EXERCISE 6.6

To study drag, we describe three related experiments.

For the first, brew a cup of tea and leave some tea leaves to steep. Once they have settled to the bottom of the cup, stir the cup. The leaves are manifestly denser than the water: this is why they sink to the bottom, and in doing so they will obey Fig. 6.10(a), so that the larger and heavier leaves settle more rapidly, while the smaller and lighter ones remain entrained in the circulating water. More than this, though, you will find that when you stir the cup, the tea

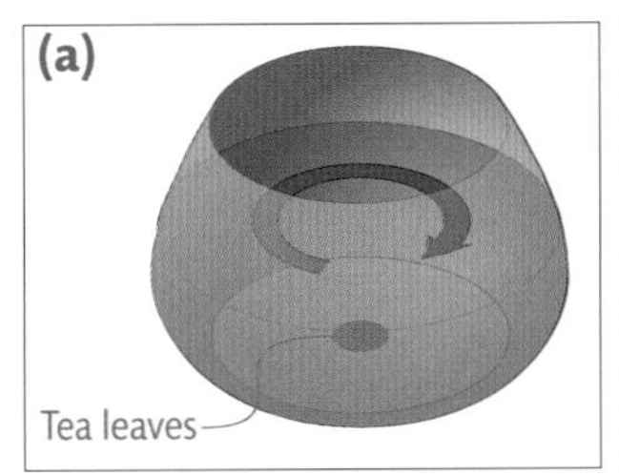

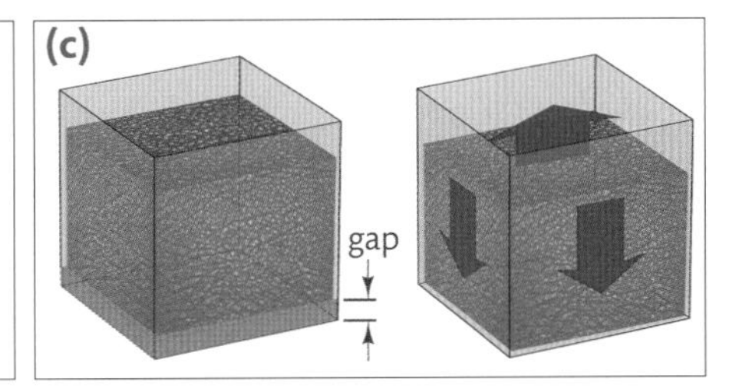

Fig. 6.11 (a) Tea leaves (green) settle to the bottom of a cup of tea (Exercise 6.6), and move to its *center* when the tea is stirred. (b) Guinness beer bubbles move *down* the outside of the glass after it has first been poured (Exercise 6.7) (PHTGRPHER_EVERYDAY/Shutterstock.com). (c) When a container of sand is vibrated (Exercise 6.8), two things happen. Left: when the container moves downward faster than gravity, the grains cannot keep up, and a gap opens beneath the bed. Right: the gap closes when the container moves back up, closing the gap: this produces upward motion in the center and downward motion along the container sides as indicated by blue arrows.

leaves will rapidly aggregate at the center of the cup bottom, as sketched in Fig. 6.11(a). But shouldn't the heavier leaves be centrifuged *outward* as we discussed earlier in the chapter on vertical flows? Perform the experiment and explain why they move *inward* instead.

This is a famous problem first studied by Einstein: try to solve the problem on your own, but if you cannot, the original solution is in Einstein.[16]

EXERCISE 6.7

A second and related problem involves the rising of bubbles rather than the sinking of tea. For this exercise, pour freshly opened Guinness stout into a tall glass, and watch the bubbles. Eventually, the bubbles will form a thick "head" at the top, but shortly after pouring the stout, you will see that the bubbles move *down* the sides of the glass, as indicated in Fig. 6.11(b). The bubbles are lighter than the liquid—which is why they ultimately rise. Explain why they move down in your experiment.

If possible, do the experiment and work out what is going on yourself. If you are legally, religiously, or otherwise unable to do this experiment, you will find movies of the effect online. Whether you resort to the movies or not, again try to solve the problem yourself before looking for existing solutions. If you do look for solutions, you will find that there are two, though one of these is difficult to find. In any event, explain what you think accounts for the downward motion of the bubbles.

Guinness, incidentally, has an important history in the development of modern science. A chemist and mathematician named William Gossett (1876–1937) was hired as part of Guinness' forward-looking policy of hiring scientists to improve its processes. While studying how to evaluate the quality of barley for stout, Gossett invented the now famous "Student t-test" that permits the estimation of uncertainties from limited quantities of data. Unfortunately, Guinness' forward-looking hiring policy was accompanied with a protectionist intellectual policy, and consequently Gossett had to negotiate with his employer to publish his statistical findings at all, and ultimately was permitted to publish only under the pseudonym "Student."[17]

EXERCISE 6.8

A final exercise dealing with intuitively challenging aspects of drag involves what is often termed vibrofluidization of grains (which is exactly as it sounds: fluidization resulting from vibration). As shown in Fig. 6.11(c), if a container of grains is vibrated vertically, the grains will convect upward in the center of the container, and downward along its sides. Two clues to the mechanism are as follows. (1) When the container moves down faster than gravity, the grains cannot keep up, and a gap appears beneath the granular bed (left in Fig. 6.11(c)). When the container moves up again, the gap closes, and then the cycle is repeated. (2) In a clever set of experiments, Knight, Jaeger, and Nagel[18] showed that the direction of convection reverses—so grains move up the sides and down the center of the bed—if the container is larger at the top than at the bottom. Using these two clues, explain how the convection shown in Fig. 6.11(c) is produced.

Recalling Fig. 6.10(a), we can go further by calculating when the downward force F_{grav} of a falling body will equal the upward drag force, F_{drag}. When this occurs, the body will reach a terminal velocity, V_t. The calculation is simple: the gravitational force on a sphere is:

$$F_{grav} = \frac{4}{3}\pi R^3(\rho_s - \rho_f)g, \qquad [6.33]$$

where ρ_s is the density of the solid and ρ_f is the density of the fluid. We note that the difference in density is what produces downward force: typically ρ_f is small, but not always. For example, if the fluid is a liquid, the density is never negligible, and even in gases the density can be important—for example, many parlor tricks can be found on internet movies using sulfur hexafluoride, which is much denser than air.

We have already calculated the drag force, and at V_t this is:

$$F_{drag} = 6\pi\mu R V_t. \qquad [6.34]$$

Setting Eq. [6.33] and [6.34] equal, we can calculate the terminal speed:

$$\begin{aligned} V_t &= \frac{1}{6\pi R\mu}\left(\frac{4}{3}\pi R^3 g\right)(\rho_s - \rho_f) \\ &= \frac{2R^2 g}{9\mu}(\rho_s - \rho_f). \end{aligned} \qquad [6.35]$$

This can be expressed slightly differently, in terms of the force applied to the sphere divided by its speed:

$$\begin{aligned} \frac{mg}{V_t} &= 4\pi R^3(\rho_s - \rho_f)g\frac{9\mu}{2R^2 g(\rho_s - \rho_f)} \\ &= 12\pi R\mu. \end{aligned} \qquad [6.36]$$

This provides a measure of the bang per buck—that is, the force needed to reach a given speed. We emphasize that this applies in the low Reynolds number regime, and we will see in Chapter 8 that different behaviors are seen at higher Reynolds number.

Example Falling sphere viscometer

An alternative expression uses Eq. [6.33] and [6.34] to provide the fluid viscosity:

$$\mu = \frac{2R^2 g}{9V_t}(\rho_s - \rho_f). \qquad [6.37]$$

So if we drop a ball in a fluid and measure its terminal speed, we can calculate the fluid viscosity from easily determined constants. This is another method of viscometry: not, however, the most accurate of available possibilities. First, this is only valid for small Reynolds number, and second it turns out that whenever the ball approaches a wall, the side of the ball nearest the wall slows, causing (1) the ball to slow, and (2) the ball to rotate. As sketched in Fig. 6.12(a), this rotation in turn entrains a layer of fluid (gray in the figure), so that the speed of fluid near the wall is increased, and the speed further from the wall is decreased. The resulting speed differential produces a "Magnus" force, drawing the ball even closer to the wall, so that particles dropped into a container of fluid end up falling near the container walls, as indicated in Fig. 6.12(b). The Magnus force refers to forces on spinning objects, and was first studied by H. Gustav Magnus (1802–1870), a German chemist. We will have more to say about this when we analyze the Bernoulli effect in the next chapter; for our purposes we remark that the Magnus effect causes falling spheres to move toward walls, and produces significant overestimation of the viscosity using Eq. [6.37]: even when the ratio of container to sphere diameter is 100, the error is 2%.

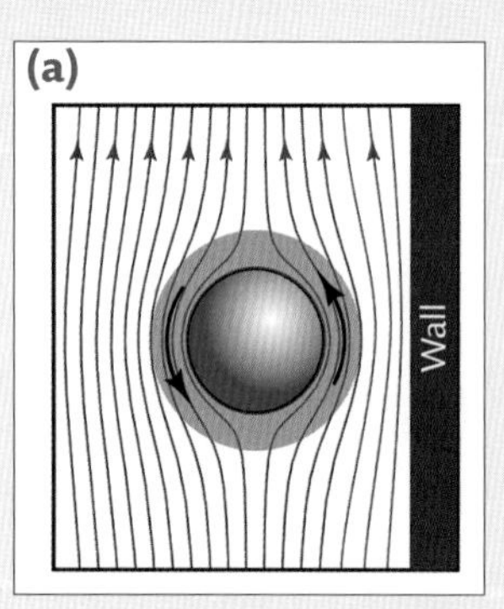

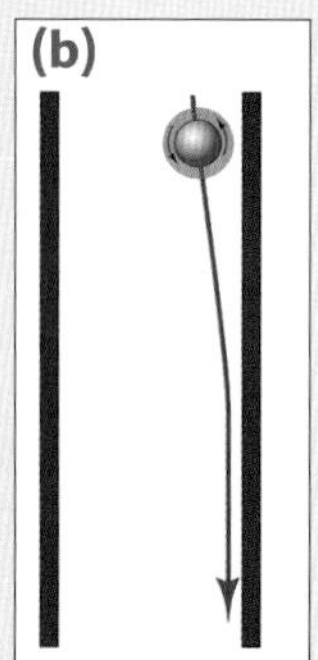

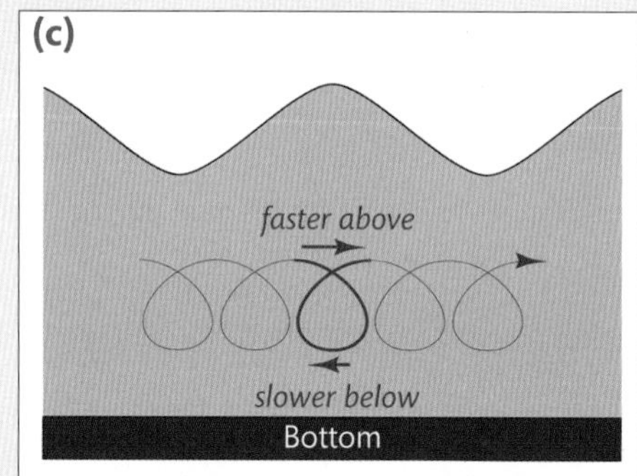

Fig. 6.12 (a) A particle falling in a fluid will rotate when it is near a wall because the side of the particle near the wall exerts higher drag than the side further from the wall. This rotation (black arrows) increases the relative fluid speed near the wall, and decreases the speed further from the wall. (b) This speed differential produces a "Magnus force" that draws the particle closer to the wall. (c) By comparison, Stokes drift causes water and entrained particles in shallow, wavy, water to move shoreward (right in this cartoon). This occurs because water near the surface feels less drag than water near bottom. Consequently, circular wave trajectories become distorted as shown, which causes shoreward transport.

Example Stokes drift

A related effect concerns the motion of fluid and particles in shallow, wavy water—an effect known as Stokes drift. Ordinarily, surface waves cause a small packet of fluid to move in circular trajectories. Whenever the packet approaches the surface, drag from the bottom is reduced, and whenever the packet approaches the bottom, drag is increased. Consequently, each wave period, packets are moved shoreward, to shallower water, as illustrated in Fig. 6.12(c). This leads to important sedimentation effects on lake and ocean shores.

6.3 **Streamfunction**

We've made considerable headway since defining the velocity, Eq. [6.25], and pressure, Eq. [6.28], around a moving sphere. Now that we've enjoyed our dessert, it's time to return to eat our cabbage—specifically, to define where the velocity came from. As we mentioned, we need some tools to do this. Specifically, we need to define the "streamfunction" and the "biharmonic equation." We start with the streamfunction.

For the reader who recalls Chapter 5's discussion of area preserving maps, this will be a breeze. Area preserving maps are algebraic expressions that conserve area: streamfunctions perform the same function in differential equations.

We begin with the continuity equation for incompressible flows:

$$\nabla \cdot \vec{V} = 0, \qquad [6.38]$$

which in two dimensions is the same as:

$$\frac{\partial V_x}{\partial x} + \frac{\partial V_y}{\partial y} = 0. \qquad [6.39]$$

Now suppose I invent a function $\psi(x, y)$ with the following property—leave it to me to worry about how $\psi(x, y)$ is defined or what it should be, just suppose it exists:

$$\begin{aligned} V_x &= -\frac{\partial \psi}{\partial y} \\ V_y &= \frac{\partial \psi}{\partial x}. \end{aligned} \qquad [6.40]$$

Note the minus sign. It doesn't matter which of the two equations contains the minus sign, just so it appears in one of these—by convention, it is put in the V_x term. If we plug Eq. [6.40] into Eq. [6.39], we get

$$\frac{\partial V_x}{\partial x} + \frac{\partial V_y}{\partial y} = -\frac{\partial}{\partial x}\frac{\partial \psi}{\partial y} + \frac{\partial}{\partial y}\frac{\partial \psi}{\partial x}, \qquad [6.41]$$

and because of the minus sign, the right side of this equation is identically zero. So if I can create a $\psi(x, y)$ with the property (Eq. [6.40]), it will define velocities that identically conserve area: incompressibility is automatic. $\psi(x, y)$ is called a streamfunction, and it's called that for a special reason.

To see why, note that along a fluid trajectory, displacements over a time Δt are given by:

$$\begin{aligned} \Delta x &= V_x \Delta t \\ \Delta y &= V_y \Delta t, \end{aligned} \qquad [6.42]$$

which is equivalent to:

$$\begin{aligned} \Delta t &= \Delta x / V_x \\ \Delta t &= \Delta y / V_y. \end{aligned} \qquad [6.43]$$

So:

$$\frac{\Delta x}{V_x} = \frac{\Delta y}{V_y} \Rightarrow \Delta x \cdot V_y - \Delta y \cdot V_x = 0. \qquad [6.44]$$

Hold that thought. Now what is the variation of $\psi(x, y)$ with changes of x and y? This is

$$\Delta\psi = \Delta x\frac{\partial\psi}{\partial x} + \Delta y\frac{\partial\psi}{\partial y}. \quad [6.45]$$

But wait: from Eq. [6.40], the quantity in red is V_y, and the quantity in blue is $-V_x$, so

$$\Delta\psi = \Delta y{\cdot}V_y - \Delta y{\cdot}V_x, \quad [6.46]$$

which is zero along a fluid trajectory, by Eq. [6.44].

So what does this mean? Along a fluid trajectory, $\Delta x \cdot V_y - \Delta y \cdot V_x = 0$. And the variation of the streamfunction is precisely $\Delta\psi = \Delta x \cdot V_y - \Delta y \cdot V_x$. So *along a trajectory, the variation of the streamfunction is zero.* Meaning if we have a streamfunction, we can both calculate velocities, using Eq. [6.40], and what is better, *if we plot lines on which the streamfunction is a constant, these will be precisely the lines of fluid trajectories.* So ψ directly provides the trajectories of a stream, hence its name.

This is simply magical: all we have to do is to plot contour lines of a streamfunction, and we have the fluid trajectories. No Euler integration, no integration errors, no integration at all! And what is more, it is enormously easier to find a streamfunction than it is to find the velocities. Why is this? To find velocities, we have to solve an equation like Stokes equation, one term of which was shown in Eq. [6.22] to be completely impenetrable. To find a streamfunction, we'll show next that we have to solve a much simpler equation—and we get both V_r and V_ϑ from ψ. That equation is called the biharmonic equation.

Before we turn to that equation, for completeness we mention that given a streamfunction, we can produce Cartesian velocities, Eq. [6.40], or equivalently we can produce spherical or cylindrical velocities using the formulas:

$$V_r = -\frac{1}{r^2\sin(\vartheta)}\frac{\partial\psi}{\partial\vartheta}, \qquad V_\vartheta = \frac{1}{r\cdot\sin(\vartheta)}\frac{\partial\psi}{\partial r} \quad \text{in spherical coordinates,} \quad [6.47]$$

or:

$$V_r = \frac{1}{r}\frac{\partial\psi}{\partial\phi}, \qquad V_\phi = -\frac{\partial\psi}{\partial r} \quad \text{in cylindrical coordinates.} \quad [6.48]$$

Two notes. First, these equations use the coordinate systems shown in Fig. 6.8(b), and so the "r" variables are different in each equation as mentioned earlier. Second, the minus signs are a matter of convention, and different texts use different signs. This shouldn't be a cause of worry: obviously in Eq. [6.47], one can change the sign of each equation and change the sign of ψ without affecting the velocities.

6.4 **Biharmonic Equation**

Let's ask what the Stokes equation looks like if we substitute streamfunction equations, Eq. [6.40], for the velocities. Stokes equation again is:

$$\nabla^2 \vec{V} = \frac{\vec{\nabla} P}{\mu}, \tag{6.49}$$

and inserting V_x and V_y from Eq. [6.40] gives for the left hand side:

$$\nabla^2 \vec{V} = -\left(\frac{\partial^3 \psi}{\partial x^2 \partial y} + \frac{\partial^3 \psi}{\partial y^3}\right)\hat{x} + \left(\frac{\partial^3 \psi}{\partial x^3} + \frac{\partial^3 \psi}{\partial x \partial y^2}\right)\hat{y}. \tag{6.50}$$

This result takes a line or two of algebra, which the reader can confirm by writing the vector Laplacian in Cartesian coordinates as its equivalent $\nabla^2 \vec{V} = \hat{x}\nabla^2 V_x + \hat{y}\nabla^2 V_y$. Now let's do a sleight of hand: we next take the curl of Eq. [6.49]. It's not hard to show that the curl of a gradient is identically zero, so the pressure term (as well as any simple external forces) vanishes straight away, and taking the curl makes the right hand side of Eq. [6.49] zero. We haven't seen much of the curl yet (we'll discuss it further in Chapter 7), but explicitly the curl of a 2D vector is written as:

$$\nabla \times (F \cdot \hat{x} + G \cdot \hat{y}) = \left(\frac{\partial G}{\partial x} - \frac{\partial F}{\partial y}\right)\hat{z}, \tag{6.51}$$

or:

$$\begin{aligned}
\nabla \times \nabla^2 \vec{V} &= \left[\frac{\partial}{\partial x}\left(-\frac{\partial^3 \psi}{\partial x^3} - \frac{\partial^3 \psi}{\partial x \partial y^2}\right) - \frac{\partial}{\partial y}\left(\frac{\partial^3 \psi}{\partial x^2 \partial y} + \frac{\partial^3 \psi}{\partial y^3}\right)\right]\hat{z} \\
&= -\left[\frac{\partial^4 \psi}{\partial x^4} + 2\frac{\partial^4 \psi}{\partial x^2 \partial y^2} + \frac{\partial^4 \psi}{\partial y^4}\right]\hat{z} \\
&= -\left[\nabla^2 \psi\right]^2 \hat{z} \\
&\equiv -[\nabla^4 \psi]\hat{z}.
\end{aligned} \tag{6.52}$$

But as we just said, the right side of Eq. [6.49], and so of this equation, is zero, which gives us the biharmonic equation:

$$\nabla^4 \psi = 0. \tag{6.53}$$

Suddenly life is good. We began this chapter with a rat's nest of equations (Eq. [6.22] being only one), and we have arrived at a single concise equation that actually concentrates *more* information than the original equations. First of all, ψ is a scalar function, which has got to be (and is) easier to solve for than separate equations for V_r and V_ϑ that additionally have somehow to be chosen to meet the constraint $\nabla \cdot \vec{V} = 0$. Second, ψ provides V_x and V_y (and equivalently, we will show, V_r, and V_ϑ). And third, ψ lets us plot exact trajectories without even having to do any integration. Magic, simply magic.

6.5 **Solution at Last**

So let's apply the biharmonic equation to flow past a sphere. In Cartesian coordinates, the biharmonic equation expanded in terms of its individual derivatives appears in the second line of Eq. [6.52]. In spherical coordinates, this becomes:

$$\nabla^4\psi = \left[\frac{\partial^2}{\partial r^2} + \frac{\sin(\vartheta)}{r^2}\frac{\partial}{\partial\vartheta}\left(\frac{1}{\sin(\vartheta)}\frac{\partial}{\partial\vartheta}\right)\right]^2\psi = 0, \quad [6.54]$$

which we need to solve subject to boundary conditions at the sphere. These are respectively that flow won't penetrate into the sphere (i.e. in the $\hat{r}$ direction), and that flow won't slip along the sphere (in the $\hat{\vartheta}$ direction):

$$V_r = -\frac{1}{r^2\sin(\vartheta)}\frac{\partial\psi}{\partial\vartheta} = 0 \text{ at } r = R \quad [6.55]$$

$$V_\vartheta = \frac{1}{r\cdot\sin(\vartheta)}\frac{\partial\psi}{\partial r} = 0 \text{ at } r = R. \quad [6.56]$$

Also, as $r\to\infty$, as we have said before (Eq. [6.24]), $V_r(\vartheta)\to V_\infty\cos(\vartheta)$ and $V_\vartheta(\vartheta)\to -V_\infty\sin(\vartheta)$. It is easily confirmed that these conditions are met by:

$$\psi \to -\frac{1}{2}V_\infty r^2\sin^2(\vartheta) \text{ as } r \to \infty. \quad [6.57]$$

This motivates us to expect a solution with angular dependence that goes as $\sin^2(\vartheta)$, so let's try a separation of variables solution of the form:

$$\psi = f(r)\cdot\sin^2(\vartheta). \quad [6.58]$$

Plugging Eq. [6.58] into Eq. [6.54] (and skipping some highly uninteresting algebra) leads us to:

$$\left(\frac{\partial^2}{\partial r^2} - \frac{2}{r^2}\right)\left(\frac{\partial^2}{\partial r^2} - \frac{2}{r^2}\right)f(r) = 0. \quad [6.59]$$

As we described in Eq. [6.11]–[6.13], whenever we see equations like Eq. [6.59] in which the power of all the terms is the same, we should look for a power law solution, like:

$$f(r) = a\cdot r^b. \quad [6.60]$$

Fair enough: so let's try this solution in Eq. [6.59]. Doing so, we get:

$$\begin{aligned} 0 &= \left(\frac{\partial^2}{\partial r^2} - \frac{2}{r^2}\right)\left(\frac{\partial^2}{\partial r^2} - \frac{2}{r^2}\right)ar^b \\ &= \left(\frac{\partial^2}{\partial r^2} - \frac{2}{r^2}\right)[ab(b-1) - 2a]r^{b-2} \\ &= [(b-2)(b-2-1) - 2][b(b-1) - 2]ar^{b-4} \\ &= [(b-1)(b-4)][(b+1)(b-2)]ar^{b-4}. \end{aligned} \quad [6.61]$$

So either $a = 0$ (meaning there is no flow) or $b = 1, 4, -1$, or 2. This means our solution is:

$$f(r) = a_1 r^{-1} + a_2 r^1 + a_3 r^2 + a_4 r^4, \quad [6.62]$$

where constants a_1, a_2, a_3, and a_4 have to be determined from the boundary conditions (Eq. [6.55]–[6.57]).

Let's apply the boundary condition of Eq. [6.57] first. As $r \to \infty$, we know right away that $a_4 = 0$, else ψ would grow as r^4, whereas we know it grows only as r^2. Also we know that a_3 must be $-V_\infty/2$ to satisfy Eq. [6.57]. So at this point, we have:

$$\begin{aligned} \psi &= f(r) sin^2(\vartheta) \\ &= \left(a_1 r^{-1} + a_2 r^1 - \frac{V_\infty}{2} r^2\right)\sin^2(\vartheta). \end{aligned} \quad [6.63]$$

We can now apply boundary conditions at $r = R$, Eqs. [6.55] and [6.56]:

$$\begin{aligned} 0 &= V_r\big|_R \\ &= \frac{-1}{r^2\sin(\vartheta)}\frac{\partial\psi}{\partial\vartheta}\bigg|_R \\ &= \left(-a_1 R^{-3} - a_2 R^{-1} + \frac{V_\infty}{2}\right)2\cos(\vartheta) \end{aligned} \quad [6.64]$$

and:

$$\begin{aligned} 0 &= V_\vartheta\big|_R \\ &= \frac{1}{r\sin(\vartheta)}\frac{\partial\psi}{\partial r}\bigg|_R \\ &= (-a_1 R^{-3} + a_2 R^{-1} - V_\infty)\sin(\vartheta). \end{aligned} \quad [6.65]$$

These hold all around the sphere, that is, at all ϑ, so we have two equations in two unknowns, a_1 and a_2, which we can solve to obtain:

$$\begin{aligned} a_1 &= -\frac{V_\infty R^3}{4} \\ a_2 &= \frac{3V_\infty R}{4}. \end{aligned} \quad [6.66]$$

This gives at last:

$$f(r) = -V_\infty\left(\frac{R^3}{4r} - \frac{3Rr}{4} + \frac{r^2}{2}\right), \quad [6.67]$$

or:

$$\psi = -V_\infty\left(\frac{R^3}{4r} - \frac{3Rr}{4} + \frac{r^2}{2}\right)\sin^2(\vartheta). \quad [6.68]$$

As advertised, we can differentiate this using Eq. [6.47] to obtain the dessert that we were given earlier, the velocity field (Eq. [6.25]).

EXERCISE 6.9

Plot contours of the streamfunction (Eq. [6.68]) and compare with the Euler integrated solutions that you obtained in Exercise 6.5. As we have shown (Eq. [6.46]), the streamfunction provides exact trajectories for fluid motion. We saw in Exercise 5.2 of Chapter 5 that Euler integration suffers a specific deficiency: show this in your comparison plots.

Example Another parlor trick

It is simple enough to test our confidence with our solution. As shown in Fig. 6.13, blowing air past a drinking glass should produce the flow plotted in Fig. 6.8(a), diverging around the upwind side, and converging again around the downwind side. This means that we should be able to blow out a candle on the other side of the glass. In fact, the reader can confirm that it is simple to blow out a candle on the other side of *two* glasses, as shown in Fig. 6.13. This needn't be true: for large Reynolds number flows, there can actually be an upstream flow behind an obstacle, but for the Reynolds numbers that you can generate by blowing, the candle effect shown is easily confirmed.

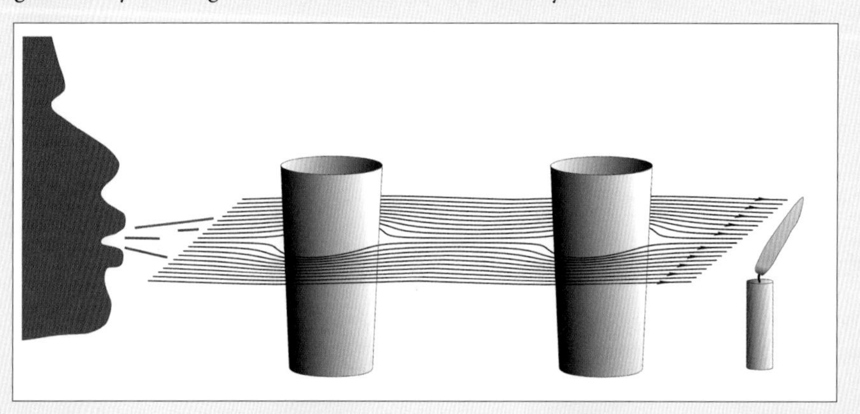

Fig. 6.13 Solution to flow obtained in this chapter implies that a candle behind two drinking glasses can be blow out: a fact that the reader can easily confirm.

Example Falling bodies

We close this chapter with a final set of examples on the theme, mentioned several times already, that fluid dynamics is richer in intricate and captivating behaviors than could reasonably be expected. We saw in an earlier example that suspensions of numerous particles produce phenomena such as the Boycott effect and colloidal stratification. Even very small numbers of particles exhibit unexpected and intriguing behaviors—many of which even today remain poorly understood.

As examples, in Fig. 6.14 we show several behaviors from the lab of Daniel Joseph (1929–2011), a mechanical engineer who started his academic career as a sociology major, and ended up having written a prodigious 400 research articles and seven books. A small part of his work involved studies of interactions between falling bodies as shown in the figure. At low Reynolds number, Fig. 6.14(a), pairs of spheres falling through water execute a dynamical dance that Joseph called "draft, kiss, and tumble." During drafting, the downstream sphere feels less drag than the upstream one, and hence the spheres approach one another until they kiss. The kissing configuration isn't stable, however, and the two spheres tumble and separate side-to-side as shown. The wake structures here are made visible by producing lines of hydrogen bubbles using a wire that is subjected to pulsed current.[19] The advected bubbles are then illuminated with a laser.

At higher Reynolds number, this dance ends, and the spheres adopt the stable configuration shown in Fig. 6.14(b). The wakes still interact, but the spheres draft off of one another in a steady manner resembling flying geese. At still higher Reynolds number, the nearly streamwise-aligned stable configuration shown in Fig. 6.14(b) becomes unstable, and instead a nearly spanwise aligned state shown in Fig. 6.14(c) appears.

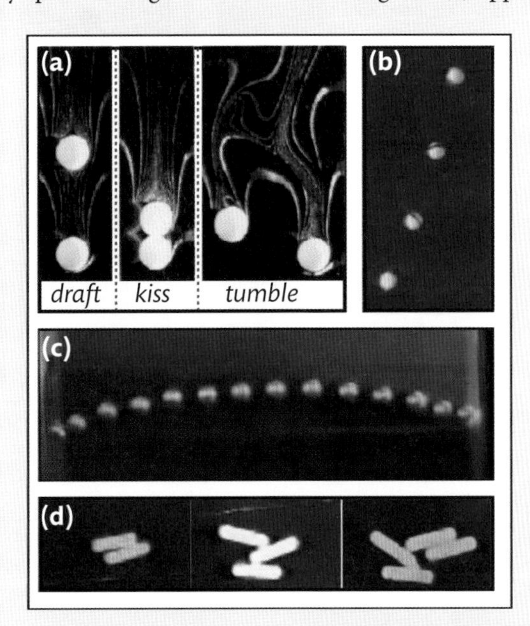

Fig. 6.14 Examples of curious phenomena that appear as particles fall through a Newtonian liquid. (a) Two spheres falling at Re = 8.5 "draft" with one downstream of the other, then "kiss," then "tumble" to fall side-by-side. Wake structures are identified using hydrogen bubbles illuminated by pulsed laser light. (b) A steady line of spheres forms spontaneously when spheres fall at $22 < Re < 43$. As in panel (a), each sphere "drafts" in the wake of the one upstream, but at this higher speed, the drafting is stable. (c) A stable arch of particles forms at $Re = 67.2$. (d) Particle shape affects structure formation as well: here elongated particles form stable two, three-, and four-particle clusters: these fall indefinitely essentially without change. (From D.D. Joseph, AEM, University of Minnesota; panel colors altered to match.)

Joseph also identified stable states of non-spherical particles, for example, Fig. 6.14(d) shows stable states of two-, three-, and four-particle falling clusters. Using Newtonian fluids, many other states can be found as well, including T-shaped structures and stable rafts of spanwise-aligned particles. Using non-Newtonian fluids, another taxonomy of falling configurations has been found. Some of these behaviors are well understood (e.g. the draft-kiss-tumble state and the stable drafting states shown in Figs. 6.14(a) and (b)); others remain to be fully analyzed.

A final example of surprising behaviors of falling particles can be seen in Fig. 6.15, where we show an experiment (panel (a)) and a simulation (panel (b)) of a heavy intruder falling into a silo of light particles.[20] In both experiment and simulation, the intruder falls with a terminal velocity as if in a fluid. Meaning it would penetrate infinitely far through a bed of solid particles! The simulation is color coded by stress, so that the darker beads are under more stress than the lighter ones. This reveals two important features of particle beds.

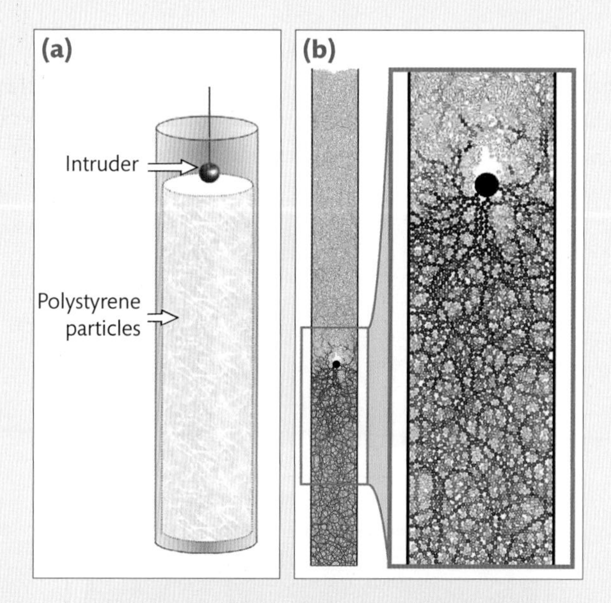

Fig. 6.15 (a) Experiment in which a heavy intruder (a ping pong ball filled with steel beads) is dropped into a 6 m high silo containing polystyrene beads. The intruder acquires a terminal velocity, and falls through the entire bed, as measured by tracking the trailing thread shown. (b) Two dimensional simulation of stresses on beads as the intruder falls. Darker beads are under greater stress than lighter beads. Note that a subset of the beads bears most of the load, and that the pressure doesn't grow with depth (Janssen's law described in text). (From: F. Pacheco-Vázquez et al (2011). Infinite penetration of a projectile into a granular medium. *Physical Review Letters 106*. By permission of the American Physical Society.)

First, the load is borne by so-called "stress chains:" a subset of highly stressed grains. The chains are unstable, so as the intruder falls, one chain will break and transfer its load to other chains: this same phenomenon leads to erratic slip events in granular beds ranging from abrupt avalanches when the other chains cannot bear the load to periodic and chaotic stick-slip motion of stressed beds.

Second, the stress—that is, the pressure—doesn't grow with depth into the silo. This is termed "Janssen's law,"[21] which obviously differs from fluids, in which pressure grows with depth, and occurs with or without an intruder. The cause of Janssen's law is known to be that the stress chains transfer downward stress to the walls, which support the bed above and relieve the underlying bed of the weight of overlying material.

REFERENCES

1. Wereley, S.T. and Lueptow, R.M. (1999) Inertial particle motion in a Taylor–Couette rotating filter. *Physics of Fluids, 11*, 325–34.
2. Andereck, C.D., Liu, S.S. and Swinney, H.L. (1986) Flow regimes in a circular Couette system with independently rotating cylinders. *Journal of Fluid Mechanics, 164*, 155–83.
3. Hoffman, C., Altmeyer, S., Pinter A. and Lücke, M. (2009) Transitions between Taylor vortices and spirals via wavy Taylor vortices and wavy spirals. *New Journal of Physics, 11*, 053002.
4. Taylor, G.I. (1923) Stability of a viscous liquid contained between two rotating cylinders. *Transactions of the Royal Society of London, 223*, 289–43.
5. Yeow, Y.L., Ko, W.C. and Tang, P.P.P. (2000) Solving the inverse problem of Couette viscometry by Tikhonov regularization. *Journal of Rheology, 44*, 1335–52.
6. Boycott, A.E. (1920) Sedimentation of blood corpuscles. *Nature, 104*, 532.
7. Richardson, J.F. and Zaki, W.N. (1954) Sedimentation and fluidisation: Part I. *Transactions of the Institution of Chemical Engineers, 32*, 35–53.
8. Mueth, D.M., Crocker, J.C., Esipov, S.E. and Grier, D.G. (1996) Origin of stratification in creaming emulsions. *Physical Review Letters, 77*, 578–81.
9. Muzzio, F.J., Unger, D.R., Liu, M., Bramble, J., Searles, J. and Fahnestock, P. (1999) Computational and experimental investigation of flow and particle settling in a roller bottle bioreactor. *Biotechnology and Bioengineering, 63*, 185–96.
10. Skloot, R. (2010) *The Immortal Life of Henrietta Lacks*. New York: Random House.
11. Gold, E.R. (1997) Body parts: property rights and the ownership of human biological materials. *Harvard Journal of Law and Technology, 10*, 369–75.
12. Allen, M.J., Powers, M.L.E., Gronowski, K.S. and Gronowski, A.M. (2010) Human tissue ownership and use in research: what laboratories and researchers should know. *Clinical Chemistry, 56*, 1675–82.
13. Anon. (2009) The ethics of egg manipulation. *Nature, 460*, 1057.
14. Ceck, E.A. (2014) Embed social awareness in science curricula. *Nature, 505*, 477–8.
15. Zaric, B., Perin, B., Becker, H.D., Herth, F.F., Eberhardt, R., Djuric, M., Djuric, D., Matijasevic, J., Kopitovic, I. and Stanic J. (2011) Autofluorescence imaging videobronchoscopy in the

detection of lung cancer: from research tool to everyday procedure. *Expert Review of Medical Devices, 8*, 167–72.

16. Einstein, A. (1955) *Essays in Science*. New York: Philosophical Library, pp. 85–91.

17. Student. (1908) The probably error of a mean. *Biometrika, 6*, 1–25.

18. Knight, J.B., Jaeger, H.M. and Nagel, S.R. (1993) Vibration-induced size separation in granular media: the convection connection. *Physical Review Letters, 70*, 3728–31.

19. Merzkirch, W. (1987) *Flow Visualization*. Orlando, FL: Academic Press.

20. Pacheco-Vázquez, F., Caballero-Robledo, G.A., Solano-Altamirano, J.M., Altshuler, E., Batista-Leyva, A.J. and Ruiz-Suárez, J.C. (2011) Infinite penetration of a projectile into a granular medium. *Physical Review Letters, 106*, 218001.

21. Janssen, H.A. (1895) Getreidedruck in Silozellen. *Z. Ver. Dt. Ing., 39*, 1045–9 (English translation in *Granular Matter, 8*, (2006) 59–65).

7 Intermezzo: Effects of Increasing Reynolds Number

7.1 Introduction

We have covered several topics in low Reynolds number flows, and we turn next to higher Reynolds numbers. Before we get involved in the mathematics of these new problems, let us describe some of the phenomenological issues involved in increasing the Reynolds number. A good place to start is a 1976 lecture[1] by Edward Purcell (1912–1997) entitled "Life at low Reynolds number." Purcell was an intellectual powerhouse, who shared the 1952 Nobel Prize for discovering nuclear magnetic resonance, now largely renamed magnetic resonance imaging to avoid popular misunderstanding associated with the term "nuclear." His discussion of low Reynolds number physics is reminiscent of Julia Child's recipe for scrambled eggs: a simple recipe as a gift from the expert to the layperson.

Purcell's discussion raised several important issues starting from a single initial observation, namely that Stokes equation is reversible. That is, Stokes equation reads:

$$\nabla^2 \vec{V} = \frac{\vec{\nabla} P}{\mu}, \tag{7.1}$$

and if we change the sign of the driving force, here the pressure gradient, this will do nothing but change the sign on the velocity, thus exactly reversing the motion of the fluid. In a famous demonstration of this effect, G.I. Taylor (mentioned Chapters 5 and 6) built a small Couette device, and showed (Fig. 7.1(a)) that after injecting a blob of dye, he can rotate one of the cylinders several turns clockwise to disperse the dye, but then when he rotates the cylinder the same number of turns counterclockwise, the dye is restored to its original shape and location. The reader can confirm that this must be so by scrutinizing Eq. [6.16] in Chapter 6: changing the signs of the wall motions Ω_{inner} and Ω_{outer} exactly changes the sign of the fluid velocity, and so any motion produced by a given sign of Ω_{inner} and Ω_{outer} will be exactly reversed by switching signs.

Purcell emphasized the fact that because of their tiny size, Reynolds numbers must be very low for microorganisms, and so their motions must be reversible. He terms this the "scallop theorem," stating that a scallop may move in one direction when it opens its hinge, but must reverse that movement when it closes the hinge, so a simple scallop cannot move forward. This is illustrated in Fig. 7.2(a).

Biomedical Fluid Dynamics: Flow and Form. Troy Shinbrot.

DOI: 10.1093/oso/9780198812586.001.0001

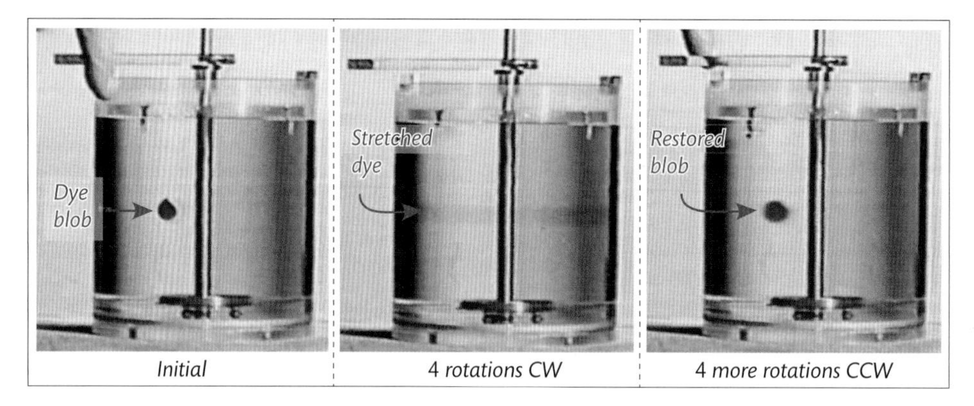

Fig. 7.1 Frames from G.I. Taylor's movie showing reversibility at low Reynolds number in a Couette cell. A blob of dye injected into glycerin (left panel) is stretched after four clockwise rotations of the inner cylinder (center panel), and is then restored to its original position after four counterclockwise rotations (right panel). (© Education Development Center.)

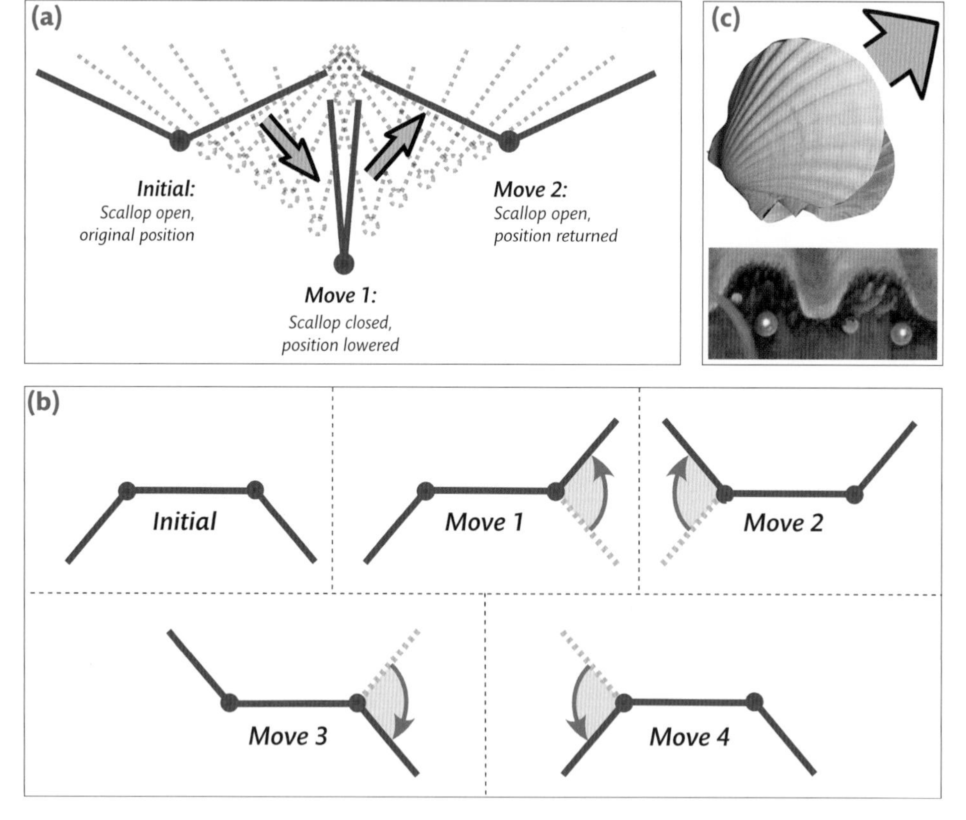

Fig. 7.2 Purcell's "scallop theorem." (a) In this illustration, an initially open scallop will move down when the scallop hinge closes (Move 1), but then must move up by precisely the same distance when the hinge reopens (Move 2). (b) Purcell's example of a device with *two* hinges that can make net motion by repeating the four movements shown. (c) Actual scallop, indicating direction of motion during swimming (upper panel), and scallop eyes (blue) at leading edge of shell (lower panel).

EXERCISE 7.1

Purcell remarks that despite the constraint imposed by the scallop theorem, microbes do swim, and he proposes a couple of mechanisms for this. First, he remarks that the scallop theorem applies in two dimensions (2D), but three-dimensional (3D) effects, produced for example by spiraling flagella in microorganisms such as *Escherichia coli*, can produce net motion. Second, he proposes that non-reciprocal actions produced in two dimensions by more than one hinge can produce motion on the microbial scale.

In Fig. 7.2(b) we show Purcell's hypothetical device with two hinges that execute a four move sequence. In this exercise:

(1) Determine what direction this device moves (up, down, left, or right) after a complete cycle of the four moves shown in Fig. 7.2(b).

(2) Identify several mechanisms that creatures use to swim at low Reynolds numbers. There are at least 10 distinct mechanisms, possibly more: find as many as you can.

(3) Real scallops do swim despite having only one hinge. Explain how they do this and why the scallop theorem doesn't apply to real scallops. Note that the direction that scallops move (Fig. 7.2(c), top) may not be the intuitive one: scallops even have rudimentary eyes at the edges of their shells pointed in the direction of their motion (Fig. 7.2(c), bottom).

Importantly, reversibility does *not* hold at higher Reynolds numbers. To see why, consider the nonlinear term in Navier–Stokes, which we recall dominates when $Re \ll 1$:

$$\vec{V} \cdot \nabla \vec{V} = \frac{\vec{\nabla} P}{\mu}. \quad [7.2]$$

In this case, changing the sign of the velocity has no effect on the sign of $\vec{\nabla} P$ (nor vice versa). This is a first indication that things must change qualitatively as the Reynolds number increases. A couple of parlor tricks, and one trick that is definitely not done in the parlor, will illustrate some qualitative effects of higher Reynolds number.

Examples More on flow around spheres

We know from the previous chapter how fluids flow around spheres at low Reynolds number ($Re \ll 1$). Let's consider flow around spheres at moderate Reynolds number ($Re \sim$ 10–100).

In Fig. 7.3, we show two simple examples involving the flow of air around a ping-pong ball. Figs. 7.3(a) and (b) show air being blown forcefully into a funnel containing a ping-pong ball. The reader can confirm that if this is done, the ball will not be ejected from the funnel. Moreover, if the funnel is turned to face downward, blowing into the funnel will produce a paradoxical force holding the ball up, *in the funnel*, against the force of gravity. This unexpected behavior is associated with the Bernoulli effect, and results from the high speed flow of air between the funnel wall and the ball.

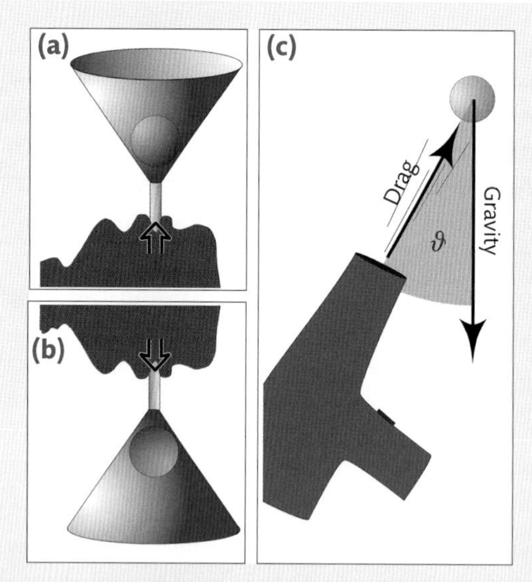

Fig. 7.3 (a) Blowing air into a funnel containing a ping-pong ball will keep the ball in the funnel. (b) This effect will even hold the ball in the funnel against the force of gravity. (c) Challenge question: what holds the ping-pong ball shown from falling when the hair drier isn't directly below the ball?

We will analyze the Bernoulli effect next, but what it says is that—all other things being equal—the pressure is lowest where the fluid speed is highest. Since the gap between the ball and the funnel wall starts out being small, air has to move rapidly to squeeze through that gap, and consequently the pressure drop near the gap is stronger than elsewhere—even stronger than the pressure drop driving the flow.

EXERCISE 7.2

Fig. 7.3(c) illustrates another ping-pong ball experiment, in which a ball is held aloft using the airflow from a hair drier. More dramatically, a beach ball can easily be held in place several meters above the ground using a leaf blower. It should not be surprising that objects as light as ping-pong or beach balls can be suspended by flow: we calculated drag forces last chapter, and in Chapters 1 and 6 we have discussed fluidized beds in which particles are suspended in upwelling flows. What may be surprising is that the ball can be held at a considerable angle to the vertical, as shown in Fig. 7.3(c). Gravity clearly tends to pull the ball down and out of the airflow. Explain what holds the ball in place.

EXERCISE 7.3

Before we leave the topics of Reynolds number and swimming, we mention that, historically, contemporaries Isaac Newton (1642–1727) and Christiaan Huygens (1629–1695) disagreed on the basic question of how an object's speed affects viscous drag.[2] This disagreement persisted for over 300 years, until 2004 when a team of chemical engineers examined the issue experimentally by comparing the speeds of 10 competitive swimmers in an ordinary swimming pool and in a pool filled with guar gum, which doubled the viscosity. Documentary evidence of the experiment is shown in Fig. 7.4; as an exercise, we ask the reader to reason which is faster, compare their answer with measurements given by Gettelfinger and Cussler,[3] and explain what accounts for the measured results. A second question that the reader may want to consider is that the Reynolds number during swimming is of order 100,000; furthermore, the swimmers shown partially leave the water during parts of their stroke. Would the result reported by Gettelfinger and Cussler change at low *Re*, or if the swimmers swam the breaststroke completely submerged? Again, answer this question based on what you know, and then compare your answer with measurements that you will be able to find for the algae *Chlamydomonas reinhardtii*.

Fig. 7.4 Swimmers in water (left) and guar gum solution (right).

7.2 Bernoulli Equation

The Bernoulli equation is named for Daniel Bernoulli (1700–1782), the most prolific of the family of famous Bernoulli mathematicians mentioned in Chapter 4. This equation can be obtained starting from Navier–Stokes equation at high Reynolds number:

$$\frac{\partial \vec{V}}{\partial t} + \vec{V} \cdot \vec{\nabla} \vec{V} = -\frac{1}{\rho} \vec{\nabla} P + \frac{\vec{F}_{ext}}{\rho}. \qquad [7.3]$$

If we assume the flow is steady and neglect external forces, we obtain:

$$\vec{V} \cdot \vec{\nabla} \vec{V} + \frac{1}{\rho} \vec{\nabla} P = 0. \qquad [7.4]$$

The leftmost term can be expanded using the (little remembered, but easily looked up) identity:

$$\vec{V}\cdot\vec{\nabla}\vec{V} = \frac{1}{2}\vec{\nabla}(\vec{V}\cdot\vec{V}) - \vec{V}\times(\vec{\nabla}\times\vec{V}). \quad [7.5]$$

Here we see again the curl, $\vec{\nabla}\times\vec{V}$, introduced briefly last chapter. The curl is a measure of circulation, or "vorticity:" a flow that circulates around a point will have curl perpendicular to the direction of circulation, and a flow without vorticity will have zero curl. So if we assume that the flow has no vorticity (termed an "irrotational" flow) we can get rid of the last term in Eq. [7.5], in which case Eq. [7.4] becomes:

$$\frac{1}{2}\vec{\nabla}(\vec{V}\cdot\vec{V}) + \frac{1}{\rho}\vec{\nabla}P = 0. \quad [7.6]$$

But look, this is the same as:

$$\vec{\nabla}\left(\frac{V^2}{2} + \frac{P}{\rho}\right) = 0, \quad [7.7]$$

where we abbreviate $V^2 = \vec{V}\cdot\vec{V}$. We can integrate Eq. [7.7], and by the fundamental theorem of calculus for line integrals (again easy to look up if you don't recall it), we obtain Bernoulli's equation:

$$\frac{V^2}{2} + \frac{P}{\rho} = constant. \quad [7.8]$$

7.3 **Vorticity Equation**

With a little thought, Eq. [7.8] can explain the effects shown in Fig. 7.3, as well as several other effects that we will discuss later. Explicitly, the equation says that the pressure goes as $-V^2$, so the pressure is lowest (most negative) where the velocity is highest. We reiterate that Bernoulli's equation requires that the flow must be irrotational (or more generally, that $\vec{V}\times(\vec{\nabla}\times\vec{V}) = 0$).

Many flows are not irrotational, however, so let's briefly address what happens in that case. We start again with Navier–Stokes equation, and to identify effects that are relevant to rotation, we take the curl of each term:

$$\vec{\nabla}\times\frac{\partial\vec{V}}{\partial t} + \vec{\nabla}\times(\vec{V}\cdot\vec{\nabla}\vec{V}) = -\frac{1}{\rho}\vec{\nabla}\times\vec{\nabla}P + \nu\,\vec{\nabla}\times\nabla^2\,\vec{V} + \vec{\nabla}\times\frac{\vec{F}_{ext}}{\rho}. \quad [7.9]$$

Let's simplify things a little by recalling that the curl of a gradient is identically zero so the green term vanishes, and if we neglect external forces, we can ignore the red term as well. We also make use of one more obscure identity that tells us the blue term is the same as:

$$\vec{\nabla}\times(\vec{V}\cdot\vec{\nabla}\vec{V}) = \vec{V}\cdot\vec{\nabla}(\vec{\nabla}\times\vec{V}) - (\vec{\nabla}\times\vec{V})\cdot\nabla\vec{V} + (\vec{\nabla}\cdot\vec{V})(\vec{\nabla}\times\vec{V}). \quad [7.10]$$

In an incompressible flow, the magenta term is zero, so Eq. [7.9] becomes:

$$\vec{\nabla} \times \frac{\partial \vec{V}}{\partial t} + \vec{V} \cdot \vec{\nabla} (\vec{\nabla} \times \vec{V}) - (\vec{\nabla} \times \vec{V}) \cdot \nabla \vec{V} = \nu \cdot \vec{\nabla} \times \nabla^2 \vec{V} . \quad [7.11]$$

We are always free to change the order of differentiation, so if we define the "vorticity" to be $\vec{\omega} = \vec{\nabla} \times \vec{V}$, Eq. [7.11] can be written as:

$$\frac{\partial \vec{\omega}}{\partial t} + \vec{V} \cdot \vec{\nabla} (\vec{\omega}) - \vec{\omega} \cdot \nabla \vec{V} = \nu \cdot \nabla^2 \vec{\omega} . \quad [7.12]$$

Almost there: this will be a little simpler if we write the left two terms as the substantive derivative:

$$\frac{D\vec{\omega}}{Dt} = \vec{\omega} \cdot \vec{\nabla} \vec{V} + \nu \cdot \nabla^2 \vec{\omega} . \quad [7.13]$$

This is the vorticity equation, which tells us how the vorticity, $\vec{\omega}$, of a parcel of fluid behaves as it is advected with velocity $\vec{V}$.

So now we have yet another equation: what does it tell us? We focus on two things that are revealing for future discussions: first, the relation between vorticity and viscosity, and second, something that we will call "vortex stretching."

7.3.1 **Vorticity and Viscosity**

In the limit that the viscosity vanishes (called an "inviscid" flow), Eq. [7.13] reduces to:

$$\frac{D\vec{\omega}}{Dt} = \vec{\omega} \cdot \vec{\nabla} \vec{V} . \quad [7.14]$$

Eq. [7.14] tells us that if $\vec{\omega} = 0$ for a packet of fluid, then the substantive derivative of $\vec{\omega}$ will be zero, so $\vec{\omega}$ in that packet will never be able to change. That is, if $\vec{\omega} = 0$ ever, then $\vec{\omega} = 0$ forever. This may seem a little strange, but consider: if a packet of fluid has zero vorticity, how would it acquire vorticity? Answer: by rotation applied by neighboring packets. But applying rotation involves transmitting shear, and without viscosity shear cannot be transmitted as we saw in Chapter 1, viscosity is the *mechanism* for transmitting shear.

This highlights the essential influence of viscosity on fluid motion at high Reynolds number, an effect famously summarized in *rhyme* by Lewis Frye Richardson (1881–1951):

"*Big whirls have little whirls that feed on their velocity,*
and little whirls have lesser whirls and so on to viscosity."

EXERCISE 7.4

Last chapter, we discussed ethics and ownership of biological materials. Richardson raised a different ethical issue: he was a mathematician who, among other things, was the first to propose applying mathematical methods to weather prediction, and who introduced a central notion leading to the study of fractals, namely that the length of a coastline (or other natural object) grows as the measuring stick gets shorter. Importantly, he was also a pacifist who refused to fight in World War I, and subsequently refused to work on any scientific topic that could be used for war. In a noteworthy example, Richardson worried that meteorology was being used to determine when and where to deploy poison gas—recall this was at a time when tens of thousands of people had been killed or horribly injured by chlorine, phosgene and mustard gas. Consequently, Richardson withdrew from contributing further to the study of weather prediction—a topic that he helped to found.

This may seem extreme to some readers, but deciding where to draw an ethical line in scientific research is a problem that arises often. This situation was described by Einstein in a famous quote:

> *"Perfection of means and confusion of goals seem, in my opinion, to characterize our age."*

Two opposing positions regarding where to draw an ethical line were taken by Hans Bethe (1906–2005) and Edward Teller (1908–2003). Bethe and Teller worked in competing groups to develop different designs for the first atomic bomb (Fig. 7.5). As many as a quarter million people, mostly civilians, were killed by the two bombs dropped on Japan in World War II.

Fig. 7.5 Faces of the role of ethics and morality in scientific decision-making. (a) Hans Bethe, an advocate for arms control, and (c) Edward Teller, a proponent of ever more destructive weaponry. Both men were involved in the design of nuclear weapons, but their views of the role of ethics in science differed profoundly. Between the two, panel (b) provides a snapshot of a nuclear test of the type opposed by Bethe and promoted by Teller.

After the war, the work and views of these two men could not have been more different. Hans Bethe's career chiefly involved academic study of nuclear theory and astrophysics, for which he received the Nobel Prize in 1967. He campaigned with Einstein against nuclear weapon development and proliferation, and he urged all scientists to take a Hippocratic oath:[4]

> "*. . . to cease and desist from work creating, developing, improving and manufacturing further nuclear weapons—and, for that matter, other weapons of potential mass destruction such as chemical and biological weapons.*"

Edward Teller, on the other hand, championed the development of the hydrogen bomb—a weapon thousands of times more powerful than the atomic bombs dropped on Hiroshima and Nagasaki. Teller's stated view on the specific topic of the development of this weapon was:

> "... *the pursuit of knowledge and the expansion of human capabilities are intrinsically worthwhile.*"

Teller rejected objections to the use of nuclear weapons on the grounds that these were "merely moral." Later in life, Teller proposed broad use of nuclear bombs, for example, to extract oil from the Alberta tar sands and to excavate land for projects such as the creation of artificial harbors. After successfully engineering hydrogen bombs, he led the development of the neutron bomb, and went on to advocate the use of satellite-borne nuclear weapons.

With this background, where would you draw your ethical line? Do you feel that this discussion has any place in a textbook: are scientists and engineers obliged to consider the ethics and morality of their work, or should scientists and engineers determine what *can* be done, leaving others to decide what *should* be done? Would you ever argue that scientists should do even work that they feel is wrong on the grounds that, "if they don't do it, someone else will?" Are there weapons, technologies or scientific topics that you would refuse to work on? Finally, would you sign Bethe's Hippocratic oath, or do you side with Teller in his view that:

> "*There is no case where ignorance should be preferred to knowledge* – especially *if the knowledge is terrible*"?

Give concise but reasoned arguments to support your views.

Returning to Richardson's rhyme, the explicit idea that it poses is that there is a cascade of vorticity from large to small scales that continues until the scale becomes small enough that viscosity can dissipate the motion into heat. This leads to vorticities at all length scales in turbulent flow, and is known as the "Kolmogorov cascade"—an important topic discussed in the turbulence literature. The attentive reader will notice that this is another aspect of mode coupling discussed in Chapter 4. Indeed, as we discussed in that chapter, it is the nonlinear term in Navier–Stokes that leads to mode coupling, and it is precisely the same term expressed in the vorticity equation that produces coupling between large and small scale vortices.

The presence of vortices across multiple length scales can be seen in almost any turbulent system—for example, we saw this near Jupiter's Great Red Spot in Chapter 4, and in Fig. 7.6(a) we show multiscale vortices in a computational shearing flow.

Underlying Richardson's description is an implicit recognition that it is viscosity that allows the transfer of vorticity from one scale to another. Without viscosity, vortices can't transfer energy: a situation that is actually seen in superfluids which have truly zero viscosity—again, a fascinating topic well discussed in other literature.

7.3.2 **Vortex Stretching**

As we have mentioned, the vorticity equation tells us that as the viscosity goes to zero, or equivalently as the scale becomes very large, a body of fluid that initially lacks vorticity will always lack vorticity. So how does vorticity develop? To answer this question, let's consider a

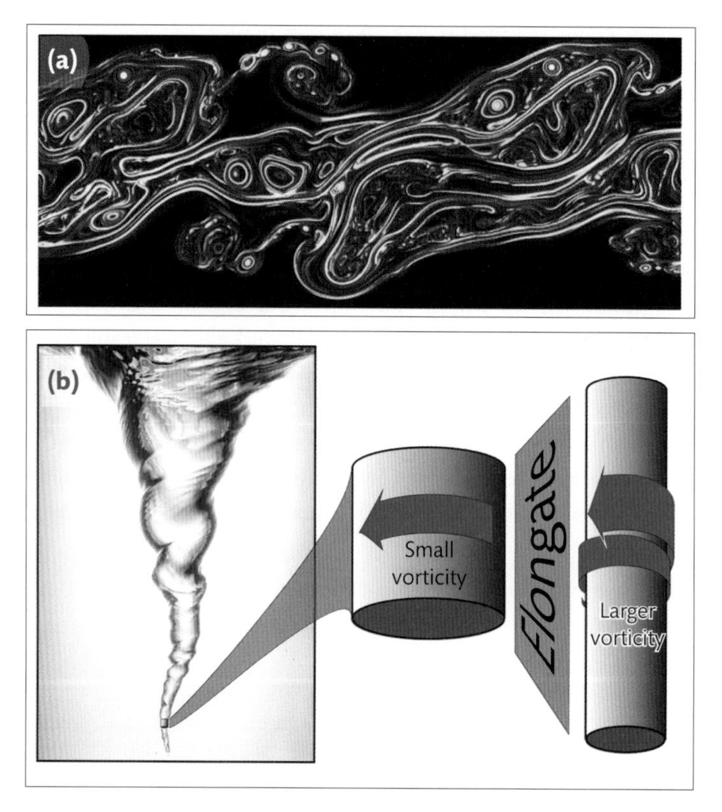

Fig. 7.6 Effects of vorticity. (a) Vortices across multiple length scales produced by a turbulent shearing flow. (From Karimabadi, H., Roytershteyn, V., Wan, M., Matthaeus, W.H., Daughton, W., Wu, P., Shay, M., Loring, B., Borovsky, J., Leonardis, E., Chapman, S.C. and Nakamura, T.K.M. (2013) Coherent structures, intermittent turbulence, and dissipation in high-temperature plasmas. *Physics of Plasmas, 20*(1). By permission of AIP Publishing LLC.) (b) Vortex stretching in a waterspout. Enlargement to the right shows a small parcel of rotating fluid: when the parcel is elongated vertically, its rotational speed necessarily increases according to Eq. [7.15] (Valentina Proskurina/Shutterstock.com).

small packet of fluid instantaneously spinning with about the z-axis, so that $\vec{\omega} = \omega_0\hat{z}$. Eq. [7.14] at that instant is:

$$\begin{aligned}\frac{D\vec{\omega}}{Dt} &= \vec{\omega}\cdot\vec{\nabla}\vec{V} \\ &= \omega_0\cdot\frac{\partial V_z}{\partial z}.\end{aligned} \qquad [7.15]$$

So if $\partial V_z/\partial z > 0$, the vorticity of the packet will grow. And what does $\partial V_z/\partial z$ represent? $\partial V_z/\partial z$ is a measure of the rate of change of z-velocity in the z-direction. That is, if the vertical velocity increases with height, then the vorticity will increase. Question: how can that occur? Answer: if a packet of fluid is stretched in the z-direction.

This means that if a packet of fluid is stretched along the axis about which it is rotating, then the packet's vorticity will increase. Just like a ballerina who draws her limbs closer to her spinning axis, a fluid that is stretched vertically—and so must by continuity be compressed horizontally—will spin faster. This mechanism is termed "vortex stretching," and is illustrated in Fig. 7.6(b).

An example of vortex stretching is spin-up in a funnel cloud preceding a tornado. In a funnel cloud, a spinning region of fluid is stretched vertically by a strong pressure difference between the air in the cloud and the air beneath. Eq. [7.15] says that as the region is stretched, its vorticity must increase. This is what leads to strong winds in a tornado: wind speeds of hundreds of kilometers per hour can be generated from overlying clouds with relative wind speeds an order of magnitude smaller. Another example is found in phalaropes, a type of sandpiper that swims rapidly in circles while feeding. Vortex stretching allows these small shore birds to produce a 50 cm long vortex that brings small crustaceans up from the shallow waters where it feeds.[5]

Examples of vortex stretching are found not only in fluids, but also in literature dealing with fusion plasma dynamics and dynamo physics—for example, every few hundred thousand years, the Earth's core is known to switch its direction so as to exchange the north and south magnetic poles. This switching is intimately related to the stretching of magnetic fluid in the earth's core, which causes vorticity to grow until a tipping point is reached, causing the direction of magnetic alignment to switch abruptly.

ADVANCED EXERCISE 7.5

Tornados travel over land, but typically die when they encounter large bodies of water: this occurs because the tornado's energy is expended churning and lifting water. Hurricanes, on the other hand, travel over water, but typically die when they encounter large bodies of land. What accounts for this difference?

7.4 **Matching Solutions: Rankine Vortex**

In Fig. 7.6(b) we show a snapshot of a waterspout: it turns out that what we know so far can be applied to solve for its shape. We begin by recalling that the curl is the measure of vorticity, and in Cartesian coordinates, the component of curl about the z-axis can be written as:

$$\omega_z = \frac{\partial V_y}{\partial x} - \frac{\partial V_x}{\partial y}, \qquad [7.16]$$

where as usual V_x and V_y are velocity components in the x- and y-directions. So the reader can confirm that uniform rotation can be defined by:

$$\begin{aligned} V_x &= -\omega_o y \cdot \hat{x} \\ V_y &= \omega_o x \cdot \hat{y}, \end{aligned} \qquad [7.17]$$

where ω_o is an amplitude defining how fast the rotation occurs. In the spirit of this chapter's exploration of inviscid flow, let's assume that the Reynolds number is high (which it must be to maintain a waterspout), so the equation we need to solve is Eq. [7.4] plus the external force of gravity:

$$\vec{V} \cdot \vec{\nabla} \vec{V} = -\frac{1}{\rho} \vec{\nabla} P + \frac{F_g}{(volume)}, \qquad [7.18]$$

where F_g is the gravitational force on a unit volume. In the presence of the uniform rotation defined in Eqs. [7.17], we can simply expand the dot product to get:

$$\left(-\omega_o y \cdot \frac{\partial}{\partial x} + \omega_o x \cdot \frac{\partial}{\partial y}\right)(-\omega_o y \cdot \hat{x} + \omega_o x \cdot \hat{y}) = -\frac{1}{\rho}\vec{\nabla}P + \vec{\nabla}f. \quad [7.19]$$

Here we have defined a new quantity, f, which is the creature whose gradient is $F_g/(volume)$. And what is f? Let's try $f = -g\rho z$. There is no x or y component to this trial function, so its gradient is easily calculated:

$$\begin{aligned} -\vec{\nabla}\, g \cdot \rho \cdot z &= 0 \cdot \hat{x} + 0 \cdot \hat{y} - g \cdot \rho \frac{\partial z}{\partial z}\hat{z} \\ &= -g\rho \cdot \hat{z}. \end{aligned} \quad [7.20]$$

And $-g\rho$ is the same as $-gm/(volume)$, where m is the mass of $(volume)$ of fluid. Which is to say that $-g\rho$ is $F_g/(volume)$, as desired.

So evidently the trial function $f = -g\rho z$ is the creature we seek. We can now carry out the derivatives in Eq. [7.19] to obtain:

$$-\omega_o^2(\vec{x} + \vec{y}) = -\frac{1}{\rho}\vec{\nabla}P - \vec{\nabla}g\rho z, \quad [7.21]$$

making use of the trivial relation that $\vec{x} = x \cdot \hat{x}$. Now we can write the left side of the equation as a gradient:

$$-\frac{\omega_o^2}{2}\vec{\nabla}(x^2 + y^2) = -\frac{1}{\rho}\vec{\nabla}P - \vec{\nabla}g\rho z, \quad [7.22]$$

as we did when we derived the Bernoulli equation (Eqs. [7.6]–[7.8]).

The astute reader should notice that something is wrong, for our flow is definitely *not* irrotational, however the left side of Eq. [7.22] holds nevertheless because the velocity $\vec{V} = \vec{x} + \vec{y}$ satisfies $\vec{V} \times (\vec{\nabla} \times \vec{V}) = 0$. It is worthwhile for the reader to confirm this as a quick exercise to review the definitions of curl and cross product.

Finally, we make use of the only physically meaningful part of this calculation: we observe that at the water surface, the pressure is constant. Pressure will certainly grow with depth beneath the surface, but at the surface itself, pressure is just atmospheric. This fact allows us to solve this so-called "free boundary" problem: notice that here we are providing the flow (Eqs. [7.17]) and solving for the shape of the boundary, rather than providing the boundary conditions and solving for the flow as we have done in previous problems.

Since the pressure is constant, after applying the fundamental theorem of calculus to Eq. [7.22], we are left with:

$$-\frac{\omega_o^2}{2}(x^2 + y^2) = \text{Const}_1 - g\rho z, \quad [7.23]$$

or:

$$z = \text{Const}_2 + \frac{\omega_o^2}{2g\rho}(x^2 + y^2). \quad [7.24]$$

Apparently, a uniformly rotating body of water has a free surface that is just parabolic. This is very simple, but unfortunately it is manifestly wrong. As x and y grow, the height of the surface must grow quadratically, and so must approach infinity—and rather rapidly at that. The arithmetic is correct, so where did this error creep in to our problem?

The answer is very straightforward: we have solved for the free surface of a *uniformly rotating* body of water. Eqs [7.17] produce an infinite velocity as x and y grow, and indeed an infinite velocity will produce a surface with infinite height. So how can we fix this problem? The problem lies with Eqs [7.17], so presumably the solution lies there also. Let's alter Eqs [7.17] as follows:

$$\begin{aligned} V_x &= -\omega(r)y \cdot \hat{x} \\ V_y &= \omega(r)x \cdot \hat{y}. \end{aligned} \qquad [7.25]$$

Here we have changed the angular speed, ω, into some function (to be determined), which presumably diminishes with radius to prevent the unfortunate divergence in velocity—and so in height.

We know something that will help us find $\omega(r)$: we know that although flow at the waterspout may be rotating rapidly, far away the flow ought to be irrotational: there is no reason for the ocean height near New Zealand to depend on a waterspout around Easter Island. Which is to say that for large r:

$$\begin{aligned} 0 &= \frac{\partial V_y}{\partial x} - \frac{\partial V_x}{\partial y} \\ &= \left[\omega(r) + x\frac{\partial \omega(r)}{\partial x}\right] - \left[-\omega(r) - y\frac{\partial \omega(r)}{\partial y}\right]. \end{aligned} \qquad [7.26]$$

We could now write $r = \sqrt{x^2 + y^2}$ and grind through the derivatives. Alternatively, we can observe that the vector $\vec{r} = \vec{x} + \vec{y}$, so that Eq. [7.26] is the same as:

$$0 = 2\omega(r) + \vec{r} \cdot \vec{\nabla}\omega(r). \qquad [7.27]$$

And now we perform a cheap trick: we note that the equation that we started with was written in Cartesian coordinates, but now no remnant of x or y remains: Eq. [7.27] depends only on r, and not on ϑ. So if we switch midstream to polar coordinates, nothing would change in Eq. [7.27] except that we could write it as the scalar equation:

$$0 = 2\omega(r) + r\frac{\partial \omega(r)}{\partial r}, \qquad [7.28]$$

whose solution is just:

$$\omega(r) = \frac{\omega_1}{r^2} \qquad [7.29]$$

for constant ω_1. Splendid, so now we know from Eq. [7.25] that at large r:

$$\begin{aligned} V_x &= -\frac{\omega_1 y}{r^2} \cdot \hat{x} \\ V_y &= \frac{\omega_1 x}{r^2} \cdot \hat{y}. \end{aligned} \qquad [7.30]$$

We can now do just as we did before: to find the free surface given this velocity field, we observe that the free surface occurs at constant pressure. Eq. [7.21] in this case becomes:

$$-\frac{{\omega_1}^2}{r^4}(\vec{x}+\vec{y}) = -\frac{1}{\rho}\vec{\nabla}P - \vec{\nabla}g\rho z, \quad [7.31]$$

We can then repeat the derivation following Eq. [7.21] to obtain:

$$\frac{V^2}{2} = \mathrm{Const}_3 - g\rho z, \quad [7.32]$$

where using Eq. [7.30]:

$$\begin{aligned} V^2 &= \frac{{\omega_1}^2}{r^4}(x^2+y^2) \\ &= \frac{{\omega_1}^2}{r^2}, \end{aligned} \quad [7.33]$$

leaving:

$$z = \mathrm{Const}_4 - \frac{{\omega_1}^2}{2g\rho(x^2+y^2)}. \quad [7.34]$$

Another quick exercise: there is a minus sign in the leftmost term of Eqs. [7.20] and [7.22]: why is there no minus sign in that term for Eqs. [7.32] and [7.34]?

So now we have two solutions: Eq. [7.24] from a flow that we assume rotates uniformly, and Eq. [7.34] that assumes the fluid doesn't rotate at all. As we mentioned, Eq. [7.24] can't be right, since it diverges at infinity, and it is equally evident that Eq. [7.34] can't be right, because *it* diverges at the origin, $x = y = 0$. What can we do?

The answer was provided by Scottish engineer William Rankine (1820–1872). Principally known for his work in thermodynamics, Rankine proposed that Eqs. [7.17] and [7.24] define the flow and free surface near the center of the waterspout where the water rotates rapidly, and that Eqs. [7.30] and [7.34] define these quantities further away, where the flow becomes nearly irrotational. Rankine achieved a global solution by matching an outer solution, Z_{outer}, defined by Eq. [7.34] with an inner solution, Z_{inner}, defined by Eq. [7.24] at a particular radius, a. The choice of a is somewhat arbitrary, but as Exercise 7.6 will show, the combined "Rankine vortex" solution looks largely the same for many choices of a.

EXERCISE 7.6

Match both the functional values *and* the derivatives, dz/dr, at $r = a$ (for your choice of a) and plot Z_{inner} and Z_{outer} as defined above. Your problem will have four parameters: $Const_2$, $Const_4$, ω_1, and ω_2 (assume $g = \rho = 1$ for simplicity). The requirement that the values of Z_{inner} and Z_{outer} must agree at $r = a$ gives you one constraint on the parameters: this will produce a continuous curve like Fig. 7.7(a), rather than a discontinuous one like Fig. 7.7(b). You will also need to match the derivatives, dz/dr: otherwise, you'd obtain something like Fig. 7.7(c), which is non-physical. You'll need two other constraints: what two things would you define in the physical problem?

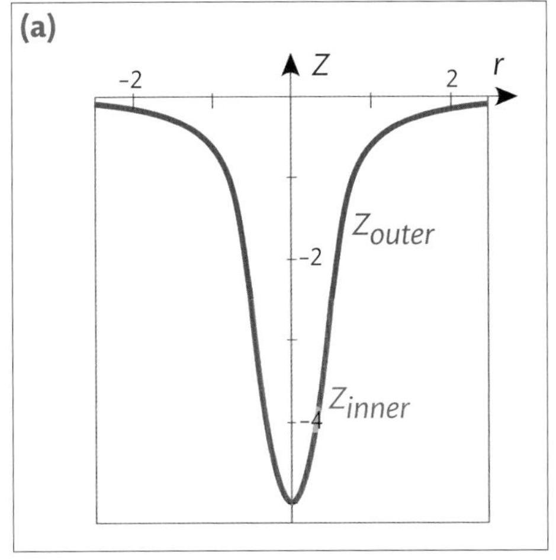

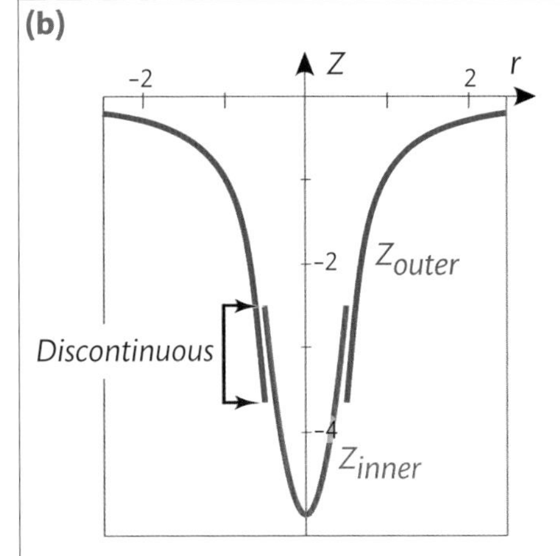

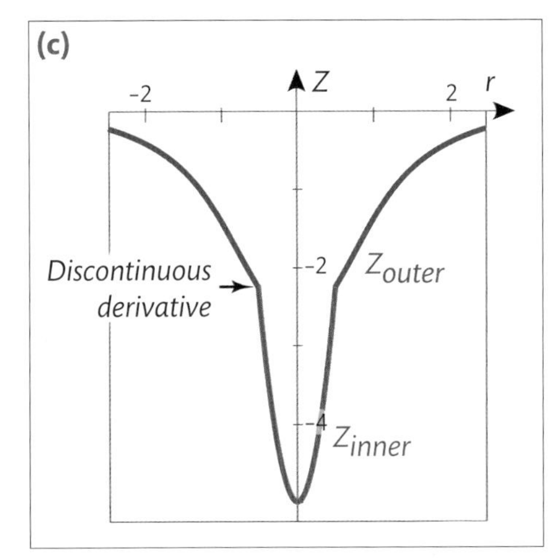

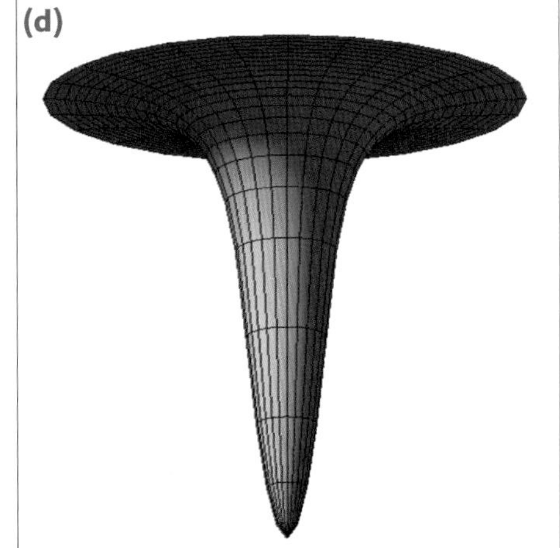

Fig. 7.7 Rankine solution to free surface of rotational flow. (a) Correct cross-sectional solution, in which inner solution (red: from Eq. [7.24]) and outer solution (blue: from Eq. [7.34]) match. (b) Incorrect solution, in which the values of Z_{inner} and Z_{outer} differ at $r = a$. (c) Incorrect solution, in which the derivatives of Z_{inner} and Z_{outer} differ at $r = a$. (d) 3D plot of correct solution (compare with waterspout in Fig. 7.6(b)).

When you have answered that question, solve for all remaining parameters and plot Z_{inner} and Z_{outer} versus r as shown in Fig. 7.7(a), demonstrating that the functions and their derivatives are continuous. Repeat this plot for a *different value* of a. Then plot the same thing in three dimensions, as shown in Fig. 7.7(d).

ADVANCED EXERCISE 7.7

Swirl a large clear soda bottle partially filled with water and photograph the waterspout that forms.

(1) Fit the photographed free surface with your Rankine solution.

(2) Note that the surface that you see will be distorted because the water in the bottle acts like a convex lens. Look up "index matching" and explain how you would correct for this distortion in a careful experiment. Sketch what an index-matched experimental geometry would look like.

7.5 **Return to Irrotational Flow**

We will have more to say about vorticity, but before we get to that, irrotational flows have much more to teach us. To start with, we reiterate that the curl of a gradient is identically zero. We mentioned this in the biharmonic equation section of Chapter 6, and as a quick exercise, the reader can take the curl of Eqs [7.36] (following) to confirm this identity. So any velocity written as:

$$\vec{V} = \vec{\nabla}\varphi, \qquad [7.35]$$

is guaranteed to be irrotational. Velocities that satisfy Eq. [7.35] define what we term "potential flows," where φ is the potential.

Incidentally, the reader may notice that Eq. [7.35] is similar to the definition of the electric field from the electric potential. Indeed, much of electromagnetism was developed by making analogies between the flows of fluids and of charges, and the pioneers in the development of fluid mechanics were also pioneers in the development of electromagnetic models. For example, Michael Faraday (1791–1867) studied both electrical induction and patterned surface waves in water, and James Clerk Maxwell (1831–1879) is known both for his work in the kinetic theory of gases and for Maxwell's equations of electromagnetism. Likewise we mentioned in Chapter 1 that Osborne Reynolds (1842–1912) developed a very thoughtful fluid-mechanical model for the "luminiferous ether," and even the earliest founders of electromagnetic theory thought of electricity as a kind of fluid, termed "effluvium" by William Gilbert (1544–1603).

Let's take a second to understand what Eq. [7.35] says. In two dimensions, $\vec{V} = \vec{\nabla}\varphi$ is the same as:

$$V_x = \frac{\partial \varphi}{\partial x}$$
$$V_z = \frac{\partial \varphi}{\partial z}, \qquad [7.36]$$

and as we mentioned, this implies that $\vec{\nabla} \times \vec{V} = 0$. In two dimensions, this is simply:

$$\frac{\partial V_z}{\partial x} - \frac{\partial V_x}{\partial z} = 0. \qquad [7.37]$$

By comparison, in Chapter 6, we introduced the stream function, ψ, where in x–z coordinates:

$$V_x = -\frac{\partial \psi}{\partial z}$$
$$V_z = \frac{\partial \psi}{\partial x} \qquad [7.38]$$

were obtained by assuming incompressibility, $\vec{\nabla} \cdot \vec{V} = 0$, which in two dimensions is:

$$\frac{\partial V_x}{\partial x} + \frac{\partial V_z}{\partial z} = 0. \qquad [7.39]$$

Note the minus signs in red in Eqs. [7.37] and [7.38]: these signs are what make the streamfunction, ψ, and the potential function, φ, do work for us to solve for velocities. In the irrotational formulation, Eqs. [7.36] and [7.37], there is a minus sign *in the curl* that makes any choice of φ produce an irrotational flow, and in the incompressible formulation, Eqs. [7.38] and [7.39], there is a minus sign in the *definition* of ψ that makes any choice of ψ produce an incompressible flow.

This observation suggests that probably there is a relation between ψ and φ: they are both functions that define the velocity field, the only difference is that one is produced from a curl-free assumption [7.37] with a minus sign in the curl of φ, and the other is produced from a divergence-free assumption [39] with a minus sign in the definition of ψ. Indeed, there is a straightforward relation between ψ and φ. To obtain it, we need only set like terms equal in Eqs. [7.36] and [7.38]:

$$\begin{aligned} \frac{\partial \varphi}{\partial x} &= -\frac{\partial \psi}{\partial z} \\ \frac{\partial \varphi}{\partial z} &= \frac{\partial \psi}{\partial x}. \end{aligned} \qquad [7.40]$$

These are known as the Cauchy–Riemann equations, named after French and German mathematicians Augustin-Louis Cauchy (1789–1857) and Bernhard Riemann (1826–1866). Each of these men was enormously influential in his own right, and together they provided the equations above, which we will see in the next chapter are really unreasonably powerful, and provide a simplified and broadly applicable way of describing fluid flow—as well as elasticity, electricity and magnetism, quantum mechanics, and other fields that are discussed elsewhere.

For the time being, let's explore what potential flows (Eqs. [7.36]) are all about. We begin by remarking that although we have chosen to assume that the flows of interest will be irrotational ($\vec{\nabla} \times \vec{V} = 0$), there is nothing to prevent them from also being incompressible ($\vec{\nabla} \cdot \vec{V} = 0$). Incompressibility of a potential flow can be written as:

$$\begin{aligned} \vec{\nabla} \cdot \vec{V} &= \vec{\nabla} \cdot \vec{\nabla} \varphi \\ &= \nabla^2 \varphi = 0. \end{aligned} \qquad [7.41]$$

So an irrotational and incompressible flow is defined by a potential function that obeys Laplace's equation, $\nabla^2 \varphi = 0$.

This is really a remarkable and unexpected result, because we started, after all, with a *non*linear *partial* differential equation, $\vec{V} \cdot \vec{\nabla} \vec{V} = \vec{\nabla} P / \mu$, which we noted suffers from mode coupling, multi-scale energy cascades, richly variable dynamics (recall the Couette phase diagram in Chapter 6) and so on. Yet we ended up with one of the most studied and simplest *linear ordinary* differential equations known. More than this: in the physically simpler *linear*, low Reynolds number problem, we had to solve the biharmonic equation, $\nabla^4 \psi = 0$, while in the physically more complex *nonlinear*, high Reynolds number problem, we need solve only the lower order Laplace's equation, $\nabla^2 \varphi = 0$! This is all provided that the flow is irrotational:

evidently, the complexities of the nonlinear term in Navier–Stokes equation are intrinsically tied to rotational motion—without rotation, the troublesome nonlinear term reduces to a solvable problem.

Let's look at some properties of potential flows.

7.6 **Rotation and the Oscillation Game**

A first problem that gives insight into the structure of solutions to Laplace's equation involves swirling flows like the Rankine vortex. Swirling flows are also encountered in arteries,[6] as well as in transport by microorganisms.[7,8] Additionally, swirling flows are used to refrigerate small devices using Ranque–Hilsch vortex tubes,[9] to concentration particulates in cyclone separators,[10] and in spin coating[11] and electrospinning applications[12] used, respectively, to produce thin films and tissue substrates. Moreover, in both biological[13] and non-biological[14] systems, swirling improves stability against turbulence in ways that aren't completely understood.

To shed some light on effects of swirling, let's consider a common example: flow down a drain. In that problem, water is never observed to travel radially straight toward a plughole, it invariably swirls. In part, this is simply due to vortex stretching, which amplifies any pre-existing circulation. Vortex stretching also amplifies Coriolis accelerations close to the plughole, though contrary to popular wisdom, careful experiments show that under normal circumstances the Coriolis contribution to motion in this problem is insignificant, and flow down a drain is clockwise as often as it is counterclockwise.

A more detailed look at the problem shows that swirling and spatial oscillations are closely related in a fundamental way. To understand why, we'll first ask what solutions to Laplace's equation look like in Cartesian coordinates where things are simplest; then we'll show that the same essential issue arises in cylindrical coordinates as would describe flow down a plughole. In Cartesian coordinates, Laplace's equation is:

$$\frac{\partial^2\varphi}{\partial x^2}+\frac{\partial^2\varphi}{\partial y^2}+\frac{\partial^2\varphi}{\partial z^2}=0. \qquad [7.42]$$

Let's separate variables by writing $\varphi = X(x)Y(y)Z(z)$, so that Eq. [7.42] becomes:

$$\frac{1}{X}\frac{\partial^2 X}{\partial x^2}+\frac{1}{Y}\frac{\partial^2 Y}{\partial y^2}+\frac{1}{Z}\frac{\partial^2 Z}{\partial z^2}=0. \qquad [7.43]$$

This can be satisfied by making each term constant, as follows:

$$\begin{aligned}
&\frac{1}{X}\frac{\partial^2 X}{\partial x^2}=k^2 \Rightarrow X = X_0e^{\pm kx}\\
&\frac{1}{Y}\frac{\partial^2 Y}{\partial y^2}=\ell^2 \Rightarrow Y = Y_0e^{\pm \ell y}\\
&\frac{1}{Z}\frac{\partial^2 Z}{\partial z^2}=m^2 \Rightarrow Z = Z_0e^{\pm mz}.
\end{aligned} \qquad [7.44]$$

Crucially, k^2, ℓ^2 *and* m^2 cannot all be positive, because they must add up to zero. At least one of these must be negative, meaning that *at least one of* k, ℓ *or* m *must be imaginary*. We have seen several times that an imaginary exponent signals an oscillatory flow. For example the oscillatory term could be in the z-direction, so that $m = i\mu$ would be imaginary, and:

$$Z = Z_0 e^{\pm i\mu x}. \qquad [7.45]$$

What this tells us is that structurally Laplace's equation requires both spatially "monotonic" (steadily increasing or decreasing, but not changing direction) and spatially oscillatory components. Moreover, we recall that Laplace's equation was obtained by assuming that a flow of interest is incompressible, so that $\vec{\nabla}\cdot\vec{V} = 0$. Near a drain, however, fluid leaves the system—that is, the total volume of fluid decreases. To account for this, we can either write that $\vec{\nabla}\cdot\vec{V} = 0$ and prescribe that fluid leaves at some boundary location, or we can define a loss of mass at the drain, in which case we would write that $\vec{\nabla}\cdot\vec{V} < 0$ there. In that case, near the drain the solution could only be satisfied by increasing the frequency, μ, of the oscillatory component in Eq. [7.45], making the solution oscillate faster closer to the drain.

In conclusion, simply to make Laplace's equation (Eq. [7.43]) add up to zero, there must be an essential asymmetry between at least one direction in which the solution is monotonic, and at least one other direction in which the solution is oscillatory. This tells us three things. (1) Potential flows tend to exhibit spatial oscillations. (2) Flow near a drain tends to circulate due to vortex stretching. And (3) oscillations tend to increase near a mass sink. All of these results are qualitatively confirmed by examination of Fig. 7.6(b) and similar flows.

EXERCISE 7.8

We caution that this has been a fast and dirty derivation intended to illustrate an essential feature of Laplace's equation, and can't be taken too literally to imply that all potential flows have to be oscillatory. To see why, notice that we haven't specified boundary conditions, and indeed we will shortly solve several problems with flows that are certainly not oscillatory. Explain how Eqs. [7.44] and [7.45] can be used to produce a non-oscillatory flow.

Example Parlor tricks showing obligatory oscillations

Notwithstanding the caution above, many problems do harbor an essential asymmetry that causes one direction to behave differently from others. Two simple parlor tricks that illustrate this phenomenon are shown in Figs. 7.8(a) and (b). These tricks are instructive in that they provide a geometric explanation for obligatory oscillations to complement the arithmetic one given above.

The first trick consists of simply tossing a book in the air while spinning it (it is obviously advisable to hold the book closed with a rubber band). The reader will find that spinning the book about either of the two axes indicated by green arrows in Fig. 7.8(a) will produce stable motion.

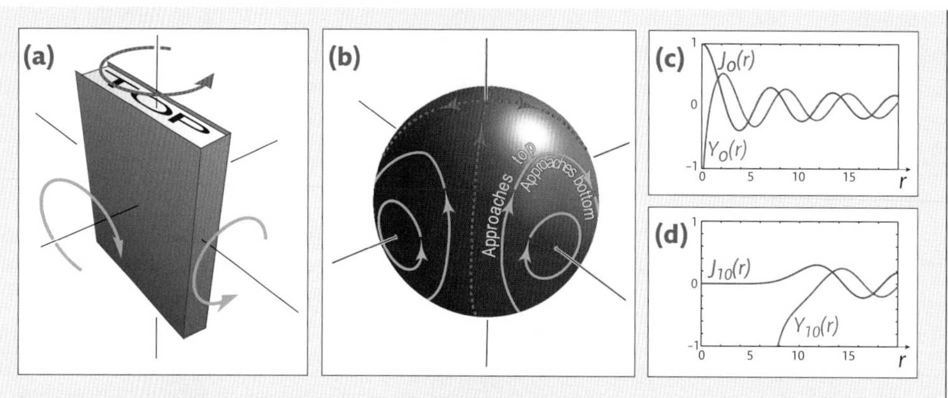

Fig. 7.8 (a) Tossing a book into the air while spinning it about two of its principal axes (green arrows) will produce stable motion, but spinning it about the third axis (pink arrow) will produce oscillations. (b) The same holds for a sphere: two axes will produce stable motion—and steady precession (green) if spun about a nearby non-principal axis, *and so* spinning near the third principal axis will produce flips from top to bottom (pink). (c) Oscillations in Bessel functions for $\nu = 0$, and (d) the same for $\nu = 10$, as described in the text.

We saw in the double pendulum example of Chapter 3 that stable motion is boring, and that maxim holds here. Something much more interesting occurs if the reader spins the book about the third axis indicated by the pink arrow. The reader can confirm that the face labeled "TOP" will periodically move to point downward and then return. We'll call this motion that oscillates between top and bottom "anti-stable." This curious effect is caused arithmetically by the need for a minus sign in Euler's equations for rigid body motion—essentially the same thing that we just discussed for Laplace's equation. For more background the reader can look up "polhode" or "Poinsot's ellipsoid" or the "Intermediate Axis Theorem."

Arithmetic is dealt with by the left brain, so let's put arithmetic aside for a moment to give the right brain a chance to get in on the action. A property of rigid body motion is that *every* such body has three unique "principal axes" that define its motion. The axes are mutually perpendicular, two of the axes are stable and the third is anti-stable. This is a mathematical theorem, which holds for books, chairs, satellites—even spheres.

The reader could well wonder how a sphere can have unique axes, and how the sphere decides which will be anti-stable. We leave that question for the reader to puzzle over, but we emphasize that this is a theorem, and is not open to doubt. While the reader is puzzling, let's look at an experiment that may explain why one axis must be anti-stable.

Consider a sphere rotating smoothly in space. An experiment can be performed by supporting a billiard ball in an air cup—an apparatus that the reader may want to construct. If the sphere is rotated about an axis near one of the stable principal axes, it will precess steadily about the nearby axis. This can be confirmed by touching a felt pen to the apparent axis of rotation: the pen mark will trace out an orbit around either of the two of the principal axes, as shown in green in Fig. 7.8(b).

But look: this means that any rotation axis near a stable axis will first approach one of the anti-stable axes, and will then recede from it and approach the other anti-stable axis, illustrated using typical trajectories in Fig. 7.8(b). The further the rotation axis is from one of the stable axes, the closer it will approach the dotted line shown in pink in Fig. 7.8(b): a line that connects one anti-stable pole at the top of the sphere to another anti-stable pole at its bottom. There is no way around this: you cannot draw successively larger circles around two perpendicular axes on a sphere without them coming together in intersecting lines as shown in pink in the figure. And these intersecting lines represent trajectories that travel from one pole to its opposite, flipping the sphere from pole to pole in an anti-stable motion.

So trajectories near a stable axis will remain near the axis, while trajectories near an anti-stable axis will oscillate between one pole to the other. This is due to simple geometry, and is a visual, right-brain, representation of the arithmetic, left-brain property commented on earlier. Right-brain learners may want to consider other visual examples of this asymmetry, for example, during twisting dives from the high board, discussed by Frohlich.[15]

Rigid body motion provides a first example of a geometric constraint on flow: we will mention another in a later chapter. Here the "flow" is the trajectory of a spinning axis; in other examples, the flow will be the trajectory of physical packets of fluid. The work of Hassan Aref (mentioned in Chapter 5) and others have conclusively shown that there is a close parallel between trajectories of dynamical systems such as a rotating sphere and of fluid systems.

Holding onto that thought, let's return to Laplace's equation governing potential flow of a fluid. We looked at the generalized problem in Cartesian coordinates following Eq. [7.42]; now let's make the connection closer to flow down a drain by considering the equation in cylindrical coordinates centered at the drain. In that case, Laplace's equation acting on a potential defined to be $\varphi = R(r) \cdot \Phi(\phi) \cdot Z(z)$ is:

$$\frac{1}{R}\left[R''(r) + \frac{1}{r}R'(r)\right] + \frac{1}{r^2}\frac{\Phi''(\phi)}{\Phi(\phi)} + \frac{Z''(z)}{Z(z)} = 0. \qquad [7.46]$$

Now let's play a game. The goal of the game is to force every part of this equation to be non-oscillatory. If we win, we will be able to define a flow that monotonically travels toward a drain—in which case we'll have to explain the universal observation that flow never does this. If we lose, we will have to accept there is an intrinsic need for solutions to Laplace's equation to oscillate in cylindrical, as well as Cartesian, coordinates. We'll see in cylindrical coordinates that this need additionally implies a connection between rotation and oscillation, especially so that near a drain, flow must rotate, or oscillate, or both, but cannot monotonically approach the drain.

Starting with the red term in Eq. [7.46], $\Phi(\phi)$, must be periodic in φ, simply because there is no difference between $\phi = 0$ and $\phi = 2\pi$. So if we set the red term to be a constant, that constant cannot be positive—otherwise $\Phi(\phi)$ will be an exponential, whose values at $\phi = 0$ and $\phi = 2\pi$ won't agree. So let's set the red term equal to $-\nu^2$ where ν is real, so that:

$$\Phi''(\phi) = -\nu^2\Phi(\phi)\,. \qquad [7.47]$$

This produces solutions for $\Phi(\phi)$ that oscillate unless $\nu = 0$.

OK, so the only way to prevent oscillations of the red term is to choose $\nu = 0$. Let's turn next to the blue term. We can set it constant as well: this constant can be real and positive, so call it k^2:

$$Z''(z) = k^2 Z(z), \quad [7.48]$$

which has exponential solutions. Good: so neither red nor blue terms need to oscillate. Now to complete the game, we have to keep $R(r)$ from oscillating. Eq. [7.46] can now be written:

$$R''(r) + \frac{1}{r}R'(r) + \left[k^2 - \frac{\nu^2}{r^2}\right]R(r) = 0. \quad [7.49]$$

This is a version of our old friend the Bessel equation, first introduced in Chapter 4. This is to be expected, for Bessel functions naturally arise in cylindrical geometries. Eq. [7.49] is second order, and so has two solutions, Bessel functions $Y_\nu(r)$ and $J_\nu(r)$. In our case, $\nu = 0$, but both of these functions oscillate for any value of ν; as a reminder, in Fig. 7.8(c) we plot $Y_0(r)$ and $J_0(r)$.

Apparently then, a solution to Laplace's equation that is not allowed to oscillate in the ϕ or z directions has no choice but to oscillate in the r direction. We can in principle continue to play the game of suppressing oscillations by pushing them away from the origin: this can be done by using larger ν. For example for $\nu = 10$, $Y_{10}(r)$ and $J_{10}(r)$ are plotted in Fig. 7.8(d), both of which only oscillate far from the origin. But recall that we chose $\nu = 0$ to prevent oscillations in the ϕ direction: if ν isn't zero, there will be both flow and oscillations in the ϕ direction.

This example shows that no matter how one plays the game, oscillations are bound to appear somewhere: if they are suppressed in one direction, they will simply pop up in another. In concrete terms, either there is no motion in ϕ, in which case there are oscillations in r, or there are no oscillations in r, in which case there is motion in ϕ. Put another way, monotonic radial flows tend to swirl, and swirl-free flows tend to have radial fluctuations.

7.7 Potential Flow around a Cylinder

In Chapter 6, we solved for incompressible Stokes flow around a sphere. We now compare that solution with incompressible and irrotational flow around a sphere. We show the idea by starting with flow around a cylinder, which may be easier to visualize, since it is essentially 2D. We already have the differential equation (Eq. [7.41]), so all we need are boundary conditions. As was the case at low Reynolds number, the first of these is no penetration:

$$\hat{\perp} \cdot \nabla\varphi = \frac{\partial\varphi}{\partial\perp} = 0, \quad [7.50]$$

where $\perp$ is the vector perpendicular to the boundary surface (the radius of the cylinder).

In problems that also obey a no slip condition, we would have:

$$\hat{\|} \cdot \nabla\varphi = \frac{\partial\varphi}{\partial\|} = 0, \quad [7.51]$$

where $||$ is the vector parallel to the surface. In this case derivatives of φ in both perpendicular and parallel directions are zero, so φ can only be a constant:

$$\varphi = \text{constant at a surface.} \qquad [7.52]$$

In our problem, there is no viscosity and so the no slip condition won't hold, however in some problems Eq. [7.52] is used as an approximation, and it is worth knowing about.

Returning to the problem of inviscid flow past a cylinder, we have a second boundary condition at ∞ as before, namely that as the radial displacement $\rho \to \infty$, flow becomes constant in the flow direction (which we take to be $\hat{x}$):

$$\vec{V} = \vec{\nabla}\varphi \to V_\infty \hat{x}. \qquad [7.53]$$

It is easily confirmed (e.g. by differentiating with respect to x and y) that condition [7.53] is met by:

$$\varphi_1 = V_\infty x, \qquad [7.54]$$

or in cylindrical coordinates:

$$\varphi_1 = V_\infty \rho \cdot \cos(\phi). \qquad [7.55]$$

Here with apologies for the continuing proliferation of symbols, we write the azimuthal coordinate as ϕ to distinguish it from the flow potential, φ.

Now we perform a trick of sorts. Laplace's equation (Eq. [7.41]) is linear, so if φ_1 and φ_2 are each solutions to the equation, then $\varphi_{total} = \varphi_1 + \varphi_2$ is also a solution. We have a function, Eq. [7.55], that meets the boundary condition at ∞, and it is easily confirmed that this function also satisfies Laplace's equation. So if we can find a second solution, φ_2, that goes to zero at ∞ and makes φ_{total} meet the condition at the surface of the cylinder, then we can just add the two solutions, and we're done. This is a bit like the Rankine vortex solution, except that we will add the two solutions together rather than choosing constants to make them match as we did in the Rankine problem.

Confused? No worries, let's try the following:

$$\varphi_2 = V_\infty \frac{R^2}{\rho} \cdot \cos(\phi), \qquad [7.56]$$

where R is the radius of the cylinder. It is easily confirmed that this solves Laplace's equation, and it is simple to see that this goes to zero as $\rho \to \infty$. Boundary condition [7.50] says at the surface of the cylinder:

$$\frac{\partial\varphi}{\partial\perp} = 0, \qquad [7.57]$$

where on a cylinder, the perpendicular direction is ρ. So if we try solution [7.55], we get:

$$\frac{\partial\varphi_1}{\partial\perp} = V_\infty \cos(\vartheta) \; at \; \rho = R, \qquad [7.58]$$

and if we try solution [7.56], we get:

$$\begin{aligned}\frac{\partial \varphi_2}{\partial \perp} &= -V_\infty \frac{R^2}{\rho^2}\cos(\phi) \\ &= -V_\infty \cos(\phi)\ at\ \rho = R.\end{aligned} \qquad [7.59]$$

So $\varphi_{total} = \varphi_1 + \varphi_2$ satisfies the boundary condition at the surface of the cylinder as well as at ∞.

To summarize,

$$\varphi_{total} = \varphi_1 + \varphi_2 = V_\infty\left(\rho + \frac{R^2}{\rho}\right)\cos(\phi) \qquad [7.60]$$

is a solution that the reader can confirm satisfies Laplace's equation and meets both boundary conditions. We can finally obtain the velocities using formulas for the gradient:

$$\begin{aligned}V_\rho &= \frac{\partial \varphi}{\partial \rho} \\ V_\varphi &= \frac{1}{\rho}\frac{\partial \varphi}{\partial \varphi}\end{aligned} \quad \text{in cylindrical coordinates,} \qquad [7.61]$$

so that:

$$\begin{aligned}V_\rho &= V_\infty\left(1 - \frac{R^2}{\rho^2}\right)\cos(\phi) \\ V_\varphi &= -V_\infty\left(1 + \frac{R^2}{\rho^2}\right)\sin(\phi).\end{aligned} \qquad [7.62]$$

7.8 Potential Flow around a Sphere

In spherical coordinates we use the same idea, but we take flow to be in the $\hat{z}$ direction (instead of $\hat{x}$ as in the cylindrical case), and we measure radial coordinate, r, and angle, ϑ, from the origin rather than from the centerline of a cylinder. Eqs. [7.55] and [7.56] then become:

$$\begin{aligned}\varphi_1 &= V_\infty r \cdot \cos(\vartheta) \\ \varphi_2 &= V_\infty \frac{R^3}{r^2}\cos(\vartheta).\end{aligned} \qquad [7.63]$$

Again we can confirm that these potential functions satisfy Laplace's equation and that their sum satisfies the boundary conditions. We use spherical gradient formulas to obtain the velocities:

$$\begin{aligned}V_r &= \frac{\partial \varphi}{\partial r} \\ V_\vartheta &= \frac{1}{r}\frac{\partial \varphi}{\partial \vartheta}\end{aligned} \quad \text{in spherical coordinates,} \qquad [7.64]$$

so that:

$$V_r = V_\infty\left(1 - \frac{2R^3}{r^3}\right)\cos(\vartheta)$$
$$V_\vartheta = -V_\infty\left(1 + \frac{R^3}{r^3}\right)\sin(\vartheta). \qquad [7.65]$$

So we find that high Reynolds number (*Re*) flow around either a cylinder or a sphere has simple solutions. We will return to this observation in the next chapter; for now, we note that the same is not true at low *Re*. There we can obtain a simple solution for flow past a sphere (Chapter 6, Eq. [6.66]), but surprisingly not for flow past a cylinder. The reader may want to puzzle through the root cause of this famous result, known as the "Stokes paradox," for it provides a clue about the limits of analytic solutions.

EXERCISE 7.9

In the next chapter, we will compare the potential function with the streamfunction in some detail. As a prelude to that comparison, we remark that a streamfunction that produces the same velocities as are defined in Eqs. [7.62] is:

$$\psi = V_\infty\left(\rho - \frac{R^2}{\rho}\right)\sin(\phi). \qquad [7.66]$$

For this exercise, first differentiate Eq. [7.60] (using Eqs. [7.62]) to produce Eqs. [7.63]. Then confirm that differentiating Eq. [7.66] gives the same velocity field. For this second calculation, use cylindrical coordinates (Eqs. [6.52] in Chapter 6 with ϕ replaced with ϑ). Spherical coordinates require a different and more complicated streamfunction. The outcome will be the same, but the algebra is more unpleasant.

Next, overlay plots of contours of both ψ and φ, and report what you observe: how are the contours related? Can you explain this observation either using the definitions of the stream and potential functions or using the observation that Eq. [7.35] resembles an equation from electrostatics? That is, if φ were the electric potential instead of the fluid flow potential, what could ψ represent, and what do you know about φ and ψ in electrostatics?

7.9 Flow over a Wavy Streambed: Some Useful Lessons

We will revisit the observations that you make in Exercise 7.9; for now, let's examine another potential flow: flow over a wavy streambed. This problem was first studied by Lord Kelvin (whom we mentioned in Chapter 1) in 1866, and it provides several instructive

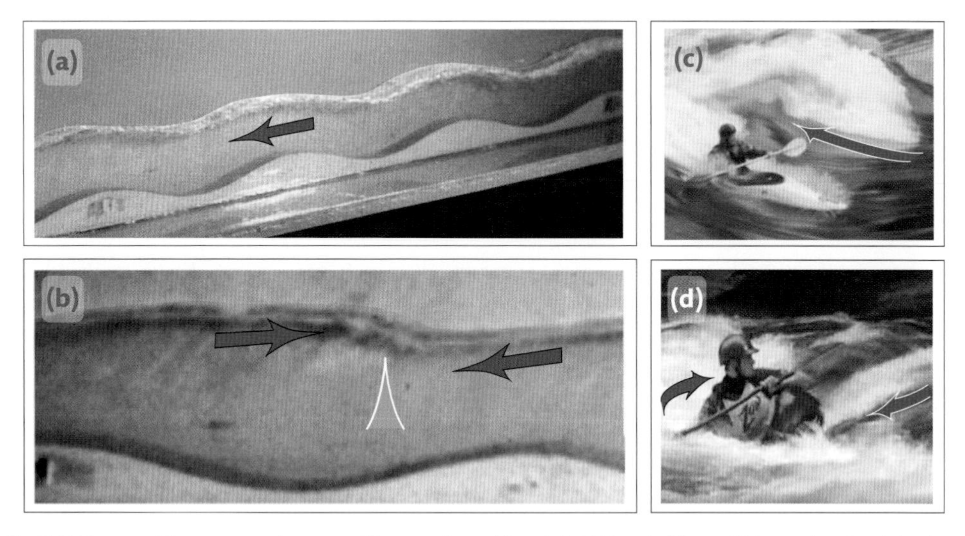

Fig. 7.9 Flow transition over wavy bottom—all flows are from right to left. (a) "Regular" flow at high speed: surface is in phase with bottom. (b) "Antiregular" flow at lower speed: surface is out of phase with bottom, and upstream flow appears along with a hydraulic jump (cyan arrowhead—see next example). (c) Kayak in a "standing wave:" a misnomer used in the kayaking community to describe nearly regular flow over subsurface obstacles, and (d) kayak trapped in a "hole" formed by antiregular flow. Flow from behind pushes the kayak upstream, while flow ahead of the kayak pushes it downstream. A transition from regular to antiregular flow occurs abruptly at a critical Froude number as described in the text.

elements. Consider then a sinusoidal streambed, as shown in Fig. 7.9(a). The height of the free surface, z, obeys the same relation as in the Rankine vortex problem:

$$\frac{V^2}{2} = \text{Const} + g \cdot z, \qquad [7.67]$$

where V is the fluid speed at the surface. Solving for V is a little involved, and performing the algebra isn't very informative; the interested reader is referred to details and additional background in Lamb.[16] A few points are worth emphasizing, however, for these involve methods that the reader may benefit from. In outline, the problem can be solved through the steps underlined below.

7.9.1 **Small Parameter Expansion**

The first step is to assume that the shape of the streambed can be written as something like $z_{bottom} = -h + \epsilon \cdot \cos(kx)$, where h is the mean depth of the water, k is the wavenumber $(= 2\pi/wavelength)$ of the wavy bottom, and ϵ is small. Whenever you see a problem formulated in this way, you can be sure that it will lead to a so-called "small parameter" or "perturbative" expansion. The idea is simple: if ϵ is small, maybe 1/10 the depth of the water, then we can expand whatever we are looking for, maybe the potential, in a power series like:

$$\text{Whatever} = c_0 + c_1\epsilon + c_2\epsilon^2 + \ldots, \qquad [7.68]$$

and then argue that terms like ϵ^2, ϵ^3, etc. must be of sizes 1/100 and 1/1000, and so can be neglected. There is a large literature on small parameter expansions: many scientists make an entire career out of looking for a small parameter, expanding the problem in terms of it, solving the simplified problem that results, and moving on to the next problem.

The attentive reader will recognize that it isn't necessarily the case that smaller terms can be neglected. For example, if you are steering a car at a high speed, small fluctuations applied to the steering wheel can lead to large effects. Similarly and more germane to this book, vortex stretching mentioned earlier in this chapter is one mechanism that can generate large effects from small causes. Likewise, the Kelvin–Helmholtz instability shown in Fig. 5.4, Chapter 5, starts out as barely noticeable waviness and builds to produce rolls that dominate the flow. Notwithstanding this legitimate worry, in many problems (including this one) ϵ remains small, and the perturbative solution works well to minimize algebraic complications by removing all but the simplest terms in the analysis.

7.9.2 **Hyperbolic Functions**

The second step is to take Laplace's equation for the potential, φ, separate variables, and solve for horizontal and vertical components. That is, we write $\varphi = X(x)Z(z)$, and the solution works out to be

$$\varphi = V_0 \cdot \{-x + [A \cdot \cosh(kz) + B \cdot \sinh(kz)] \cdot \sin(kx)\}, \qquad [7.69]$$

where V_o is the mean flow speed, A and B are constants, and cosh and sinh are the hyperbolic cosine and sine:

$$\begin{aligned} \cosh(kz) &\equiv \frac{e^{kx} + e^{-kx}}{2} \\ \sinh(kz) &\equiv \frac{e^{kx} - e^{-kx}}{2}. \end{aligned} \qquad [7.70]$$

These functions are pronounced "cosh" and "sinch" respectively. The reader may recall that we saw a hyperbolic secant in Chapter 4, which is logically enough 1/cosh. It is a common game to interchange exponentials with hyperbolic functions, just as we saw in Chapter 3 that it is common to interchange imaginary exponentials with sines and cosines. The choice of whether to use exponentials or sines and sinhs is purely a matter of algebraic convenience: the solutions are indistinguishable. As a quick exercise, show that this is true by solving for A and B in terms of C and D, and *vice versa*, in the following relation:

$$A \cdot \cosh(kz) + B \cdot \sinh(kz) \equiv C \cdot e^{kx} + D \cdot e^{-kx}. \qquad [7.71]$$

7.9.3 **Boundary Conditions**

The third step is to apply a no penetration condition at both the top and the bottom of the streambed to provide two relations for A and B in Eq. [7.69]. We can then differentiate φ to find the velocity, which we can square and solve for z from Eq. [7.67]. After all is said and done, the solution for z at the free surface is:

$$z = \frac{\epsilon}{\cosh(kh) - (gh/V_0{}^2)\sinh(kh)}\cos(kx). \quad [7.72]$$

Let's pause to understand what Eq. [7.72] says. Note in particular that the denominator changes sign when:

$$\frac{\sinh(kh)}{\cosh(kh)} = \frac{V_0^2}{gh}. \quad [7.73]$$

When this occurs, the solution goes from cosine (times some stuff) to *minus* cosine (times other stuff), meaning that the free surface flips upside down. This is shown in Fig. 7.9(a) for a comparatively high-speed flow, and in Fig. 7.9(b) for a lower speed flow. The first solution, in which the surface wave is in phase with the bottom wave, is termed "regular" flow, and the second solution, in which the surface is out of phase with the bottom, is termed "antiregular."

Notice that the formation of antiregular flow coincides with an *upstream* flow as indicated by the leftmost arrow in Fig. 7.9(b). This reversal can be understood by examination of the critical value determining when this transition will occur, termed the "Froude number:"

$$Fr = \frac{V_0{}^2}{gh}, \quad [7.74]$$

mentioned also in Chapter 2. *Fr* was named after naval architect William Froude (1810–1879), who was interested in building scale models to predict the stability of ships. *Fr* is sometimes expressed as the square root of Eq. [7.74], and is also sometimes called the "Boussinesq number," after French mathematician Joseph Boussinesq (1842–1929).

As with other dimensionless groups, *Fr* represents a competition between effects, here the kinetic energy of the flow (which goes as the velocity squared) and its potential energy (which goes as gravity times depth). When $Fr > 1$, the flow is dominated by kinetic energy, which drives the fluid downstream along the most direct path—a regular flow. When $Fr < 1$, however, kinetic energy is no longer strong enough to push the downstream fluid aside, and the fluid's potential energy causes it to back up downstream of each crest in the streambed. This is indicated by the left arrow in Fig. 7.9(b). Between the upstream and downstream flows is a "hydraulic jump," indicated by the cyan arrowhead: more on this next.

The transition between regular and antiregular flows has practical implications that are easily observed in any real streambed. For example, as shown in Fig. 7.9(c), downstream regular flow accelerates into the troughs, but slows near the peaks, so although a kayaker has to paddle continually to travel upstream in this flow, she can rest in the white water near the peak. On the other hand, a boat in fully antiregular flow, shown in Fig. 7.9(d), will be trapped in what kayakers call a "hole," between accelerating flow upstream (right arrow) and backflow downstream (left arrow). This "hole" has been termed a drowning machine for its effect on trapped swimmers.[17] By judging the amount and location of white water, kayakers learn where they can and cannot safely travel. Interested readers may want to look up "eddy turns," "s-turns," and "peeling in and out," all of which are techniques used by kayakers to take advantage of flow reversals downstream of obstacles.

Similarly, the reader can provoke this transition by hiking through a (not too rapid!) creek. Look for a boulder, say a meter in diameter, with regular flow downstream, as shown in Fig. 7.10 (a). If you stand on the boulder (red arrowhead in Fig. 7.10(a)) so as to reduce the flow speed downstream, you will find that you can often produce a transition to antiregular flow, which occurs as a downstream hydraulic jump (cyan arrowhead) moves upstream against the flow.

Example Kitchen hydraulic jump

Hydraulic jumps appear wherever liquids slow from high ($Fr > 1$) to low ($Fr < 1$) Froude numbers. As one might imagine, this can be very common. One doesn't even need to hike through a creek: the reader has doubtless seen these jumps many times. To produce a jump, all that is needed is to direct flow from a faucet onto a plate, as shown in Fig. 7.10(b). Near the point where the water strikes the plate, the flow is rapid and so is dominated by

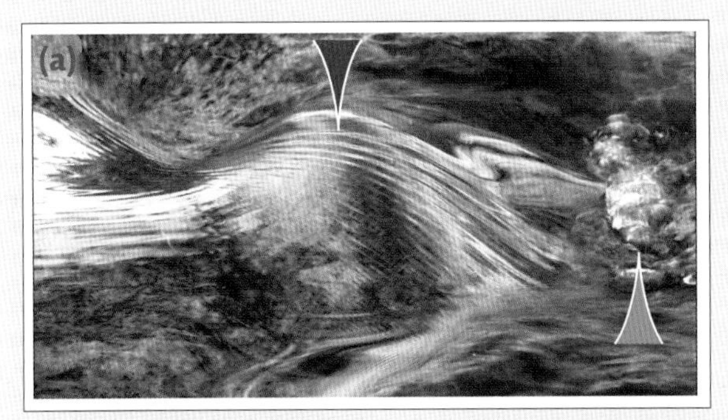

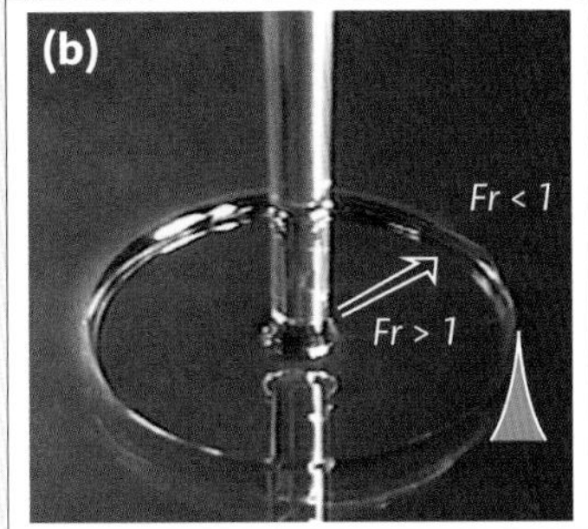

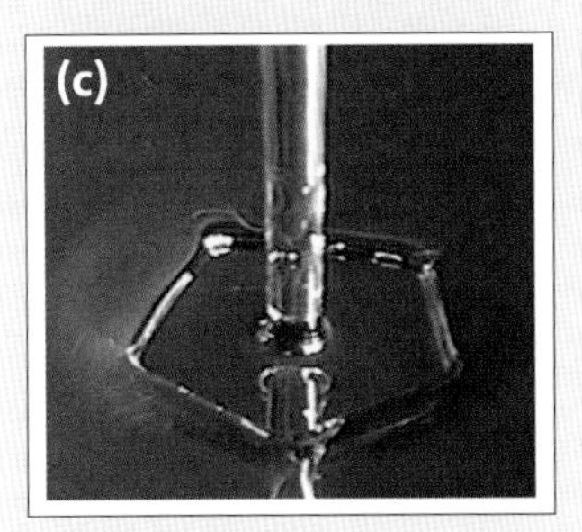

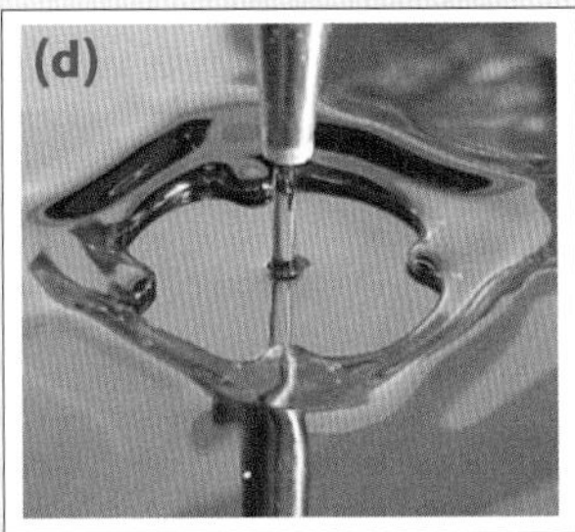

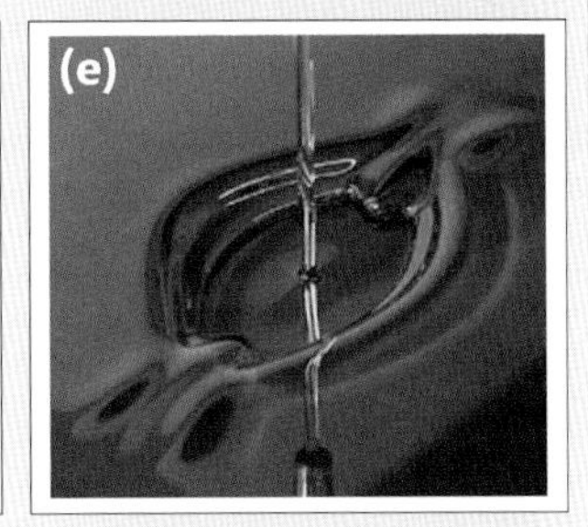

Fig. 7.10 Hydraulic jumps. (a) Hydraulic jump (cyan arrowhead) downstream of regular flow in a creek (flow here is left to right). Partially blocking the flow upstream, for example by standing on the rock near the magenta arrowhead, will slow the flow and cause the jump to travel upstream, producing antiregular flow. (From: https://www.la.utexas.edu/users/bump/images/Krause%20Springs/Creek%20Water%20over%20Rocks.jpg) (b) Classic circular hydraulic jump (arrowhead) in the kitchen sink: inside the circular radius, flow is dominated by kinetic energy, so $Fr > 1$; outside the radius, flow is dominated by potential energy, so $Fr < 1$. (c)-(e) Unusual shape hydraulic jumps produced by flowing fluids with varying viscosities and surface tensions. (Panels (b) and (c) from Ellegaard, C., Hansen, A.E. Haaning, A., Hansen, K., Marcussen, A., Bohr, T., Hansen J.L. and Watanabe, S. (1998) Creating corners in kitchen sinks. *Nature, 392*, 767–8. By permission of Springer Nature. Panels (d) and (e) from JWM Bush, JM Aristoff & AE Hosoi. An experimental investigation of the stability of the circular hydraulic jump. *J. Fluid Mech* 558 (2006) 33–52.)

kinetic energy—Fr is large here. Further away, the flow unavoidably slows to produce a higher potential energy, lower Fr, body of fluid. And where kinetic and potential energies are equal—at $Fr = 1$—a hydraulic jump appears.

The classic hydraulic jump is circular, as shown in Fig. 7.10(b). This is where our story would end, but for the observation by Shinya Watanabe and colleagues that higher viscosity fluids produce geometrically complex jumps. In Fig. 7.10(c), for example, is a polygonal jump produced using ethylene glycol in place of water,[18] and in Fig. 10(d) and (e) are "cloverleaf" and "cat's eye" shapes—a few of many patterns produced using glycerin.[19] This is another example, of many illustrated throughout this book and elsewhere, that fluid flows are more intricate and captivating than we have any right to expect.

7.10 Stability, Instability, and Transitions

In the problem of flow over a wavy bottom, we saw that flow can undergo a distinct transition—here from regular to antiregular flow. This is not the first example that we have seen: in Chapter 2 we saw a "pitchfork bifurcation" in a model for bleb growth, in Chapter 3, we discussed stability and instability of the double pendulum, and in Chapter 6 we saw that numerous qualitatively different flows appear in the Couette apparatus as rotation speeds are varied. What all of these problems have in common is that one behavior became unstable (e.g. Taylor vortex flow in the Couette system) and a different behavior became stable (e.g. wavy vortex flow). We began a discussion of stability in Chapter 3, in our "brief encounter with complex analysis," where we explained that stability is characterized by exponential decay and instability by exponential growth. Stability, instability, and transitions between qualitatively different behaviors are themes that we will return to, and we close this chapter with a few illustrative examples.

Example Stability and instability in the rin

As a first example, consider the "rin," or "Tibetan singing bowl," shown in Fig. 7.11(a). This is a metal bowl that rings like a bell when struck. Its ringing, incidentally, is prolonged because rins are made from low impurity metals that form large crystal lattices, and vibration travels in these lattices with minimal disturbance from impurities that would scatter sound.

An idealized signal that one would expect from this ringing is shown in Fig. 7.11(b), where we plot oscillatory vibrations that decay exponentially in time. We can write a simple model for this signal as something like:

$$Amplitude = A_0 e^{(-|k|+i\omega)\cdot Time}, \quad [7.75]$$

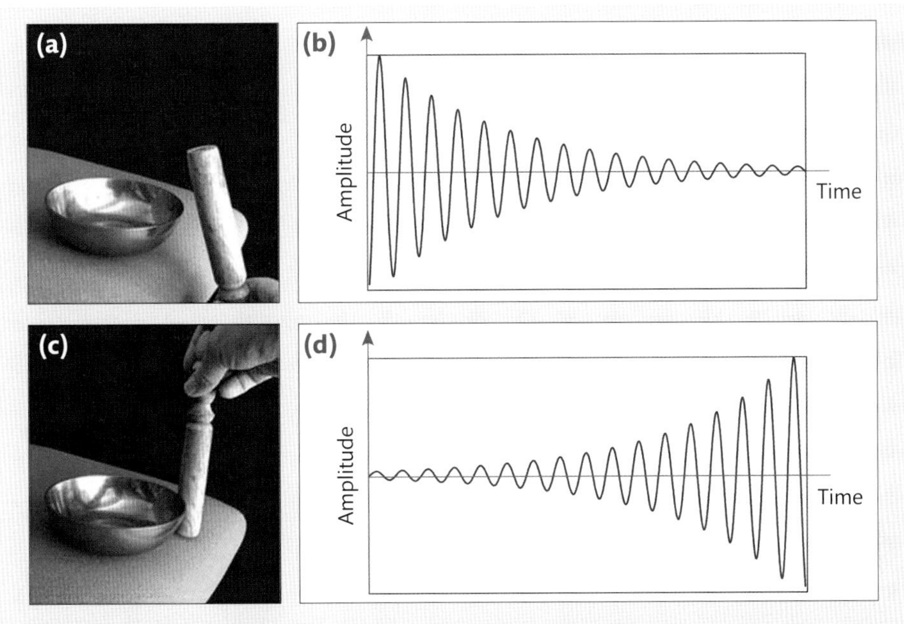

Fig. 7.11 Stability and instability in a singing bowl. (a) Illustration of striking the bowl, leading to (b) oscillations that diminish in time: this behavior is stable. (c) Illustration of rubbing the bowl with a wooden rod. This leads to (d) oscillations that grow in time: this behavior is unstable.

where A_0 is an initial amplitude. We recall from Chapter 3 that the real part of the exponential argument defines whether a disturbance grows (if positive) or decays (if negative), and the imaginary part defines an oscillation. Accordingly, k is an exponential rate of decay, and ω is a frequency of oscillation.

So far, so good. If, however, instead of striking the rin, we rub its perimeter with a wooden rod, as shown in Fig. 7.11(c), the rin will again ring—and this time, the amplitude will *grow* while it is being rubbed. An idealized plot of this growth is shown in Fig. 7.11(d). In this case, the simple model for the signal would become

$$Amplitude = A_0 e^{(|k|+i\omega)\cdot Time}, \qquad [7.76]$$

where the growth of the signal is codified by a positive rate, $|k|$.

The reader should rightly wonder why the rin rings when rubbed: we will leave this as a puzzle for the reader to solve. The point of this example is that the bowl is a very, very simple object, yet it exhibits either stable or unstable oscillations depending on how it is treated. In this sense, the rin shares features with more complicated systems. For example in the Couette apparatus, Taylor vortices become unstable and wavy vortices become stable when the inner cylinder speed is increased. So while the rin and the Couette device are physically dissimilar, the mathematics of their dynamical transitions are both described by a change of sign in a term like $|k|$ in Eqs. [7.75] and [7.76].

In flow over a wavy bottom, the change in sign depends on straightforward physical quantities: the kinetic divided by the potential energy (Eq. [7.74]). In the Couette system, the change depends on the relative strength of centrifugal acceleration. And in the rin, the change depends on something else, which as we say we leave for the reader to unveil.

Example Stability of flow-driven structures

Transitions between stable and unstable behaviors are exceedingly common, and can in all cases be described in terms of a change of sign of a critical parameter in an exponent. Some illustrative examples are shown in Fig. 7.12. Fig. 7.12(a) shows the infamous example of the 1940 Tacoma narrows bridge. This bridge crossed the Columbia gorge: a location known for its strong winds. These winds caused torsional oscillations in the bridge as shown, leading to its ultimate collapse, only 4 months after it was opened to traffic. Crucially, the bridge was stable to these oscillations until the wind speed exceeded a critical value. At this point, the bridge surface twisted to present a larger area to the wind, which amplified the growing oscillations. This mechanism leads to the growth of surface waves described at the start of this book, and in exactly the same way, vibrations are produced by blowing on a blade of grass or an oboe reed held between the lips.

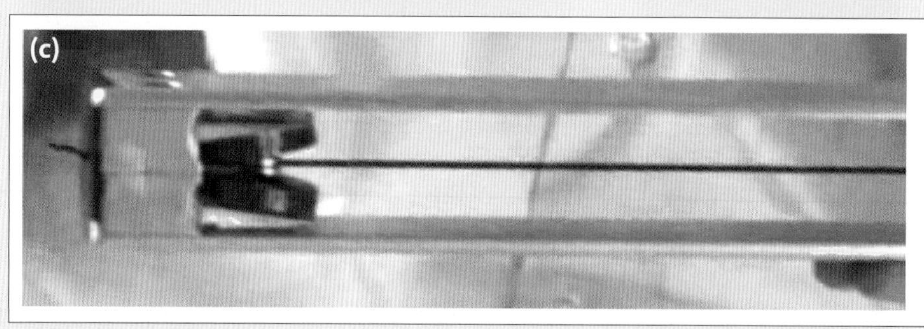

Fig. 7.12 (a) Instability of Tacoma Narrows bridge before its collapse in 1940. Strong winds through the Columbia gorge caused the torsional oscillation shown. (From film by Barney Elliott.) (b) Many structures are stabilized by tuned mass dampers, such as these on the Severn Bridge linking England to Wales. (Photo © Ian Taylor, cc-by-sa/2.0.) (c) Wind-driven instability has been exploited to generate electricity in the "windbelt," a plastic strip (black) that oscillates in response to wind, driving a magnet back and forth between solenoid coils. (From: https://www.youtube.com/watch?v=AMojRXK14jU.)

Many strategies are employed to make unstable structures stable. Although the strategies differ in many important ways, they can all be described in terms of changing a positive exponent into a negative one. Some strategies involve active control, as with the suspensions of cars developed by Lotus—indeed, we execute active control when we balance on one foot, which is plainly an unstable configuration. Others use passive control, for example, so-called "tuned mass dampers" used on skyscrapers—such as the 700 ton steel ball mounted near the top of Taipei 101. In Fig. 7.12(b), we show tuned mass dampers (also called Stockbridge or dog-bone dampers) mounted on cables supporting the Severn Bridge that connects England to Wales. The role of these odd-shaped devices is to distribute the energy of damaging low frequency modes of oscillation into more complex higher frequency modes that will ultimately dissipate the energy away as heat—essentially the mechanism described by Richardson earlier in this chapter.

A final clever and practical application appears in Fig. 7.12(c), which shows the "windbelt" device. This is a very inexpensive and low-tech invention that converts wind energy to AC electricity for use in underserved communities. The device consists of a plastic ribbon (black at the center of the photograph) that vibrates in response to wind, attached to a magnet (silver) between two wire coils at the left of the photo. As the ribbon vibrates, the magnet moves between the coils, generating alternating current sufficient to power a radio or light a small room.

Example Blowup in finite time

As a final example, we note that the transition expressed in Eq. [7.72] has the unusual feature that the denominator goes through zero, and consequently, the surface waveform formally diverges. In this particular problem, the divergence is simply a mathematical artifact that can be traced to the approximation involved in using a small parameter, but in practice, actual divergences can occur as well, leading to startling results.

A simple case of a divergence appears in the "Euler disk" shown in Fig. 7.13(a). This is a heavy metal disk that rolls as it settles onto a smooth surface. As the disk settles, its frequency of rotation grows, and formally the frequency becomes infinite in finite time—a behavior known as "blowup in finite time." The reader can confirm that the disk's frequency grows dramatically to an abrupt climax before it comes to rest (these devices are easily purchased or seen online).

Problems involving divergences in fluid or fluid-like systems are shown in Figs. 7.13 (b)–(d). Fig. 7.13(b) shows a jet of water being ejected from a cylindrical container. This jet is produced by oscillating the cylindrical boundaries up and down: each oscillation produces a wave that concentrates energy at a single point in the center of the container, resulting in an energy density that, again formally, is infinite. In practice, this produces a very fine, sharply pointed jet of water as shown.

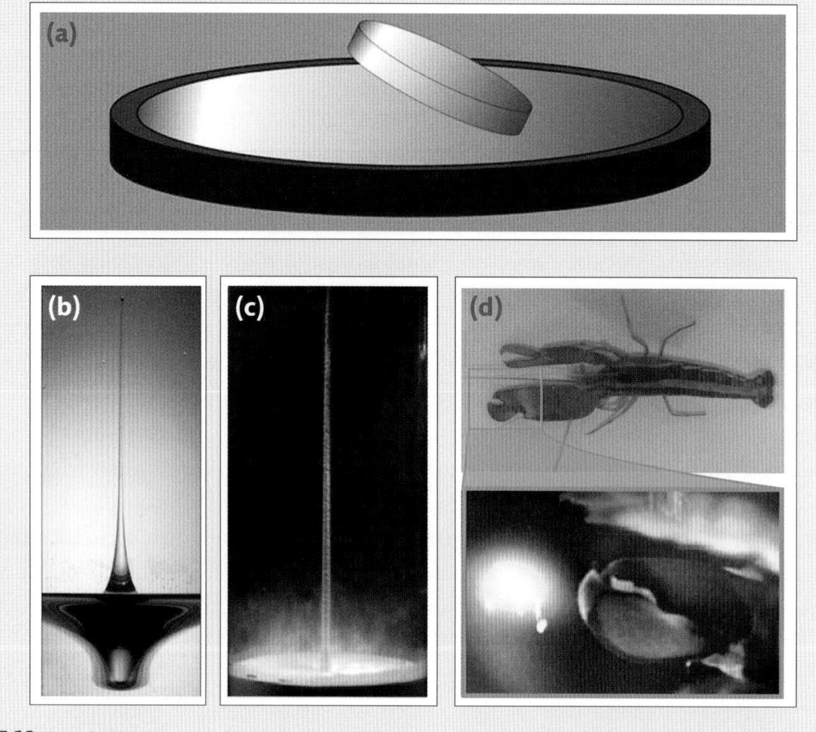

Fig. 7.13 Singularities associated with finite time blowup. (a) "Euler disk," whose wobbling frequency formally reaches infinity before coming to rest. (b) Spike forms in a cylindrical container of water whose boundaries are oscillated: the oscillation becomes focused at the center of the container, leading to formally infinite energy density. (From: Zeff, B., Kleber, B., Fineberg, J., Lathrop, D. Singularity dynamics in curvature collapse and jet eruption on a fluid surface. *Nature* 403, 401-404. By permission of Springer Nature.) (c) Granular jet that forms when a heavy ball is dropped into a container of grains. The jet forms after the crater produced by the ball collapses to a singular point. (From: J. R. Royer, E. I. Corwin, A. Flior, M.-L. Cordero, M. L. Rivers, P. J. Eng, and H. M. Jaeger. (2005). Formation of granular jets observed by high-speed X-ray radiography. *Nature Physics* 1, 164-167. By permission of Springer Nature.) (d) A trigger shrimp, which produces sonoluminescence (inset) by suddenly closing its claw. The rapidity of this motion ejects cavitation bubbles that collapse to a point, producing temperatures that have been measured to be higher than at the surface of the sun. (KP Aquatics LLC.)

The same phenomenon can be implemented by dropping a large marble or billiard ball into a bucket of fine, loose sand, as shown in Fig. 7.13(c). With care, the ball can be made to produce an outgoing wave that collapses symmetrically, generating a singularity that results in a jet of air and sand being ejected.

In these water and sand jet experiments, energy is focused to a point, but the collapse is principally in the horizontal plane, and material is therefore ejected vertically. A 3D collapse, logically enough, concentrates energy even more effectively, as was discovered in 1934 following research into cavitation of bubbles emanating from ship propellers. At that time, it was observed that collapsing bubbles emit light—a process known as "sonoluminescence." Detailed measurements have shown that the concentration is so complete that temperatures inside a collapsing bubble can reach 10,000 K—nearly twice the temperature at the surface of the Sun! This striking effect is used in nature by pistol shrimp (hence their name), as shown in Fig. 7.13(d). These animals, only about 4 cm long, can shut one of their pincers so rapidly that it ejects a cavitating stream powerful enough to kill small fish several centimeters away.

REFERENCES

1. Purcell, E.M. (1977) Life at low Reynolds number. *American Journal of Physics, 45*, 3–11.
2. Newton, I. (c.1687) Motion of fluids, and the resistance made to projected bodies. In: *Principia Mathematica, Book II.*
3. Gettelfinger, B. and Cussler, E.L. (2004) Do humans swim faster or slower in syrup? *AIChE Journal, 50*, 2646–7.
4. See also Wiener, N. (1947) A scientist rebels. *Atlantic Monthly, 179*, 46.
5. Obst, B.S., Hamner, W.M., Hamner, P.P., Wolanski, E., Rubega, M. and Littlehales, B. (1996) Kinematics of phalarope spinning. *Nature, 384*, 121.
6. Paul, M.C. and Larman, A. (2009) Investigation of spiral blood flow in a model of arterial stenosis. *Medical Engineering and Physics, 31*, 1195–1203.
7. Leptos, K.C., Guasto, J.S., Gollub, J.P., Pesci, A.I. and Goldstein, R.E. (2009) Dynamics of enhanced tracer diffusion in suspensions of swimming eukaryotic microorganisms. *Physical Review Letters, 103*, 198103.
8. Drescher, K., Leptos, K.C., Tuval, I., Ishikawa, T., Pedley, T.J. and Goldstein, R.E. (2009) Dancing Volvox: hydrodynamic bound states of swimming algae. *Physical Review Letters, 102*, 168101.
9. Eiamsa-ard, S. and Promvonge, P. (2008) Review of Ranque–Hilsch effect in vortex tubes. *Renewable and Sustainable Energy Reviews, 12*, 1822–42.
10. Huard, M., Briens, C., Berruti, F. and Gauthier, T.A. (2010) A review of rapid gas–solid separation techniques. *International Journal of Chemical Reactor Engineering, 8*(1).
11. Pichumani, M., Bagheri, P., Poduska, K.M., González-Viñasa, W. and Yethiraj, A. (2013) Dynamics,crystallization and structures in colloid spin coating. *Soft Matter, 12*, 3220–9.
12. Schiffman, J.D. and Schauer, C.L. (2008) A review: electrospinning of biopolymer nanofibers and their applications. *Polymer Reviews, 48*, 317–52.
13. Stonebridge, P.A. and Brophy, C.M. (1991) Spiral laminar flow in arteries. *Lancet, 338*, 1360–1.
14. Syred, N. and Beér, J.M. (1974) Combustion in swirling flows: a review. *Combustion & Flame, 23*, 143–201.
15. Frohlich, C. (1980) The physics of somersaulting and twisting. *Scientific American, 242*, 155–64.
16. Lamb, H. (1932) *Hydrodynamics*, 6th ed. Cambridge: Cambridge University Press, p. 409.
17. Gioia, G., Chakraborty, P., Gary, S.F., Zamallao, C.Z. and Keane, R.D. (2011) Residence time of buoyant objects in drowning machines. *Proceedings of the National Academy of Sciences, 108*, 6361–3.
18. Ellegaard, C., Hansen, A.E., Haaning, A., Hansen, K., Marcussen, A., Bohr, T., Hansen, J.L. and Watanabe, S. (1998) Creating corners in kitchen sinks. *Nature, 392*, 767–8.
19. Bush, J.W.M., Aristoff, J.M. and Hosoi, A.E. (2006) An experimental investigation of the stability of the circular hydraulic jump. *Journal of Fluid Mechanics, 558*, 33–52.

8 Inviscid Flows

8.1 Introduction

We have seen that either the streamfunction, ψ, or the potential function, φ, can be used to produce velocity fields for inviscid, irrotational flows. In this chapter, we apply our knowledge to analyze flows in a variety of flow geometries. Let's start by re-examining the problem of flow past a sphere, analyzed in Exercise 7.9 of the last chapter, which holds both lessons and surprises for understanding the transition between low and high Reynolds number behaviors.

To recap, we have derived two different solutions for flow past a sphere. At low Reynolds numbers, we obtained the viscous stream function (Chapter 6, Eq. [6.68]):

$$\psi = -V_\infty \left(\frac{R^3}{4r} - \frac{3Rr}{4} + \frac{r^2}{2} \right) \sin^2(\vartheta), \qquad [8.1]$$

while at high Reynolds numbers, we presented (Chapter 7, Eq. [7.63]) the irrotational potential function:

$$\varphi = V_\infty \left(r + \frac{R^3}{r^2} \right) \cos(\vartheta). \qquad [8.2]$$

To illustrate how viscosity affects flow, in Fig. 8.1(a) we show a superposition of both of these flows, which look largely similar. Detailed scrutiny (inset to right) reveals that close to the sphere, the irrotational flow (Eq. [8.2]) executes more rapid changes in direction than the viscous flow (Eq. [8.1]). On the other hand, far from the sphere the viscous streamline is perturbed more from a straight path than the irrotational streamline. Both of these results can be explained with a little thought: viscosity penalizes shear—so rapid changes in direction (which produce shear) are more difficult to execute in a viscous than in an inviscid fluid. Similarly, viscosity provides a mechanism for fluids to transmit stresses far from the sphere that inviscid systems lack.

These effects are subtle; the effect of viscosity is more evident in a comparison between time courses of fluid trajectories, as shown in Fig. 8.1(b). In that panel, we plot trajectories by integrating the velocities obtained from the stream functions, Eqs. [8.1] and [8.2]. We use "fourth order Runge–Kutta" integration (described in the Appendix), which tracks trajectories more accurately than the "Euler integration" method discussed in Exercise 5.3, Chapter 5. Logically enough, viscosity slows the fluid with respect to the spherical intruder, and consequently both near and far from the sphere, fluid trajectories reach shorter distances in the viscous than in the inviscid case.

Biomedical Fluid Dynamics: Flow and Form. Troy Shinbrot.

DOI: 10.1093/oso/9780198812586.001.0001

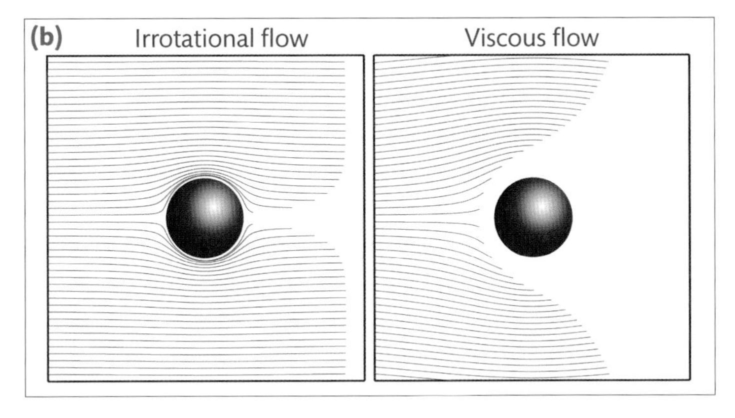

Fig. 8.1 Comparisons between irrotational and viscous streamlines. (a) Superposition of the flows themselves, along with enlargement in inset. (b) Low and high viscosity trajectories, both plotted using the same fourth order Runge–Kutta algorithm (see Appendix) for the same length of time.

EXERCISE 8.1

Reproduce Fig. 8.1(a), and confirm the differences between irrotational flow and viscous stream functions. Notice that the trajectories shown in Fig. 8.1(a) are nearly equally spaced: how can you achieve that using the stream function of Eq. [8.1] (give some thought to the fact that the contours are obtained from a surface with varying slope and height).

8.2 **Stokes Paradox**

We mentioned the Stokes paradox in Chapter 7: this is the curious fact that a simple expression for flow around a cylinder is available at high Reynolds numbers (*Re*), but not at low. This is part of the theme that we mentioned earlier that the trick of restricting attention to the irrotational limit (Chapter 7, Eq. [7.6]) permits solutions to be obtained in a way that is in some ways *simpler* at high *Re* than at low. Even as a mathematical sleight of hand it is remarkable that we can reduce the nonlinear Navier–Stokes equation to Laplace's equation, $\nabla^2\varphi = 0$, at the high *Re* limit, while the linearized equation at the low *Re* limit obeys the higher order biharmonic equation, $\nabla^4\psi = 0$.

This remarkable feat comes at a cost, and indeed to obtain Laplace's equation we have disregarded very real nonlinear effects that lead at their simplest to exponential growth of disturbances (cf. the sine flow of Chapter 1) and at their most complex to a wealth of turbulent phenomena (see Van Dyke[1]). By the same token there is a cost to approximating Navier–Stokes at either high or low *Re*, we ignore one term or another, and although doing so permits us to obtain analytic solutions, we should understand what we lose in the bargain.

We can view the high *Re* limit as allowing viscosity to go to zero, and in this case as we have discussed we lose the ability to capture shear transmission. As we described in Chapter 7, this means that vortices cannot be produced or destroyed. It also means that we cannot say anything about flow near boundaries, where we know shear gradients must be strong.

What do we lose in the low *Re* limit? In Chapter 7, we saw what happens in the Rankine vortex example: the high *Re* solution fails at small distances, and the low *Re* solution fails at large distances. In the Rankine problem, the failure at small distances was physically due to the fact that shear is strong close to the origin—and the high *Re* solution has no way of transmitting shear. At the other extreme, the failure at large distances was due to the fact that the low *Re* solution allows stresses—and so speeds—to grow indefinitely with distance. More formally speaking, $Re = \langle V\rangle \cdot \langle L\rangle / \nu$, and as the characteristic distance, $\langle L\rangle$, grows, it is impossible for *Re* to remain small. So effectively we can define a high Re solution only if we ignore behaviors at small $\langle L\rangle$ (close to an obstacle) and we can define a low Re solution only if we ignore behaviors at large $\langle L\rangle$ (far away).

EXERCISE 8.2

After separation of variables in cylindrical coordinates, the biharmonic equation (Chapter 6, Eq. [6.52]) becomes:

$$\left[\frac{\partial^2}{\partial\rho^2} + \frac{1}{\rho}\frac{\partial}{\partial\rho} - \frac{1}{\rho^2}\right]^2 R(\rho) = 0. \quad [8.3]$$

Prove that there is no solution to this equation for flow past a cylinder. To do this, first show that Eq. [8.3] has solution:

$$R(\rho) = a_1\rho^3 + a_2\rho\ln(\rho) + a_3\rho + a_4\rho^{-1}, \quad [8.4]$$

where the constants $a_1 \ldots a_4$ must be determined from boundary conditions. Second, show that these conditions forbid the solution in Eq. [8.4], as follows. Use a cylinder aligned with the z-axis and flow in the *x*-direction. Far from the cylinder, show that the velocity must go as $V_o\rho\sin(\vartheta)$, and at the cylinder surface, obtain no-penetration and no-slip conditions. Then show what this implies for the constants in Eq. [8.4]. From this, you should be able to demonstrate that Eq. [8.4] cannot be reconciled with the known boundary conditions for flow past a cylinder.

ADVANCED EXERCISE 8.3

Use the low *Re* solution for flow past a sphere (Chapter 6, Eq. [6.25]) to estimate the magnitudes of growth of inertial ($\vec{V} \cdot \nabla \vec{V}$) and viscous ($\nabla^2 \vec{V}$) terms with distance. Show that the ratio of these terms grows with radius, *r*. This implies that even for a sphere, inertia grows with distance, so the solution obtained in Chapter 6 must only be valid sufficiently close to the sphere.[2]

As we described last chapter, development of the fields of fluid mechanics and electrostatics went hand-in-hand: Maxwell proposed that electrostatics could be explained as the flow of an imaginary incompressible and inertia-less fluid.[3] Use this fact to estimate how much more rapidly the velocity around a cylinder must grow with distance than that around a sphere. This gives a heuristic explanation for why a solution can be obtained for viscous flow around a sphere, but not a cylinder.

In this chapter, we'll look more deeply into the effects of viscosity on fluid flow, and we'll begin by examining high Reynolds number flows. For this task, it will be useful to carefully consider what stream and potential and functions are and how they are related. Recall Exercise 7.9, Chapter 7, in which you plotted velocity and potential functions for flow past a sphere. The potential function in that case was Eq. [7.60], which we reproduce here:

$$\varphi = V_\infty \left(r + \frac{R^2}{r} \right) \cos(\vartheta). \qquad [8.5]$$

Using Eqs. [8.2] and [8.5], you should have obtained a result something like Fig. 8.2(a), whose inset illustrates the central point that the stream and potential contours are everywhere perpendicular.

This can be proven analytically by considering the two sets of equations that define these functions. For the stream function, we know that:

$$\begin{aligned} V_x &= -\frac{\partial \psi}{\partial y} \\ V_z &= \frac{\partial \psi}{\partial x}, \end{aligned} \qquad [8.6]$$

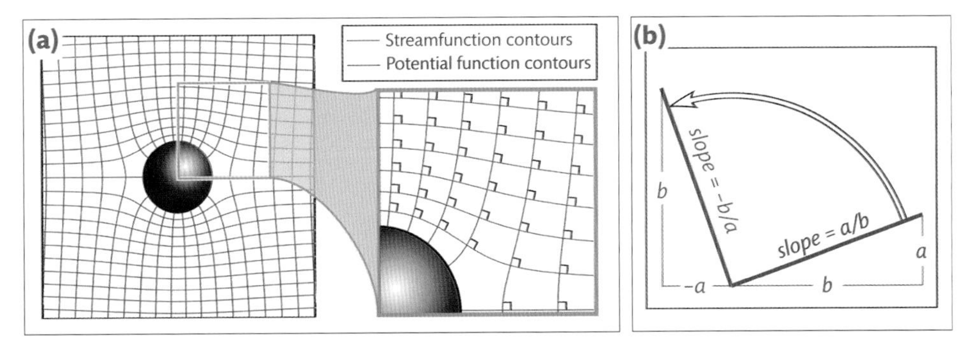

Fig. 8.2 Angles between stream and potential function contours. (a) Angles between these contours for irrotational flow past a sphere, with inset highlighting the observation that the contours are perpendicular everywhere. (b) Illustration that rotating a line by 90° (indicated by the arrow, rotating the blue line to the red one) changes the slope from a/b to $-b/a$, so that the product of the two slopes is -1.

and we saw earlier that a line of constant ψ is defined by:

$$\begin{aligned} 0 &= \frac{\partial\psi}{\partial x}\Delta x + \frac{\partial\psi}{\partial y}\Delta y \\ &= -V_y\Delta x + V_x\Delta y. \end{aligned} \quad [8.7]$$

So for constant ψ:

$$\frac{\partial y}{\partial x} = \frac{V_y}{V_x}. \quad [8.8]$$

Similarly for the potential function:

$$\begin{aligned} V_x &= \frac{\partial\varphi}{\partial x} \\ V_z &= \frac{\partial\varphi}{\partial z} \end{aligned}, \quad [8.9]$$

so a line of constant φ is defined by:

$$\frac{\partial y}{\partial x} = -\frac{V_x}{V_y}. \quad [8.10]$$

Eqs. [8.8] and [8.10] imply that the product of slopes of constant ψ and of constant φ is -1. The reader may recall from an analytic geometry course that this in turn implies that the two slopes are perpendicular. Since minutiae of this kind are easily forgotten, we illustrate the identity in Fig. 8.2(b), where we show that rotating a line of slope a/b by 90° produces a new line of slope $-b/a$.

This tells us that for an inviscid and irrotational flow, streamlines and equipotential lines are always perpendicular. This geometrical fact is central to the Cauchy–Riemann equations obtained last chapter:

$$\begin{aligned} \frac{\partial\varphi}{\partial x} &= -\frac{\partial\psi}{\partial z} \\ \frac{\partial\varphi}{\partial z} &= \frac{\partial\psi}{\partial x}, \end{aligned} \quad [8.11]$$

which say in essence that it is equivalent to know either ψ or φ: one can be derived from the other, and the two functions are always perpendicular to one another. Hold on to these thoughts, which we will return to after we learn some more about how real flows change with Reynolds number.

EXERCISE 8.4

Since φ can be obtained from ψ, doesn't this mean that streamlines obtained for viscous flow can also be described in terms of a potential function? If not, why not, and if so, does this mean that viscous flows are irrotational (since the curl of $\nabla\varphi$ is identically zero)?

Additionally, a solution, f, to Laplace's equation, $\nabla^2 f = 0$, is also a solution to the biharmonic equation, $\nabla^4 f = 0$, so both high and low viscosity flows obey the biharmonic equation. This being the case, what accounts for the difference between the solutions shown in Fig. 8.1? Mathematically speaking, shouldn't the solution to the same differential equation (here the biharmonic equation) in the same geometry (flow past a sphere) be unique? If not, why not?

8.3 Drag as a Function of Reynolds Number and Roughness

A good way to understand how flows change with Reynolds number is to focus on drag on obstacles in a moving stream: as we will show, this problem reveals a number of surprising aspects that provide essential clues to fluid behaviors in a variety of contexts. We saw in Eq. [2.27] of Chapter 2 that based on dimensional arguments, motion of fluid past a body can be characterized by a drag coefficient, which we traditionally write as the nondimensionalized ratio of force applied to the body divided by fluid kinetic energy:

$$C_d = \frac{F}{(\rho \cdot V^2 \cdot A)/2}. \qquad [8.12]$$

Here ρ is the fluid density, V is the speed of the body through the fluid, and A is the body's cross-sectional area. The literature on drag sometimes instead mentions a "friction factor," f: this is synonymous with C_d. The distinction between these terms is merely a matter of context: typically C_d is used to describe external flow, around an obstacle, while f is used for internal flow, through a pipe. To complicate matters further, in engineering texts, f is sometimes referred to as a "Fanning" friction factor, after John Fanning (1837–1911), the "Darcy" friction factor, after Henry Darcy (1803–1858), or the "Moody" friction factor, after Lewis Moody (1880–1953). These are not all precisely the same thing—for example, the Darcy f is four times the Fanning f—and in technical discussions it may be necessary to discriminate whose friction factor is under discussion. We'll discuss drag for flow through pipes in Chapter 9; for our purposes at the moment, we will attempt to limit confusion—as well to avoid compounding the semantic problem that f has nothing to do with friction—by only speaking about the drag coefficient, C_d.

We can expand Eq. [8.12] by making use of the force acting on a sphere at low Reynolds number (Eq. [6.34] of Chapter 6):

$$F = 6\pi\mu RV, \qquad [8.13]$$

so:

$$C_d = \frac{12\nu}{V \cdot R}, \qquad [8.14]$$

where we have used the cross-sectional area of a sphere, $A = \pi R^2$. But look, if we write Eq. [8.14] in terms of the diameter, $D = 2R$, of the sphere, we get

$$C_d = \frac{24}{Re}. \qquad [8.15]$$

So at low Reynolds number, the drag coefficient should simply depend on the Reynolds number. This is true more generally as well: if we perform a Buckingham pi analysis (described in Chapter 2), we find that drag acting on an obstacle depends on four quantities: viscosity (μ), density (ρ) speed (V). and radius (R), in three units: length (L), time (T). and mass (M). Consequently, the drag can be defined in terms of only one dimensionless group, and any other dimensionless group (e.g. C_d) must be expressible in terms of that single group. We may as well choose that group to be *Re*—which we know something about already—so drag can apparently be expressed only in terms of Reynolds number.

In Fig. 8.3(a), we therefore show a plot of the drag coefficient versus Reynolds number, for flows past a cylinder and a sphere. Both of these plots confirm the expected drag (Eq. [8.15]) at low *Re*, but both produce very different behavior as *Re* grows. Empirically, it is found for a sphere that between $Re = 500$ and $Re = 3 \times 10^5$, C_d approaches 0.44, after which C_d decreases by a factor of 3, and then grows again at very high *Re*. Similar behaviors are seen for cylinders and other cross-sectional shapes as shown in Fig. 8.3(b).

This is very strange, right? The growth in drag compared with Eq. [8.15] isn't surprising: as *Re* grows, other stuff, of some kind, is probably going on, but why does C_d dip so suddenly above $Re \sim 10^5$? And put in practical terms, can this dip in C_d be used to deliberately reduce drag?

We'll answer these questions next, but to get a flavor for what "other stuff" is going on, let's consider an implication of Fig. 8.3(b). This figure shows that a cylinder of circular cross section (solid green curve) has the same drag coefficient as an airfoil *ten times* its

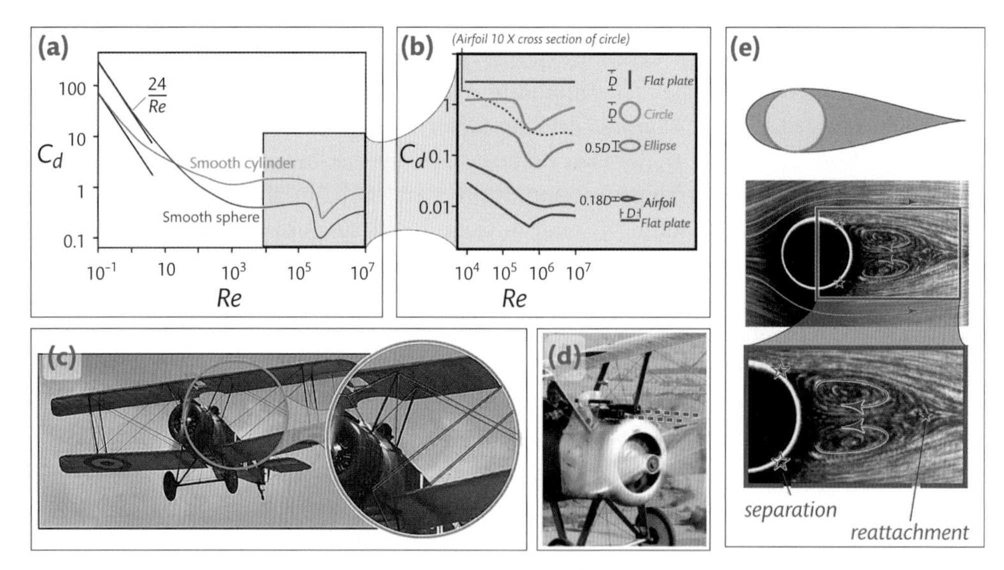

Fig. 8.3 (a) Drag coefficient plotted as a function of Reynolds number for a sphere and a cylinder. Notice two key features: at small *Re*, C_d scales like 24/*Re*, and at about $Re = 5 \times 10^5$, C_d executes a dramatic dip. (b) Drag coefficients depend strongly on shape, here plotted for several two-dimensional cross sections—note especially that an airfoil has much lower drag than a cylinder. (From: *Fundamentals of Fluid Mechanics*, Third Edition by Bruce R. Munson, Donald F. Young, and Theodore H. Okiishi (1998). By permission of John Wiley & Sons.) (c) Early biplane, with guy wires identified in inset as red lines. As mentioned in the text, the drag on a cylinder is about the same as the drag on an airfoil with 10× the diameter. (MRYsportfoto/Shutterstock.com.) (d) "Sopwith Camel" biplane with machine guns that fire between propeller blades, as indicated by red dashes. (http://imgproc.airliners.net/photos/airliners/7/1/9/1473917.jpg?v=v40.) (e) Comparison between airfoil and circle: airfoil shown has 1/10 drag of circle. Flow beneath, at $Re = 26$, exhibits trailing vortices due to pressure inversion described in text. Source for panels (c)-(e): MRYsportfoto/Shutterstock.com. (Taneda, S. 1956. *J. Phys. Soc. Jpn.* 11:1104–1108.)

cross-sectional profile (dotted magenta curve). Looked at another way, as shown at the top of panel (e), if you take a cylinder and *add material* to it as shown in pink, you will *reduce* the drag by an order of magnitude! This is completely counterintuitive—as if using a bigger saddle could make a horse run faster.

Example Airplane design

Historically, the counterintuitive finding that a cylinder has the same drag as an airfoil ten times its diameter had an interesting implication. The first airplanes were designed to maximize wing area, because the idea of flight was thought to be so extraordinary that surely large wings must be necessary. As a consequence, as shown in Fig. 8.3(c), early airplanes used two, or even three, wings, which of mechanical necessity were lashed together with numerous struts and guy wires. But as we've said, Fig. 8.3(b) tells us that the drag imposed by these rigging members was equivalent to that due to airfoils ten times their diameter. So considering the enlargement shown to the right of the figure, it is evident that the designers unknowingly produced the same drag as if the entire area of attack of the biplane were one large airfoil.

Thus most of the energy consumed by early planes went to defeating drag, and comparatively little went toward providing sufficient lift, which can easily be achieved with a single wing. Indeed any beginning pilot will report that today's small planes—with much smaller wing area than early models—take off at very low speeds, and a pilot's first challenge is commonly to learn how to keep the plane *down* during landing, as it tends to become airborne again after each bounce off the tarmac.

Drag was only one of many unanticipated technical challenges involved in early plane design—for example, fighter planes built during WWI typically accommodated only one propeller and one pilot, and so the machine guns were mounted where the pilot could reach it: directly behind the propeller. As shown in Fig. 8.3(d), this meant that the bullets had to pass through the path of the propeller blades. So how could the pilot fire the gun without damaging the propellers? This design challenge has a clever solution that the reader may find amusing.

8.3.1 **Mechanisms**

The curious order-of-magnitude reduction in drag that airfoils produce is important for us because it gives a first clue to what governs the drag coefficient at high Reynolds number. To see why, consider the flow behind a cylinder, shown at the bottom of Fig. 8.3(e) at $Re = 11$. The reader may recall from Chapter 5 (Fig. 5.5) that flows can separate near a boundary due to a pressure inversion. For flow behind an obstacle, the inversion is caused by trailing flow that pulls fluid away from the obstacle. Above $\mathrm{Re} \sim 1$, the pressure inversion becomes large enough to create a backflow, producing separation and reattachment points identified as red stars in Fig. 8.3(e), accompanied by recirculating vortices shown as red arrows in the figure.

So there is apparently both high-pressure ahead of an obstacle—caused by the incoming fluid flow pushing the obstacle—and low pressure behind—caused by trailing fluid flow pulling fluid away from the obstacle. The difference between these pressures causes what is termed "form drag." This is drag produced by pressure alone, as distinguished from what the literature terms "skin friction," which (like the friction factor mentioned earlier) isn't friction at all, but is drag produced by shear stress—that is, viscosity.

Airfoils reduce form drag by filling in the trailing low-pressure region to combat the pressure reduction. Fluid then follows smooth streamlines, and consequently form drag is dramatically diminished. So apparently how fluid flows—smoothly to fill in a space, or breaking apart to produce a pressure drop—strongly affects drag.

The process of separation and reattachment becomes elaborate at higher Reynolds numbers, and understanding what drives that process has lessons for both high and low Reynolds number flows, so in Fig. 8.4 we depict schematics of flows as a function of Reynolds number.

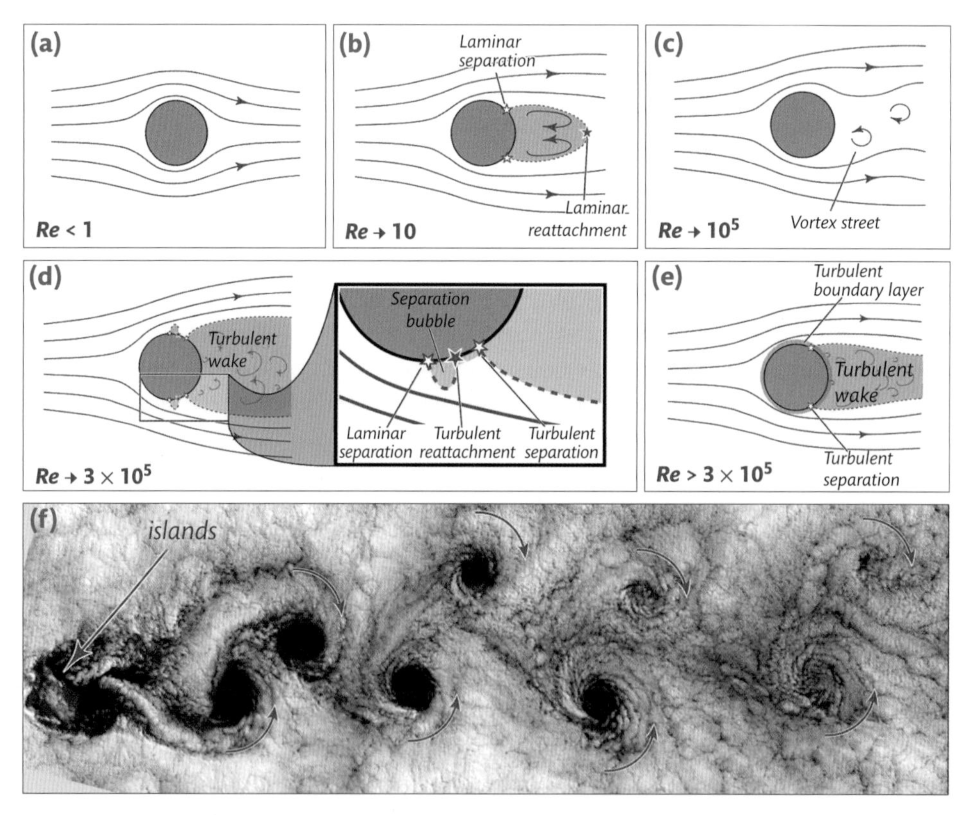

Fig. 8.4 Sequence of sketches of flows past a cylinder as Reynolds number is increased. (a) Laminar, low *Re*, flow described in Chapter 6. (b) Between $Re \sim 1$ and $Re \sim 10$, low pressure behind a cylinder produces bound recirculating vortices. (c) Above $Re \sim 47$, the downstream stress pulls vortices away from the cylinder to produce a "von Karman vortex street" (see also panel (f)). (d) At $Re \sim 3 \times 10^5$, a separation bubble forms that enlarges the effective cross-sectional area of the obstacle, producing a large turbulent wake. (e) At slightly larger *Re*, the separation bubble washes downstream, the wake dramatically decreases in size, and drag abruptly drops. (f) Landsat 7 image of street of counter-rotating vortices trailing Juan Fernandez Islands, $Re \sim 10^5$. (Source for panel (f): USGS/NASA Landsat.)

As shown in Fig. 8.4(a), at low Reynolds numbers there is no separation or re-attachment: flow is purely laminar as described in Chapter 6. As Re approaches 10, inertia of the trailing fluid literally pulls the fluid trajectories apart, and the trailing low-pressure produces a separation bubble (light blue). In the figure we identify separation and reattachment points as being "laminar," which in this context means steady. Above $Re \sim 10$, shear stress from the surrounding fluid becomes strong enough to detach the trailing vortices from the cylinder, leading to unsteady flow as one vortex after another detaches from the cylinder. This creates a "von Karman vortex street," named after Theodore von Karman (1881–1963), a Hungarian-American aerospace engineer who co-founded NASA's famous Jet Propulsion Laboratory. Streets of detached vortices appear over a large range of Reynolds numbers, and can be seen in laboratory, oceanic and atmospheric flows—for example, a dramatic snapshot of a vortex street is included in Fig. 8.4(f), here in clouds trailing behind islands off the coast of Chile. These vortices can last several minutes and persist over distances of 5 km or more: it is the persistence of these vortices that limits the rate at which planes that can take off from an airfield. Vortices trailing behind a large plane can easily flip over another plane attempting to take off or land too close behind.

Above $Re \sim 3 \times 10^5$, the abrupt dip in drag shown in Fig. 8.3(a) and (b) appears, and in Figs. 8.4(d) and (e) we plot a caricature of what happens to the flow in this range of Reynolds numbers. Close to 3×10^5, a separation bubble is seen, sketched in the enlarged inset to Fig. 8.4(d). This bubble impinges outward into the fluid flow, and so enlarges the effective size of the obstacle. This enlargement in turn accentuates the downstream pressure drop, and leaves behind a large turbulent wake. These two effects—the increase in pressure drop and the growth of downstream turbulence—increase the form drag—or more precisely, reduce the rate of decline shown in Figs. 8.3(a) and (b). Above 3×10^5, the same thing happens as we saw with von Karman vortices: an increase in shear stress drags the separation bubble downstream, forcing the three separation-reattachment points shown in Fig. 8.4(d) to move downstream, where they merge into a single separation point. This is sketched in Fig. 8.4(e) for $Re > 3 \times 10^5$.

This intricate interplay between small-scale separation points and large-scale surrounding flow features is associated with two things. First, since the separation bubble vanishes, the effective cross-sectional area is reduced, and the turbulent wake becomes much smaller. This is the immediate cause of the reduction in drag shown in Figs. 8.3(a) and (b).

Second, and related, at the surface of the obstacle, the upstream flow merges downstream into a turbulent wake at a separation point (actually a line, since we are talking about a cylindrical obstacle). Unlike the case at $Re \sim 3 \times 10^5$, however, there is no separation bubble insulating the laminar flow upstream from the turbulent flow downstream, and near the obstacle surface, the turbulence invades the formerly laminar upstream flow.

What these observations tell us is that small changes at the surface of an obstacle can produce large effects on downstream fluid behavior, and consequently on drag. It turns out, for instance, that 80% of the turbulent kinetic energy produced by aircraft is associated with turbulent bursts that occur when hairpin-shaped vortices break free from the surface, enlarging the cross-sectional area and generating downstream turbulence. For this reason, there has been extensive research in both aircraft and marine industries in developing control surfaces that limit turbulence (more on this shortly). Likewise in biomedical applications, "bruits" are noises generated in carotid arteries that are caused by turbulence downstream of obstacles produced by plaques. The formation of plaques is in this respect believed to be a self-fulfilling mechanism: a small plaque produces a trailing separation bubble, which holds blood in place, leading often to growth in the plaque as well as the shedding of emboli.

The details of the mechanism leading to changes in drag change considerably with obstacle geometry—for example, for a cylinder there is a well-defined symmetry axis that trailing vortices align with. For a sphere, on the other hand, the trailing flow lacks this symmetry and so is more complex. Similar qualitative behavior is seen for a variety of geometries that we showed in Figs. 8.3(a) and (b).

8.3.2 **Surface Features**

Since downstream fluid behavior—and so drag—depends strongly on flows at the surface of an obstacle, let's look at surface characteristics in a little more detail. In Fig. 8.5(a), we plot how the drag coefficient, C_d, of a sphere changes with surface roughness. Roughness can be measured in terms of the height of surface asperities, ϵ, divided by the sphere

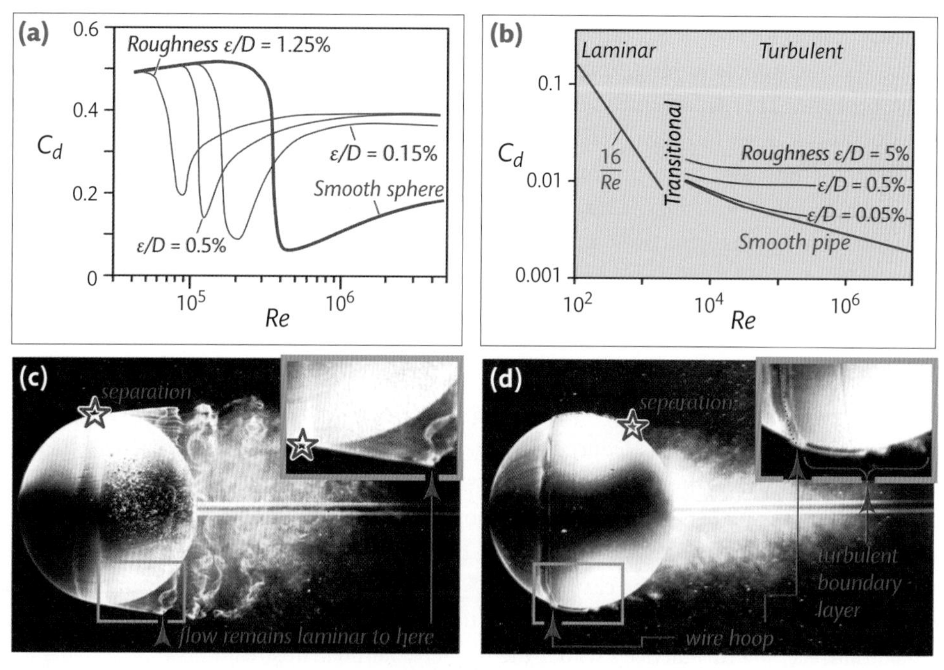

Fig. 8.5 (a) Drag coefficients for rough and smooth spheres. (b) Drag coefficient for smooth and rough pipes. Flow behind spheres for (c) smooth sphere at $Re = 15,000$ and (d) artificially roughened sphere at $Re = 30,000$. Artificial roughness is introduced by attaching a wire hoop forward of the equator (red arrow).[4] Notice in panel (c) that flow separates from the sphere upstream of the equator, remains laminar for some distance downstream (see enlarged inset), and so the effective cross-sectional diameter presented by the sphere and its attached boundary layer is larger than in case (d). Inset highlights turbulent boundary layer trailing behind wire hoop. (Panels (c) and (d) from Van Dyke, M. (1982) *An Album of Fluid Motion*, reprinted here: (c) http://www.iahrmedialibrary.net/turbulent-flow/; (d) http://www.iahrmedialibrary.net/laminar-flow/)

diameter, D. Fig. 8.5(a) shows that as ϵ/D grows, the dip in drag coefficient moves to lower Reynolds number.

Hey, this could be useful, right? If we're working at $Re = 10^5$, Fig. 8.5(a) tells us that we can minimize drag by tuning the roughness to about $\epsilon/D = 0.0125$. To fully appreciate and take advantage of this behavior, let's consider further how roughness affects drag.

Example Drag and roughness

The connection of between drag and roughness has an eclectic history. In sports, it has long been known that rough balls fly differently than smooth ones. Early golf balls, for instance, were smooth: a seventeenth century leather ball stuffed with feathers and a nineteenth century gutta percha (a type of natural rubber) ball are shown in Fig. 8.6(a). Golfers sought out older, scratched and dented, balls because they found that they could be driven further. Dimples were introduced in the late nineteenth century, and were found to allow golf balls to be driven 50–100% further than comparable smooth balls. It turns out that dimples do this in two ways: first, they reduce drag through a mechanism that we'll discuss next, and second they entrain air near the boundary, so that they can produce more lift if they are given backspin when struck. Similarly, "hop-up" uses backspin to increase the range of air rifles. This entrainment also accentuates the bane of golfers, "hook" or "slice" on a poorly struck ball; we'll discuss curved trajectories later in this chapter; for more on the aerodynamics of sports balls; see, for example, Mehta.[5]

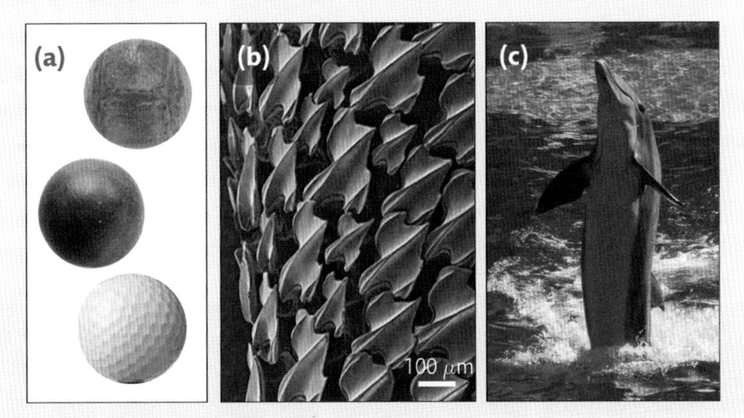

Fig. 8.6 (a) Top to bottom: leather-feather, gutta percha, and modern dimpled golf balls. (Gutta percha: Reproduced by kind permission of The Royal and Ancient Golf Club of St Andrews.) (b) "Riblets" on sharkskin, believed to reduce drag. Scale bar approximate (Eye of Science/Science Photo Library). (c) Dolphin holding itself up using muscle power not accurately accounted for in Gray's paradox (Courtesy of Loren Sztajer).

On another front, in naval engineering, a zoologist named James Gray (1891–1975) concluded in 1936 that dolphins achieve higher speeds with 1/7th the effort than can be explained using drag understood to act on smooth bodies. This conclusion, termed

"Gray's paradox," led to the finding that sea creatures like sharks have "riblets," shown in Fig. 8.6(b), on their skins that reduce drag, and in both naval and aeronautic circles, work continues to this day to develop both passive (stationary) and active (moving) control surfaces that reduce drag. Indeed, the Speedo company has promoted "fastskin" swim suits claimed to reduce drag by emulating sharkskin. Similarly, some marine paints have been marketed to reduce drag by modifying surface characteristics, and the scalloped leading edges of whale fins have been found to significantly increase lift and reduce drag.

Notwithstanding the real reductions in drag that Gray first observed, it was found in 2008 that Gray's calculations were wrong. Dolphins can deliver substantially more force than Gray believed—indeed, they are well known to be able to hold themselves out of the water by the force of their tail fins, as shown in Fig. 8.6(c). These newer calculations reveal that although dolphins, sharks and other creatures are optimized by body shape and surface properties for low drag, there is no underlying paradox: the drag that they overcome is exactly accounted for by the energy that they expend.

EXERCISE 8.5

We've been discussing the interface between an animal and its environment. Fig. 8.7 shows several different examples of animals that manipulate that interface to live on a surface. For each case, identify the animal and explain what it manipulates to stay at the surface. Note that surface tension is involved in most of these cases, but isn't sufficient by itself. For example, the animal in case (a) would sink if it stayed still, and sticks shaped like the animal in case (b) would puncture the water surface and sink as well. In detail then, what forces are involved, and what keeps each of these animals at the surface? Some of these answers are not yet known, by the way: do your best at a minimum to identify the forces involved.

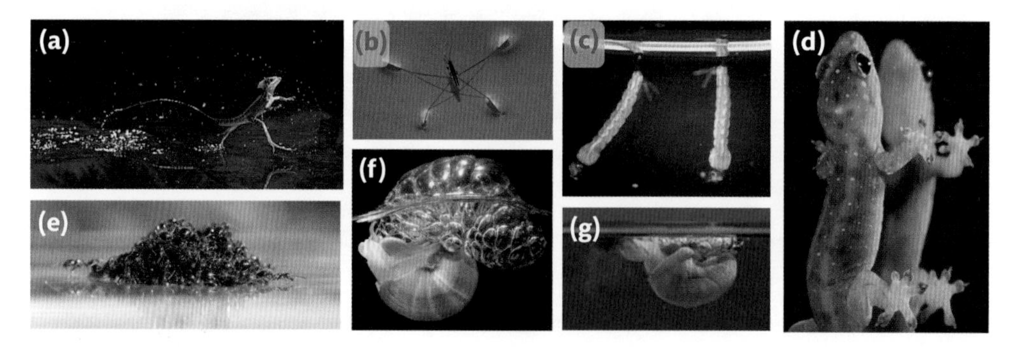

Fig. 8.7 Surface dwelling animals. (Sources: (a) © Stephen Dalton / naturepl.com. (b) mjf99/Shutterstock.com. (c) © CDC/ Dr. Pratt. (d) Phonpipat Jampatip/Shutterstock.com. (e) From: Mlot, N., Tovey, C., Hu, D. (2011). Fire ants self-assemble into waterproof rafts to survive floods. *Proceedings of the National Academy of Sciences 108* (19) 7669–7673. By permission of PNAS. (f) Denis Riek. (g) © NHPA/Photoshot.)

Let's return to examining precisely how roughness affects drag. Figs. 8.4(d) and (e) tell us that drag can be reduced for a *smooth* obstacle by increasing the Reynolds number, thereby pushing the separation points downstream (asterisks in the figures). Moving the separation point downstream in turn reduces the size of the trailing wake structure, which we've seen ultimately reduces form drag.

By comparison, in Fig. 8.5(a) we see that adding roughness reduces the Reynolds number at which the dip in drag occurs. For smooth obstacles this dip is associated with washing the separation point downstream and shrinking the trailing wake. So does roughness cause the same to occur?

To answer this question, in Figs. 8.5(c) and (d) we compare trailing wakes of dye for a smooth sphere (Fig. 8.5(c)) and a sphere with artificial roughness introduced by encircling the sphere with a wire hoop (Fig. 8.5(d)). The second of these snapshots shows that the wire triggers both the downstream motion of the separation point and the contraction of the wake at Reynolds numbers an order of magnitude lower than would be seen for a smooth sphere. In the enlarged inset to Fig. 8.5(d), we highlight turbulence downstream of the wire. This turbulence in the boundary layer insulates the flow outside of the layer from the sphere itself, and as a result the sphere can exert less stress on the surrounding fluid, and the shearing flow more easily drags the separation point downstream. Thus adding roughness causes turbulence at the surface that reduces drag by effectively lubricating the obstacle from the external shear flow. Sharks, eels, and some fish produce the same lubricating effect by using specialized mucus (discussed later in this book).

It is curious that adding turbulence *decreases* drag in flow past an obstacle. In flow through a pipe, on the other hand, turbulence also appears, but in that case it produces a very different, and in a sense more expected, result. The dependence of C_d on Re is shown in Fig. 8.5(b) for pipe flow (to be discussed again in Chapter 9). At low Reynolds number, the drag coefficient goes as $1/Re$, which can be derived using a simple analysis similar to that shown in Eqs. [8.12]–[8.15]. At about 2×10^3, the coefficient abruptly jumps, and above that Reynolds number, the drag coefficient reaches an asymptote. The jump in drag is directly caused by turbulence, and in pipe flow, turbulence, and drag, *increase* with roughness.

So in pipe flow, turbulence produces the response that one might expect: turbulence produced by roughness fills the pipe, reduces the area available for laminar flow, and so *increases* drag. It turns out that theoretically, a perfectly smooth pipe should exhibit no turbulence at all. Of course no real pipe is perfectly smooth, and there are important practical intricacies involved in the development of turbulence in a pipe as well; see, for example, Nikuradse[6] and Gioia and Chakraborty.[7] Notwithstanding these intricacies, the overarching message is that the boundary layer turbulence caused by roughness in *exterior* flow past an obstacle reduces drag, while the same in *interior* flow within a pipe increases drag. We'll have more to say about drag in pipes in Chapter 9.

EXERCISE 8.6

Figs. 8.5(c) and (d) identify a flow of interest using simple dye. It is important that the dye moves as a "passive" marker, meaning that it isn't buoyant, it doesn't contain heavy particles that could fail to take a turn rapidly (more on this in a later section), it doesn't react or change the viscosity of the fluid, etc. Many methods have been introduced to track flows, some in two dimensions, some in three dimensions, some imposing structure on the flow, some using esoteric techniques, etc. Identify and briefly summarize as many methods for the visualization of fluids and their flows as you can: at least a dozen distinct methods have been used in the literature. As part of your summary, explain how dyes used for tracking flows are made neutrally buoyant. Bear in mind that many dyes are made by adding a powder in a fluid: shouldn't this increase the fluid density?

8.4 **Curve Balls**

Now that we know something about how an object interacts with its boundary, let's consider what happens to a spinning ball to make it curve. Isaac Newton in 1672 appears to have been the first to record that tennis balls curve in response to spin. He explained this in a picturesque and personified way typical of his time, by saying that the faster moving side of the ball would "beat the contiguous air more violently than the other [side]," and so would excite a "reluctancy and reaction of the air." Newton proposed this picture as part of an appeal to explain the refraction of light, and didn't remark further on curving of balls.

Example Rifling and projectile design

Nearly 200 years elapsed until curve balls were meaningfully studied: the next significant contribution came from H.G. Magnus (1802–1870), mentioned in Chapter 6's discussion of the falling sphere viscometer. Magnus studied the effect of rifling on the trajectories of artillery shells for the German military. It had been known at least since the fifteenth century that the accuracy of arrows, bullets, and cannon shells could be improved by rifling: cutting helical grooves to cause a projectile to rotate in flight—similar to spin on an American football. The effectiveness of rifling was minimal through the eighteenth century, in large part because the explosives used (chiefly black powder) left large amounts of sooty residue that fouled rifling grooves. So-called "smokeless explosives" developed in the nineteenth century largely eliminated this problem.

Enter H.G. Magnus, who produced equations to relate projectile spin to the amount of drift, or curvature, of its trajectory. We'll discuss the underlying mechanical effects that he uncovered shortly, but in this example we focus on rifling and its consequences. Rifling has undergone a great deal of study since the nineteenth century, one aspect of which was the work of Eugene Stoner (1922–1997), who developed the M16 rifle and ammunition that has been used for at least 50 years by the US military. This weapon has a diabolically clever design: it uses a high rifling aspect ratio to make the bullet spin very fast, and it fires a bullet that is intentionally tail-heavy.

This design was developed to cause the bullet to fly straight—because it spins rapidly – but to tumble energetically when it strikes soft tissue—because the fast-spinning bullet is unstable and its tail tends to overtake its nose. As a result, it produces extreme tissue damage using a small caliber bullet. This approach was adopted in part because the US was signatory to the 1907 Hague Convention, which specifies that "... it is especially forbidden [in international warfare] ... to employ arms, projectiles, or material calculated to cause unnecessary suffering." For this reason, hollow-point and similar bullets designed to expand or flatten on impact were avoided in favor of the M16 design—likewise for the Russian AK-74. In a similar way, weapons using plastic shrapnel to injure but not kill—and to be difficult to detect by x-ray—were designed during the Vietnam War, because the strategic decision was made that this would maximize the burden of injuries on the Viet Cong. Thus designs by engineers both intentionally increased suffering and provided a stratagem to circumvent the letter of international law. These issues may bring to the reader's mind Chapter 7's discussion of scientific ethics.

Returning to Newton's picturesque description, let's try to construct a meaningful mathematical description by using Bernoulli's equation, Eq. [7.8] of Chapter 7:

$$\frac{V^2}{2} + \frac{P}{\rho} = constant. \qquad [8.16]$$

This tells us that if we define V_{top} and V_{bottom} to be the speeds of fluid at the top and bottom of a body, then the difference in pressure between these surfaces is:

$$P_{bottom} - P_{top} = \frac{\rho}{2}\left(V_{top}{}^2 - V_{bottom}{}^2\right). \qquad [8.17]$$

It's useful to separate the right side of this equation into two parts as follows:

$$\begin{aligned} P_{bottom} - P_{top} &= \rho\frac{(V_{top} + V_{bottom})}{2}(V_{top} - V_{bottom}) \\ &= \rho \cdot \langle V \rangle \cdot (V_{top} - V_{bottom}), \end{aligned} \qquad [8.18]$$

where we have defined $\langle V \rangle$ to be the average velocity, $\langle V \rangle = (V_{top} + V_{bottom})/2$. So the lift (if the orientation is as shown in Fig. 8.8), or drift (if the picture is on its side), acting on an obstacle is simply proportional to the mean speed, $\langle V \rangle$, and the difference in speeds, $V_{top} - V_{bottom}$.

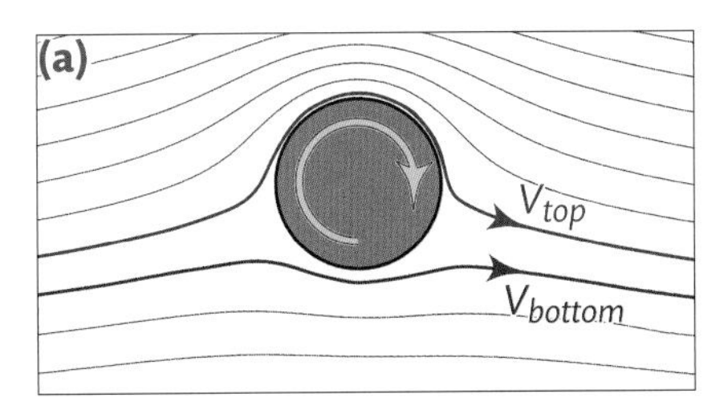

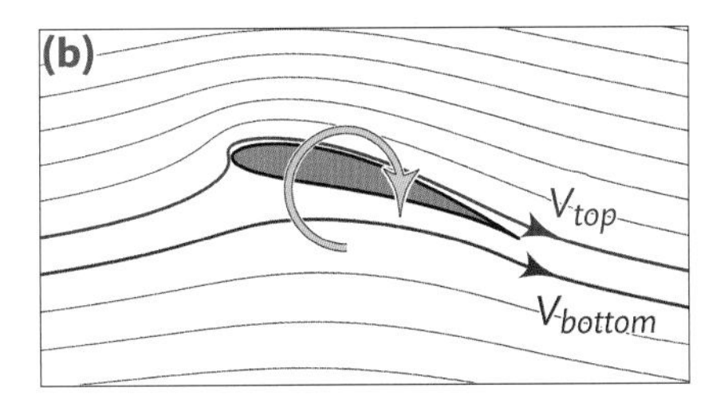

Fig. 8.8 (a) Streamlines of flow around spinning cylinder; (b) streamlines of flow around static airfoil. Cyan arrows indicate circulation.

As we have indicated in Fig. 8.8, lift can be produced either by spinning a symmetrical body—as shown in Fig. 8.8(a)—or by making the body asymmetrical so that the fluid takes different paths on top and bottom—as shown in Fig. 8.8(b). We'll have more to say about the analogy between spinning bodies and airfoils in a moment, but first let's calculate what the actual force acting on a body is due to the pressure difference shown in Eq. [8.18]. This is easily done: we need only integrate Eq. [8.18] over the surface area:

$$\begin{aligned} F_{lift} &= L\int_{left}^{right}(P_{bottom} - P_{top})dx \\ &= \rho L\langle V\rangle\int_{left}^{right}(V_{top} - V_{bottom})dx, \end{aligned} \qquad [8.19]$$

where the integral is from left to right as drawn in Fig. 8.8, and L is the total length of the cylinder or airfoil. But look, if we switch the order of integration, we can remove the minus sign to get:

$$F_{lift} = \rho L\langle V\rangle\left(\int_{left}^{right} V_{top}dx + \int_{right}^{left} V_{bottom}dx\right), \qquad [8.20]$$

which means that we are integrating left to right along the bottom, and then right to left along the top. This is the same as integrating in a circle around the obstacle. We write this as:

$$F_{lift} = \rho L\langle V\rangle\oint V dx, \qquad [8.21]$$

and the integral around the obstacle is defined to be the "circulation," Γ:

$$\Gamma = \oint V dx. \qquad [8.22]$$

This couldn't be simpler, right? Putting aside the density, ρ, and length, L (which are just fixed numbers defined for each problem), the lift acting on a body is simply the mean speed times the circulation. So the faster an airplane goes, the greater the lift: this seems self-evident. But also to *design* the greatest lift, the only thing that matters is maximizing the circulation. The shape of the wing? The size of the fuselage? The height of the rudder? None of these matter to lift except insofar as it affects the circulation—that is the only thing that needs to be controlled.

Eqs. [8.21] and [8.22] are of obvious importance in aeronautical engineering, and are known as the Kutta–Joukowski theorem, after German mathematician G.W. Kutta (1867–1944) and Russian N.Y. Joukowski (1847–1921), sometimes called the father of modern aerodynamics. We'll have more to say about contributions of both of these scientists.

Example Lift and circulation

We'll examine next what the circulation depends on, but first let's consider some unexpected applications of Eq. [8.21]. A first example is the Flettner rotor, shown in a modern freighter in Fig. 8.9(a). Anton Flettner (1885–1961) was a German engineer who worked chiefly on helicopter design. Flettner recognized the parallel between lift produced by circulation on a rotating cylinder and on a wing shown in Figs. 8.8(a) and (b). He reasoned that vertical rotating cylinders could propel a boat forward in a crosswind, and he designed and launched the first sailboat using this idea in 1925. The rotors require a motor, typically diesel, and so the ship isn't as energy efficient as a conventional sailboat, but the rotors have the advantages that the forward lift can be selected by simply setting the circulation of the rotors. It also turns out that Flettner rotors can tack more tightly into the wind than traditional sails, an idea possibly as counterintuitive as adding turbulence to reduce drag (Fig. 8.5).

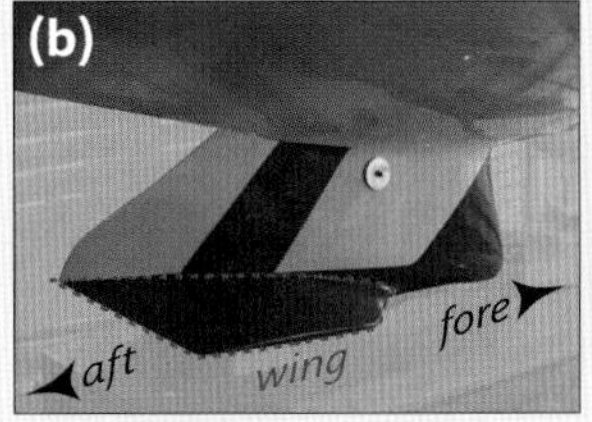

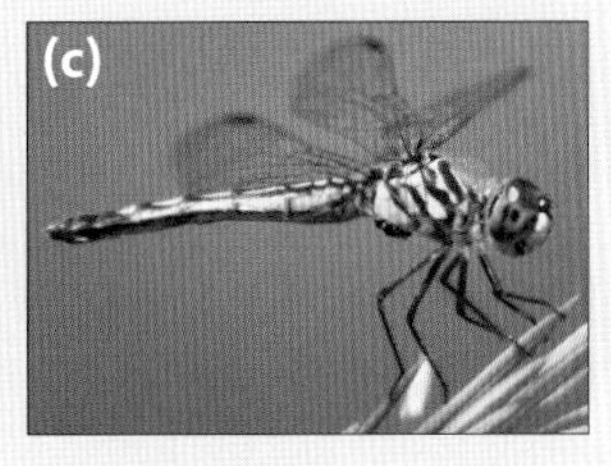

Fig. 8.9 (a) E-ship 1, using Flettner rotors as sails to produce lift (Alan Jamieson/CC-BY-2.0). (b) Winged keel on Australia II sailboat to stabilize the keel by producing negative lift (Adam Fuller/CC BY 2.0). (c) Dragonfly, which produces and sheds vortices downward to generate lift upward (R.A. Nonenmacher/CC BY-SA 3.0).

Another application of Eq. [8.21] to sailing appears in Fig. 8.9(b), which shows the first "winged keel" on the Australia II, winner of the 1983 America's Cup. This keel incorporates an airfoil with long side down (i.e. Fig. 8.8(b) constructed upside-down). This design caused vigorous controversy at the time, as it kept the keel vertical in rapid tacking maneuvers and allowed the team from Perth to break the New York Yacht Club's 132 year undefeated record in the competition. Thus paradoxically, adding vorticity allowed the Australia II to gain speed over a traditional design.

A final example of a clever connection between circulation and lift has been found in insect flight.[8] In Fig. 8.9(c), we show a dragonfly, which generates a vortex with its forewing, passes it to its hindwing, and discharges it beneath to generate upward thrust. Thus the upward momentum imparted to keep the fly aloft is provided by downward momentum of the vortices being shed.

EXERCISE 8.7

Since Flettner worked on helicopters, perhaps you can deduce an important consequence of his work in that application. The fastest helicopter built, the Westland Lynx, can travel at a little over 300 km/hr—impressive, but a *hundred times* slower than the fastest airplane, the SR-71 Blackbird. Based on Flettner's work on the rotor, what limits helicopter speed? If you look up the answer, you'll see that, not surprisingly, lots of factors that govern the speed, but there is one simple over-riding limitation that you should be able to work out just by thinking about the geometry of helicopter blades.

To solve for flow around rotating bodies, let's start with what we know: the potential function (Eq. [8.5]) and stream function (Eq. [7.64] of Chapter 7) for flow past a sphere:

$$\varphi = V_\infty\left(r + \frac{R^2}{r}\right)\cos(\vartheta), \qquad [8.23]$$

$$\psi = V_\infty\left(r - \frac{R^2}{r}\right)\sin(\vartheta). \qquad [8.24]$$

These are based on 2D calculations, and in two dimensions a sphere and a cylinder are indistinguishable, so the same functions apply to flow past a cylinder, where it is understood in both geometries that ϑ must be the angle to $\vec{V}_\infty$.

It turns out that analysis is simplest if we use some shorthand from complex analysis. What follows is much like the nabla, $\vec{\nabla}$, introduced at the start of the book: it is merely a shorter way of writing things: nothing substantive is being done, but writing things as we do will make the algebra simpler. This said, let's define a "complex potential," that contains both the potential and stream functions in one, as follows:

$$\Phi(z) = \varphi + i\cdot\psi, \qquad [8.25]$$

where z is the complex position, $z = x + iy = r\cdot e^{i\vartheta}$. We acknowledge that rewriting perfectly good equations in yet another way, involving complex numbers for no apparent reason, seems like a perverse thing to do, but we ask the reader's indulgence: it really does make things easier.

If we now plug Eqs. [8.23] and [8.24] into Eq. [8.25] and make use of the relations $x = r\cdot\cos(\vartheta)$ and $y = r\cdot\sin(\vartheta)$, we get

$$\begin{aligned}\Phi(z) &= V_\infty x\left(1 + \frac{R^2}{r^2}\right) + i\cdot V_\infty y\left(1 - \frac{R^2}{r^2}\right)\\ &= V_\infty\left[(x + i\cdot y) + R^2\frac{(x - i\cdot y)}{r^2}\right].\end{aligned} \qquad [8.26]$$

But $r^2 = (x + i\cdot y)(x - i\cdot y)$, so:

$$\begin{aligned}\Phi(z) &= V_\infty\left[(x + i\cdot y) + \frac{R^2}{(x + i\cdot y)}\right]\\ &= V_\infty R\left[\frac{z}{R} + \frac{R}{z}\right].\end{aligned} \qquad [8.27]$$

As advertised, this is a very compact way of describing flow past a cylinder. This can be made simpler yet if we take for granted that Complex Analysis invariably makes hard things easy. In this case, we might be tempted to write *the* simplest imaginable thing, that the velocity too is complex: $V_{complex} = V_x + i \cdot V_y$, and in the best of possible worlds, since the ordinary velocity goes as $\partial\varphi/\partial\vec{x}$, we might imagine that:

$$\begin{aligned} V_{complex} &= \frac{\partial \Phi}{\partial z} \\ &= V_\infty\left(1 - \frac{R^2}{z^2}\right). \end{aligned} \quad [8.28]$$

EXERCISE 8.8

Confirm that Complex Analysis does give us the best of all possible worlds: solve for $V_{complex}$ from Eq. [8.28] and confirm that the velocities are the same as those obtained in Chapter 7 (Eq. [7.62]). You can grind out the answer using $z = x + i \cdot y$, but you will get the same answer much more easily if you use $z = re^{i\vartheta}$. Trigonometric double-angle identities will aid this calculation.

Eq. [8.27] gives us a tidy description of potential flow past a sphere. What about circulation? As we have mentioned, a very nice property of stream and potential functions is that we can choose any function at all for either of these, and we can be sure that it will describe some flow that will conserve area (for a stream function) or vorticity (for a potential function). We can therefore experiment with all kinds of functions and investigate what flows each function defines. The same is true for the complex potential (Eq. [8.25]), but in that case once we choose a function, *both* φ and ψ are prescribed: that is, the flow is required to conserve both area and vorticity.

In the spirit of experimentation, we'll examine several functions shortly; for now, let's experiment with a particular function that turns out to have an intimate connection with vorticity:

$$\Phi = -i\frac{kV_\infty}{2\pi}ln(z). \quad [8.29]$$

The logarithm is nice because when we substitute $z = r \cdot e^{i\vartheta}$, it naturally simplifies:

$$\begin{aligned} \Phi &= -i\frac{kV_\infty}{2\pi}ln(r \cdot e^{i\vartheta}) \\ &= -i\frac{kV_\infty}{2\pi}[ln(r) + i\vartheta] \\ &= \frac{kV_\infty}{2\pi}[\vartheta - i \cdot ln(r)]. \end{aligned} \quad [8.30]$$

From Eq. [8.25], φ must go as ϑ, and ψ must go as $\ln(r)$. This seems simple enough: it means that the potential function simply consists of lines of constant ϑ—that is, lines outward from the origin. This is fine for a circulating flow, because the perpendiculars are lines of constant r, encircling the origin as illustrated in Fig. 8.10. Makes perfect sense for a circulating flow, right?

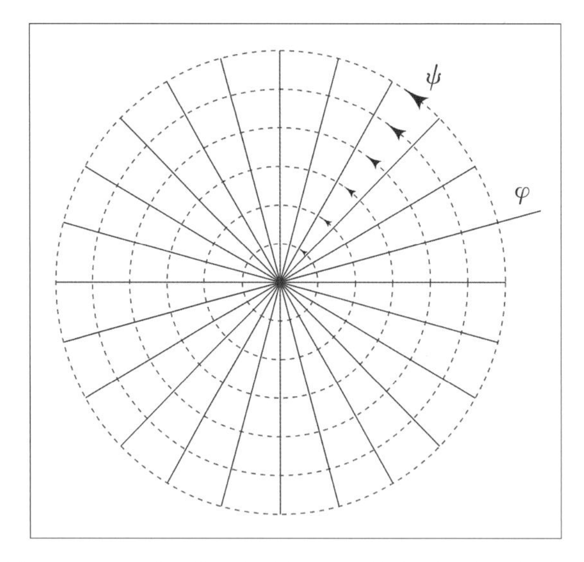

Fig. 8.10 Equipotential lines and streamlines for circulating flow. Notice that in order for the streamlines (red) to symmetrically encircle the origin, the equipotentials (blue) must be lines of constant ϑ, and as explained in the text, this implies in turn that the stream function can only go as ln(r).

But wait: this means that in order for the potential function to be lines outward from the origin, the radial dependence is somehow *prescribed*: the dependence *has* to be logarithmic in r. The same result is obtainable by using the Cauchy–Riemann equations to derive ψ from φ.

EXERCISE 8.9

Something is very peculiar—and revealing—about this result. Let's rewrite Eq. [8.30] in Cartesian coordinates. In this case, we get:

$$\Phi = \frac{kV_\infty}{2\pi}\left[arctan(y/x) - \frac{i}{2}ln(x^2 + y^2)\right]. \tag{8.31}$$

So exactly the same velocities are prescribed for the *arctan* and for the *logarithm*. Perform the necessary differentiations to confirm this and then explain how these two very different functions are so closely related.

Additionally, refer back to the Rankine vortex solution in Chapter 7. In that problem, two different circulatory solutions were obtained: an inner solution near the origin, and an outer solution far from the origin. You should find that the velocities you obtain are exactly one of these two solutions. This is reassuring, but what about the *other* Rankine velocity? The present exercise confirms that *a flow with streamlines encircling the origin can only have stream function that is logarithmic in radius.* Explain why a circulating flow can only be logarithmic and explain whether or not the Rankine solution is therefore erroneous.

This exercise identifies a relation between two functions that otherwise might have gone unnoticed. The presence of related functions will hold for any potential flow: any potential function that you choose will always prescribe a second, orthogonal, function defined by the Cauchy–Riemann equations. And both of these functions together will define a third, complex, potential function.

To recap, we now have one complex potential, Eq. [8.27], that defines flow past a cylinder, and a second complex potential, Eq. [8.29], that defines circulation. We've seen this situation before: in Chapter 7, we obtained potential flow past a sphere by adding two potential functions together (Eq. [7.60])—which we can do since Laplace's equation is linear. Laplace's equation has remained linear since last chapter, so we can add Eqs. [8.27] and [8.29] to get circulatory flow around a cylinder:

$$\Phi(z) = V_\infty R\left[\frac{z}{R} + \frac{R}{z} - \frac{ik}{2\pi R}ln(z)\right]. \quad [8.32]$$

To plot the flow, we need to extract the stream function, which we do next. We separate the real (φ) and imaginary (ψ) parts of Φ, by writing $z = r \cdot e^{i\vartheta}$ as usual, to get

$$\begin{aligned}\Phi &= V_\infty\left\{r \cdot e^{i\vartheta} + \frac{R^2}{r}e^{-i\vartheta} - \frac{ik}{2\pi}[ln(r) + i\vartheta]\right\} \\ &= V_\infty\left\{r \cdot [\cos(\vartheta) + i \cdot \sin(\vartheta)] + \frac{R^2}{r}[\cos(\vartheta) - i \cdot \sin(\vartheta)] - \frac{ik}{2\pi}[ln(r) + i\vartheta]\right\} \\ &= V_\infty\left[r \cdot \cos(\vartheta) + \frac{R^2}{r} \cdot \cos(\vartheta) + \frac{k\vartheta}{2\pi}\right] + iV_\infty\left[r \cdot \sin(\vartheta) - \frac{R^2}{r} \cdot \sin(\vartheta) - \frac{k}{2\pi}ln(r)\right].\end{aligned} \quad [8.33]$$

The real and imaginary parts of this are then:

$$\varphi = V_\infty\left[\left(1 + \frac{R^2}{r^2}\right) \cdot r \cdot \cos(\vartheta) + \frac{k\vartheta}{2\pi}\right], \quad [8.34]$$

$$\psi = V_\infty\left[\left(1 - \frac{R^2}{r^2}\right) \cdot r \cdot \sin(\vartheta) - \frac{k}{2\pi}ln(r)\right]. \quad [8.35]$$

EXERCISE 8.10

Plot streamlines as the strength of circulation, k, is increased. You should get results that look like Figs. 8.11(a)–(d). Those plots use $k < 0$, which produces clockwise rotation and positive lift as in Fig. 8.8; you can use either positive or negative k for your plots.

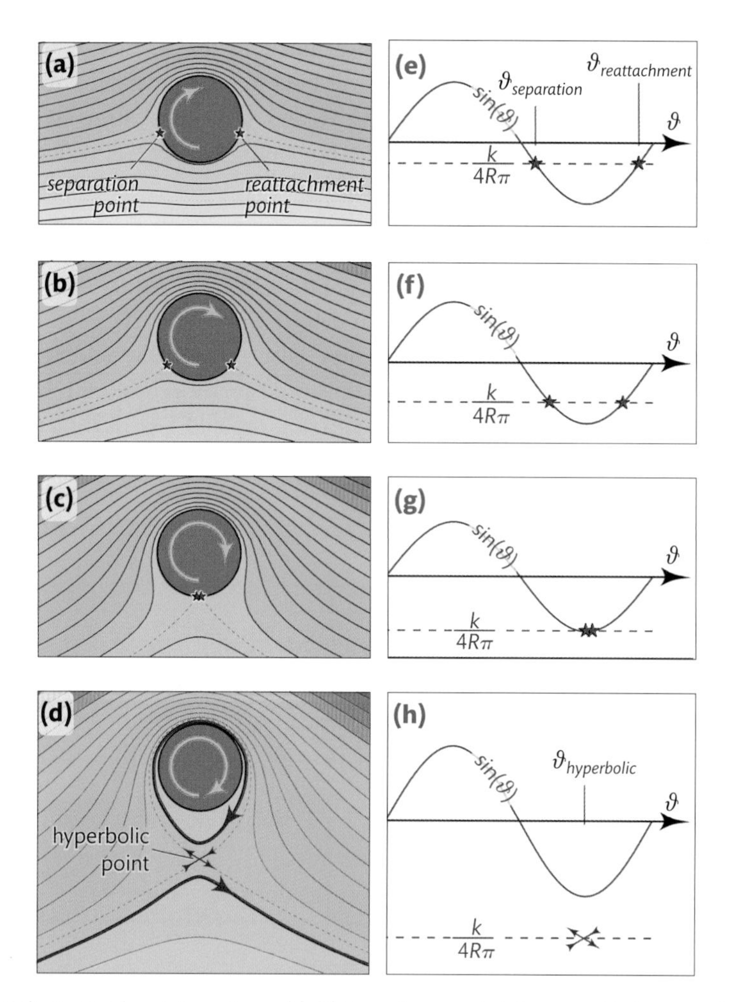

Fig. 8.11 Streamlines around a rotating cylinder. (a)–(d) Streamlines as spin is increased. Notice separation and reattachment points (red asterisks) converge as spin increases. (e, f) Corresponding comparison between $\sin(\vartheta)$ and $k/4R\pi$. Notice a hyperbolic point appears when $\sin(\vartheta) = k/4R\pi$.

Notice that the separation and reattachment points (red asterisks in Fig. 8.11) move closer together as k is increased, and at a certain value of k the points merge and leave the surface of the cylinder. The locations of these points can be calculated by observing that the angular speed at the cylinder surface is zero where the flow separates. The angular speed at $r = R$ is:

$$\begin{aligned} V_\vartheta &= \frac{1}{r}\frac{\partial\varphi}{\partial\vartheta}\bigg|_{r=R} \\ &= V_\infty\left[-2R\cdot\sin(\vartheta) + \frac{k}{2\pi}\right], \end{aligned} \qquad [8.36]$$

and setting $V_\vartheta = 0$ gives the criterion:

$$\sin(\vartheta) = \frac{k}{4R\pi}. \qquad [8.37]$$

Figs. 8.11(e)–(h) show plots of $\sin(\vartheta)$ in blue, and $k/4R\pi$ in red (recall that $k < 0$). The separation and reattachment points are where these two curves are equal. Confirm in your streamline plots that the angles at which separation occurs agree with Eq. [8.37] *within several significant digits*. You'll need to give some thought to how you can accurately estimate the separation points—you can do it graphically, but you'll need to produce streamlines very near the separation point, perhaps straddling the point using what is called a "shooting" method. Or you can do it numerically, for which you'll have to use "Newton's method." Or perhaps you can construct a different approach of your own.

We make a final observation to presage a topic that we'll bring up in a later chapter: notice that when $| k/4R\pi | > | \sin(\vartheta) |$, Eq. [8.37] has no solution, meaning that there are no points on the cylinder surface that satisfy $V_\vartheta = 0$. The separation and reattachment points are referred in the mathematics literature as "parabolic" points, and when they combine and leave the cylinder surface, the topological convergence of trajectories shown by a red X in Fig. 8.11(d) is referred to as a "hyperbolic" point. Note that two parabolic points turn into one hyperbolic point: this is guaranteed by a topological theorem constraining what changes a flow can possibly undergo. This is an elegant result that at its heart states that the sum of elliptic and hyperbolic points in any well behaved flow cannot change. So, if you stir a reactor faster, you can spawn new elliptic points, but they will always be accompanied by a hyperbolic point (or a pair of parabolic points) elsewhere. Details of this lovely result are unfortunately beyond the topic of this book; however the reader can find more information by investigating "topological index theorems."

8.5 **Conformal Mappings**

We've obtained stream and potential functions that define flow and circulation past a circular obstacle, Eqs. [8.33]–[8.35]. This is nice, but of course most obstacles aren't this simple. Is there a way to use a simple solution, like Fig. 8.8(a)), to obtain stream and potential functions for more complicated geometries, like Fig. 8.8(b)?

Here's an idea: since we know that any stream function we choose can be used to define a flow, maybe we can take our existing stream function and distort it somehow to conform to a more complicated geometry of our choosing. How might this be done?

Let's start with the canon that Complex Analysis makes hard things easy. What's the easiest thing that could possibly be done? Well, we could take an existing solution, for example flow past a cylinder, and apply any old random function to transform the solution that we know into a new solution. And in the best of all possible worlds, the result would solve Laplace's equation, and would produce new stream and potential functions that would solve a new flow problem.

The world is not quite that fantastic, but it's awfully close. We have seen that a stream function and a potential function both are solutions to Laplace's equation, and we know that these two functions must be perpendicular to one another. So for a transformed solution to also be a solution, the transformed stream and potential functions must still be perpendicular at every point. A transformation that preserves angles in this way is called "conformal," and almost any well-behaved function turns out to produce a conformal transformation, or "mapping." Technically, well-behaved means twice differentiable: any twice differentiable function will take one flow and transform it into a new one. Practically speaking, there are tables that provide conformal mappings for any reasonably simple geometry. Moreover, conformal mappings can be produced for very complicated geometries using what are called "Schwartz–Christoffel" transforms: there are texts and online computer programs to achieve this. An excellent resource for these problems is Mathews and Howell.[9]

In the rest of this chapter, we'll look at several examples of conformal mappings that allow us to obtain flows in a variety of useful geometries. We'll start by obtaining flow around an ellipse by applying a simple function to the solution for flow around a circle (Eqs. [8.33]–[8.35]), and we'll then embellish this solution to provide flow around more complicated airfoils, as shown in Fig. 8.8(b). Then we'll turn to other geometries, such as flows in or around wedges (cf. Chapter 3, Fig. 3.7), out of pipes, and so forth.

8.5.1 **Flow Around an Ellipse**

Let's begin. As proposed, we'll start with flow around a circle, and apply a function that maps the flow past the circle onto flow past an ellipse. This function turns out to be:

$$w = z + \frac{\lambda^2}{z}. \qquad [8.38]$$

EXERCISE 8.11

Confirm graphically that Eq. [8.38] makes ellipses from circles directly, by plotting circles in the complex plane where $z = x + i \cdot y$, then applying Eq. [8.38] and plotting the result in the new complex plane where $w = u + i \cdot v$, as shown in Fig. 8.12. Confirm (again graphically, by zooming in on a few cases) that the perpendiculars to circles in the z-plane remain perpendicular to the ellipses in the w-plane. Choose λ to be a convenient number, for example, $\lambda = 2$ in the figure.

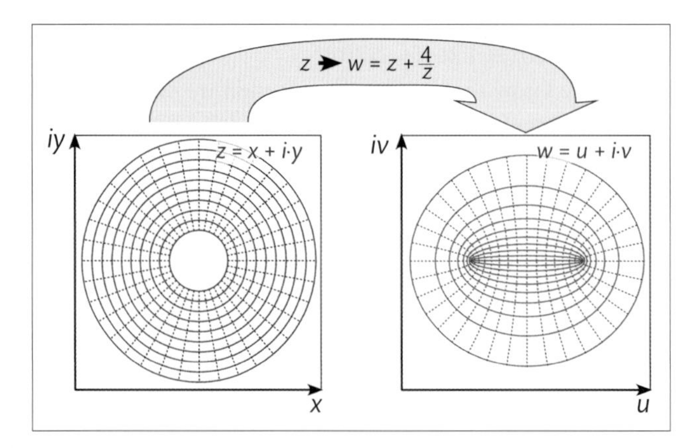

Fig. 8.12 Conformal mapping (Eq. [8.38]) that takes circles to ellipses. Notice that the perpendiculars to the circles remain perpendicular to the ellipses after application of the conformal mapping (perpendiculars are dotted lines in both plots).

After you've reconstructed Fig. 8.12, try a slightly different mapping:

$$w = z + \frac{\lambda^2}{z^3}. \qquad [8.39]$$

Before you make a plot of what this does to a set of circles, guess what you think the mapping may do. Was your guess correct? What do you suppose will happen if the z^3 is changed into a z^4?

We can also confirm algebraically that the mapping of Eq. [8.38] takes a circle to an ellipse. We know that $z^2 = x^2 + y^2 = 1$ defines a circle for radius 1, while $x^2/a^2 + y^2/b^2 = 1$ defines an ellipse of radii a and b. So for points on a circle, $z^2 = 1$, does w^2 for those points produce $x^2/a^2 + y^2/b^2 = 1$?

We can solve for w^2 as follows. We know that:

$$\begin{aligned} w &= z + \frac{\lambda^2}{z} \\ &= x + i \cdot y + \lambda^2 \frac{x - i \cdot y}{x^2 + y^2} \\ &= x\left(1 + \frac{\lambda^2}{r^2}\right) + i \cdot y\left(1 - \frac{\lambda^2}{r^2}\right), \end{aligned} \qquad [8.40]$$

so:

$$w^2 = \left[x\left(1 + \frac{\lambda^2}{r^2}\right)\right]^2 + \left[y\left(1 - \frac{\lambda^2}{r^2}\right)\right]^2. \qquad [8.41]$$

When $w^2 = 1$, this is exactly the equation for an ellipse $w^2 = x^2/a^2 + y^2/b^2 = 1$ with semi-minor and semi-major lengths:

$$\begin{aligned} a &= \left(1 + \frac{\lambda^2}{r^2}\right)^{-1} \\ b &= \left(1 - \frac{\lambda^2}{r^2}\right)^{-1}. \end{aligned} \qquad [8.42]$$

Here λ is termed the "eccentricity" of the ellipse: when $\lambda = 0$, the ellipse is simply a circle; when $\lambda = 1$, the ellipse compresses completely into a flat plate.

EXERCISE 8.12

We now know that Eq. [8.38] indeed maps circles to ellipses—and since $f(z) = z + \lambda^2/z$ is twice differentiable, it is conformal (which you also confirmed graphically in Exercise 8.11). So, confident that Complex Analysis makes everything beautiful, we can hope that flow past a cylinder, Eq. [8.27], can be mapped into flow past an ellipse by using Eq. [8.38]. Confirm this graphically by producing a plot like Fig. 8.13.

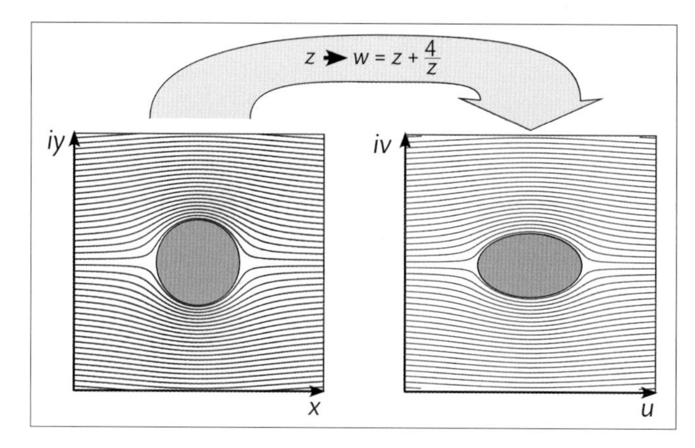

Fig. 8.13 Conformal mapping (Eq. [8.38]) applied to streamlines obtained from Eq. [8.27].

EXERCISE 8.13

In the spirit of continuing to sneak up on increasingly complicated flows, we can perform a cute trick—which turns out to work unreasonably well. The idea behind this trick is as follows. It is difficult to figure out how to rotate the ellipse to the right of Fig. 8.13 so as to produce a variable "angle of attack" on an obstacle of nontrivial shape. On the other hand, it is very easy to rotate the angle of attack on the circle to the left of Fig. 8.13. Could we perhaps take the ellipse, transform it back into a circle, rotate the flow around the *circle*, and then transform the result forward again to produce flow around an ellipse at a rotated angle of attack? Nutty though this idea is, the answer is yes, as shown in Fig. 8.14. Your task in this exercise is to make plots similar to those shown, using a few different eccentricities and angles of attack.

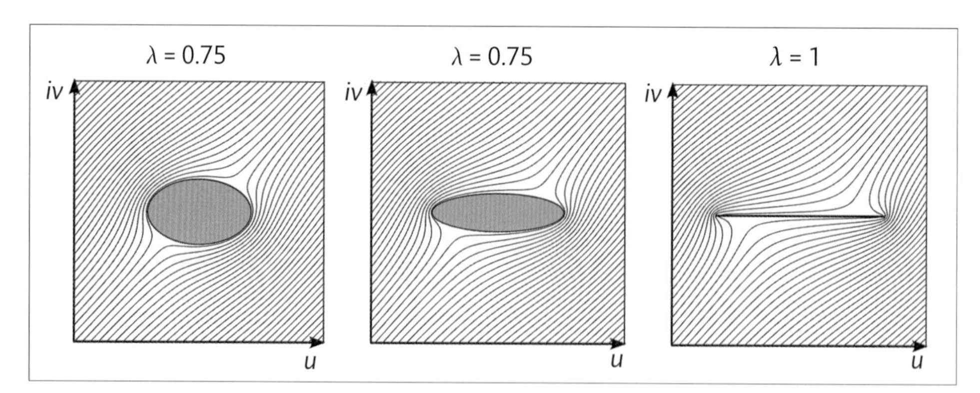

Fig. 8.14 45° angle of attack of flow on three different eccentricity ellipses. Notice that this approach works even for the limiting case of an ellipse, $\lambda = 1$: a flat plate.

To do this, consider the following question: what is the difference between a conformal mapping and a complex potential? Both are complex functions. Both keep perpendicular lines perpendicular: the conformal mapping takes one perpendicular grid and maps it to a new perpendicular grid (see Fig. 8.12), while the complex potential maps x- and y-directions (which are always perpendicular) to ϕ and ψ directions (likewise). Indeed, aside from interpretation there is no difference between distorting spaces with a conformal mapping and applying a complex potential function: the mapping distorts space, while the potential tells us how fluid moves—in a distorted space.

With this in mind, break the problem of a variable angle of attack on an ellipse into manageable pieces. First, start with an ellipse and apply a conformal mapping to get back a circle. Second, rotate the circle, and third apply a conformal mapping to get back to the ellipse. To keep things clear, in the following details of these steps, we'll use the variable w to denote space around the original ellipse, z to denote space around the circle, ζ to denote space around the rotated circle, and ω to denote space around the rotated ellipse.

(1) **Conformally map from ellipse to circle:** do this by inverting Eq. [8.38], the mapping that takes a circle to an ellipse. Show that this gives you:

$$z = \frac{w \pm \sqrt{w^2 - 4\lambda^2}}{2}. \tag{8.43}$$

(2) **Rotate by an angle ϑ_0:** to do this, just apply the mapping:

$$\zeta = z \cdot e^{i \cdot \vartheta_0}. \tag{8.44}$$

It should be clear that this rotates by ϑ_0: by applying it to a vector, $z = r \cdot e^{i\vartheta}$, we get $\zeta = r \cdot e^{i \cdot (\vartheta + \vartheta_0)}$.

(3) *Conformally map back from circle to ellipse*: do this by applying Eq. [8.38] again:

$$\omega = \zeta + \frac{\lambda^2}{\zeta}. \tag{8.45}$$

This should get you streamlines that look like Fig. 8.14. How? Recall that a conformal mapping can be viewed as a complex potential—and the imaginary part of a complex potential is a stream function, right? So the imaginary part of Eq. [8.45] is the stream function, whose contours are streamlines that determine fluid trajectories.

8.5.2 **Joukowski Transform**

A final example of flow past an obstacle involves an airfoil, as shown in Fig. 8.8(b). Remember N.Y. Joukowski, who brought us the relation between circulation and lift, Eqs. [8.21] and [8.22]? His interest was in lift on airfoils, which brought him to experiment with different complex potentials, and we have learned that these are the same as conformal mappings. Eq. [8.32] gave us flow past a circle with circulation; Joukowski developed a variation of that equation as follows:

$$\Phi(z) = z + \frac{e^{i \cdot \vartheta_0}}{z - s} - 2i \cdot \sin(\vartheta_0) \cdot \ln(z). \tag{8.46}$$

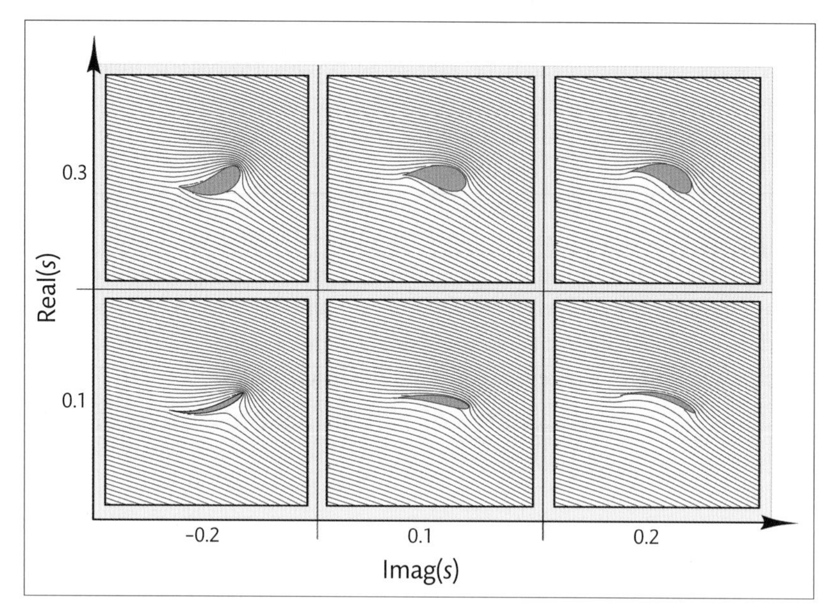

Fig. 8.15 Joukowski streamlines as real and imaginary parts of the shape parameter, s, are varied at fixed angle of attack, $\lambda_0 = 20^\circ$.

In this formula, ϑ_0 controls the angle of attack as before, and s affects the shape of the airfoil. The parameter s has both a real and an imaginary part; essentially the real part controls the thickness of the airfoil, and the imaginary part controls the amount that the airfoil bends, as shown in Fig. 8.15. It's remarkable that flows past such complicated objects can be obtained by merely plotting contours of a straightforward function. We'll see more examples after a summary.

8.5.3 **Summary and Other Conformal Geometries**

We summarize our discussion of conformal mappings by remarking that if we know some flow in one complex geometry, $z = x + i \cdot y$, then if we deform that geometry into a new one, $w = u + i \cdot v$, in an angle-preserving way, this necessarily defines a new flow. This idea is sketched pictorially in Fig. 8.16, where on the left we show a simple potential flow of flow past a couple of cylinders, and on the right we show a deformed space with irregular boundaries and obstacles. Notice we aren't saying how a conformal mapping from left to right can be found, but we are saying that if there is a potential flow in one, and a conformal mapping can be found from one to the other, then the potential flow in the other is just the first flow acted on by the mapping.

We close this chapter with descriptions of several other conformal mappings that may be useful for constructing flows in practical geometries. Many more mappings exist and as we mentioned earlier these can be found in tables. Additionally, Schwartz–Christoffel transforms will generate conformal mappings onto more complicated polygonal geometries. Each of the examples chosen here involves a specific lesson summarized in its title.

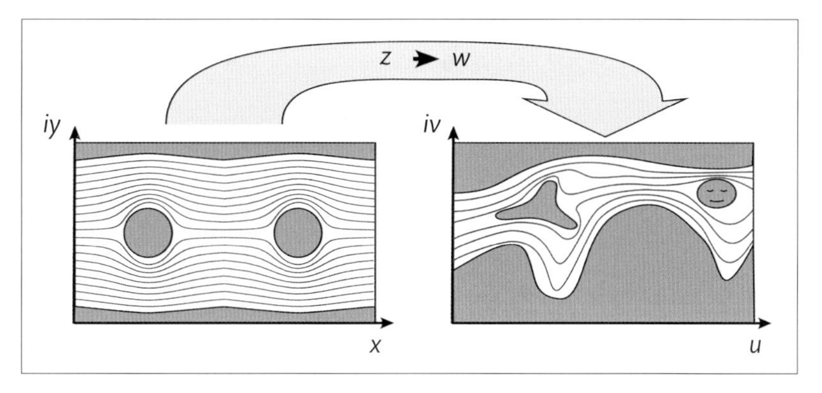

Fig. 8.16 Schematic of the idea underlying conformal mappings: a simple potential flow on the left, in $z = x + i \cdot y$ space, can be transformed into a new potential flow in a more complicated geometry, $w = u + i \cdot v$, on the right if a conformal mapping from z to w can be found.

8.5.4 **Wedge Flows—Lesson: Nature and Causes of Divergences**

A conformal mapping that can be completely analyzed and that is easily understood is

$$w = z^n, \qquad [8.47]$$

where n is a constant. Let's rewrite this in polar coordinates, so that:

$$\begin{aligned} w &= r^n e^{in\vartheta} \\ &= r^n \cos(n\vartheta) + i \cdot r^n \sin(n\vartheta). \end{aligned} \qquad [8.48]$$

Cosine and sine of course are periodic, so notice that if $2n$ is an integer, the flow will have a $2n$-fold symmetry. That is, whatever the flow looks like at ϑ, it will look the same at $\vartheta + 2n\pi$, so there will be $2n$ copies of the flow as the angle is varied from 0 to 2π.

This is somewhat abstract, so to see how this works in a more concrete way, recall again that a conformal mapping is equivalent to a complex potential, and that the imaginary part of the complex potential is the streamfunction. So the streamfunction of the mapping in Eq. [8.47] is:

$$\psi = r^n \sin(n\vartheta). \qquad [8.49]$$

Let's look at a particular example, $n = 2$, that we can solve exactly. In this case:

$$\begin{aligned} \psi &= r^2 \sin(2\vartheta) \\ &= 2[r \cdot \sin(\vartheta)][r \cdot \cos(\vartheta)] \\ &= 2xy. \end{aligned} \qquad [8.50]$$

This defines hyperbolas, as shown in Fig. 8.17(a), right? As x goes to 0, y has to go to ∞ to keep ψ constant, and the same for x as $y \to 0$. And since $n = 2$, there are $2n = 4$ copies of hyperbolas, as shown in the figure. Similarly, for $n = 3/2$, there are $2n = 3$ copies, shown in Fig. 8.17(b), and for $n = 5/2$, there are five copies, as shown in Fig. 8.17(c).

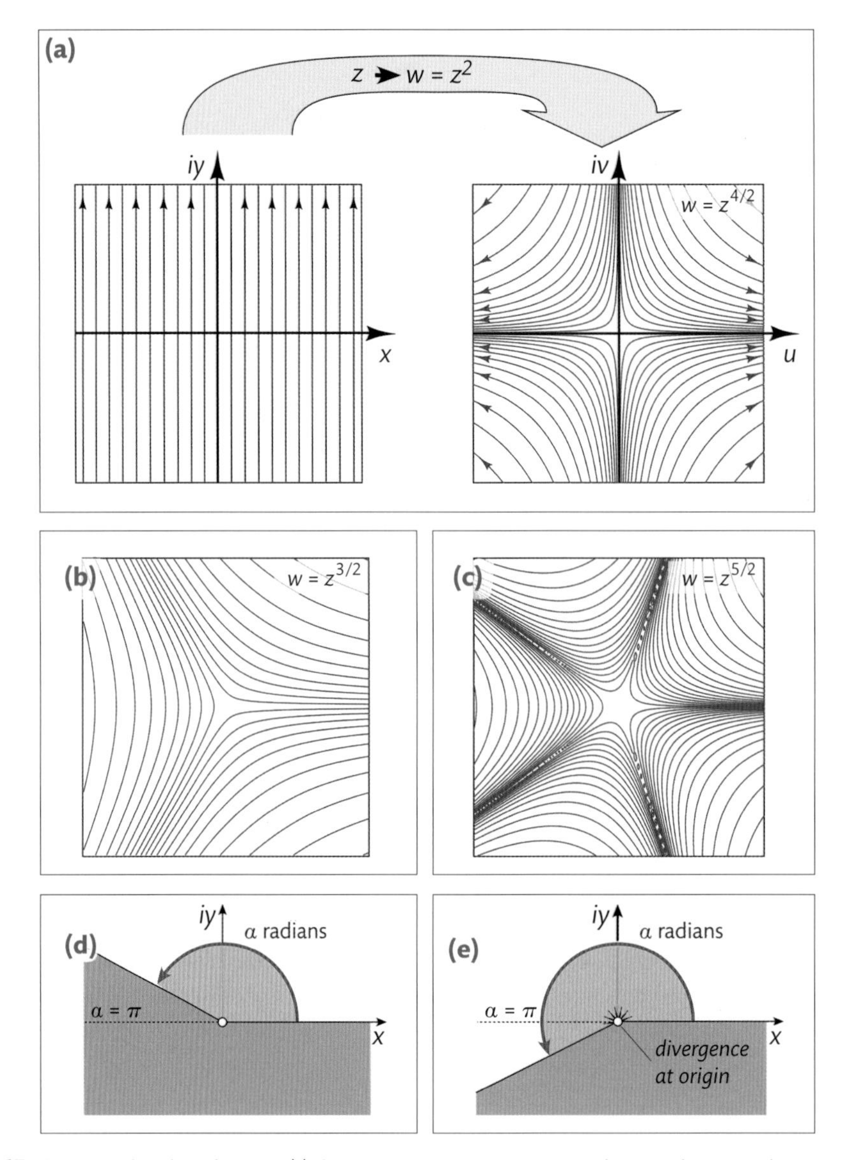

Fig. 8.17 Flows in wedge-shaped regions. (a) The Cartesian space $z = x + i \cdot y$ can be viewed as a complex potential with stream functions $\Psi = y$– that is, vertical lines of flow at any x, as shown on the left. On the right, the mapping $w = z^2$ produces fourfold symmetric hyperbolas, as described in the text. (b, c) Similarly, the mappings $w = z^{3/2}$ and $w = z^{5/2}$ produce three- and five-fold symmetries respectively. (d) The open angle of the resulting wedge can be expressed as $\alpha = \pi/n$ in radians. (e) When $\alpha > \pi$, a divergence appears, as described in the text.

EXERCISE 8.14

Given the stream function of Eq. [8.49], it is a simple matter to simulate flow in and out of wedges of various angles. Do so and reproduce Figs. 8.17(b) and (c). Then find a table of conformal mappings, and identify one that describes flow past a forward-facing step, as shown in Fig. 8.18. Reproduce this flow as well.

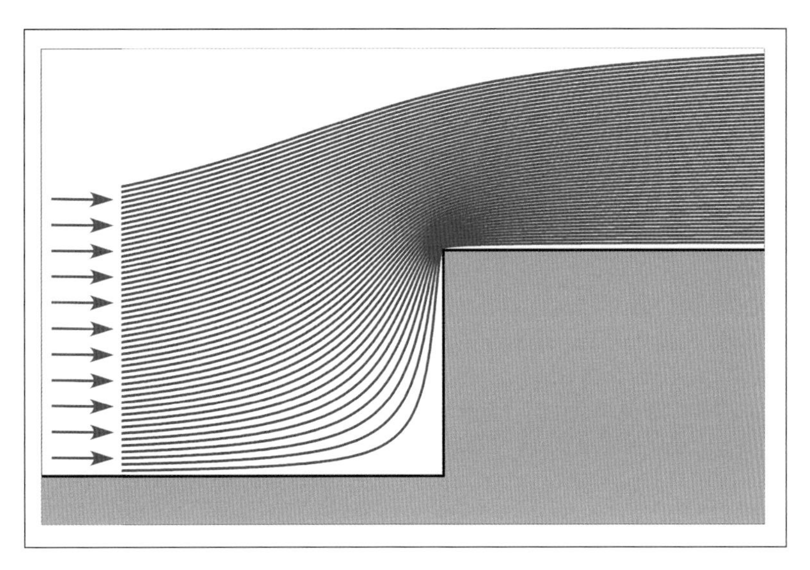

Fig. 8.18 Flow over a forward-facing step.

Apparently we can analyze and understand flow defined by the conformal mapping of Eq. [8.47]. This is good, but consider what happens if we rewrite the mapping in terms of $\alpha = n/\vartheta$. We'll reinforce the idea that conformal mappings and complex potentials are equivalent by writing Eq. [8.47] as a complex potential:

$$\Phi = r^{\pi/\alpha} e^{i\pi\vartheta/\alpha}. \quad [8.51]$$

Last time we had a go at this equation, we looked at the *stream* function, Eq. [8.49]. To likewise reinforce the complementarity of stream and potential functions, this time we extract the *potential* function, which again is the real part of the complex potential:

$$\varphi = r^{\pi/\alpha} \cos(\pi\vartheta/\alpha). \quad [8.52]$$

Let's now calculate the velocities, and we'll find something surprising:

$$\begin{aligned} V_r &= \frac{\partial\varphi}{\partial r} = \frac{\pi}{\alpha} r^{\left(\frac{\pi}{\alpha}-1\right)} \cos\left(\frac{\pi\vartheta}{\alpha}\right) \\ V_\vartheta &= \frac{1}{r}\frac{\partial\varphi}{\partial\vartheta} = -\frac{\pi}{\alpha} r^{\left(\frac{\pi}{\alpha}-1\right)} \sin\left(\frac{\pi\vartheta}{\alpha}\right). \end{aligned} \quad [8.53]$$

Simple enough, right? Note, though, the red terms: these tell us that *the exponent becomes negative* when $\alpha > \pi$. Which means that close to the origin, $r = 0$, *the velocities diverge*! We mentioned divergences briefly in Chapter 7, Fig. 7.13—here they arise again, but in an unexpectedly benign-looking context. Concave wedges, shown in Fig. 8.17(d), produce simple and well-behaved flows. But as soon as a wedge becomes convex, as shown in Fig. 8.17(e), flow speeds diverge at the apex of the wedge (now a step).

EXERCISE 8.15

Divergences of this kind appear from time to time; this one is known as the backward-facing step problem. If you were unaware of this issue and tried to perform a simulation of flow past a backward-facing step using a computational package, you would be confronted with erratic results close to the corner of the step. Knowing the mathematical cause of the problem, how might you correct it in your computations?

In the physical problem of flow past a backward-facing step, the fluid speeds surely don't go to infinity (Do they? Recall sonoluminescence and pistol shrimp in Chapter 7, Fig. 7.13!). Is this then a purely mathematical artifact without physical implications, or are there measurable consequences? What physical effect might be produced? Finally, give some thought to the fact that potential flows are defined by Laplace's equation: is there another problem defined by Laplace's equation in which a corner could produce real physical divergences?

8.5.5 Hose–Lesson: a First Example of an Implicit Model

A conformal mapping that illustrates a new technique involves flow from a hose. So far in this book, our problems have all been explicitly solvable, meaning we could write down a velocity, stream function, or other relevant object as some function of location and system parameters. This isn't always the case—consider for example the following conformal mapping:

$$z = \Phi + e^{\Phi}. \qquad [8.54]$$

Here it isn't possible to write down $\Phi(z)$ explicitly. Yet given z, Φ is completely prescribed (as we'll see). So how do we find Φ?

The answer is to turn to a so-called "implicit" method. There are many implicit approaches; we describe one here. Let's separate Φ into its real and imaginary parts:

$$\begin{aligned} z &= \varphi + i \cdot \psi + e^{\varphi + i \cdot \psi} \\ &= [\varphi + e^{\varphi}\cos(\psi)] + i[\psi + e^{\varphi}\sin(\psi)]. \end{aligned} \qquad [8.55]$$

Now we recall that a streamline is defined to be where the stream function is constant—so let's treat ψ as if it were constant. Then we can write:

$$z = [\varphi + e^{\varphi}C_1] + i[C_2 + e^{\varphi}C_3], \qquad [8.56]$$

where C_1, C_2, and C_3 are constants: $C_1 = \cos(\psi)$, $C_2 = \psi$, and $C_3 = \sin(\psi)$. We can similarly separate z into its parts to get:

$$\begin{aligned} x &= \varphi + e^{\varphi}C_1 \\ y &= C_2 + e^{\varphi}C_3. \end{aligned} \qquad [8.57]$$

These are parametric definitions of a curve in x–y space, which you may recall from Analytic Geometry means that for any constant ψ (i.e. along any streamline), we can change φ to vary x and y, thereby tracing out a curve.

For example, if $\psi = 0$, $C_1 = 1$, and $C_2 = C_3 = 0$, so that:

$$\begin{aligned} x &= \varphi + e^{\varphi} \\ y &= 0, \end{aligned} \qquad [8.58]$$

and as φ goes from negative to positive numbers, a line from left to right is traced out along $y = 0$, shown in green in Fig. 8.19(a). We can do the same thing for any other value of ψ–for instance, $\psi = \pm\pi$ produces

$$\begin{aligned} x &= \varphi - e^{\varphi} \\ y &= \pm\pi. \end{aligned} \qquad [8.59]$$

This produces curves shown in red in Fig. 8.19(a) that lie along $y = \pm\pi$, and as φ is increased, the trajectories travel in the $+x$ direction up to $x = -1$, and then reverse direction. Between $\psi = -\pi$ and $\psi = +\pi$, trajectories follow the lines shown in Fig. 8.19(b).

EXERCISE 8.16

Reproduce Fig. 8.19(b). This doesn't look like flow out of any hose that I've seen—is something wrong? Has something been overlooked? Explain.

8.5.6 **Cavity Flow–Lesson: Limitations of Fancy Tools**

We leave discussions of conformal mappings with a word of warning. These mappings are very handy tools that provide solutions for a variety of 2D flows with very little actual work. You just write down the stream function for a geometry that looks like your flow, plot its contours (or differentiate it to get numerical values of velocities), and you're done. As we've mentioned, there are even online tools to accommodate convoluted geometries.

These mappings, however, are deceptively svelte, and can produce flows that look very nice, and do satisfy Laplace's equation as advertised, but are not actually flows that Nature chooses. This is, it is worth stressing, not at all restricted to conformal mappings: many analytic tools produce pretty, and seemingly realistic, results that do not in fact occur. We'll see other examples of this caution in the remaining chapters of this book—for example, when we look at

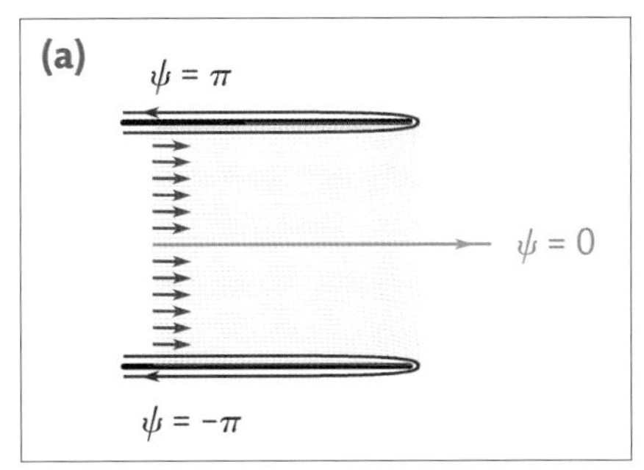

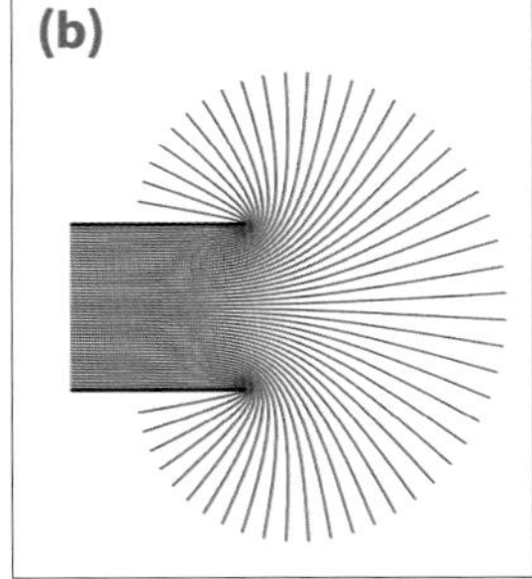

Fig. 8.19 Flow out of a hose. (a) Geometric construction described in text; (b) parametric plot of constant stream function lines.

some surprising behaviors of blood and other bodily fluids. Because surprises are universal in practical fluid problems, it is imperative that all models be validated by experiment. Flow through a cavity is a case in point.

A conformal mapping that appears to describe flow through a cavity, shown in Fig. 8.20(a), is

$$\Phi = \sqrt{z^2 - 1} + \arcsin(1/z). \qquad [8.60]$$

Unlike prior examples, there is no simple way of knowing that this mapping defines this particular flow, but thankfully generations of mathematicians have experimented with innumerable combinations of functions, and this turns out to define flow in a cavity. Cavity flows are important for a number of reasons: if you are growing cells at the bottom of a multi-well plate, you'll want to know how rapidly oxygen and other nutrients are being replenished in the well; if you are cleaning a textured surface, you'll need to know whether the cleaning flow will reach into grooves; and if you are devising a lab on a chip or pouring polymer to make a mold, you'll need to know where fluid travels and where it stagnates.

The streamlines in Fig. 8.20(a) seem ready-made to address all of these issues, right? Unfortunately, they do not describe the flow that is actually seen. In Fig. 8.20(b), we show flows produced by more careful and detailed "direct numerical simulation," which consists of dividing the fluid volume into tiny boxes, evaluating stresses on each box, and moving the fluid from box to box as defined by Newton's laws of motion and of viscosity. At $Re = 1000$, several asymmetric secondary vortices appear as shown, none of which mix with one another, and each of which travels about an order of magnitude slower than the one above.

If you were incautious, you could easily convince yourself from Fig. 8.20(a) that flow trajectories deliver nutrients all the way to the bottom of the cavity, which they don't, that the cavity would eventually be cleaned by continual flow, which it wouldn't, and that there are

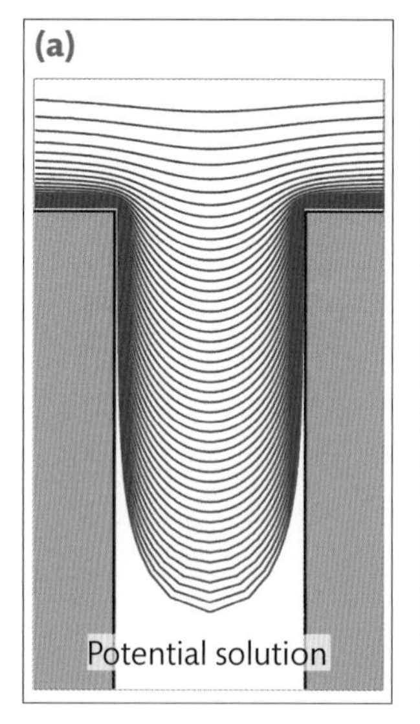

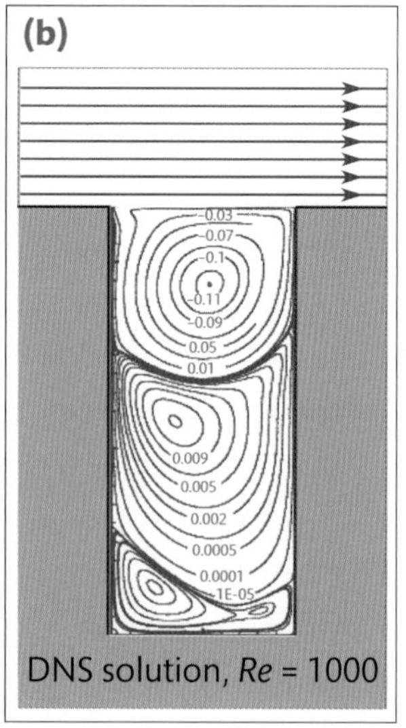

Fig. 8.20 Flow through a cavity produced (a) by the potential function (Eq. [8.60]) and (b) from direct numerical simulation at $Re = 1000$, from Cheng and Hung.[10]

no stagnated regions of flow in the cavity, which also is wrong. We therefore end this chapter with a caution that all computational and analytic tools need to be compared with experiment before they are trusted.

EXERCISE 8.17

Reproduce Fig. 8.20(a) using the mapping of Eq. [8.60]. Explain why a flow that solves Laplace's equation in the geometry of interest can disagree with more careful analysis. We remark that Eq. [8.60] actually has no bottom boundary: a conformal mapping that includes a bottom boundary can also be found: this involves elliptic functions that are difficult to work with, so we don't use that solution here. Take it for granted, though, that those solutions look very similar to what you see in Fig. 8.20(a), and explain why it is that the solution to Laplace's equation differs so much from the actual flow. What is missing from our analysis?

REFERENCES

1. Van Dyke, M. (1998) *An Album of Fluid Motion*. Stanford, CA: Parabolic Press.
2. Oseen, C.W. (1910) Über die Stokessche formel, und über eine verwandte Aufgabe in der Hydrodynamik. *Arkiv för Matematik, Astronomi och Fysik, 6*, 1–20.
3. Maxwell, J.C. (1890) On Faraday's lines of force. In: Niven, W.D. (ed.), *The Scientific Papers of James Clerk Maxwell*. Reprinted by Cambridge University Press, 2011.
4. Van Dyke, M. (1998) *An Album of Fluid Motion*. Stanford, CA: Parabolic Press, pp. 34–5.
5. Mehta, R.D. (1985) Aerodynamics of sportsballs. *Annual Review of Fluid Mechanics, 17*, 151–89; see also http://www.aviation-for-kids.com/aerodynamics-in-sports.html.
6. Nikuradse, J. (1933) Strömungsgesetze in rauhen Rohren. *Forschung auf dem Gebiete des Ingenieurwesens, Ausgabe B, Band 4*. Translated into English in NACA Technical Memorandum 1292.
7. Gioia, G. and Chakraborty, P. (2006) Turbulent friction in rough pipes and the energy spectrum of the phenomenological theory. *Physical Review Letters, 96*, 044502.
8. Wang, Z.J. (2005) Dissecting insect flight. *Annual Review of Fluid Mechanics, 37*, 183–210.
9. Mathews, J.H. and Howell, R.W. (2012) *Complex Analysis: For Mathematics and Engineering*. Sudbury, MA: Jones & Bartlett; see also http://math.fullerton.edu/mathews/complex.html.
10. Cheng, M. and Hung, K.C. (2006) Vortex structure of steady flow in a rectangular cavity. *Computers & Fluids, 35*, 1046–62.

9 Rheology in Complex Fluids 1

9.1 Introduction

We ended the last chapter with a caution that theoretical solutions need to be compared with experiment. In the next two chapters, we will look at experiments involving complex fluids. We will see that suspended materials, such as particles, proteins or cells, strongly affect how fluids, such as blood and synovial fluid, behave. In this chapter, we begin our discussion of experimental flows by describing the mechanical influences that suspended materials can have on fluids. This involves the study of "rheology," meaning the response of a fluid or a plastic to deformation. In Chapter 10, we'll consider more complex behaviors that biological fluids are prone to.

Example Strange suspensions

Suspensions can behave in very unexpected ways—for example, consider the difference between ketchup and cornstarch suspended in water. Ketchup is an example of a "yield stress" fluid, also referred to as a "Bingham" plastic, after chemist Eugene Bingham (1878–1945). Bingham materials are easily understood as consisting of suspended solids or droplets that support a load until a critical stress is reached, at which the solids can slip past one another (see Casson model, Chapter 10). At this point, the material suddenly yields—as ketchup does when poured from a bottle. For ketchup, the suspended material is pulp from the tomato, but the same is seen with mayonnaise, where the suspended material is oil droplets, or paint, where latex spheres are suspended. We'll have more to say about these materials later, but the essential point is that they behave like a solid under low stress, and they flow like a fluid at higher stress. The ability of a yield stress fluid to support a load like a solid is an extreme example of so-called "shear thinning," which refers to a reduction in viscosity with an increase in shear stress.

At the other extreme is cornstarch in water, sometimes called "oobleck,"[1] which flows easily at low stress, but becomes solid-like at high stress.[2] The general behavior in which viscosity increases with shear stress is termed "shear-thickening." We'll talk more about both shear thinning and shear thickening later; for now we invite the reader to make a dense suspension of cornstarch and experiment with it. If you haven't played with cornstarch and water before, you'll find it to be very strange stuff (see Fig. 9.1).[3]

Biomedical Fluid Dynamics: Flow and Form. Troy Shinbrot.

DOI: 10.1093/oso/9780198812586.001.0001

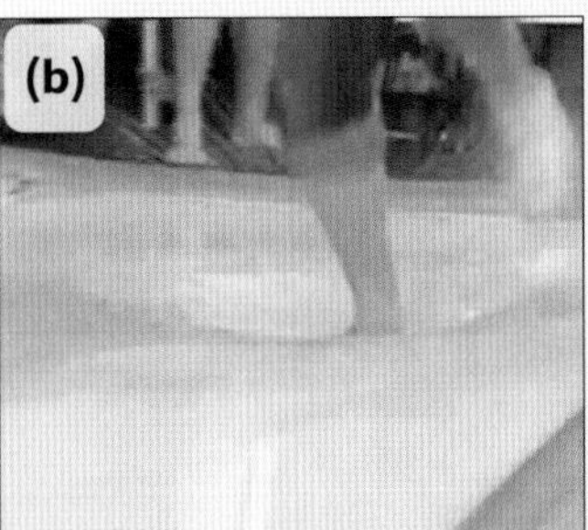

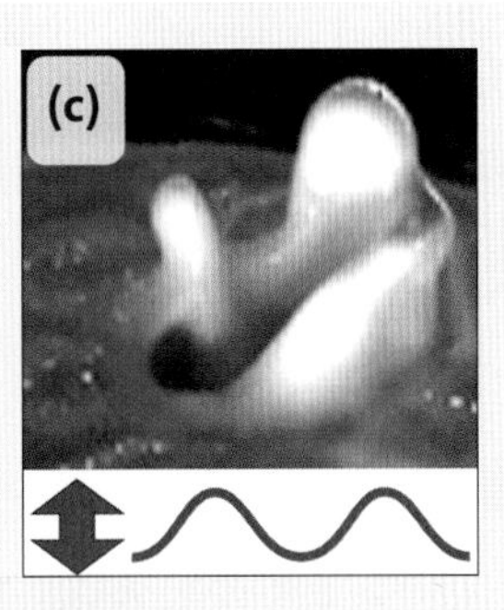

Fig. 9.1 Strange behaviors of cornstarch in water. (a, b) Running across a pool of cornstarch, from Spanish television program, Hormiguero.[6] (© El Hormiguero/Antena 3.) (c) Undulating fingers rising from pool of vibrated cornstarch.[7] (From Merkt, F.S., Deegan, R.D., Goldman, D.I., Rericha, E.C. and Swinney, H.L. (2004) Persistent holes in a fluid. *Physical Review Letters, 92*. By permission of the American Physical Society.)

Unlike suspended tomato particles, which flow in response to increased stress, cornstarch particles in water flow at *low* stress, but at high stress bind together to stiffen the surrounding area—so much so that a related suspension of silica particles in polyethylene glycol (PEG) is being studied as a flexible, but knife-proof, material.[4] Curiously, cornstarch in alcohol does not exhibit shear thickening.[5] It turns out that tough fibers, such as Kevlar™, can stop bullets by distributing the impact along the two longitudinal directions of the fibers, but the fibers themselves are more easily pushed aside in any one dimension, as by a knife. By incorporating materials similar to cornstarch, textiles with both bullet and knife resistance can be made because the cornstarch holds the Kevlar fibers together. Future designs have even been proposed to prevent needle sticks.

So it appears that real suspensions can choose, seemingly on their own, whether to be shear thinning or shear thickening. Additionally, we saw in Chapter 6 (Fig. 6.14) that individual grains in suspension interact in unexpectedly complex ways. A delightful review of some of these behaviors can be found in Walker.[8] In this section, we'll analyze the simplest behaviors of suspensions, with the recognition that these expectations cannot by themselves account for the variety of behaviors seen. We'll then briefly examine how simple fluids interact with grains in tubes or pipes, and finally we'll describe behaviors seen in so-called "non-Newtonian" fluids, typically consisting of entangled proteins or polymers in suspensions. Armed with this background knowledge, we'll conclude this chapter with mention of several extraordinary effects seen in suspensions, which will set the stage for a deeper analysis of blood and other bodily fluids, presented in Chapter 10.

9.2 Viscosity of Simple Suspensions

9.2.1 Einstein Model for Dilute Suspensions

We discussed one of Einstein's contributions to fluid mechanics in Chapter 6 (Fig. 6.11) when we considered secondary flow in a teacup. Here we discuss a second contribution that he

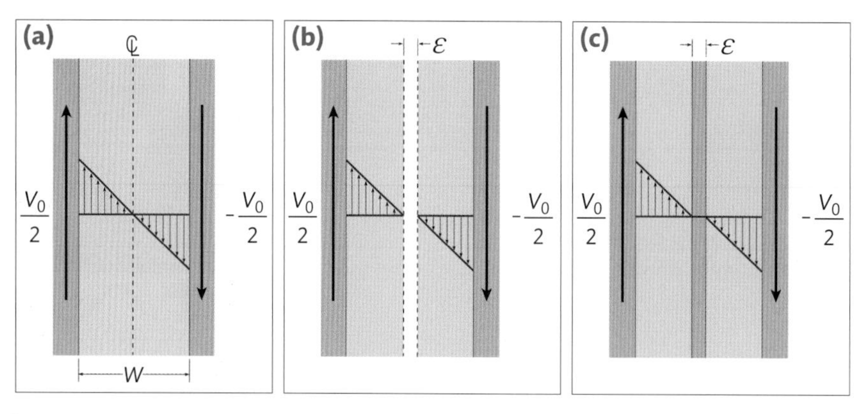

Fig. 9.2 Einstein's thought experiment for analysis of suspension viscosity. (a) Fluid (blue) between shearing plates (gray). An initially horizontal line is transported by the shear flow as shown in red. (b) The same shear problem with the fluid separated at the centerline by a gap of width ϵ. (c) Since the flow speed at the centerline is zero, we can place a solid wall into the gap without changing the flow.

made, and in Chapter 11, we'll discuss a third contribution dealing with particle diffusion. Einstein's model for the behavior of suspensions involves the following very simple thought experiment. Suppose we consider a fluid trapped between two shearing plates, each moving with speed $V_0/2$, as sketched in Fig. 9.2(a). The stress, τ, acting on the fluid is given by Newton's law of viscosity, so for two plates separated by distance w:

$$\begin{aligned}\tau_{fluid} &= \mu_{fluid}\frac{\partial V}{\partial x}\\ &= \mu_{fluid}\frac{V_0}{w},\end{aligned} \quad [9.1]$$

where μ_{fluid} is the viscosity of the fluid.

Next, Einstein observed that at the centerline between the plates, the velocity is zero, so the fluid between the left wall and the centerline would behave in precisely the same way if the centerline were replaced by a stationary wall. The same is true for the right half of the fluid, so the sheared fluid is the same as the two sheared fluids shown in Fig. 9.2(b), separated by a gap of thickness ϵ. But since the flow speed is zero at the dashed lines in Fig. 9.2(b), we can insert a solid wall into the gap without changing the flow, as shown in Fig. 9.2(c).

The stress on the fluid in the thought experiment shown in Fig. 9.2(a) is identical to that in Fig. 9.2(c). On the other hand, an experimenter who did not know that there was solid material between the left and right walls, would measure the separation between the walls to be $w + \epsilon$, would measure the wall speeds to be $V_0/2$, and would conclude that the stress on the system must be:

$$\tau_{fluid+solid} = \mu_{fluid}\frac{V_0}{w+\epsilon}. \quad [9.2]$$

But we would get precisely the same answer if we changed the viscosity in Eq. [9.1] to:

$$\mu_{fluid+solid} = \mu_{fluid}\frac{w}{w+\epsilon}, \quad [9.3]$$

meaning that there is no difference between an experiment in which we insert a fraction ϵ/w of solid into a pure fluid of viscosity μ_{fluid}, and an experiment in which we use a fluid of viscosity $\mu_{fluid} w/(w+\epsilon)$. If the fraction of solids, $\phi = \epsilon/w$, is small, we can do a little algebraic fidgeting by expanding Eq. [9.3] as follows. First, we define a function $f(1+\phi)$:

$$f = \frac{1}{1+\phi}, \tag{9.4}$$

which we expand as a Taylor series in ϕ:

$$\mu_{fluid+solid} = \mu_{fluid}\left[1 + \frac{\partial f}{\partial \phi}\cdot\phi + Order(\phi^2)\right]. \tag{9.5}$$

Since ϕ is small, we ignore terms of order ϕ^2, leaving

$$\begin{aligned}\mu_{fluid+solid} &= \mu_{fluid}\left[1 + \frac{1}{1-2\phi}\cdot\phi\right] \\ &\approx \mu_{fluid}[1+\phi].\end{aligned} \tag{9.6}$$

This was Einstein's simple and elegant 1905 result, which stood until German mathematician, Heinz Hopf (1894–1971) pointed out a central flaw. Perhaps this flaw has occurred to you as well: a flat vertical wall doesn't change the flow at all—this was the point of Einstein's thought experiment. But suspended solids such as spheres *do* change the flow, because fluid has to travel around the spheres. Hopf's observation led Einstein to write a correction including this effect, giving:

$$\mu_{fluid+solid} \approx \mu_{fluid}\left[1 + \frac{5}{2}\phi\right], \tag{9.7}$$

a prediction that holds reasonably well for small solids concentrations: $\phi < 5\%$. At higher particle fractions, particles interact with one another, and it is these interactions that produce the disparate behaviors ranging from yield stress (ketchup) to shear thickening (oobleck) suspensions. We cannot fully explain these behaviors here, in large part because they are not fully understood[5], but we will next provide a couple of phenomenological models of how the viscosity changes with concentration for suspensions, after which we will present very brief motivational descriptions of microscopic effects that strongly influence macroscopic fluid behaviors.

9.2.2 Models for Dense Suspensions

Models to extend Einstein's result are based on the observation that when solids become close packed, like a stack of cannonballs, they support loads like a single solid body, and the viscosity must then become infinite. A regular lattice like stacked cannonballs has volume fraction $\phi = 74\%$, but particles in a real suspension do not stack into a regular lattice like cannonballs. Rather, they reach a so-called "random close packed" limit, whose value depends on particle type and on precisely how the particles are prepared. For example, simply dumping spheres into a container of water will produce $\phi_{rcp} = 60\%$, whereas suspensions of spheres that are repetitively sheared back and forth produce $\phi_{rcp} = 68\%$, and randomly packed ellipsoids[9] shaped like M&M candies, can exceed $\phi_{rcp} = 74\%$. Commonly a value of $\phi_{rcp} = 64\%$ is used in the literature for spheres.

For any model, when the density approaches ϕ_{rcp}, the viscosity must diverge, so the viscosity ratio, $\mu_{fluid+solid}/\mu_{fluid}$, must follow a curve as shown in red in Fig. 9.3(a), reaching the Einstein limit at small ϕ, and approaching infinity at $\phi = \phi_{rcp}$. Many functional forms for curves go between these extremes; two that are commonly used are the Mooney and the Krieger–Dougherty models. The Mooney approximation was developed by Melvin Mooney (1893–1968), who used his model to predict the viscosity of rubber—which consists of isoprene particles suspended in water. His model is:

$$\mu_{fluid+solid} = \mu_{fluid}\, exp\left(\frac{5/2\phi}{1-\phi/\phi_{rcp}}\right), \qquad [9.8]$$

and has the merit that as ϕ becomes small, the viscosity approaches the Einstein solution (Eq. [9.7]): this can easily be seen by writing the first two terms in the Taylor expansion for the exponential. Although simple and analytically irreproachable, the Mooney model turns out not to be accurate in detail and lacks the ability to include features of suspended particles such as shape or polydispersity that are known to significantly affect viscosity.

A second and more accurate model, which was produced by I.M. Krieger and T.J. Dougherty in 1959, looks like:

$$\mu_{fluid+solid} = \mu_{fluid}\left(1-\phi/\phi_{rcp}\right)^{-B\cdot\phi_{rcp}}, \qquad [9.9]$$

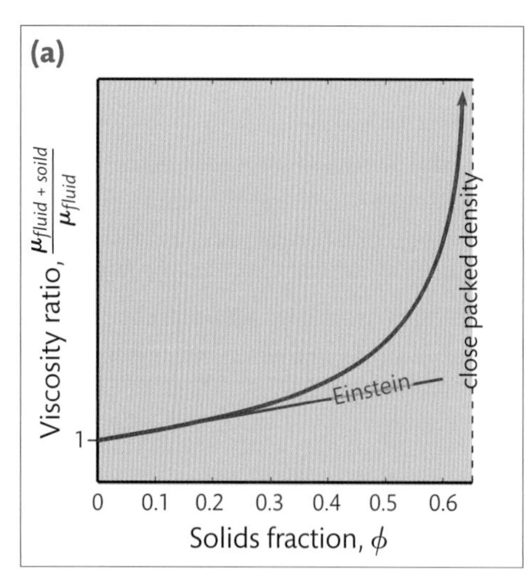

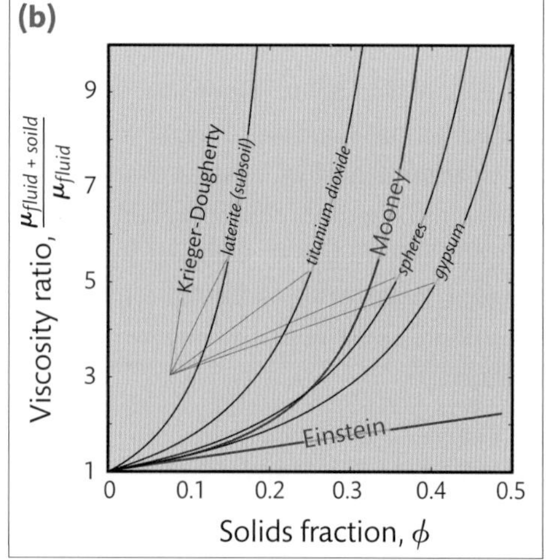

Fig. 9.3 Growth of viscosity with solids fraction in suspensions. (a) Einstein model applies at low ϕ, and viscosity zdiverges as ϕ approaches close packed density, so we qualitatively expect behavior that matches these two endpoints, shown in red. (b) Comparison between Einstein, Mooney, and Krieger-Dougherty models for suspension viscosity. Einstein's model is the simplest, but only applies at very low particle concentrations; Mooney is the next most accurate model, and Krieger-Dougherty accurately describes viscosities for many suspended particle types.

where B is an empirical constant that depends on particle shape and material. Values for B and ϕ_{rcp} can be found in the literature, for example, in Bird et al.,[10] and we plot viscosity ratios for several materials in Fig. 9.3(b).

9.2.3 Lubrication Theory and Colloidal Dynamics

All of the behaviors that we have analyzed so far have been Newtonian—that is, the stress has been assumed to be proportional to the rate of strain, and in the past few pages we have only observed that the constant of proportionality depends on solids concentration. This is so because we have assumed that suspended particles interact in very simple ways, essentially just gumming up the fluid. But we know from the ketchup and cornstarch examples that this can't be all that takes place: sometimes a fluid will go from completely static to freely flowing. In this section, we'll begin to look at effects of interactions between suspended particles, and in later sections we'll discuss more exotic behaviors.

EXERCISE 9.1

To get a preliminary feel for how, qualitatively, viscosities can vary with solids fraction, suppose you performed viscosity measurements on a suspension of solids, and you obtained the results shown in Table 9.1. Find the best fit of these points to a Mooney and to a Krieger–Dougherty model—for the Mooney fit, note that multiple different values of the parameters 5/2 and ϕ_{rcp} can be found in the literature: use ones that fit the data best. Use any tools that you like; Matlab's polyfit and logarithm functions may help (see Appendix for description of polyfit).

Table 9.1 Sample values of viscosity ratio and solids fraction to fit with Mooney and Krieger–Dougherty models.

Viscosity Ratio	Solids Fraction
1.3	0.06
2.0	0.15
2.9	0.22
4.5	0.28
6.5	0.31
8.3	0.33
10	0.35

As a place to start understanding how the, somewhat tame, fluid mechanics described so far lead to more interesting effects, let's ask how fluid trapped between two spheres behaves. Consider the spheres shown in Fig. 9.4, and suppose we are trying to pull the spheres apart. In order for them to move, fluid must squeeze into the gap between them. But if the spheres are very close together, with a gap of size h, then the Reynolds number must be very small, since $Re = Vh/\nu$. And we learned in Chapter 1 that a small Re can equally be viewed as a small scale or as a large viscosity. Which is to say that it will be very difficult, and it will take a very long time, to squeeze fluid into, or out of, narrow gaps between particles. And naturally, the closer the particles get to one another, the longer that will take. Of course the problem is compounded by the fact that in a collection of spheres, for fluid to *get* to the gap between any contacting pair, it has to percolate through interstices between spheres—and these interstices get smaller both as the spheres get closer and as the sphere size diminishes.

For this reason, particles—especially small particles—packed closely together strongly resist being moved with respect to one another. If a suspension is slowly strained, there may be time for the fluid to move into the gaps between particles—and so the apparent viscosity may be given by a Krieger–Dougherty or similar model. But if the suspension is strained rapidly, the effective fluid viscosity will be better represented by:

$$\nu_{eff} = \nu \frac{L}{h}, \qquad [9.10]$$

where ν is the viscosity of the pure fluid, L is a characteristic length scale for the problem (say a diameter of pipe), and h is a typical gap between suspended particles. Obviously if h is on the order of a micron and L is on the order of a millimeter, viscosities several orders of magnitude larger than that of the pure fluid can be expected.

This simple caricature goes by the name of a "lubrication" model, because of its similarity to the lubrication of oil between bearings. The caricature accounts broadly for the growth in viscosity that some suspensions undergo as the rate of shear is increased – so called "shear thickening" fluids. We remark that this is only one of very many intriguing behaviors seen in so-called "colloidal" suspensions: suspensions involving fine particles, typically under a micron or so in diameter. For example, we mentioned briefly stratified emulsions in Chapter 6, and we'll talk a little about colloids in a couple of chapters when we consider entropic ordering: the spontaneous self-assembly of structures of colloids. For our purposes, we'll just mention that the same effect that prevents suspensions of fine particles from being rapidly moved also prevents dense suspensions from being formed in the first place: India inks, for

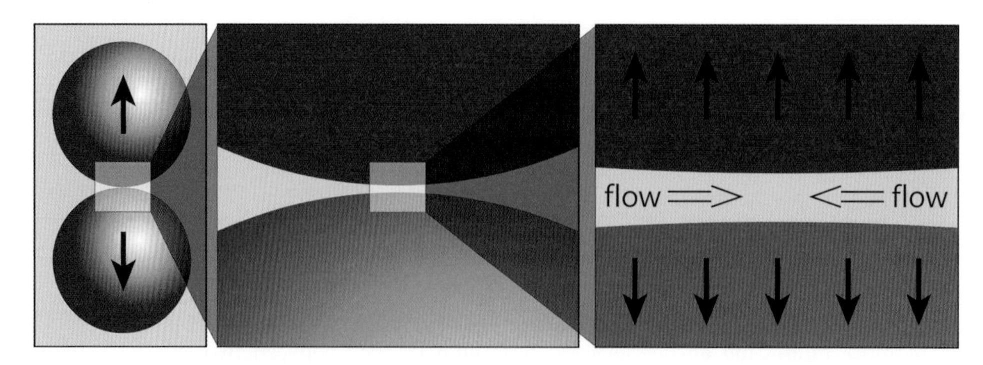

Fig. 9.4 Sketch of a gap between two spheres. Notice that at increasing magnification, the gap looks like two parallel plates, and fluid must flow out of and into the gap as the spheres are moved toward and away from one another. Since the gap between the spheres is small, the effective Reynolds number must also be small, which is equivalent to flow of a high viscosity fluid in a larger gap.

example, are often aged for months to years to allow very fine soot particles to equilibrate with one another—a process that takes such a long time precisely because of the lubrication effect depicted in Fig. 9.4.

EXERCISE 9.2

It has probably occurred to you that the caricature of Fig. 9.4 has significant shortcomings. First, it only applies to smooth solids: asperities on the surfaces of suspended solids cause friction and prevent gaps between the solids from becoming very small. Second, a chain is only as strong as its weakest link, so fluid won't need to move between spheres separated by small gaps if there are larger gaps elsewhere that can accommodate the applied strain. Thus non-uniformities in suspensions can lead to sudden slip events, and consequently the uniformity of a suspension is central to the stress–strain behavior that is seen.

On the other hand, by the same token larger gaps can cause the suspension to slip, and if the suspended solids themselves are relatively weak—like tomato pulp—the solids themselves can fail, causing an abrupt decrease in viscosity when the particle yield stress is reached. By considering effects such as these, this simple lubrication model can in principle be made to qualitatively describe the behavior of suspensions that shear thin,[3] as well as shear thicken.

In slightly more detail, we note that it is observed experimentally that the total force needed to deform a fixed volume of suspension grows as the size of the suspended particles diminishes: fine cornstarch in water is shear thickening, but coarse cornmeal is not. In this exercise, you will determine whether the lubrication model accounts for this observation.

Consider Fig. 9.5, which depicts a suspension being sheared. A densely packed suspension is intrinsically interlocked, and so in order for the suspension to move, particles need to get out of one another's way, or "dilate" as shown in Fig. 9.5(b), to make space for the motion to occur. We will discuss "dilatency" later in this chapter. Dilatency requires particles to separate as sketched in Fig. 9.4, so in this exercise, you'll calculate what this depends on.

This can be done in many ways, to a greater or lesser degree of accuracy: here, we are only interested in asking whether to first order the lubrication idea agrees or disagrees with the

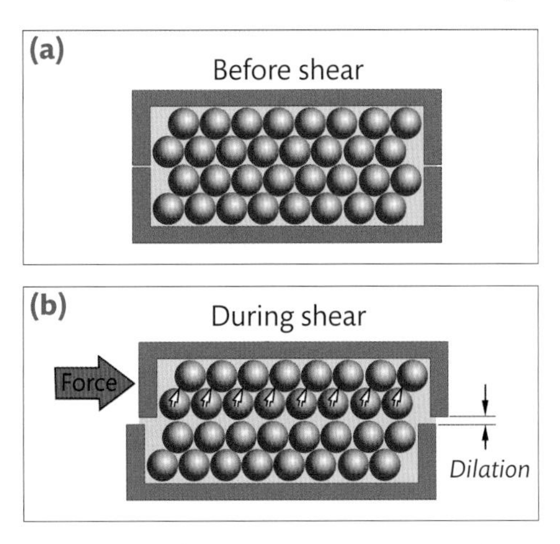

Fig. 9.5 Dilatency of a suspension of spheres. (a) A densely packed suspension of particles is interlocked, and cannot move without generating space between the particles. (b) By applying shear, particles are forced apart, and "dilate," as shown, so that a shear band of mobile particles can form.

observation that smaller particles in a suspension resist motion more strongly than larger particles. Make any assumptions that you choose, and estimate how the apparent viscosity varies with particle size. For example, you may want to solve the problem in two dimensions, for infinitely long cylinders, or you may want to assume that interstices between particles are tubes that can be approximated using Poiseuille flow, or you may want to assume that spaces between particles are perfect plates. In any case, state what your assumptions are, and calculate the viscosity dependence on particle size. You should obtain an approximate like:

$$viscosity \sim size^{power}, \quad [9.11]$$

and if the power is negative, you can conclude that lubrication accounts qualitatively for the experimental observations; if it is positive, you must conclude that something else is at work. In this case, propose some possibilities.

In more detail still, it takes time for particles to dilate, and if a suspension is sheared slowly enough to permit this, it can therefore develop slip planes between shearing layers, as sketched in the upper left inset to Fig. 9.6. The presence of these slip planes reduces the effective

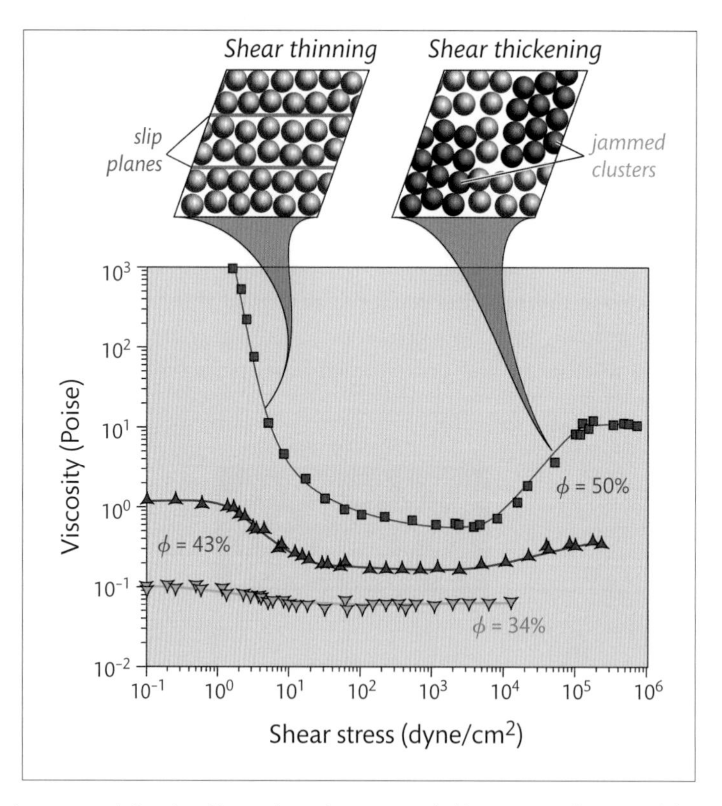

Fig. 9.6 Main plot: measured viscosity of latex microspheres suspended in water as a function of shear stress for various solids concentrations, ϕ. Insets above main plot: simplified caricatures of behavior of suspended spheres, showing that shear thinning occurs when time is allowed for slip planes to develop, and shear thickening occurs when particles jam into clusters that are produced when particles are sheared too rapidly to slide past one another. From Wagner and Brady.[11]

viscosity, and so leads to shear thinning. This is seen in data taken using dense suspensions of latex microspheres in water, as shown to the left of the main plot in Fig. 9.6.

On the other hand, if shearing occurs more rapidly than dilation, particles will be forced into near contact by applied shear, and this leads to compact, "jammed," clusters in which no particle can move at all. This is depicted in the upper right inset to Fig. 9.6. Jamming is itself a very complex phenomenon which leads in this case to an increase in viscosity—that is, shear thickening, as shown in experimental data to the right of the main plot of Fig. 9.6.

Evidently, changes in viscosity of suspensions can be quite complex—often too complex to treat analytically. In the next sections, we will describe some phenomenology seen in dense suspensions, after which we will introduce some tools to analyze some of the less complicated non-Newtonian phenomena of suspensions and other complex fluids.

Example Lubrication

Lubrication theory is important for many biomedical applications, both old and new. Later in this chapter, we will describe remarkable properties of synovial fluid, which acts to lubricate and protect joints. New biomedical implants also propose to use ideas from lubrication theory, as we discuss shortly.

A clever engineering approach to lubrication involves "Oilite™" bearings, introduced by Chrysler Motor Company in the 1950s to lubricate clutches, leaf springs, and other moving parts. An Oilite bushing is shown in Fig. 9.7(a): this is an example of "powder metallurgy," in which the bushing is made from metal powder that is sintered (partially melted) to produce a solid object with holes where interstices between the powder particles once were. In an Oilite bearing, the sintered metal is infused with oil, so that oil is extruded where the bearing is under load, and is reabsorbed where the load is released. As shown in Fig. 9.7(b), this results in a bearing that produces oil at the leading edge—where lubrication is needed—and then recovers the oil again at the trailing edge. Thus the bearing continuously lubricates itself, and rarely if ever needs oil replenishment.

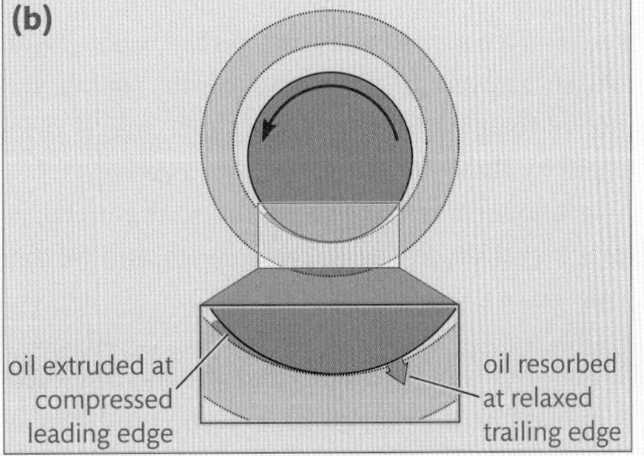

Fig. 9.7 Oilite bearing: a creative example of powder metallurgy. (a) An Oilite bushing: note the evident pores that can be used to infiltrate oil (© Oilite). (b) Infiltrated oil is extruded at the leading edge, where stress is high, and is reabsorbed at trailing edge, where stress is relaxed.

Sintered metals have been widely used and proposed for biomedical devices ranging from infusion filters to fluidized bed plenums to implants that either elute drugs or provide a porous substrate intended to promote osteointegration. Using lasers to perform the sintering also enables rapid prototyping of metal implants, and new sintering technologies are being used to develop nanostructured materials for applications ranging from ceramic armor to third generation photovoltaics.

9.3 **Return to Drag in Pipes: Empirical Relations**

We've touched briefly on a few topics that arise in suspensions. We'll examine other behaviors in suspensions and other complex fluids later in this chapter, but let's first consider the limiting case of what happens when the solids density is so high that the solids become jammed and cannot move at all. This situation arises in filters and liquid chromatography columns, and is often referred to as a "packed bed." After we have covered this topic, we'll return to behaviors of complex and biological fluids—blood in particular.

In Chapter 8 (Fig. 8.5(b)), we saw that the drag experienced by a simple fluid traveling through a pipe exhibits two regimes: one for $Re < 10^3$ where the drag coefficient, C_d, goes as $16/Re$, and another for $Re > 10^4$, where C_d decreases with a much shallower slope (we'll see that $C_d \sim 1/Re^{1/4}$ in that regime). Let's recap what we know of drag of simple fluids through pipes, after which we'll compare the situation with flow through pipes packed with grains.

9.3.1 **Drag in Smooth Pipes**

We begin with the definition we started with for the drag coefficient (Chapter 8, Eq. [8.10]):

$$C_d = \frac{F}{\frac{1}{2}\rho \cdot \langle V \rangle^2 \cdot A}, \qquad [9.12]$$

where F is the streamwise force on the walls, ρ is the fluid density, $\langle V \rangle$ is the mean fluid speed in the streamwise direction, and A is the surface area of the pipe walls. This may be made more intuitive for pipe flow by noting that this is the same as the stress on the walls, $\tau = F/A$, divided by the unit kinetic energy, $\frac{1}{2}\rho\langle V \rangle^2$.

We know something about the terms in Eq. [9.12], right? The surface area of a pipe of radius R and length L is just $A = 2\pi RL$, and we can calculate the force, F, on the walls in either of two ways.

(1) The first way is to note that the force needed to slow a fluid flowing through a pipe is applied by the pipe walls, and the fluid flow is generated by the pressure drop from one end of the pipe to the other, so at equilibrium the two have to be equal. This implies that the force at the walls must be:

$$F = \pi R^2 \Delta P, \tag{9.13}$$

where ΔP is the pressure drop along the length of pipe. This makes Eq. [9.12] become:

$$\begin{aligned} C_d &= \frac{\pi R^2 \Delta P}{\frac{1}{2}\rho \cdot \langle V \rangle^2 \cdot 2\pi RL} \\ &= \frac{1}{4}\frac{2R}{L}\frac{\Delta P}{(1/2)\rho \cdot \langle V \rangle^2} \\ &= \frac{\Delta P \cdot R}{L \cdot \rho \cdot \langle V \rangle^2}. \end{aligned} \tag{9.14}$$

At low *Re*, we can find the average velocity, $\langle V \rangle$, from the Poiseuille solution in Chapter 1, Eq. [1.47], which gives:

$$Q = \frac{\pi \Delta P R^4}{8L\mu}. \tag{9.15}$$

Moreover, $\langle V \rangle$ is the volumetric flow rate per unit area:

$$\begin{aligned} \langle V \rangle &= \frac{\pi \Delta P R^4}{8L\mu} \bigg/ \pi R^2 \\ &= \frac{\Delta P R^2}{8L\mu}, \end{aligned} \tag{9.16}$$

and inserting this into Eq. [9.14] gives:

$$\begin{aligned} C_d &= \frac{\Delta P \cdot R}{L \cdot \rho \cdot \langle V \rangle \left[\frac{\Delta P R^2}{8L\mu}\right]} \\ &= \frac{8\mu}{\rho \cdot \langle V \rangle R} \\ &= \frac{16}{Re}. \end{aligned} \tag{9.17}$$

So just as we showed in Chapter 8 that drag on an obstacle goes as $24/Re$, here for pipe flow we find that drag goes as $16/Re$. Both of these results apply only at low *Re*. Which brings us to the second way of calculating the force, without restricting the Reynolds number.

(2) This second way is to evaluate the viscous stress exerted by the fluid on the pipe. At equilibrium this has to give the same result as the first way, but calculating the viscous stress is more general, and provides helpful insight, as we'll see.

In this case, the force is the integrated viscous stress on the wall:

$$F = \int_0^{2\pi} \int_0^L \left(-\mu \frac{\partial V_z}{\partial r}\right)_{r=R} R \cdot d\vartheta dz, \tag{9.18}$$

where the terms in parentheses are the viscosity times the shear strain rate as defined in Chapter 1. So here we are assuming that the fluid is Newtonian—remember this when we examine blood, which we will see is not at all Newtonian!

We can then insert Eq. [9.18] into Eq. [9.12] to obtain:

$$C_d = \frac{\int_0^{2\pi}\int_0^{L}\left(-\mu\frac{\partial V_z}{\partial r}\right)_{r=R} R\cdot d\vartheta dz}{½\rho\cdot V^2\cdot 2\pi RL}. \qquad [9.19]$$

To understand what this tells us, it's useful to nondimensionalize and perform a little algebra to obtain:

$$C_d = \frac{2R}{\pi L\cdot Re}\int_0^{2\pi}\int_0^{L/2R}\left(-\frac{\partial \widetilde{V}_z}{\partial \widetilde{r}}\right)_{\widetilde{r}=½} d\vartheta d\widetilde{z}, \qquad [9.20]$$

where the nondimensionalized variables are:

$$\widetilde{V}_z = \frac{V_z}{V} \qquad \widetilde{r} = \frac{r}{2R} \qquad [9.21]$$

$$\widetilde{z} = \frac{z}{L} \qquad Re = \frac{2RV\rho}{\mu}.$$

Finally, we note two things about Eq. [9.20]. First, if the fluid obeys Navier–Stokes equations, the nondimensionalized velocity in Eq. [9.20] can in principle depend only on Reynolds number and coordinates, $\widetilde{r}, \widetilde{\vartheta}$ and $\widetilde{z}$. But $\widetilde{r}$ is fixed at the wall (where $\widetilde{r} = ½$), and furthermore $\widetilde{\vartheta}$ and $\widetilde{z}$ are integrated over. So the integral cannot depend on coordinates at all, and must only depend on *Re*. Consequently, Eq. [9.20] tells us that the drag coefficient can only depend on aspect ratio, $L/2R$, and Reynolds number. Moreover for well-developed flow far from the entrance of long pipes, $L/2R$ will be very large and won't vary, so C_d can only depend on *Re*! This applies at all *Re*: at low flow speeds C_d is particularly simple, as shown in Eq. [9.17], but *at all speeds, C_d depends only on Re.*

In turbulent flow, for example, an empirical formula for the time averaged drag coefficient gives:

$$C_d = \frac{0.0791}{Re^{1/4}}, \qquad [9.22]$$

known as the "Blasius" formula, after German physicist P.R.H. Blasius (1883–1970). Corrections to account for roughness and other factors can also be found in the literature. Bottom line: flow resistance in pipes must depend only on *Re*, but that dependence can vary dramatically, from Re^{-1} (Eq. [9.17]) to $Re^{-1/4}$ (Eq. [9.22]).

EXERCISE 9.3

In Chapter 8, we used a Buckingham pi analysis of flow past an obstacle to demonstrate that flow in that problem likewise depends only on *Re*. This shouldn't be too surprising, since the dimensionless Navier–Stokes equation also is defined only in terms of *Re* (see Chapter 1). Yet drag clearly does depend on other things as well, for example obstacle shape (Chapter 8, Fig. 8.3(b)) and pipe roughness (Chapter 8, Fig. 8.5(b)). How can this be? Is the Buckingham Pi analysis flawed? Is something else wrong?

The fact that C_d is prescribed by *Re* tells us also that we can compute the pressure drop needed to achieve a desired flow rate. To do this, let's return to Eq. [9.14], which we can rewrite in terms of the flow rate, *Q*:

$$\frac{\Delta P}{L} = \frac{\rho \cdot C_d}{Area^2 R} Q^2, \qquad [9.23]$$

where *Area* is the cross-sectional area of the pipe. We write this equation in this way because not all pipes are round, and to include irregular pipe shapes it is often useful to write an effective Reynolds number as:

$$\begin{aligned} Re_{eff} &= \frac{4R_h \langle V \rangle \rho}{\mu} \\ &= \frac{Q \cdot \rho}{\pi \cdot R_h \cdot \mu}, \end{aligned} \qquad [9.24]$$

where R_h is the "hydraulic radius::

$$R_h = \frac{Area}{Perimeter}. \qquad [9.25]$$

We can use Eq. [9.23] combined with an experimentally obtained plot of C_d vs. Re_{eff} to obtain the pressure per unit length needed to sustain a desired flow rate, *Q*. As an example, suppose we have a tube with hydraulic radius $R_h = 1$ cm, and we want to sustain a flow rate of $Q = 1000$ cc/sec (a fairly fast flow for a reason that will become evident in a moment). Using typical values for density and viscosity of water (which are 1 and 0.01 respectively in cgs units), we obtain $Re_{eff} = 32{,}000$. We can then look up C_d for that Re_{eff} in a plot like Fig. 9.8(a) (reproduced from Chapter 8, Fig. 8.5(b)), and we obtain $C_d \sim 0.005$ for a smooth pipe, or $C_d \sim 0.01$ for a moderately rough one. If we plug the appropriate value for C_d into Eq. [9.23], we obtain a pressure per unit length of:

$$\begin{aligned} \frac{\Delta P}{L} &= \frac{\rho \cdot C_d}{8\pi {R_h}^5} Q^2 \\ &= \begin{cases} 200\, dyne/cm^3 \, for\ a\ smooth\ pipe \\ 400\, dyne/cm^3 for\ a\ rough\ pipe \end{cases}. \end{aligned} \qquad [9.26]$$

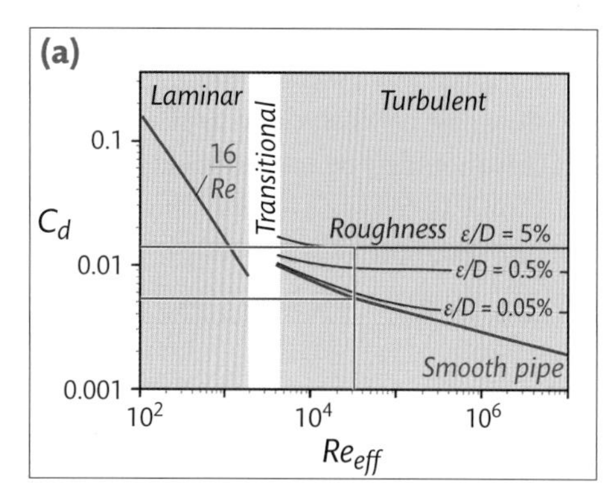

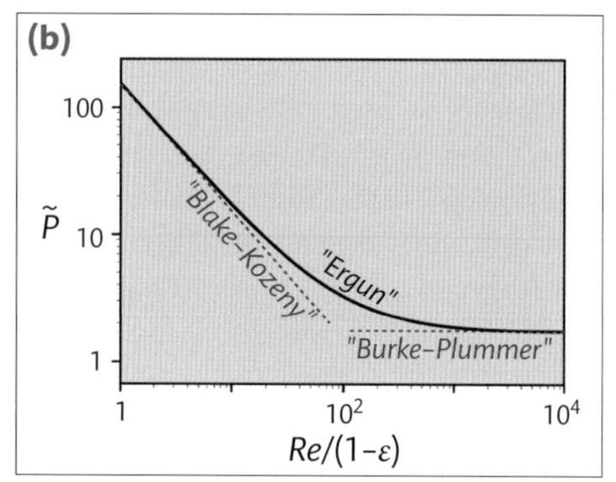

Fig. 9.8 (a) Plot of drag coefficient vs. effective Reynolds number, Re_{eff}, in simple pipes. For the example problem described in the text, $Re_{eff} = 32{,}000$, and the magenta lines show corresponding C_d values for smooth or moderately rough pipes. (b) Plot of nondimensionalized pressure vs. modified Reynolds number for packed columns (pipes filled with beads). Here, $\tilde{P} = (\Delta P/L)(D_p/\rho\langle V\rangle^2)$, $Re = \rho\langle V\rangle D_p/\mu$, ϵ is the void fraction, and D_p is an effective particle diameter.

9.3.2 **Drag in packed columns**

In many applications—reactors, liquid chromatography, filters, and so on—flow through a pipe filled with particles is of interest. We summarize some results in this field here. Before doing so, we remark that the results that we will include deal only with pipes completely filled with static beds of particles. Particles that can move are studied in "fluidized bed" research, which is a rich and complex topic in its own right. Additionally, we neglect pathological states such as "ratholes" or "channels" that can develop when unobstructed tubes of flow develop within a bed of grains: these can be both intricate, as the following example illustrates, and important, as they substantially alter flow, but they are beyond the thrust of this section, which deals with drag through *uniform* beds of grains.

Example Ratholes and channels

In Fig. 9.9, we show snapshots that reinforce the theme that nature is invariably more intricate than we may expect. For example, in Fig. 9.9(a), we show a "rathole" that has developed in an ore stockpile. These holes can appear when grains are drained from the bottom of a pile, and their appearance causes industrial problems ranging from flow blockages to particle segregation (notice that particles lining the rathole are finer than particles surrounding it). Ratholes appear in dry, though cohesive, grains. In wet grains, complicated anastomosing channels commonly develop, as shown in Fig. 9.9(b) from an experiment intended to mimic inland river deltas.

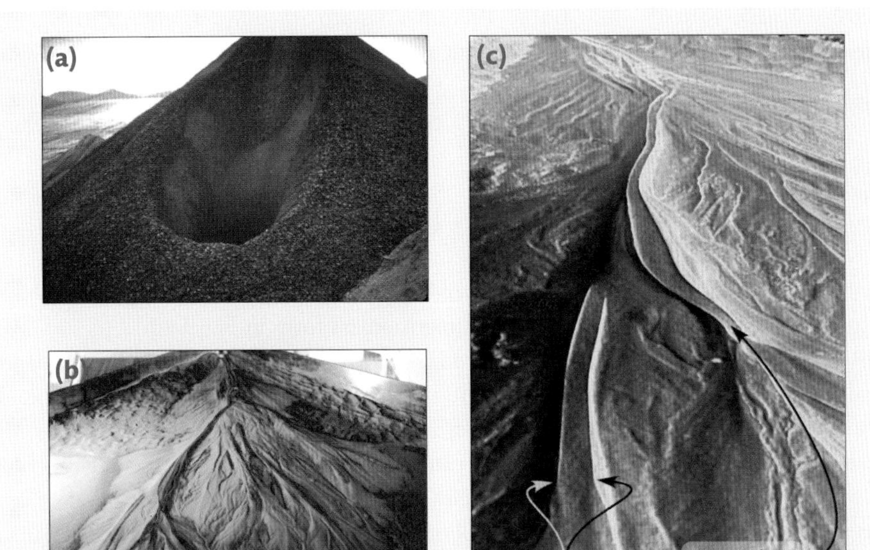

Fig. 9.9 Nonuniform structures that develop in dry and wet grains. (a) "Rathole" in dry ore stockpile that has been drained from beneath. Flow commonly stops once this tubular hole reaches the surface of the pile. (From: http://jenike.com/files/2012/10/Rathole-InPile.jpg.) (b) False colored laboratory experiment,[12] illustrating the formation of a so-called "inland" delta. (c) Enlargement, showing that flow occurs in levy-bounded channels made from deposited sediment, *above* the ambient surface. (Panels (b) and (c) from Mehdi Doumi.)

Familiar deltas, such as the Mississippi or the Nile, form where water runs from mountains or highlands to the sea. Less well known deltas, termed inland deltas, form as water flows from the other side of the mountains into the desert, as in the Mojave or the Okavango. Both types of deltas exhibit surprising features; for instance inland deltas carry sediment downstream, and as a result the accumulated sediment causes the water to flow in channels as much as tens of meters above the desert floor.

With the recognition that complex and unexpected things can happen in real particle beds, the first thing that can be said of drag in packed columns is that as far as a passive fluid is concerned, it doesn't care if it is flowing through a pipe or through interstices between grains, and so a natural approximation to the drag coefficient can be obtained by simply rewriting Eq. [9.14] for drag in pipe flow:

$$C_d \approx \frac{1}{4}\frac{D_p}{L}\frac{\Delta P}{\frac{1}{2}\rho \cdot \langle V \rangle^2}, \qquad [9.27]$$

where we have written an effective particle diameter D_p in place of the pipe diameter. This isn't a terrible lowest order approximation, but not surprisingly it isn't quantitatively accurate. Experimentally, as with simple pipes, separate low and high *Re* regimes are observed: these are termed "Blake–Kozeny" and "Burke–Plummer" approximations, as shown in Fig. 9.8(b). Empirical expressions for the pressure gradient as a function of mean flow speed are as follows:

$$\frac{\Delta P}{L} = 150 \frac{\mu \langle V \rangle}{D_p^2} \frac{(1-\epsilon)^2}{\epsilon^3} \quad \text{Low } Re \text{ "Blake–Kozeny" equation,} \tag{9.28}$$

$$\frac{\Delta P}{L} = \frac{7}{4} \frac{\rho \langle V \rangle^2}{D_p} \frac{(1-\epsilon)}{\epsilon^3} \quad \text{High } Re \text{ "Burke–Plummer" equation.} \tag{9.29}$$

Here, ϵ is the void fraction (i.e. the ratio of void volume in the packed bed to total volume in the pipe). These expressions are plotted in dimensionless form in Fig. 9.8(b).

This provides us with a central lesson of importance both to this topic and to science in general. These two complicated and difficult to understand equations are *empirical fits* to data. They are named after four different people, they include numbers like 150 and 7/4 whose origins are totally abstruse, and embellishments that can be found in the literature include other terms to account for things like asphericity of particles, differences in packing density, and so forth.

It is very easy to become intimidated by all of these complications. The lesson is: don't. All that has happened is that people observed that filling pipes with stuff slows down the flow. We can't calculate how much the flow should be slowed down from first principles, so measurements have been taken, and by and large—if we ignore fluidization, channeling, particle shape and so on—the data fit two lines, one at low *Re*, and another at high *Re*. Nothing deeper than that is going on, despite the complicated equations and names.

Moreover, these two low- and high-Reynolds number fits can be combined in yet another equation, this time named after Turkish chemical engineer Sabri Ergun (1918–2006):

$$\frac{\Delta P}{L} = 150 \frac{\mu \langle V \rangle}{D_p^2} \frac{(1-\epsilon)^2}{\epsilon^3} + \frac{7}{4} \frac{\rho \langle V \rangle^2}{D_p} \frac{(1-\epsilon)}{\epsilon^3} \quad \text{"Ergun" equation.} \tag{9.30}$$

This is a famous and important equation for engineers working with flow through packed beds, but again, if you took the lesson to heart not to be intimidated by Blake–Kozeny or Burke–Plummer, you *really* shouldn't be intimidated by the Ergun equation: it just consists of the sum of the other two equations, right? This isn't to say that the equation isn't useful or that it didn't involve dedicated and careful measurements. It is a very useful equation, but like the Bessel equation that we introduced in Chapter 4, it is very easy to feel overwhelmed by fancy names and complicated formulas. Invariably something very basic underlies these complications, and recognizing this will save a great deal of confusion and angst.

9.4 **Power Law Fluids**

9.4.1 **Phenomenology**

We've seen that flows of suspensions, and flows through suspensions, generate complexities spanning the range from shear thinning to shear thickening. There are numerous behaviors that cannot be easily analyzed, but focusing just on this range allows us to fairly readily define a phenomenological model. To do so, let's recall where we started this book, with Newton's law of viscosity:

$$\tau = -\mu \cdot \frac{\partial V_x}{\partial y}, \tag{9.31}$$

where μ is the viscosity and τ is the stress on a volume of fluid. When we first introduced this law, we emphasized that this was the *simplest* possible model for fluid response to applied forces, and as defined here, it only pertains to flow in the x-direction, V_x, that changes in the y-direction.

More generally, we can write Eq. [9.31] as a tensor relation:

$$\tau_{ij} = -\sum_k \sum_\ell \mu_{ijk\ell} \frac{\partial V_k}{\partial x_\ell}. \tag{9.32}$$

Tensors refer to objects (such as stresses, τ_{ij}) that depend on two or more coordinates (here i and j which can vary over 1, 2, or 3 to represent directions x, y, and z). Tensors are the higher order generalization of vectors: vectors can be defined in the x, y, or z directions, whereas tensors are defined in terms of two directions. So the derivative in Eq. [9.31] depends both on x (the direction of the velocity) and y (the direction of its change). The convention used for stress tensors is that the second index defines the direction in which the stress acts, and the first index defines the direction of change. So τ_{yx} (or equivalently, τ_{21}) defines a stress in the x-direction that acts because of a change in velocity in the y-direction.

We don't work much with tensors in this book; for a more complete tensorial treatment of fluid behaviors, see Bird et al.[13] We use tensors here because they make explicit a relevant feature of viscosities, namely that in principle they have 81 components: one for each of three directions of velocity, each varying over three directions of variation of velocity, and each of these nine components can affect three directions of stress that, like the velocity, can themselves vary over three directions.

Example Tensors and liquid crystals

Eighty-one components to a viscosity probably seems excessive—and as we'll see, this number can be dramatically reduced in most applications, but if you think about it, you can convince yourself that this many components really are possible. Consider for example a box containing a slanted rod, as depicted in Fig. 9.10(a). If the box is compressed along the z-direction, the rod will transmit stress along the y-direction—that is, the top of the box will move horizontally as shown by the blue arrow. The compression can be provided by changing the velocity in the z-direction, V_z, as z itself changes. This would require a strain term due to z-components: $\partial V_z/\partial z$, and a stress term, in the y-direction that changes with z: τ_{yz}, and so the viscosity to account for this must be μ_{yzzz}.

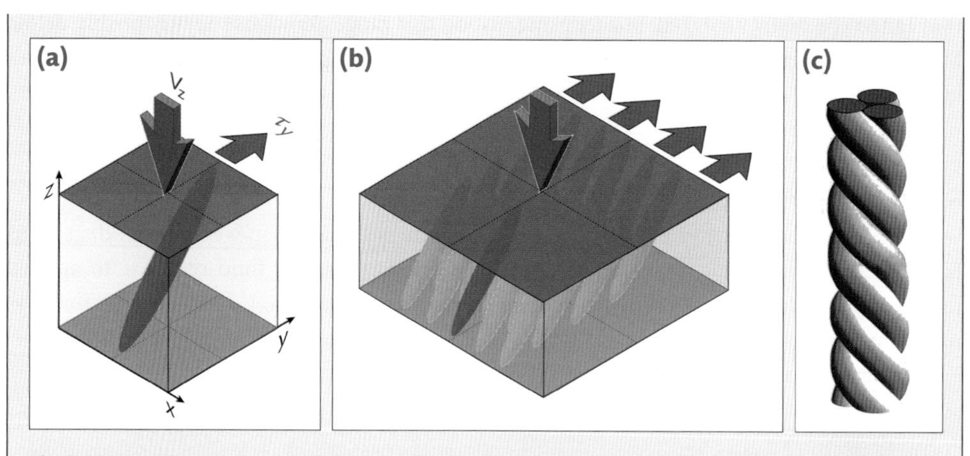

Fig. 9.10 Coupling between orthogonal stresses and strain rates produced by structured material suspended within a fluid. (a) A single rod-like particle will transform a strain rate in the z-direction (V_z) into a stress in the y-direction (τ_y). (b) When numerous rods are suspended in a fluid, this is called a liquid crystal, and when aligned as shown, the state is called "smectic C:" smectic to indicate that all rods lie in a single plane, and C to indicate that the rods are uniformly tilted. (c) Similarly, a twisted rope or cable couples vertical stretching with rotation, thus involving all of x, y, and z vectors in a complex way that can be described by a tensor Young's modulus (for a solid) or viscosity (for a liquid).

This term is actually seen in liquid crystals: suspensions of rods aligned in a so-called "smectic C" arrangement shown in Fig. 9.10(b). This arrangement is only one of many ordered states that liquid crystals can adopt, and their ability to take on multiple different states accounts for their importance in engineering: liquid crystals are most likely used in your computer display, and they are also found in lasers, detergents, thermometers, and other devices.

It isn't at all hard to create other combinations of strain rates and resulting stresses that require large tensors to express. For example, consider a twisted rope. The rope shown in Fig. 9.10(c) will attempt to straighten under tension, and so will produce both x- and y-components of stress in response to a strain in z. There are chiral phases in liquid crystals, as well as in the motions of suspended bacteria, that produce measurable optical and mechanical effects; a delightful result of this coupling can be seen in the Wilberforce pendulum: a spring-driven pendulum that trades energy between rotational and vertical modes in an unexpected way.

So in principle there are 81 terms in the viscosity tensor. The tensor itself looks like a matrix, with diagonal terms that represent stresses in the x-direction that change only in the x-direction due to velocities in the x-direction that also change only in the x-direction, and similarly for the y- and z-directions, which is a long way of describing pressure. The off-diagonal terms represent a stress, or its change, that occurs due to a velocity, or its change, in a different direction—as illustrated in the examples of Fig. 9.10.

For most fluids, many of these terms are the same, so for example fluids without suspended rods or the like have no preferred direction. In this case, nothing changes if an experiment is rotated by 90°, or if i is exchanged with j, or k is exchanged with ℓ. It can be shown using nothing more than algebraic manipulation that any strain rate that is symmetric under exchange between k and ℓ can be written in one of only two possible forms:

$$\left(\frac{\partial V_i}{\partial x_j}+\frac{\partial V_i}{\partial x_j}\right) \text{ or } \left(\frac{\partial V_1}{\partial x_1}+\frac{\partial V_2}{\partial x_2}+\frac{\partial V_3}{\partial x_3}\right), \quad [9.33]$$

so the 81 possible terms reduce to only *two*:

$$\tau_{ij} = -\mu\left(\frac{\partial V_i}{\partial x_j}+\frac{\partial V_i}{\partial x_j}\right) + \beta\left(\frac{\partial V_1}{\partial x_1}+\frac{\partial V_2}{\partial x_2}+\frac{\partial V_3}{\partial x_3}\right)! \quad [9.34]$$

And indeed it is easily seen that this is symmetric as required: $\tau_{ij} = \tau_{ji}$. The term in the second parenthesis is readily recognized as being the divergence, and of course for an incompressible fluid, this term vanishes. Moreover for a compressional fluid like a gas, Eq. [9.34] is often broken down in a different way, writing $\beta = 2\mu/3 - k$:

$$\tau_{ij} = -\mu\left(\frac{\partial V_i}{\partial x_j}+\frac{\partial V_i}{\partial x_j}\right) + \left(\frac{2}{3}\mu - k\right)\left(\frac{\partial V_1}{\partial x_1}+\frac{\partial V_2}{\partial x_2}+\frac{\partial V_3}{\partial x_3}\right), \quad [9.35]$$

where k is the "dilational viscosity."

We talked a little about dilatency in Exercise 9.2, and as its figure (Fig. 9.5) shows, a dilational fluid doesn't conserve volume—not because it compresses—rather, it *expands* in response to shear.

Materials that expand in all dimensions when stretched in one are more generally called "auxetic," a term borrowed from cell biology, where it refers to growth without division. Several applications have been developed using auxetic growth. For example, notice in Fig. 9.11(a) that the surface of wet sand becomes *dry* when deformed by squeezing—because the sand must dilate in order to deform, and this dilation produces more space to hold water. This observation has led to an amusing application: a diaper or absorbent towel with embedded grains.[14] Most diapers have the unfortunate property that they exude liquid when squeezed: a diaper with embedded grains does the opposite, and absorbs *more* liquid when it is compressed.

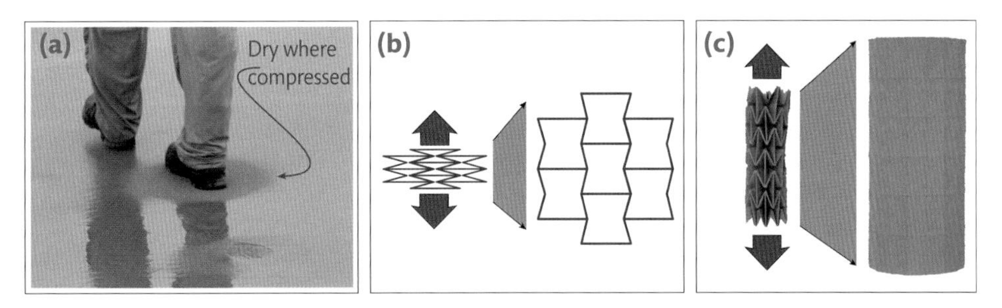

Fig. 9.11 (a) Dilatency causes wet sand to expand, thus holding more water, so that the surface of the sand appears dry where it has been deformed. (Courtesy of Alan Holyoak.) (b) Geometrical construction illustrating how pulling an auxetic material apart can cause growth in all directions. (c) Proposed stent design using Miura-Ori folds that grows auxetically when pulled.

Other applications involve expansion more directly: consider, for example, the mechanism depicted in Fig. 9.11(b). By geometric construction, the folded structure on the left enlarges its pores when pulled apart vertically, as shown to the right of the panel. Thus the enclosed area of the structure grows when stretched. Correspondingly, it has been proposed that a stent constructed from a folded sheet as shown on the left of Fig. 9.11(c) will grow to enclose a larger volume when stretched as shown on the right. The folds here are known as "Miura–Ori" folds, after Japanese astrophysicist, Koryo Miura, and are widely used to construct compact maps as well as to dramatically reduce the size of solar panels and other large surfaces in extraterrestrial satellites. The folds are also seen naturally, for example, in the leaves of trees such as beech or hornbeam.

Dilatent materials can be found in specialized applications as we've discussed, but by and large, most fluids—especially biological fluids—are incompressible. A common and more compact way of writing Eq. [9.35] is:

$$\bar{\bar{\tau}} = -\mu\left[\vec{\nabla}\vec{V} + \left(\vec{\nabla}\vec{V}\right)^{\dagger}\right] + \left(\frac{2}{3}\mu - k\right)(\vec{\nabla}\cdot\vec{V})\bar{\bar{\delta}}, \qquad [9.36]$$

where the double overbars represent a tensor, the dagger, $(\,)^{\dagger}$, represents the transpose, and $\bar{\bar{\delta}}$, is the "Kronecker delta," a matrix containing ones along the diagonal, and zeros elsewhere:

$$\bar{\bar{\delta}} \equiv \begin{bmatrix} 1 & 0 & 0 \\ 0 & 1 & 0 \\ 0 & 0 & 1 \end{bmatrix}. \qquad [9.37]$$

Otherwise, term-by-term the matrices in Eq. [9.36] are defined to be composed of the corresponding expressions in Eq. [9.35]. Written in this way, it is evident that symmetric and incompressible fluids that obey $\vec{\nabla}\cdot\vec{V} = 0$ must be described by:

$$\bar{\bar{\tau}} = -\mu\left[\vec{\nabla}\vec{V} + \left(\vec{\nabla}\vec{V}\right)^{\dagger}\right]. \qquad [9.38]$$

A final simplification and we're done: recall from Chapter 1 that Newton's law of viscosity says that the viscous stress on a fluid, here $\bar{\bar{\tau}}$, is proportional to the space derivatives of the velocity (Chapter 1, Eq. [1.27]). But the symmetric part of those derivatives is given by the terms in square brackets in Eq. [9.38]. So this equation is often abbreviated:

$$\bar{\bar{\tau}} = -\mu\dot{\bar{\bar{\gamma}}}, \qquad [9.39]$$

where $\dot{\bar{\bar{\gamma}}}$ is the symmetric rate of strain:

$$\dot{\bar{\bar{\gamma}}} \equiv \vec{\nabla}\vec{V} + \left(\vec{\nabla}\vec{V}\right)^{\dagger}. \qquad [9.40]$$

Written in this way, Eq. [9.39] is the same law of Newtonian viscosity that we have used all along, and simply says that stress is proportional to the rate of strain.

9.4.2 **Viscosity of Biological Fluids**

Real biological liquids such as synovial or amniotic fluid or respiratory mucus, however, do not respond in this simple, Newtonian, way. One way in which such fluids can behave is described by a different law:

$$\bar{\bar{\tau}} = -m\dot{\bar{\bar{\gamma}}}^n, \qquad [9.41]$$

where m and n are constants. These are called "power law" fluids, and their viscosities grow (for $n > 1$) or diminish (for $n < 1$) with shear rate, according to:

$$\mu = m\dot{\gamma}^{n-1}. \qquad [9.42]$$

For $n \neq 1$, these are termed "non-Newtonian" fluids because they do not obey the simple Newtonian law of viscosity, and as always these first examples are only the simplest cases of non-Newtonian fluids: in the next chapter we will describe more complicated non-Newtonian behaviors.

In Fig. 9.12, we show the change in viscosity and stress as functions of how rapidly a fluid is sheared for several fluids. We remind the reader that taking the log of Eq. [9.42] produces:

$$\ln(\mu) = \ln(m) + (n-1)\cdot\ln(\dot{\gamma}), \qquad [9.43]$$

so that a linear fit in a log–log plot implies that there must be a power law relation such as Eq. [9.41] or [9.42]. As Fig. 9.12 shows, water, synovial fluid, and hyaluronic acid (the ingredient in synovial fluid that most strongly affects its rheology) are all power law fluids over the range in strain rates shown. We reiterate that, as shown for example in Fig. 9.6, more complicated behaviors are possible, and fluids can behave as a power law over one range in shear rates and not over another.

Fig. 9.12 shows us that biological fluids behave in a more complex manner than does simple water. In the case of synovial fluid, it is strongly shear thinning, and we will see shortly that synovial fluid, blood, mucus, and other fluids are even more complicated than this—synovial fluid, for example, is elastic as well as being shear thinning. The shear thinning property itself is important for synovial fluid for it allows joints under rapid strain (e.g. during running) to produce low resistance to motion, while at the same time sustaining high stresses under slow strain conditions (e.g. when lifting heavy weights).

Synovial fluid contains a number of ingredients, but the shear thinning property is chiefly associated with hyaluronic acid. This is made evident by the comparison shown in Fig. 9.12(a) between synovial fluid from an ox (chosen because it has a large quantity of fluid that can be obtained for testing) and a small concentration of hyaluronic acid in water. Many joint diseases are associated with a degradation of this shear thinning property. For example, as shown in Fig. 9.12, rheumatoid arthritis ("r.a." in the figure) produces essentially Newtonian rheological response, with two unfortunate results. First, inflammatory byproducts produce a viscosity much lower than synovial fluid from healthy patients, so imposed strains must be absorbed by the surrounding tissues rather than by the fluid. And second, the rate of dissipation is the same at slow and fast strains—yet the rate of energy production is obviously

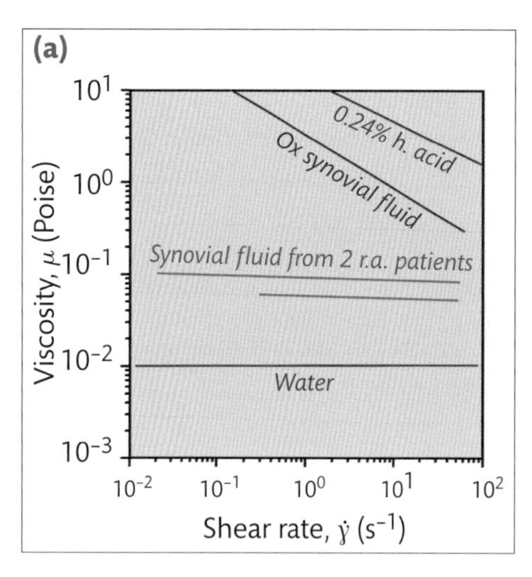

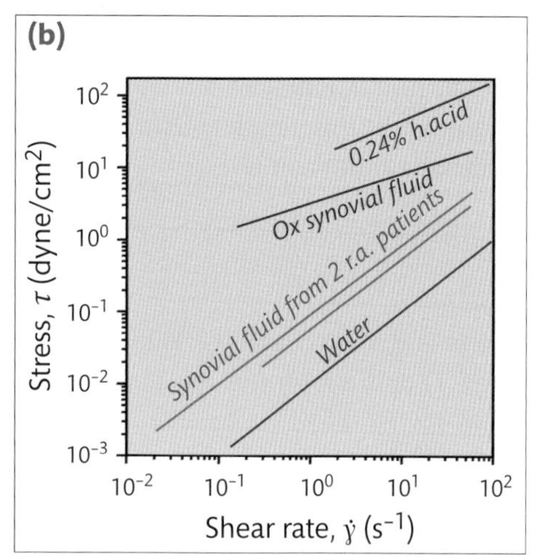

Fig. 9.12 Rheology of power law fluids. (a) Viscosities as a function of shear rate for water, synovial fluid from two rheumatoid arthritis ("r.a." in the plots) patients, synovial fluid from a healthy ox, and 0.24% hyaluronic acid ("h. acid" in the plot) in water. (b) The same data plotted in terms of stress vs. shear rate. (Source: after Lightfoot.[15])

faster at higher strain rates. As a consequence of the low viscosity, every motion of the joint produces an excess stress on the joint, and as a consequence of the lack of shear thinning, this stress gets worse during more rapid motion.

9.4.3 **Poiseuille Problem for a Power Law Fluid**

Let's illustrate how power law fluids behave by considering a simple problem, Poiseuille flow between infinite parallel plates. We first repeat the Newtonian solution, which we will then compare with a power law solution. Expressed in terms of stress, Newton's law of viscosity is:

$$\tau_{yx} = -\mu \frac{\partial V_x}{\partial y}, \tag{9.44}$$

for flow in the x-direction between fixed plates separated in the y-direction. Stokes equation prescribes that this stress is driven by the pressure gradient, and if we integrate Stokes equation once over y, we obtain:

$$\nabla P \cdot y + C_1 = -\mu \frac{\partial V_x}{\partial y}, \tag{9.45}$$

where P is the pressure gradient and C_1 is an integration constant. We can integrate this once more to obtain:

$$V_x = \frac{\nabla P}{2\mu} \cdot y^2 + C_1 y + C_2, \tag{9.46}$$

where C_2 is another integration constant. If we place the origin midway between the plates, we can make use of symmetry (as we did in Chapter 1) to set $C_1 = 0$, and if the plates are separated by a distance $2b$, we can apply non-slip boundary conditions to get:

$$V_x = \frac{\nabla P}{2\mu} \cdot b^2 \left(1 - \frac{y^2}{b^2}\right). \qquad [9.47]$$

As a quick exercise, the reader can integrate over a flow cross section of width w to calculate the mass flow rate of a fluid of density ρ, which as expected from Chapter 1 works out to be:

$$Q = \frac{2\rho\nabla P}{3\mu} \cdot b^3 w. \qquad [9.48]$$

We now compare this solution using Newton's law of viscosity with a solution using a power law viscosity. For a power law fluid, Eq. [9.44] becomes:

$$\tau_{yx} = m\left(-\frac{\partial V_x}{\partial y}\right)^n. \qquad [9.49]$$

As we described previously, for $n = 1$ the fluid is Newtonian. Eq. [9.45] is then:

$$\nabla P \cdot x + C_1 = m\left(-\frac{\partial V_x}{\partial y}\right)^n, \qquad [9.50]$$

which we take the $1/n$-th root of and integrate to obtain:

$$V_x = \left(\frac{\nabla P}{2\mu} \cdot b\right)^{1/n} \frac{b}{1 + 1/n}\left(1 - \frac{y^{1+1/n}}{b^{1+1/n}}\right). \qquad [9.51]$$

As before, the reader can perform a quick exercise of integrating to get the mass flow rate of a power law fluid flowing between parallel plates:

$$Q = \frac{2wb^2\rho}{1 + 1/n}\left(\frac{\nabla P}{m} b\right)^{1/n}. \qquad [9.52]$$

As we would expect, when $n = 1$ this reduces to the Newtonian solution (Eq. [9.48]). For a shear thickening fluid, $n > 1$, and the flow rate slows at higher pressure gradients as compared with the Newtonian case:

$$Q \sim (\nabla P)^{1/n} \qquad [9.53]$$

That is, at large n, the flow rate doesn't change much with pressure gradient: it becomes more and more difficult to force such a fluid through a gap. This occurs because the harder such a shear thickening fluid is pushed, the more it resists. The converse of course is true for a shear thinning fluid: for such fluids, small changes in pressure produce large changes in flow rate.

EXERCISE 9.4

It isn't hard to repeat these derivations to show that the rate of flow of a power law fluid through a tube goes as:

$$Q = \frac{\pi R^3 \rho}{3 + {}^1/_n} \left(\frac{\nabla P}{m} R \right)^{1/n}. \qquad [9.54]$$

Do this derivation, and plot the flow rate versus R for shear thinning, Newtonian, and shear thickening fluids. Fix the other parameters at any convenient values of your choosing. Repeat this for the parallel plate solution, Eq. [52], and comment on what differences between these solutions tell you about conditions under which a kidney dialyzer should be designed using parallel plates or tubes.

9.5 **Complex Shear Modulus**

The power law description (Eq. [9.41]) addresses a first complication to Newton's simple law of viscosity. Real biological fluids are considerably more intricate than Eq. [9.41] can describe, and indeed a full description of such fluids defies analytic treatment altogether. As before, our approach will therefore be to creep up on the problem by including one complication at a time. To understand both how complicated these material behaviors can be and what makes the behaviors so peculiar, let's take a lesson from physics and consider how polymers, including biological proteins, behave.

Example Spaghetti model of polymers

Polymers are commonly added to pharmaceuticals, foods, and other materials to modify their transport properties. For example, polymers such as PEG are used to produced timed release of drugs; likewise capsule coatings such as polyvinyl acetate phthalate (PVAP) are used to protect drugs from the acidic stomach environment and allow release in the higher pH intestine. Polymers such as carboxymethyl cellulose (CMC: described in Chapter 3, Fig. 3.5) are used to thicken ice cream, toothpaste, and other products. And perhaps the grandparents of these additives are viscosity improvers such as polymethacrylates (PMAs) and olefin copolymers (OCPs) that are added to motor oils.

Until the 1950s,[16] motor oils had to be changed every season, because in the winter engines needed low viscosity oil to lubricate the engine during startup, and in the summer engines needed high viscosity oil so that the oil would remain on surfaces at high temperatures. Aside from being a nuisance, this necessitated compromises: for instance a low viscosity oil would allow a car to start in a Minnesota winter, but would lubricate poorly once the engine warmed up.

By adding long chain polymers, so-called "multi-grade" oils were produced: these oils typically consist of a low viscosity oil with added polymers that uncoil at higher temperatures to hold the oil together. So a 10W-30 motor oil is made almost entirely from a low viscosity, 10-grade oil. At low temperatures, the polymers coil tightly, so the low viscosity oil flows easily to coat the engine surfaces. At high temperatures, the polymers unwind and act like a meshwork to boost the viscosity to a 30-grade rating when the engine warms up.

This clever mechanism has an obvious drawback: engines apply high stresses to the oil, and so the polymers are physically ripped apart over time, leaving just the low viscosity base oil to lubricate the engine. This (plus the accumulation of grunge) is a central reason that engine oils must be replaced.

So let's give some thought to what adding polymers does to a fluid. This question has been studied by many scientists, including Pierre-Gilles de Gennes (1932–2007), who won the Nobel Prize in Physics in 1991 for analysis of order-disorder transitions in many materials, including polymeric molecules. One of de Gennes' insights was that a network of polymers, such as polyvinyl alcohol (PVA) shown in Fig. 9.13(b), both looks and behaves like a collection of spaghetti. PVA is the principal ingredient in Elmer's and other white glues, and it is what makes the toys "slime" and "flubber" behave oddly. De Gennes' spaghetti model explains many of these odd behaviors.

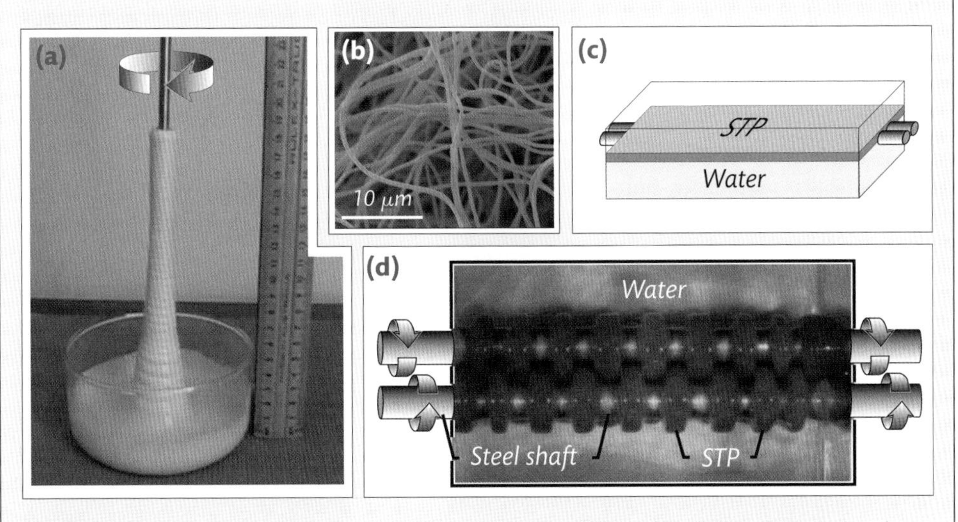

Fig. 9.13 Strange behaviors of polymers in fluids. (a) Rod climbing, or "Weissenberg," effect of Elmer's glue + borax. The polymeric liquid climbs a rotating rod driven by a drill. (From: psidot/Youtube). (b) Micrograph of polyvinyl alcohol, showing entangled, spaghetti-like tendrils.[17] (From Coburn, J.M., Gibson, M., Monagle, S., Patterson, Z. and Elisseef, J.H. (2009) Bioinspired nanofibers support chondrogenesis for articular cartilage repair. *Proceedings of the National Academy of Sciences, 109*, 10012–7. By permission of the National Academy of Sciences.) (c) STP oil treatment floating on water and driven by a pair of counter-rotating steel rods spontaneously generate interlinked "rollers" shown in overhead view in panel (d). Panel (d) retouched to identify rotating rods and subsurface water.[18] (From Joseph, D.D., Nguyen, K. and Beavers, G.S. (1986) Rollers. *Physics of Fluids, 29*(9). By permission of the National Academy of Sciences.)

Consider for example the "Weissenberg," or "rod climbing," effect shown in Fig. 9.13(a). If you mix white glue with a cross-linking agent such as borax, the glue's sticky ends attach together. The glue ceases being sticky, and instead it behaves like a consolidated, but oozing, gel. When a spinning rod is inserted into the blend, the gel climbs the rod to a considerable height. To understand why this occurs, visualize the gel as a collection of spaghetti, and the rod as a spinning fork. The spaghetti will wind around the fork, and as it does so it will tighten around the fork, squeezing inward. Spaghetti near the fork will then be extruded out, along the fork's rotational axis. Because this effect produces a compressive stress along one axis (the rotational axis) due to tensile stress along a perpendicular axis (the azimuthal axis), this is termed an effect due to a "normal stress difference" (see, for example, Bird et al.,[13] p. 237).

So viewing a polymer as spaghetti can help to explain macroscopic effects. De Gennes' chief contribution here was to a more subtle issue: the microscopic rearrangement of polymer chains in response to applied stresses. If we view the tangled mass of polymers shown in Fig. 9.13(b) as a collection of spaghetti, it will make logical sense that pulling rapidly on the tangle will produce a reversible and elastic recoil: the spaghetti will remain tangled, and will respond to being stretched like a coiled spring. On the other hand, if it is pulled more slowly, spaghetti can be disentangled and will align with a twirling fork. This disentanglement requires that pulling is done slowly enough that individual spaghetti (or polymer) strands can slide (or "reptate") with respect to one another. In this case, strands will *irreversibly* separate and align with the imposed stress. This is termed a plastic, rather than elastic, response.

Thus de Gennes' spaghetti model accounts for the fact that silly putty bounces elastically when dropped abruptly, but elongates plastically when slowly stretched. Such materials are manifestly non-Newtonian, since they do not obey Newton's law of viscosity, $\tau = -\mu\dot{\gamma}$, and they are often called "elastoplastic," or "viscoelastic." Additionally, polymeric materials depend on the history of deformation—for example, they align with previously applied stresses. We do not treat such materials in this book, for history-dependent materials do not even obey the Navier–Stokes equation—which after all has no place to include history—on which this book is based. Nevertheless, the study of these materials is a rich and complex topic, and the reader is referred to articles and texts on the topic, beginning perhaps with Walker.[7]

As a final, and extreme, example of extraordinary behaviors of polymeric materials, consider STP oil treatment. This consists of long chain polymers such as are added to make multi-weight motor oils. One reason for the creation of such materials is to treat engine wear. Internal combustion engines employ a pair of compression rings in each piston to maintain the high pressures generated by burning fuel. As engines age, the rings develop small nicks, which allow hot combustion gas to tear past the cylinder wall at high speeds. This causes grooves to be eroded along the cylinders that mechanics term "blow-by." To treat this problem, oil treatments that produce very high viscosities in response to rapid stresses were developed: these treatments coat the cylinder walls and respond very stiffly to sudden forcing, limiting the high speed escape of hot gases, but permitting the oil to lubricate the engine nonetheless.

So far, so good. However, because this material is designed to be spaghetti-extraordinaire, it behaves very strangely. For example, floating a layer of STP on water and imposing two counter-rotating rods as shown in Fig. 9.13(c) causes small sinusoidal fluctuations in thickness to be amplified, producing the "roller" structure shown in Fig. 9.13(d). These look for all the world like solid rollers attached to the rods, but in fact they are made of a thick polymeric fluid that settles back into a uniform layer, as shown in Fig. 9.13(c), when the steel rods stop rotating.

9.6 Dynamic Shear Moduli: Vibrating Rheometers

As the preceding examples make clear, polymeric fluids can exhibit both elastic behaviors—rod-climbing and roller formation are only two of these—and viscous behaviors—as in motor oils or toothpaste. Correspondingly, it is useful to quantify the magnitudes of elastic and viscous effects separately. To do this, notice that the essential feature that the spaghetti model relies on is the difference in timescales between elastic and viscous responses. To establish relative magnitudes of elasticity and viscosity, then, it would be useful to stress a sample over different timescales—which is precisely what modern rheometers do.

9.6.1 Elastic Solution

To understand how this is achieved, consider a parcel of fluid trapped between a stationary plate and a plate that is sheared back and forth at frequency ω, as shown in Fig. 9.14(a). In practice, this shearing is accomplished in a fixed geometry such as in the gap between Couette cylinders, or between parallel rotating plates, or between a rotating cone and a fixed plate, or so on. Whatever the geometry, a viscometer varies the frequency of applied strain on one surface and measures the resulting stress on another. If the parcel being sheared were purely elastic, and if the applied strain were slow enough to avoid elastic waves, then the stress would be transmitted uniformly throughout the parcel. Then as shown in Fig. 9.14(b), for a parcel of height h sheared at the top according to $\gamma = \gamma_0 \cos(\omega t)$, the displacement at a fractional height, y/h, would be:

$$\gamma_{elastic} = (y/h) \cdot \gamma_0 \cos(\omega t), \qquad [9.55]$$

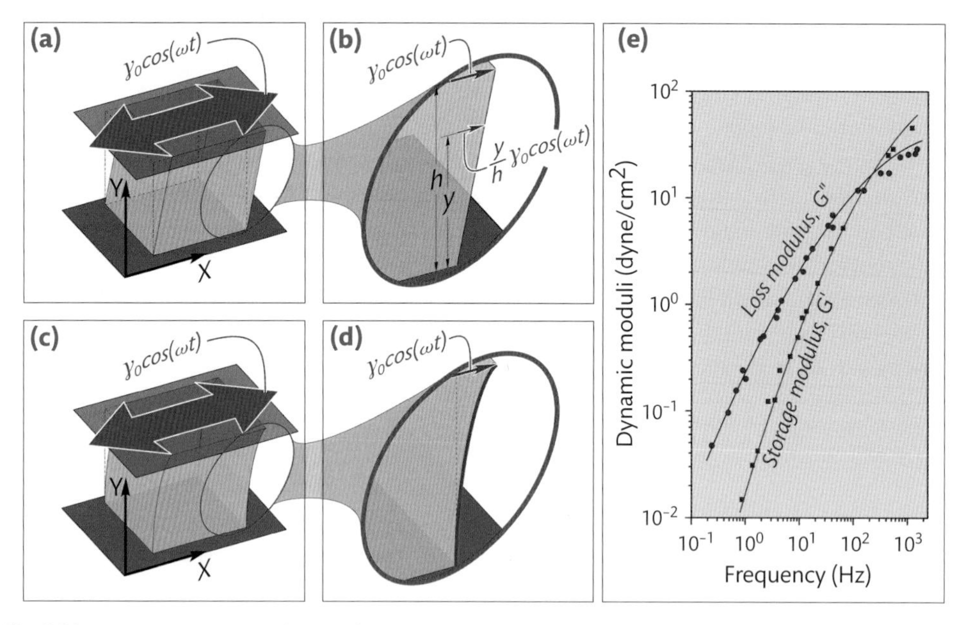

Fig. 9.14 Measurement of elastic ("storage") and viscous ("loss") moduli. (a, b) Elastic parcel of material held between two plates is sheared periodically with varying frequency, ω. (c, d) Viscous parcel sheared in the same way. (e) Storage and loss moduli measured for hyaluronic acid in water. (Panel (e) source: Lightfoot.[19])

On the other hand, if the parcel were purely viscous, the material beneath the top plate would respond lethargically, producing a displacement sketched in Figs. 9.14(c) and (d).

9.6.2 **Viscous Solution**

Figs. 9.14(a) and (b) are exact: neglecting elastic waves, a simple solid will behave purely linearly, as sketched. Figs. 9.14(c) and (d), however, are (so far) just a cartoon. We can solve analytically for how a viscous fluid would respond to shear in the idealized geometry shown as follows. We assume low Reynolds number, so that flow in the horizontal (x) direction obeys Stokes' equation:

$$\frac{\partial V_x}{\partial t} = \nu \frac{\partial^2 V_x}{\partial y^2}, \qquad [9.56]$$

with boundary conditions:

$$\begin{aligned} V_x &= 0 \qquad && at\ y = 0, \\ V_x &= \gamma_0 e^{i\omega t} \qquad && at\ y = h. \end{aligned} \qquad [9.57]$$

Let's separate variables by assuming that $V_x(y,t) = \gamma_0 e^{i\omega t} \cdot V_x(y)$, so Eq. [9.56] becomes:

$$i\omega V_x(y) = \nu \frac{\partial^2 V_x(y)}{\partial y^2}, \qquad [9.58]$$

which has solution:

$$V_x(y) = V_1 \cdot exp\left(\sqrt{\frac{i\omega}{\nu}}y\right) + V_2 \cdot exp\left(-\sqrt{\frac{i\omega}{\nu}}y\right), \quad [9.59]$$

where V_1 and V_2 are constants chosen to satisfy the boundary conditions (Eqs. [9.57]). The reader can confirm that the boundary conditions are satisfied for:

$$\begin{aligned} V_1 &= \frac{-i}{2 \cdot sin\left(\sqrt{\frac{i\omega}{\nu}}h\right)}. \\ V_2 &= -V_1. \end{aligned} \quad [9.60]$$

After a little algebra, the solution becomes:

$$V_x(y,t) = \frac{\gamma_0}{\sin\left(\sqrt{\frac{i\omega}{\nu}}h\right)} \cdot \sin\left(\sqrt{\frac{i\omega}{\nu}}y\right) e^{i\omega t}. \quad [9.61]$$

We have blithely carried along the factor of i that originated in Eq. [9.57]'s description of oscillation as a complex exponential. As we have discussed previously, this saves algebra but any solution must at some point be reconciled with the fact that only the real part of an expression can be directly measured. This reconciliation can be done, but in this case is long and tedious on account of the presence of three separate complex terms in Eq. [9.61]. Instead of carrying out that process, let's solve a slightly different problem that has a simpler, and much similar, solution.

That problem involves a single vibrating plate at the origin infinitely far from any other boundary. In this case, the general solution is the same as we obtained earlier (Eq. [9.59]):

$$V_x(y) = V_1 \cdot \exp\left(\sqrt{\frac{i\omega}{\nu}}y\right) + V_2 \cdot \exp\left(-\sqrt{\frac{i\omega}{\nu}}y\right), \quad [9.62]$$

but the boundary conditions are now (compare with Eqs. [9.57]):

$$\begin{aligned} V_x &= \gamma_0 e^{i\omega t} \qquad at\ y = 0, \\ V_x &= 0 \qquad as\ y \to \infty. \end{aligned} \quad [9.63]$$

A little work shows that these conditions imply that $V_1 = 0$ and $V_2 = 1$, so that:

$$\begin{aligned} V_x(y) &= e^{-\sqrt{\frac{i\omega}{\nu}}y} \\ &= e^{-\sqrt{\frac{\omega}{2\nu}}y}\left[\cos\left(\sqrt{\frac{\omega}{2\nu}}y\right) + i \cdot \sin\left(\sqrt{\frac{\omega}{2\nu}}y\right)\right]. \end{aligned} \quad [9.64]$$

Almost done: finally, the separation of variables ansatz, $V_x(y,t) = \gamma_0 e^{i\omega t} \cdot V_x(y)$, implies that the measurable part of the velocity is:

$$Re[V_x(y,t)] = \gamma_0 e^{-\sqrt{\frac{\omega}{2\nu}}y} \cos\left(\omega t - \sqrt{\frac{\omega}{2\nu}}y\right). \qquad [9.65]$$

EXERCISE 9.5

Fill in the missing steps between Eq. [9.62] and Eq. [9.65]. Explain why Eq. [9.61] goes as a sine in y, while Eq. [9.65] goes as a cosine.

What Eq. [9.65] tells us is that the motion of a simple fluid near an oscillating plate is damped exponentially with distance, y, from the plate, and exhibits a phase lag of $\sqrt{\omega/2\nu}\,y$ with respect to the plate's motion. The exponential damping makes sense: after all, damping is what viscosity does. The phase lag is likewise caused by viscosity, because a viscous fluid takes time to respond to a disturbance. For large ω and small ν, Eq. [9.65] reduces to Eq. [9.61], so in that limit, the solution for flow near an oscillating plate with boundaries infinitely far away agrees with the solution between an oscillating plate and a fixed boundary. The infinite and fixed boundary solutions agree at a characteristic distance of $h_0 \sim \sqrt{\nu/2\omega}$: plates separated by $h \gg h_0$ behave approximately according to Eq. [9.65]. Plates separated by less require the full solution (Eq. [9.61]).

9.6.3 Combined Elastic and Viscous Models

Whether we use the exact solution (Eq. [9.61]) or the infinite boundary solution (Eq. [9.65]), another way of expressing the phase difference between elastic and viscous parcels is to use trigonometric sum identities such as $cos(a+b) = cos(b)cos(a) + sin(b)sin(a)$ to produce:

$$\tau_{yx} = G' \cdot \gamma_0 cos(\omega t) + G'' \cdot \gamma_0 sin(\omega t), \qquad [9.66]$$

where G' determines the parcel's in-phase, elastic, contribution and G'' determines its out-of-phase, viscous, contribution. In this expression, the distance y from the vibrating plate is taken to be constant, as would be the case in an actual rheometer, where stress is measured a fixed distance away from a moving plate. A traditional shorthand is to combine these two moduli into a single complex modulus:

$$G = G' + i \cdot G''. \qquad [9.67]$$

In this shorthand, the elastic term, G', is termed a "storage" modulus (because it measures stored and released energy) and the viscous term, G'', is termed a "loss" modulus (because it measures viscous loss of energy). Typical output from a viscometer is expressed in terms of these two moduli as shown in Fig. 9.14(e) for measurements of hyaluronic acid in water. Expressed in that way, it is evident from this plot that at low frequencies a hyaluronic acid solution is predominately viscous—since the loss modulus is as much as an order of magnitude larger than the storage modulus—but at frequencies above 100 Hz, the fluid behaves elastically.

Notice that for an elastic material, stress is proportional to strain $\tau \sim \gamma$, while for a fluid, stress is proportional to strain rate: $\tau \sim \dot{\gamma}$. That is, in elastic materials, stress depends on total deformation, but not to its rate, while in fluids, faster deformation dissipates more energy, but in the limit of infinitely slow movement, no stress appears for any total deformation. Viscometers account for this distinction by either producing plots of storage and loss moduli as shown in Fig. 9.14(e), or equivalently by plotting storage and loss *viscosities*, using the complex viscosity:

$$\eta = \eta' + i \cdot \eta''. \qquad [9.68]$$

A technical note is needed here: because the viscosity depends on the *derivative* of the strain (and because historically the stress was expressed as η' times the cosine to agree with Eq. [9.66]), η' conventionally gives the viscous (loss) part of the stress, while η'' gives its elastic (storage) part. This is the opposite convention to that of the moduli (Eq. [9.67]), where G' defines the elastic and G'' gives the viscous part of the stress.

9.6.4 **Maxwell Model–An Alternative Formulation**

To summarize, for elastic materials we would write that the stress goes as the strain: $\tau \sim \gamma$, while for viscous materials we would write that the stress goes as the strain rate: $\tau \sim \dot{\gamma}$. However an elastic fluid responds *both* to the total deformation and to its rate, so there is no perfect description: either the elastic or the viscous model will leave something out. Several alternative formulations have been proposed that include both elastic and viscous effects. One of these is the so-called "Maxwell model," in which the stress and its derivative are both included:

$$\tau + \lambda \dot{\tau} = -\eta \dot{\gamma}. \qquad [9.69]$$

In this model, $1/\lambda$ gives a relaxation time that characterizes how elastic or viscous the material is. For small λ, the model reduces to Newton's law of viscosity: $\tau \sim \dot{\gamma}$, and for large λ, $\lambda\dot{\tau}$ becomes much larger than τ, so the model predicts (after time integration) an elastic response: $\tau \sim \gamma$. The reader can easily imagine more complicated models. For example, multiple relaxation times can be added to the left hand side of Eq. [9.69]: this is called a generalized Maxwell model. Or a term in the second derivative of γ can be added to the right hand side: this is called a Jeffreys model. Other formulations such as the Oldroyd model include derivatives that both move with a fluid element (as in the substantive derivative introduced in Chapter 1) and rotate with the element. For our purposes, we stick with the simpler Maxwell model and leave embellishments for specialized texts.

Armed with the Maxwell model, we can compare how flow of a Newtonian fluid compares with flow of a simple viscoelastic material. We emphasize that this analysis only applies to the simplest of viscoelastic materials, and none of the history-dependent behaviors described in Example 9.5 can be analyzed using this approach. This again is because history isn't included: Eq. [9.69] applies at an instant in time and doesn't depend on what happened to the fluid at previous times. We'll touch on some more complicated issues encountered in complex and biological fluids—especially blood—in the next chapter.

To understand the Maxwell model, let's use it to repeat the derivation of the fluid velocity surrounding an oscillating plate. We begin with the Stokes equation (Eq. [9.56]), which prescribes the rate of strain in the x direction, $\partial V_x/\partial t$. The right hand side of this equation defines how the fluid responds, which can no longer simply be the viscosity times the Laplacian, $\nu \cdot \partial^2 V_x/\partial y^2$: we got that in Chapter 1 from Newton's law of viscosity, right? More generally, the right hand side is given by the change in the stress with y (refer to the Viscosity section in Chapter 1 if you need a reminder). So for a non-Newtonian fluid, we write:

$$\frac{\partial V_x}{\partial t} = -\frac{1}{\rho}\frac{\partial}{\partial y}\tau_{yx}, \tag{9.70}$$

and our job is to figure out what the stress, τ_{yx}, is for the Maxwell model. The trouble is that Eq. [9.69] includes both the stress and its rate. So how are we going to get τ_{yx}?

To do so, we make use of a heuristic idea, namely that if the left hand side of Eq. [9.70] depends on time, the right hand side ought to as well—so presumably the stress must change in time. Rather than describing the stress at one instant in time, maybe then we ought to write it as an integral over the fluid's past history of deformation. How long a history? Well, we expect that recent events will more strongly affect the stress than older events, so perhaps we should write that the influence of history decays rapidly in time. One way of writing this is:

$$\tau_{yx} = -\int_{-\infty}^{t} \frac{\eta}{\lambda} e^{-(t-t')/\lambda} \cdot \frac{\partial V_x(y,t')}{\partial y} dt', \tag{9.71}$$

where the rate of change of strain is given by $\dot{\gamma} = \partial V_x(y,t')/\partial y$. This heuristic—and in particular the fact that it allows us to exponentially cut off strain histories—is really just a ploy to solve for what we want, but as we will see it meets our needs and produces a useful analytic description of Maxwell model rheology.

Two cautions. First, this formulation appears to include an integration over history, but in fact for the integral to be solved, $\dot{\gamma}(t')$ must vary in quite simple ways over time. So solving for complex polymeric fluids cannot typically be performed using this method. Second, this description is heuristic in that Eq. [9.71] is not directly derivable from Eq. [9.69]. It does, however, contain the essential ingredient that the stress depends on the strain rate in a way that decays in time with a time constant λ: this is the necessary ploy.

The integral defined in Eq. [9.71] is part of a more general method known as Green's functions, named after George Green (1793–1841), a British miller, who dutifully and laboriously worked on his family farm until the age of 34, when without mathematical education—or indeed any schooling at all after the age of 9—he inexplicably and astonishingly self-published a sophisticated essay on mathematical physics, and thereafter launched himself into a distinguished career as a mathematician at Cambridge. A small but important part of his work involves the formulation of problems using integrals of the form:

$$\int G(t-t') \cdot f(t')dt', \tag{9.72}$$

where f is a function of interest and $G(t - t')$ defines a "Green's function" that permits the integral to be solved. In our problem, G is simply an exponential that codifies the notion that we want to include recent, but not distant, history; in other problems choosing G can be quite involved. We don't describe this method in general, but for our problem of the vibrating plate, we use $\dot{\gamma}(t') = V_x(y, t')$ and substitute Eq. [9.71] into Eq. [9.70] to obtain

$$\frac{\partial V_x}{\partial t} = \int_{-\infty}^{t} \frac{\eta}{\rho\lambda} e^{-(t-t')/\lambda} \frac{\partial^2 V_x(y,t')}{\partial y^2} dt'. \quad [9.73]$$

As before (see Eq. [9.58]), we assume the plate oscillates according to $V_x|_{y=0} = \gamma_0 e^{i\omega t}$, and we try separation of variables using $V_x(y,t') = \gamma_0 e^{i\omega t'} \cdot V_x(y)$:

$$\begin{aligned} i\omega V_x(y)e^{i\omega t} &= \int_{-\infty}^{t} \frac{\eta}{\rho\lambda} e^{-(t-t')/\lambda} \frac{\partial^2 V_x(y)}{\partial y^2} e^{i\omega t'} dt' \\ &= \frac{\eta}{\rho\lambda} \frac{\partial^2 V_x(y)}{\partial y^2} \int_{-\infty}^{t} e^{-(t-t')/\lambda} e^{i\omega t'} dt'. \end{aligned} \quad [9.74]$$

The reader may recognize the integral to be a Laplace transform: such transforms can be solved directly or simply looked up, to produce:

$$i\omega V_x(y)e^{i\omega t} = \frac{\eta}{\rho} \frac{\partial^2 V_x(y)}{\partial y^2} \left[\frac{e^{i\omega t}}{1 + i\lambda\omega}\right]. \quad [9.75]$$

But look, this is just an ordinary differential equation of the form $A \cdot V_x(y) = V_x''(y)$. A is a function of time, but as far as the differential equation in y is concerned, A can be taken to be constant. We know the solution to this equation: it is an exponential—and after some uninformative algebra that we omit, the solution works out to be:

$$V_x(y) = V_0 e^{(\alpha + i\beta)y}, \quad [9.76]$$

where:

$$\begin{aligned} \alpha &= \sqrt{\frac{\rho\omega}{2\eta}} \left[\sqrt{1 + (\lambda\omega)^2} - \lambda\omega\right]^{1/2} \\ \beta &= \sqrt{\frac{\rho\omega}{2\eta}} \left[\sqrt{1 + (\lambda\omega)^2} - \lambda\omega\right]^{-1/2}. \end{aligned} \quad [9.77]$$

Just as we did in Eq. [9.65], we can use the separation ansatz $V_x(y,t') = \gamma_0 e^{i\omega t'} \cdot V_x(y)$ and take the real part of the result to finally obtain:

$$Re[V_x(y,t)] = \gamma_0 e^{-ay} cos(\omega t - \beta y). \quad [9.78]$$

Naturally, this approaches the Newtonian solution (Eq. [9.65]) when the relaxation time of the fluid goes to 0; more than this, we can examine how the velocity differs for a Maxwell fluid

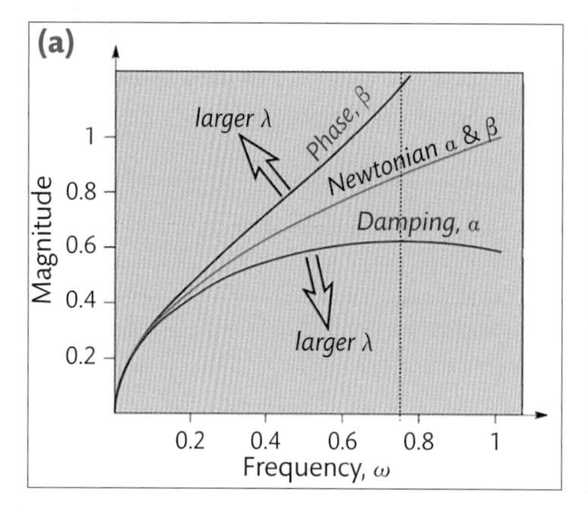

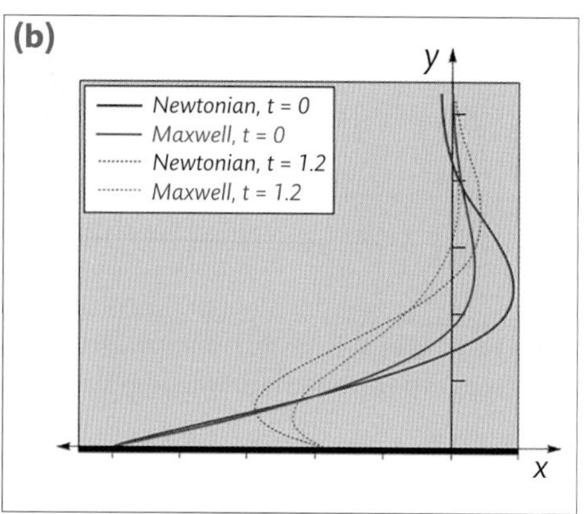

Fig. 9.15 Comparison between Newtonian and Maxwell fluids near vibrating plate. (a) Magnitudes of α and β from Eq. [9.77]. For a Newtonian fluid, $\alpha = \beta$, as indicated by the gray curve, and as relaxation time, λ, of the Maxwell fluid grows, α and β diverge from one another. (b) Typical plots of motion of fluid near vibrating plate. Plate is bold black line at bottom of plot, and locations of a vertical stripe of marker particles is shown at two different times for Newtonian fluid (gray) and Maxwell fluid using values taken from panel (a) at $\omega = 0.75$ (dotted line). All other parameters in Eqs. [9.77] and [9.78] are taken to be one.

with non-zero λ. To do this, we plot α and β as a function of ω, as shown in Fig. 9.15(a). From the figure and Eq. [9.78] it is evident that as the relaxation time, λ, grows, the damping, α, and phase, β, diverge from the Newtonian solution. We remark that β can instead be viewed as a spatial frequency, so that as λ grows, there is less damping and a higher spatial frequency, as shown in Fig. 9.15(b). So Maxwell fluids near a vibrating plate exhibit larger and more rapid fluctuations than simple Newtonian fluids.

Which makes sense, right? The whole point of the Maxwell model (Eq. [9.69]) is that a fluid can respond to an applied strain rate, $\dot{\gamma}$, with an added term associated with a rate of change of stress, $\dot{\tau}$, which of course will accelerate the fluid. So in the end, our heuristic ploy works insofar as it describes the behavior of a complex fluid with an added time dependence, producing fluctuations not present in Newtonian fluids.

We have so far seen a brief survey of some models for the simplest of non-Newtonian fluids. We turn in the next chapter to a closer examination of how biological fluids—and blood in particular—behave. We will see that the cells and proteins in blood produce surprising phenomena that are both unexpected and beautiful, requiring new and elegant tools to approach.

REFERENCES

1. After Geisel, T. (Dr. Seuss). (1949) *Bartholomew and the Oobleck*. New York: Random House.
2. Waitukaitis, S.R. and Jaeger, H.M. (2012) Impact-activated solidification of dense suspensions via dynamic jamming fronts. *Nature, 487*, 205–9.
3. Brown, E., Forman, N.A., Orellana, C.S., Zhang, H., Maynor, B.W., Betts, D.E., DeSimone, J.M. and Jaeger, H.M. (2010) Generality of shear thickening in dense suspensions. *Nature Materials, 9*, 220–4.

4. YouTube. (2009) Liquid armor. Available at http://www.youtube.com/watch?v=rYIWfn2Jz2g

5. Gálvez, L. O., de Beer, S., van der Meer, D., and Pons, A. (2017). Dramatic effect of fluid chemistry on cornstarch suspensions: Linking particle interactions to macroscopic rheology. *Physical Review E*, 95, 030602.

6. YouTube. (2006) ¡Caminaron sobre el agua! Available at http://www.youtube.com/watch?v=4KzuezE82js

7. Merkt, F.S., Deegan, R.D., Goldman, D.I., Rericha, E.C. and Swinney, H.L. (2004) Persistent holes in a fluid. *Physical Review Letters, 92*, 184501.

8. Walker, J. (1978) Serious fun with polyox, silly putty, slime and other non-Newtonian fluids. *Scientific American, 239*, 142–9.

9. Donev, A., Cisse, I., Sachs, D., Variano, E.A., Stillinger, F.H., Connely, R., Torquato, S. and Chaikin, P.M. (2004) Improving the density of jammed disordered packings using ellipsoids. *Science, 303*, 990–3.

10. Bird, R.B., Stewart, W.E. and Lightfoot, E.N. (2002) *Transport Phenomena*, 2nd ed. New York: John Wiley & Sons, p.33.

11. Wagner, N.J. and Brady, J.F. (2009) Shear thickening in colloidal dispersions. *Physics Today, 62*, 27–32.

12. Seybold, H.J., Molnar, P., Devrim, A.M., Doumi, M., Cavalcanti Tavares, M., Shinbrot, T., Herrmann, H.J. and Kinzelbach, W. (2010) The topography of an inland delta: observations, modeling, experiments. *Geophysical Research Letters, 37*, L08402.

13. Bird, R.B., Stewart, W.E. and Lightfoot, E.N. (2007) *Transport Phenomena* New York: John Wiley & Sons.

14. Goldhirsch, I. and Orszag, S.A. (2002) Composite absorbent structure and method. US Patent 6,455,114.

15. Lightfoot, E. (1974) *Transport Phenomena and Living Systems*. New York: Wiley Interscience, p.38.

16. Van Horne, W.L. (1949) Polymethacrylates as viscosity index improvers and pour point depressants. Industrial & Engineering Chemistry, 41, 952–9.

17. Coburn, J.M., Gibson, M., Monagle, S., Patterson, Z. and Elisseef, J.H. (2012) Bioinspired nanofibers support chondrogenesis for articular cartilage repair. *Proceedings of the National Academy of Sciences of the USA, 109*, 10012–7.

18. Joseph, D.D., Nguyen, K. and Beavers, G.S. (1986) Rollers. *Physics of Fluids, 29*, 2771.

19. Lightfoot, E. (1974) *Transport Phenomena and Living Systems*. New York: Wiley Interscience, p.40.

10 Rheology in Complex Fluids 2

We saw in Chapter 9 that fluids can behave oddly when mixed with particles (e.g. cornstarch), polymers (e.g. STP) and proteins (e.g. hyaluronic acid). Let's now consider a vital bodily fluid—blood. Blood behaves in some ways even more oddly than seen in previous examples, and it provides unexpectedly important insights into biophysical processes that apply to a host of other problems.

10.1 Blood Flow: Basic Phenomenology

As a first example, let's look at results that can largely be understood using tools that we have developed already. Then we will turn to more complex behaviors. One behavior that we can readily understand is how cells affect the viscosity of blood. Red blood cells (RBCs) are the largest component of cells in blood, and we might expect the viscosity of blood to increase with concentration of RBCs. Perhaps the viscosity might obey something like the Kreiger–Dougherty relation, which, as we explained in Chapter 9, describes the growth in viscosity that suspensions of solids undergo as they approach a jammed state.

In Fig. 10.1(a), we plot the effect of volume fraction on the "relative viscosity," the ratio between the suspension viscosity and the pure fluid viscosity for various suspensions. We show as a blue curve the expected Kreiger–Dougherty behavior, as defined in Eq. [9.9] of Chapter 9, using parameter values typical of spheres, $B = 2.5$ and $\phi_{rcp} = 60\%$. By comparison, we plot in red viscosity measurements for normal RBCs obtained using a Couette viscometer. For reference, we include as a dotted line a value of volume fraction for 45% RBCs. The volume fraction of RBCs is termed "hematocrit," which we abbreviate "H"—so $H = 45$ means 45% of the blood is RBCs by volume. $H = 48$ is a typical value for men; $H = 42$ is typical for women.

Plainly, viscosity does increase with volume fraction of RBCs, but by *much* less than solid spheres: notice that the vertical scale is logarithmic, so the viscosity at 45% volume fraction is reduced by nearly a factor of three for RBCs compared with solid spheres. Several effects, which account for the substantial reduction of viscosity of RBCs as compared with other suspensions,[4] are led by their deformability. This can be seen by comparison with the other curves in Fig. 10.1(a) showing suspended materials with different deformabilities. The pink curve shows that stiff and angular suspended RBCs—which appear, for example, in sickle cell disease (more on this shortly)—produce higher viscosity than either normal RBCs or rigid

Biomedical Fluid Dynamics: Flow and Form. Troy Shinbrot.

DOI: 10.1093/oso/9780198812586.001.0001

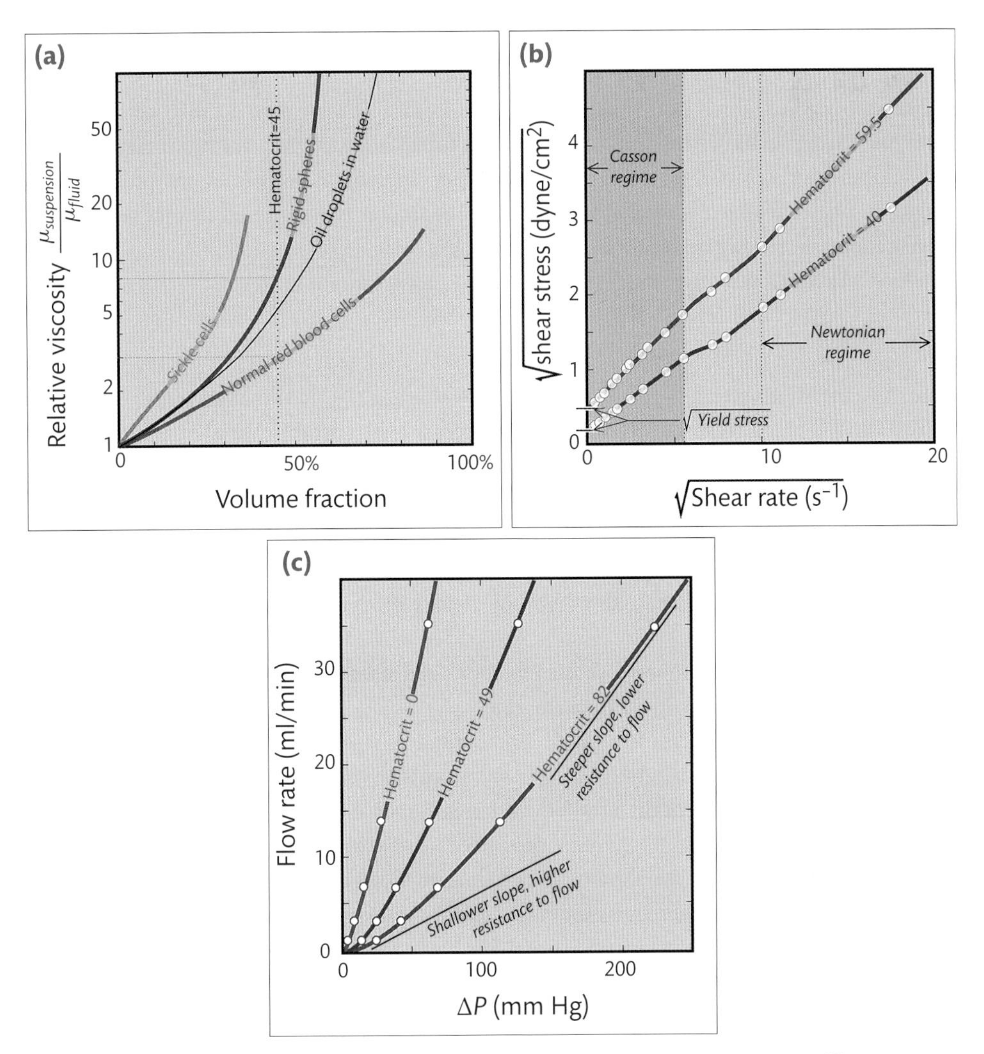

Fig. 10.1 Rheology of blood and other suspensions. (a) Effect of solids fraction on effective viscosity[1,2]. Notice that normal red blood cells have much lower viscosity than other suspensions. (b) Effect of shear rate on shear stress exhibits yield stress behavior at low shear rates. This is fit well by the Casson model (Eq. [10.1]).[3] (c) Effect of pressure drop on flow rate for suspension of washed RBCs injected into the hind paw of a dog. Here flow rate decreases as hematocrit increases, but also the slope increases at higher pressure—because the dog's blood vessels expand, permitting more rapid flow.[4]

spheres. This makes sense: like rigid spheres, these cells don't deform, and additionally their irregular shapes cause them to interlock, further interfering with shear.

On the other hand, oil droplets are more deformable than rigid spheres, and consequently droplets flow more readily than spheres in suspension as shown in the black curve in Fig. 10.1(a). This trend toward lower viscosity with increased deformability is enhanced for RBCs: these cells are biconcave (see Fig. 10.7(c)) rather than spherical because they are underinflated. Consequently (like beanbag chairs) RBCs are more deformable than firmly inflated spherical oil droplets (like exercise balls), and so suspended RBCs generate lower viscosity than oil droplets.

A second behavior that we can readily understand is closely related. We saw in Chapter 9 that suspended materials can make fluids behave like a solid so that they sustain shear up to a yield stress, beyond which point they flow. Many common materials do this: ketchup is the archetypal example, but also paints and inks don't drip because they are manufactured from small globules of colored oil or latex in a carrier solvent. Likewise as we mentioned in Chapter 1, pharmaceutical suspensions such as amoxicillin are designed to mix when vigorously shaken and to stay mixed when left to rest.

Yield stress fluids are often modeled using the Casson equation:[5]

$$\sqrt{\tau} = \sqrt{\tau_{yield}} + \sqrt{\eta_{Newtonian}}\sqrt{\dot{\gamma}}, \qquad [10.1]$$

where τ is the shear stress, τ_{yield} is the yield stress, $\eta_{Newtonian}$ is the asymptotic Newtonian viscosity, and $\dot{\gamma}$ is the shear strain rate. This model isn't strictly appropriate for blood, whose rheology as we saw in Fig. 10.1(a) differs from that of oil droplets; nevertheless as an empirical fit, the Casson model approximates bulk blood rheology well at shear rates under 20 s^{-1} ($\sqrt{\dot{\gamma}} < 4.5$). This is shown in Fig. 10.1(b) where experimental data for $\sqrt{Shear\ stress}$ is plotted as a function of $\sqrt{\dot{\gamma}} = \sqrt{Shear\ rate}$ using blood samples with two hematocrit values. Notice as indicated in the plot that these curves have a yield stress (i.e. y-intercept) that grows with hematocrit. This makes sense, since it is the suspended RBCs that sustain solid-like behavior at low strain rates. At shears above τ_{yield}, the effect of yield stress disappears, and the shear stress becomes Newtonian: proportional to shear rate.

A third behavior that we can intuitively understand is shown in Fig. 10.1(c). In that figure, we plot data in which a 5% albumin-Ringer solution with varying concentrations of washed RBCs (i.e. at varying hematocrit) was fed under different inlet pressure heads, ΔP, into the hindpaw of a dog at constant venous pressure.[6] Two things are evident here. One, by comparing the different curves shown, as hematocrit increases the flow rate decreases. This is what we expect: Fig. 10.1(a) tells us that the viscosity grows with hematocrit, so of course the flow rate will decrease at any given pressure drop. Two, as indicated by the labeled black lines in Fig. 10.1(b), as the pressure drop grows the slope increases—meaning a greater increase in flow per unit change in pressure is achieved at high than at low pressure. This involves several effects, but the dominant one is again easy to understand. The venous pressure is fixed, so increasing ΔP means that the inlet pressure has increased, and so the total pressure inside the blood vessels must increase. Since the blood vessels are elastic, they must inflate – which from the Poiseuille relation (Eq. [1.47] of Chapter 1) implies that there will be faster flow with less resistance. This explanation neglects autonomic regulation of blood vessel tone, but nevertheless the essential mechanism associated with elasticity accounts for the reduced resistance to flow at higher pressures shown in Fig. 10.1(c).

10.2 **Fåhraeus and Fåhraeus-Lindqvist Effects**

This is encouraging: flow rates decrease with increasing hematocrit and increase with increasing pressure, both in logical ways. A slight complication appears as the diameter of a vessel decreases. This was revealed in 1931 by Swedish pathologist Robin Fåhraeus (1888–1968) and

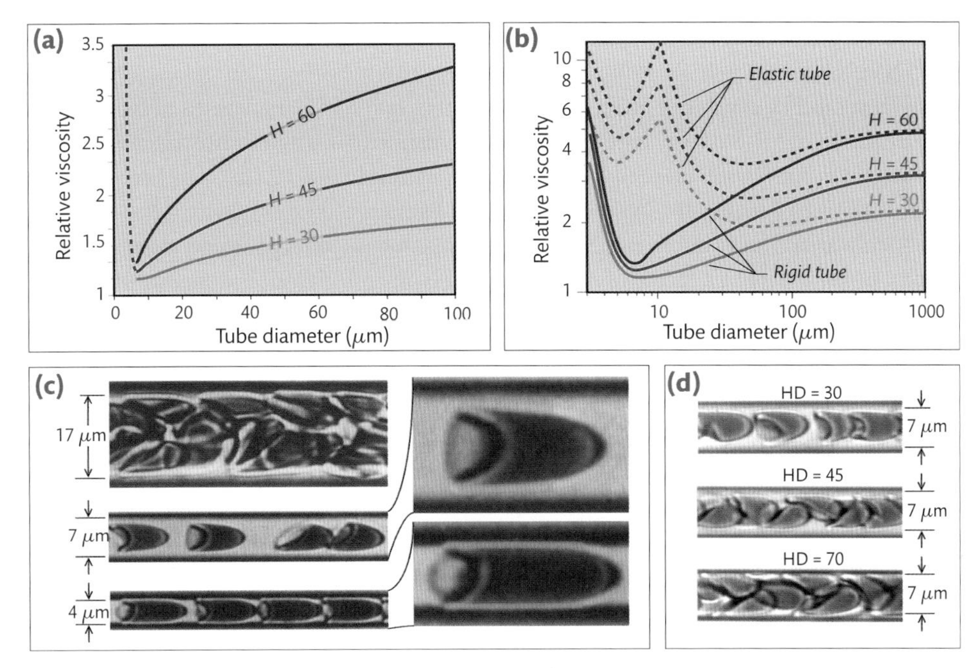

Fig. 10.2 Effect of tube diameter on measured viscosity of blood.[9] (a) Relative viscosity vs. diameter of rigid tubes on a linear-linear scale. (b) The same plot as in (a), but on a log-log scale, also including elastic tubes. (c) Red blood cells (RBCs) traveling through tubes: for larger tubes (top), multiple RBCs travel abreast, but as tube diameter decreases (center and bottom), RBCs travel single file. Cells deform strongly but are lubricated at boundaries by blood plasma (cyan in enlarged views at right).[10] (From Pries, AR., Secomb, T.W., Gaehtgens, P. and Gross, J.F. (1990) Blood flow in microvascular networks. Experiments and simulation. *Circulation Research, 67(4)*, 826–34. By permission ofthe American Heart Association.) (d) Effect of increasing hematocrit (HD is hematocrit within the tube, measured at discharge).

his young collaborator, Torsten Lindqvist (1906–2007), who measured the viscosity of blood flowing through glass tubes of differing diameters. Fåhraeus was chiefly interested in sedimentation and aggregation of RBCs (more on this later), but he and Lindqvist also performed measurements of blood flow through glass tubes.[7]

Lindqvist found the result expected from Fig. 10.1(a), namely that the viscosity of blood measured using capillary viscometry (Eq. [1.48], Chapter 1) increased with hematocrit. But he was perplexed to also find, as shown in Fig. 10.2(a), that the measured viscosity *decreased* as the capillary diameter decreased.[8] Additionally, he found that irrespective of hematocrit, the viscosity approached the plasma viscosity for tubes around 7 μm in diameter.

This, Fåhraeus and Lindqvist reasoned, was simply due to filtration: as the tube diameter approached the RBC diameter, only a fraction of the cells made it into the tube, and so the tube was largely filled with plasma—thus naturally the viscosity diminished to that of the plasma. This story was corroborated by microscopic analysis, which confirmed that finer glass tubes contained lower concentrations of RBCs than larger tubes. The exclusion of RBCs has since been named the Fåhraeus effect, and its effect in turn on viscosity, shown in Fig. 10.2(a) is termed the Fåhraeus–Lindqvist effect.

10.3 **Blood Flow: Complications Multiply**

So far, so good. But capillaries are smaller than 7 μm, and RBCs manage to make their way through them, right? So what happens at smaller diameters? The relative viscosity at very small diameters may best be seen in a log-log plot, as shown in Fig. 10.2(b), and here the first of several curiosities peculiar to RBCs is found. The Fåhraeus filtration effect, after all, would occur for any suspension, but as the diameter of a glass tube is reduced from 7 μm to 3 μm, the viscosity *jumps* by about a factor of 4. This doesn't happen for plasma alone, so RBCs aren't simply being excluded.

The cause of this growth in viscosity for diameters smaller than 7 μm is a bit more complicated. As shown in the top photograph of in Fig. 10.2(c), large vessels allow RBCs to travel in convoys multiple cells abreast. At around 7 μm, however, cells only travel through the tube single file, as shown in the center photograph in Fig. 10.2(c). Each cell is substantially deformed, and travels through the tube lubricated by a thin film of plasma, identified as cyan rectangles in the enlargements of Fig. 10.2(c). So the decrease in viscosity with decreasing tube diameter is due to Fåhraeus filtration *combined with* the lubrication of the highly deformed RBCs. Without this lubrication, the viscosity would be that of the RBCs themselves rather than that of the lubricating plasma. As the tube diameter decreases below 7 μm, the lubricating film is increasingly squeezed out,[11] as shown in the bottom enlargement in Fig. 10.2(c).

This does two things. First, it produces a thinner lubricating film, and as we have seen from the definition of Reynolds number, this is equivalent to increasing the viscosity of the film. Second, the squeezing by smaller tubes causes the viscosity measurement to be increasingly affected by the deformability—that is, the viscosity—of the RBCs rather than of the plasma. Indeed, as shown in Fig. 10.2(d), when the hematocrit increases, RBCs increasingly pack the tube, and the viscosity is consequently increasingly dominated by the RBCs. These two effects conspire to cause a rapid jump in viscosity, shown as a dotted line to the left of the linearly scaled plot Fig. 10.2(a), and shown more smoothly in the log–log plots of Fig. 10.2(b).

This is where understanding stood from Fåhraeus and Lindqvist's 1931 paper until the 1990s. At that point, a model for blood flow including effects of endothelial layer thickness—which itself varies with blood vessel diameter—was compared with data. Results from that work are shown as dotted lines in Fig. 10.2(b). Contrary to the Fåhraeus and Lindqvist result, apparent viscosity of blood in real blood vessels *increases* as the vessel diameter decreases, and reaches a maximum, rather than a minimum, at around 10 μm. These data use a fit of a model to flow measurements from flow through rat mesentary networks, and do not represent direct measurements of viscosity flow velocities; nevertheless they indicate that the endothelial layer within blood vessels produces an increase in viscosity that has been attributed to interactions between RBCs and the endothelial layer.[9,12]

Evidently the behaviors of biological materials are well characterized by Oscar Wilde's words, "The truth is rarely pure and never simple." Just measuring viscosity of blood can produce results that change by an order of magnitude as the container size, flexibility or material is varied. This is, as we will see, only the beginning of the richness of dynamics found in blood and its cells.

10.4 Shear-Induced Migration

We have seen that viscosity of blood can be strongly affected by a lubricating film highlighted in Fig. 10.2(c). For thin tubes that only admit single file flow of RBCs, it may seem self-evident that plasma will be forced to the outside of the tube. What about wider tubes? In Fig. 10.3(a), we display results of simulations of deformable RBCs in a rectangular channel:[13] these simulations show RBCs that have rapidly migrated to the center of the channel.

This inward migration was originally detected in measurements of viscosities of suspensions of spheres.[16] In these experiments, it was found that there is typically a decrease in viscosity over time, as shown in Fig. 10.3(b). This behavior—exhibited by non-spherical and deformable particles also—was found to be caused by a mechanism that is straightforward in hindsight. To describe this mechanism, at the top of Fig. 10.3(c) we depict two red particles in a Poiseuille flow. Because the flow is sheared near the boundary, the top-most particle will be overtaken by its lower, faster moving, neighbor. If the particles are sufficiently large, they will collide as indicated by the velocity vectors shown.

By this argument, larger particles in higher shear flows must collide more frequently than smaller particles in lower shear—so, for example, the blue particles at the bottom of Fig. 10.3(c) may pass by one another without undergoing a collision. Each collision will must inevitably disturb the trajectories, and so the disturbances—or diffusion—of particles will grow with both particle size and shear rate. This dependence of particle diffusion on particle size and shear rate is termed shear-induced diffusivity, and its result is termed shear-induced migration.

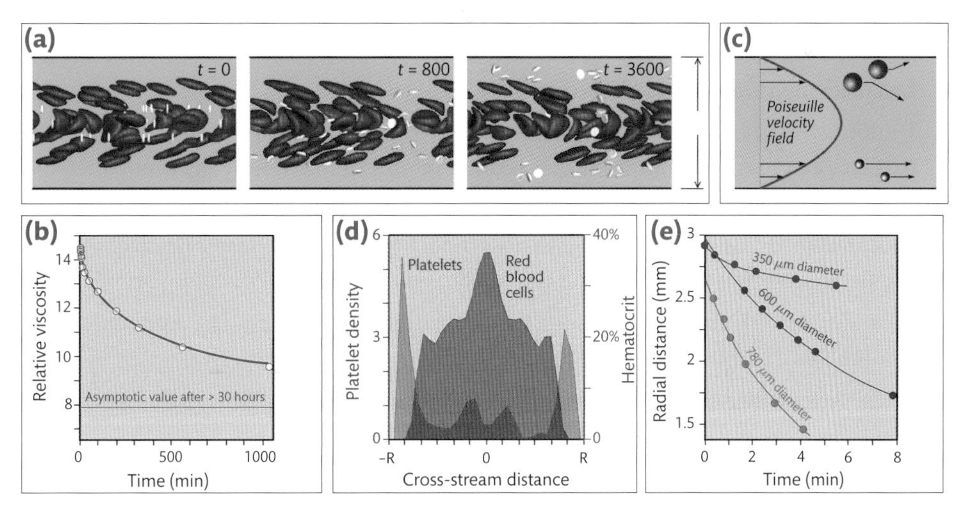

Fig. 10.3 Shear-induced migration. (a) Simulations of large RBCs (red) and small platelets (white), showing rapid migration of RBCs to center, followed by slower 'margination' of smaller particles to outside of channel. (Reproduced from Zhao, Shaqfeh & Narsimhan (2012)[13], with the permission of AIP Publishing.) (b) Measured decrease in viscosity vs. time for suspended 46 μm spherical polystyrene particles in silicone oil.[14] (c) Schematic of process leading to migration of larger particles to center of flow (see text). (d) Quantification of segregation shown in panel (a): red curve shows hematocrit after 200 computational timesteps; blue shows margination of platelets after 4800 timesteps.[15] (e) Measurements of radial migration toward center of water drops in silicone oil over time; diameters of drops shown.[1]

Notice that since shear is strongest near boundaries in flow through a tube, shear-induced diffusivity implies that large particles will diffuse faster near boundaries than at the center where the shear is smallest. Essentially, like a tin can in a roadway, particles will be continually bombarded until they reach a location away from the source of these collisions—shoulders for a highway, or the center, shear-free region, for flow through a tube.

Both simulations and experiments confirm this. Experimentally, as shown in Fig. 10.3(e), water droplets in silicone oil migrate away from the edge of a tube, and larger droplets migrate more rapidly.[17] In simulations, as shown in Fig. 10.3(a), RBCs rapidly migrate toward the center of the channel. As usual, the truth is more complex than this simple picture would indicate: the snapshot at $t = 0$ is shown after a brief and richly dynamic transient time during which those RBCs at the boundaries roll along the channel walls in a motion termed 'tank-treading', and subsequently tumble and lift away[17] to participate in shear-induced diffusivity (see, for example, Zhao et al.[13] and Dupire et al.[18] for details).

In the simulation shown in Fig. 10.3(a), white platelets are injected into the stream at $t = 0$, and as time progresses the RBCs continue to press inward, expelling the smaller platelets. This expulsion is termed "margination." Although reported in many physiology texts to be associated with chemical attraction to the vessel walls, simulations reveal that shear-induced migration is sufficient to accomplish the observed separation of large and small particles.

EXERCISE 10.1

Simulate shear-induced migration and plot the resulting distribution of particles using a simple model, as follows:

(1) Assume that flow through a tube is defined by the Poiseuille formula from Chapter 1. Calculate the shear rate as a function of radius.

(2) Uniformly distribute a number of particles in the flow—you can either distribute the particles at a fixed distance from one another or you can randomly distribute them.

(3) For each of multiple successive timesteps, give every particle a kick in a random direction (to simulate diffusion) with magnitude proportional to the shear rate that you have calculated, and transport each particle axially according to the Poiseuille formula.

(4) Show qualitatively that particles migrate to the center of the tube. The amplitudes of the random kicks and the maximum Poiseuille velocity are up to you.

(5) Calculate the distribution of particles and confirm quantitatively that they migrate inward over time.

ADVANCED EXERCISE 10.2

Given the particle density, you can compute the effective viscosity using, for example, the Kreiger–Mooney law from Chapter 9. You have the particle distribution, so you can calculate the viscosity at any point in the tube. Integrate the viscosity over the tube cross section. Does this decrease over time? Why, or why not?

ADVANCED EXERCISE 10.3

Particles in Exercise 10.1 will sometimes end up overlapping. Devise a computational strategy to prevent this. You might have each particle exert a repulsive potential, you could use a particle dynamics simulation as shown in the Appendix (to be described in Chapter 11), you could give overlapping particles random trial kicks until a trial emerges such that they no longer overlap, you could remove one of the overlapping particles and insert it at a random location—again chosen at random until the trial location doesn't overlap with any other particle—or you could invent your own strategy. You will find that there is extensive literature describing such strategies included, for example, in Guo and Sinclair Curtis.[19]

10.5 Unexpected Consequences of Diffusion: Introduction

We have seen that diffusion can both concentrate RBCs and separate them from smaller particles. This should be surprising from a number of perspectives. As we will see in greater detail, the essential mechanism underlying diffusion is randomizing collisions between particles. Indeed, this is what leads to the Second Law of Thermodynamics, which can be phrased as "disorder always increases."

On the grand scale, this is unequivocally true: any process that appears to reduce disorder is always accompanied by another process that increases disorder by a larger amount. So, for example, the ordering of RBCs and platelets shown in Fig. 10.3 only occurs so long as blood is pushed through vessels, and the disorder of the universe produced by body heat associated with doing this pushing far exceeds the ordering of the blood-borne particles. The same is true of refrigerators, which after all separate hot from cold, but do so at the cost of energy and so produce net heat—or molecular disorder.

Nevertheless, reciting the abstract tautology that the disorder of the universe always increases allows us to miss beautiful and crucial aspects of ordering that are essential to the formation of organized structures in biology and other sciences. Our very bodies are intricately organized structures that are in turn made of structures organized at all scales, ranging from organs to tissues to cells to organelles to molecules themselves. It is both deeply interesting and exceedingly important that these structures form, and if we dismiss the fact that this occurs by appealing to the net increase in disorder of the universe, we lose the opportunity to understand how order comes about and how this ordering can be turned to productive use.

In the next few chapters, we will examine both disordering and ordering in detail, but before we do so, it is useful to understand clearly how ordering occurs in mechanical, rather than biological, systems. Following these examples, we will return to biology and discuss ordering of RBCs and of structures within RBCs.

Example Mechanical ordering

In Figs. 10.4 and 10.5, we show four cases of mechanical ordering. Since we have mentioned refrigerators as a case of separating hot from cold at the expense of energy, let's start with an example of refrigeration.

In Fig. 10.4(a), we sketch a Ranque-Hilsch vortex tube. As shown, compressed air is input into a hollow tube from an off center location. This produces a swirling flow within the tube. The tube has two exit ports: one exhausting air from the center of the tube, and a second exhausting air from its periphery by means of the cone insert shown in Fig. 10.4(a). These vortex tubes aren't very efficient, but they are used in specialized applications such as cooling machine tools. The reader may want to puzzle through the details of how the temperature separation occurs, but the mechanism hinges on the creation of a temperature gradient from the inside to the outside of the tube, and this gradient is in turn produced by a Couette-like flow (Eq. [6.20] of Chapter 6 may help here).

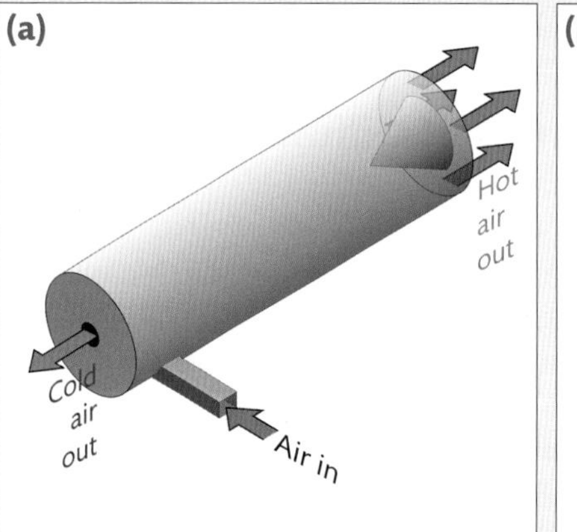

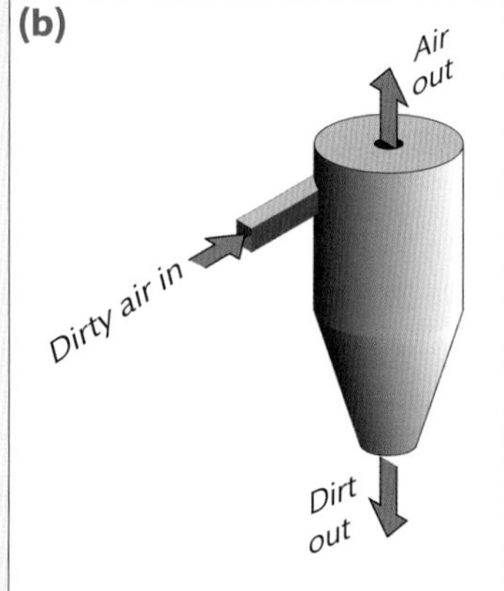

Fig. 10.4 Disorder and order. (a) A Ranque-Hilsch vortex tube takes compressed air in and delivers cold air out through a central hole and hot air out from around an conical insert. (b) A cyclone separator takes compressed particle-laden air in and separates particles from air. Neither the vortex tube nor the cyclone separator has any internal moving parts.

If we turn the Ranque–Hilsch vortex tube on its end (and remove the cone insert), we get a second example of mechanical separation: a "cyclone separator," shown in Fig. 10.4 (b). These devices are widely used to remove particulates from airflows, for example from kilns and sawmills. Here again the reader may work through the mechanism in detail, but in short, inertia keeps heavier particles to the outside of the tube where they fall out the bottom of the funnel shown, leaving air free of these particles to exit from the top of the separator. So in this geometry, centrifugal acceleration and gravity combine to produce a separating gradient.

In both the vortex tube and the cyclone separator, ordering (either of heat or of particles) occurs by virtue of mechanically establishing a gradient. This requires energy, which in both cases is produced by an external compressor, whose disordered heat always exceeds the order produced by the device itself. In these applications, the details are subtle, but the essential mechanism for producing the gradient, and the ordering, is well understood.

EXERCISE 10.4

In other applications, it is not so clear how ordering comes about, and careful thought is required. Consider the examples shown in Fig. 10.5.

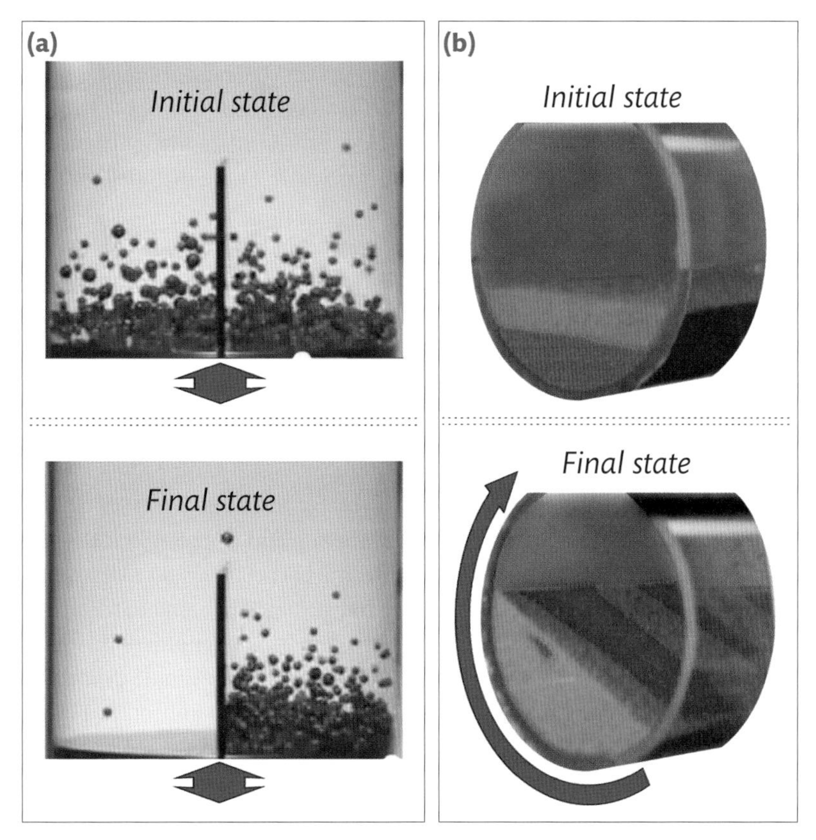

Fig. 10.5 *Disorder and order. (a) Granular Maxwell's demon consisting of a shaken container of beads with a central barrier. High energy particles can hop the barrier, leading spontaneously from a uniform state to a state with almost all particles on one side of the barrier. By adjusting the vibration amplitude, two sizes particles can also be separated by this device.*[23] (Reproduced from Mikkelsen, R., van der Weele, K., van der Meer, D., Versluis, M. and Lohse, D. (2003) Competitive clustering in a granular gas. *Physics* of Fluids, 15, S8, with the permission of AIP Publishing.) (b) Size segregation in a tumbling blender: here the red, smaller, particles form striped bands along the direction of rotation. This experiment is shown starting from a separated, layered, state, but the same occurs starting from a well blended state. (Adapted from Shinbrot and Muzzio.[21])

In Fig. 10.5(a), we show a third separation device: a large-scale, granular, version of "Maxwell's demon." The original demon was conceived by Scottish scientist James Clerk Maxwell (1831–1879: mentioned in previous chapters) to describe a thought experiment in which molecules are ordered by opening or closing a door depending on which type of molecule approaches. By opening a door whenever a hot molecule approaches it from inside a container and keeping it closed otherwise, a demon could remove all hot molecules from the container, progressively cooling it over time. It turns out that the disorder associated with detecting and responding to signals from particular molecules always exceeds the ordering achieved,[20] but nevertheless Maxwell's demon serves to clarify how ordering occurs.

Maxwell's demon was originally conceived for molecules, but its essential mechanism can be elaborated on in an analogy using large scale grains, as Fig. 10.5(a) shows.[22] In this figure, we see steel beads (2.5 mm diameter blue and 1.25 mm red) in a vertically vibrated container with a barrier (from Mikkelsen et al.[23]). As time progresses, the initially uniformly distributed beads make their way from one side of the barrier to the other. The side that the beads migrate to is random: sometimes they will choose the left, other times the right. In this granular Maxwell's demon, there is no door to open or close, instead high energy particles jump the barrier, leaving low energy ones behind. Different particle sizes are included only to illustrate the generality of the method.

For this exercise, explain what causes this to occur. In the results shown, all beads move to the right: this occurs for a particular choice of vibration frequency and amplitude. For other parameter choices, the larger (blue) beads will move to one side, leaving the smaller (red) beads on the other. Explain this as well. Finally, explain how disorder increases (which it must) despite the apparent ordering of the particles.

ADVANCED EXERCISE 10.5

A fourth and final example of mechanical ordering is shown in Fig. 10.5(b). Here two sizes of glass beads (0.8 mm red, 1.4 mm green) are shown before and after 5 minutes of tumbling at 20 rpm in a 7.5 cm diameter glass tube. In this case, the particle sizes do matter: evidently the smaller red and larger green particles form azimuthal stripes. The same outcome will result if the particles are initially well mixed rather than layered. This phenomenon is termed "axial banding," and has been known to occur since 1939,[24] in both this simple cylindrical tumbler and in more complicated designs used to blend pharmaceuticals and other powders.[25] The detailed mechanisms of demixing of grains remain controversial.[26]

For this exercise, describe two mechanisms known to be at work in axial banding, and as before explain how disorder increases. There have been many debates over which of the two mechanisms causes the banding—this is a common situation in all fields of science. Choose a side and provide evidence that your chosen mechanism matters and the other does not.

From these examples, several straightforward mechanisms can be identified that produce ordering. Because these systems are macroscopic and physically based, the mechanisms that leads to ordering are more comprehensible than will be the case in biological (and later, in chemical) examples to come. Using margination as a case in point, RBCs move to the center of a tube simply because they get ejected by colliisions near the boundaries. In the cyclone separator, heavier particles with higher momentum get centrifuged outward by the swirling flow and then fall down under gravity into the collecting funnel. The other examples require a bit more thought, but in all cases a simple underlying mechanism is present. We turn next to more intricate ordering problems, but we stress that however complex the ordering phenomenon, and however much it may seem otherwise, none of the effects to follow are magic, and there is always an underlying mechanism waiting to be unveiled.

10.6 **Ordering of Red Blood Cells**

Armed with an understanding of comparatively simple mechanisms for mechanical segregation, we return to behaviors seen in RBCs. Figs. 10.1, 10.2, and 10.3 have shown effects on RBCs in tubes; we ask next what happens to RBCs between shearing plates. As we will see, this will lead us to analyzing a new and remarkable set of issues that turn out to have relevance to a broad range of biophysical problems.

We begin with Fig. 10.6(a), which shows viscosities of RBCs measured using a cone and plate geometry. In this geometry,[3] a wide cone is rotated above a flat plate, as shown in Fig. 10.6(b). Viscosities measured in this way are similar to those obtained in a Couette viscometer, but because the cone presents a continuum of separations from the plate, this geometry avoids the fixed length scale that can influence measurement (which we saw can be important in the Fåhraeus and Lindqvist effects).

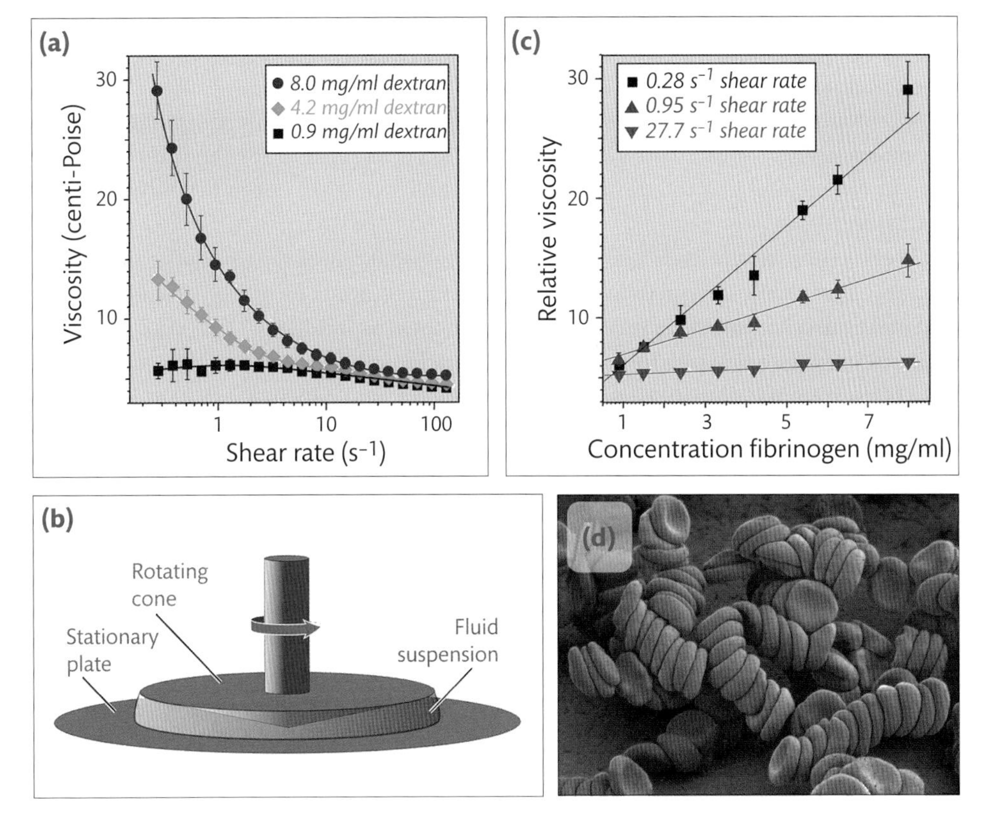

Fig. 10.6 (a) Bulk viscosity of RBC (centrifuged and washed) suspensions at 45% haematocrit as a function of shear rate for various dextran concentrations (molecular weight: 70 kDa). Viscosities measured with cone and plate,[27] shown in (b). (c) Relative viscosity of RBC suspension as a function of added fibrinogen. Both panels (a) and (c) show that RBC suspensions become shear thinning and increase their viscosity with added solute. (d) Human RBCs imaged by SEM (pseudocolor). (Sources: (a) and (c) from Chauqi Misbah and Brust et al.[27]. (d) from Thomas Deerinck and Mark Ellisman, the National Center for Microscopy and Imaging Research, UCSD.)

Fig. 10.6(a) shows a very odd thing: the fluid is a suspension of RBCs that have been washed and suspended in a saline solution free of other blood constituents. When a small amount of dextran—a sugar—is added, the viscosity is initially nearly constant with shear rate, but as the dextran concentration is increased, the RBCs make the suspension strongly shear thinning. The same is shown in Fig. 10.6(c), where the relative viscosity (the measured viscosity divided by the same for the solution without RBCs (as in Fig. 10.1(a)) is shown as a function of fibrinogen concentration at several shear rates. Dextran is a sugar, fibrinogen is a protein, and both Figs. 10.6(a) and (c) show that concentration of the additive increases the viscosity and strengthens the shear thinning behavior. Neither dextran nor fibrinogen solutions are significantly shear thinning, nor are suspensions of RBCs alone, but for some reason adding fibrinogen or dextran to RBCs makes the suspension non-Newtonian.

So what is going on? A clue can be found by visualizing RBCs in suspension, as shown in Fig. 10.6(d). Rather than remaining isolated, the RBCs group together into chains, known as rouleaux—French for rolls, like rolls of coins. Remarkably, these rouleaux form in the presence of an added solute, but do not form in a simple suspension of RBCs—and as in the macroscopic examples of Figs. 10.4 and 10.5, an ordered state of aligned cells forms spontaneously, seemingly counter to the tendency of disorder to increase as demanded by the Second Law of Thermodynamics.

Mechanically speaking, these chains of RBCs increase viscosity at low shear rates, but break apart under stronger shear. This has been shown to be responsible for the shear thinning,[28] shown in Figs. 10.6(a) and (c), through a process similar to what is seen for polymeric fluids (see Chapter 9).

So breakup of rouleaux leads to shear thinning, but what causes the rouleaux to form in the first place? In principle, several things could account for rouleaux formation: the added solutes could make the cell membranes sticky in a process known as "bridging."[29] It isn't clear why the cells should stick into tidy rolls instead of disordered clumps as they do in sickle-cell anemia or malaria (see example following), but perhaps the solutes bind only to the flattened surfaces of RBCs. Or possibly the solutes could activate internal cellular machinery causing cells to preferentially migrate toward one another, as some amoeba do.[30]

To discriminate the true mechanism from conjecture, a number of experiments have been performed.[31] Perhaps the most illuminating[32] involves the observation that many of the solutes that produce rouleaux are identical across species—so the same dextran that produces rouleaux in human RBCs does so in sheep and rabbit as well, so if cells are simply becoming sticky or are otherwise being activated, one would expect a mix of RBCs from different species to aggregate into intermixed rouleaux, combining cells from each species. On the contrary, it is found that rouleaux preferentially separate, so a mix of human and rabbit RBCs will form human rouleaux and rabbit rouleaux rather than intermixed human–rabbit rouleaux.

Evidently something new is happening—and in particular in the interspecies experiments, cells form both organized rouleaux, and the human and rabbit cells *separate*! It is truly strange that cells not only form ordered structures, but also the structures sort themselves by species: an impressive feat that we would be hard pressed to engineer on our own. We'll explain this curious behavior in the next chapter, in which we examine in detail how microscopic objects—cells, molecules, and particulates—behave through a study of so-called statistical mechanics. Before we turn to this new topic, a few vignettes about RBCs themselves and self-organization of other systems may motivate the work to come.

Example Pathologies that affect blood flow

RBCs are subject to many pathologies that affect blood flow.[3] One of the better known is sickle cell disease (SCD). SCD is an insidious mutation that substitutes a single amino acid, valine, for glutamic acid in the hemoglobin molecule. Normal hemoglobin (hemoglobin-A) binds and releases oxygen in a cooperative way, but because glutamic acid is hydrophilic, while valine is hydrophobic, when mutated hemoglobin (hemoglobin-S) releases oxygen, a hydrophobic valine is exposed. This repels water and allows adjacent hemoglobin molecules to self-assemble at the valine site, producing dimers that chain together into filaments, sketched in Fig. 10.7(a). These filaments then collect into bundles, sketched in Fig. 10.7(b).

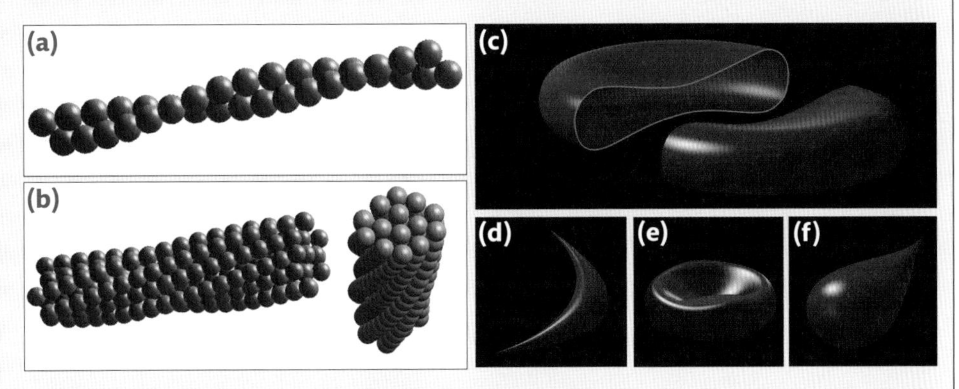

Fig. 10.7 Hemoglobin-S bundles leading to a change in RBC morphology. (a) The first process leading to bundle formation is the creation of chains of dimers. Hemoglobin is a "globular" protein, sketched here as spheres. (b) The second process is the assembly of bundles, each holding seven dimer chains. Micro-scale dynamics of hemoglobin affect macro-scale structure and function, for example, producing different shapes: (c) normal, "biconcave" red blood cell (d) sickled cell, (e) stomatocyte, and (f) teardrop shape that occurs in thalassemia major.

The bundles of polymerized hemoglobin cause three effects. First, they change the normal, biconcave, RBC shape shown in Fig. 10.7(c) into a telltale sickle shape sketched in Fig. 10.7(d). SCD is one of many disorders that affect cell shape; a couple of other variations are shown in Figs. 7(e) and (f). The mechanisms that lead from a single amino acid change to these larger scale effects hold many mysteries, and the connection between micro-scale molecular behaviors and macro-scale shapes has long been an area of active research.

Second, the bundles stiffen the cells and prevent their deformation as they pass through capillaries. Capillaries, recall, can be as small as 3–4 μm across, while RBCs are 8–10 μm along their longest dimension, so deformation is integral to blood flow. Cellular stiffening can lead to obstruction of capillaries and consequent cell and organ damage downstream of the blockage.

Third, because the binding only occurs in the deoxygenated state, repeated cycles of oxygenation and de-oxygenation as the blood flows between lungs and body deform the hemoglobin molecules, exposing iron, which in turn generates free radicals that damage the cells themselves. As a result, sickled RBCs have half-lives as short as 4 days, rather than the 40 days typical of healthy RBCs, and so patients can suffer from profound anemia. This is beyond the anemia caused by tying up hemoglobin and so preventing it from capturing and releasing oxygen in a normal way.

The history behind the study of SCD is itself fascinating: see, for example, Gormley.[33] We mention two historical tidbits. One, SCD represents the first ever discovery of a "molecular disease," whose precise genetic and molecular cause was unveiled. Later identifications of molecular diseases were made in Alzheimer's disease, hemophilia, and prion disorders. And two, it turns out that SCD occurs in homozygous patients: in the heterozygous state the mutation leading to SCD protects against malaria – which is believed to account for the longstanding observation that SCD is reported in people originating in Africa or India where malaria is endemic.

For our purposes, SCD holds a different lesson. Recall that the fibers form because valine is hydrophobic, and this single fact is what leads to the formation of dimerized chains and bundles. In detail, the hydrophobic valine in a so-called "β-subunit" on one hemoglobin molecule ends up adjacent to a hydrophobic region of an "α-subunit" of another molecule: this leads to the offset and twist shown in Figs. 10.7(a) and (b). But this again is a detail. The crucial mechanism that produces chains and bundles is not an attraction between nearby hemoglobin molecules: valine is no more attractive than glutamic acid is. The thing that forces mutated hemoglobin together in SCD is the *repulsion* by the surrounding water. As sketched in Fig. 10.8(a), the hemoglobin molecules undergo collisions with water on their sides facing away from one another, and so feel a pressure on opposing sides of each molecule. However there are no collisions on the adjoining sides—because these sides have excluded water. This phenomenon, known as self-assembly due to excluded volume, is important in other biophysical problems, as we will see.

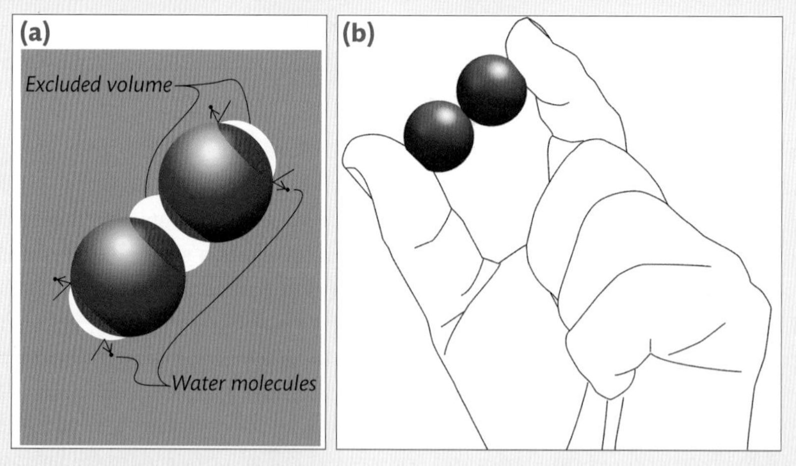

Fig. 10.8 Self-assembly of dimers. (a) Both chains and bundles self-assemble as a consequence of "excluded volume:" the white regions shown around a hemoglobin dimer exclude water due to hydrophobic groups, indicated in red. Because these regions exclude water, there are no collisions with water molecules on the inside, facing, surfaces of the hemoglobin dimer. There are, however, collisions on the outsides, so there is pressure on the outside pushing the molecules in, but not on the inside pushing the molecules out. (b) This pressure holds the molecules together as you would hold two marbles together with your fingers. If you try this, though, you will find that this arrangement is unstable, so there must be more to the story!

Excluded volumes can produce differences in molecular collision rates, and so differences in pressure, but clearly this isn't the whole picture. Pressure on the outside of a hemoglobin dimer does hold the molecules together, but what gets them aligned in the first place? After all, the molecules could approach one another in any orientation: the hydrophobic regions needn't exactly face one another, as shown in Fig. 10.8(a). And the same molecular collisions that produce the pressure holding the molecules together will surely cause continual Brownian jiggling. So how do the molecules stay together? If you try to hold two marbles, as shown in Fig. 10.8(b), you will find it a difficult proposition: this arrangement is highly unstable even without Brownian disturbances.

EXERCISE 10.6

As we have said, clearly something more is going on. To understand what this is, we will need some simple ideas from statistical mechanics, which we introduce in the next chapter. In the interim, see if you can work out this puzzle: why do unstable and seemingly well-organized arrangements form, and once they form, why do they stay together? Three tantalizing clues that you'll have to account for are as follows.

(1) You could surely imagine that structures like those shown in Fig. 10.7(a) and (b) are created by invoking additional forces—for instance, van der Waals or electrostatic attraction. For example, consider Fig. 10.9(a). This is one of several elaborate constructions that spontaneously self-assemble out of nothing more than water (more examples are shown in Chapter 13). This is a false colored electron micrograph of a crystal of rapidly freezing ice, known as rime.[34] Here the self-assembly is governed by attractive hydrogen bonds between water molecules.

(2) It turns out, however, that structures commonly self-assemble on their own without any such attraction. For example, in Fig. 10.9(b), we show another ice pattern, known as pancake ice. In this case an ice sheet breaks up under the influence of gentle water waves, and these same waves erode the ice fragments into rounded disks. Here attraction plays no role in the formation of these structures. Finally, in Fig. 10.9(c) we show two selections from a rich variety of patterns seen in wind-blown sand:[35] at the top is a dune field near Rabe Crater and at the bottom is a Barchan dune in Hellas Planitia, both on Mars. The sand, like fragments of an ice sheet, somehow self-assemble without any apparent attraction between their parts.

(3) The structures in Figs. 10.9(b) and (c) can only form in the presence of agitation. In Fig. 10.9(b), agitation is required both to break the ice sheet into fragments and to round the fragments into pancakes. In Fig. 10.9(c), agitation by wind causes sand grains to collide with one another, resulting in the structures shown. We mentioned that Brownian agitation is present in the self-assembly of hemoglobin bundles – it turns out in fact that if hemoglobin-S molecules are *cooled down*[36,37] (which reduces thermal agitation), the bundles break apart! Heating the molecules up again restores the bundles. This is the opposite of what happens in the freezing of ice or the solidification of gelatin. Why is this?

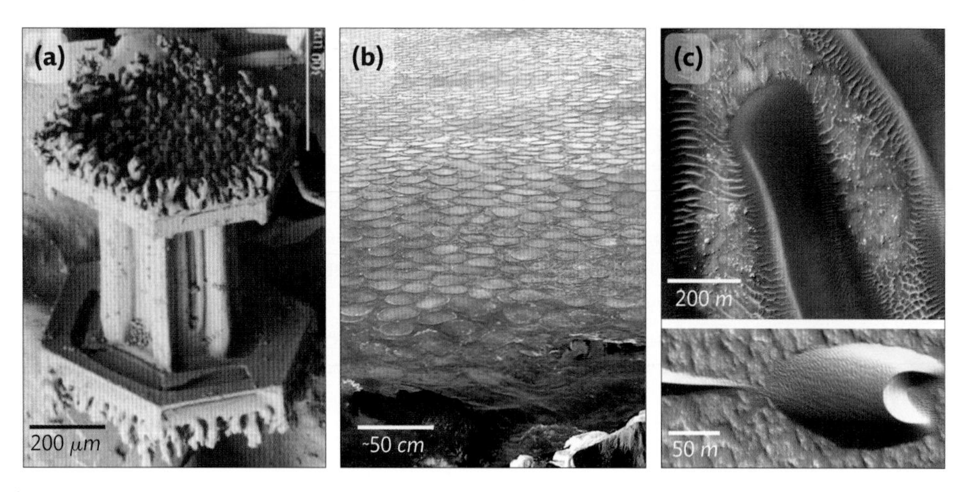

Fig. 10.9 Examples of self-assembly. (a) Spontaneously appearing structure seen in a sample of "rime" ice. (b) "Pancake ice:" a pattern that forms from ice without any attraction between elements. (c) Sampling of patterns seen in sand dunes on Mars (upper image: a variety of dunes from Noachis Terra; lower image: a barchan dune from Hellas Planitia). (Panel (c) from NASA/JPL-Caltech/University of Arizona.)

REFERENCES

1. Goldsmith, H.L. and Mason, S.G. (1967) The micro-rheology of dispersions. In: Eirich, F. (ed.) *Rheology. Theory and Applications, Vol. 4.* New York: Academic Press, pp.86–246
2. Pal, R. (2014) New models for the viscosity of nanofluids, *Journal of Nanofluids, 3*, 260–6.
3. Merrill, E.W. (1969) Rheology of blood. *Physiology Review, 49*, 863–88.
4. Skalak, R. (1971) Synthesis of a complete circulation. In: Bergel, D. (ed,) *Cardiovascular Fluid Dynamics, Vol. II* New York: Academic Press.
5. Casson, N. (1959) A flow equation for pigment-oil suspensions of the printing ink type. In: Mill, C.C. (ed.) *Rheology of Disperse Systems* Oxford: Pergamon Press, pp.84–104.
6. Benis, A.M., Usami, S. and Chien, S. (1970) Effect of hematocrit and inertial losses on pressure-flow relations in the isolated hindpaw of the dog. *Circulation Research, 27*, 1047–68.
7. Toksvang, L.N. and Berg, R.M.G. (2013) Using a classic paper by Robin Fåhraeus and Torsten Lindqvist to teach basic hemorheology. *Advances in Physiology Education, 37*, 129–33.
8. Fåhraeus, R. and Lindqvist, T. (1931) The viscosity of the blood in narrow capillary tubes. *American Journal of Physiology, 96*, 562–8.
9. Pries, A.R. and Secomb, T.W. (2005) Microvascular blood viscosity in vivo and the endothelial surface layer. *American Journal of Physiology, 289*, H2657–64.

10. Pries, A.R. and Secomb, T.W. (2008) Blood flow in microvascular networks. In: *Handbook of Physiology. The Cardiovascular System: Microcirculation*. San Diego, CA: Academic Press, sect. 2, vol. IV, pp.3–36.

11. Freund, J.B. (2014) Numerical simulations of flowing blood cells. *Annual Review of Fluid Mechanics, 46*, 67–95.

12. Pries, A.R., Neuhaus, D. and Gaehtgens, P. (1992) Blood viscosity in tube flow: dependence on diameter and hematocrit. *American Journal of Physiology, 263*, H1770–8.

13. Zhao, H., Shaqfeh, E.S.G. and Narsimhan, V. (2012) Shear induced particle migration and margination in a cellular suspension. *Physics of Fluids, 24*, 011902.

14. Leighton, D. and Acrivos, A. (1987) The shear induced migration of particles in concentrated suspensions. *Journal of Fluid Mechanics, 181*, 415–39.

15. Zhao, H. and Shaqfeh, E.S.G. (2011) Shear induced platelet margination in a microchannel. *Physical Review E, 83*, 061924.

16. Gadala-Maria, F. and Acrivos, A. (1980) Shear induced structure in a concentrated suspension of solid spheres. *Journal of Rheology, 24*, 799–814.

17. Goldsmith, H.L. and Mason, S.G. (1965) Further comments *on* the radial migration of spheres in Poiseuille flow. *Biorheology, 3*, 33–36.

18. Dupire, J., Socol, M. and Viallat, A. (2012) Full dynamics of a red blood cell in shear flow. *Proceedings of the National Academy of Sciences of the USA, 109*, 20808–13.

19. Guo, Y. and Sinclair Curtis, J. (2015) Discrete element method simulations for complex granular flows. *Annual Review of Fluid Mechanics, 47*, 21–46.

20. Ball, P. (2012) The unavoidable cost of computation revealed. *Nature, 483*,256–7.

21. Shinbrot, T. and Muzzio, F.J. (2000) Nonequilibrium patterns in granular mixing and segregation. *Physics Today, 53*(3), 25–30.

22. Eggers, J. (1999) Sand as Maxwell's demon. *Physical Review Letters, 83*, 5322–6.

23. Mikkelsen, R., van der Weele, K., van der Meer, D., Versluis, M. and Lohse, D. (2003) Competitive clustering in a granular gas. *Physics of Fluids, 15*, S8.

24. Oyama, Y. (1939) Horizontal rotating cylinders, *Bulletin of the Institute of Physical and Chemical Research (Japan), 18*, 600–39.

25. Alexander, A.W., Shinbrot, T. and Muzzio, F.J. (2001) Granular segregation in the double-cone blender: transitions and mechanisms. *Physics of Fluids, 13*, 578–87.

26. Choo, K., Molteno, T.C.A. and Morris, S.W. (1997) Traveling granular segregation patterns in a long drum mixer. *Physical Review Letters, 79*, 2975–9.

27. Brust, M., Aouane1, O., Thiébaud, M., Flormann, D., Verdier, C., Kaestner, L., Laschke, M.W., Selmi, H., Benyoussef, A., Podgorski, T., Coupier, G., Misbah, C. and Wagner, C. (2014) The plasma protein fibrinogen stabilizes clusters of red blood cells in microcapillary flows. *Scientific Reports, 4*, 4348, 1–6.

28. Snabre, P., Bitbol, M. and Mills, P. (1987) Cell disaggregation behavior in shear flow. *Biophysics Journal, 51*, 795–807.

29. Brooks, D. (1988) Mechanism of red cell aggregation. In: Platt, D. (ed.) *Blood Cells, Rheology and Aging*. Springer-Verlag.

30. Dusenbery, D.B. (1996) *Life at Small Scale: The Behavior of Microbes*. New York: Scientific American Library.

31. Wagner, C., Steffena, P. and Svetina, S. (2013) Aggregation of red blood cells: from rouleaux to clot formation. *Comptes Rendus Physique, 14*, 459–69.

32. Forsdyke, D.R. and Ford, P.M. (1983) Segregation into separate rouleaux of erythrocytes from different species. *Biochemical Journal, 214*, 257–60.

33. Gormley, M. (2003) It's the blood! A documentary history of Linus Pauling, hemoglobin, and sickle cell anemia. Available at http://scarc.library.oregonstate.edu/coll/pauling/blood/

34. ARS, USDA. (no date) Rime and graupel. Available at: http://emu.arsusda.gov/snowsite/rimegraupel/13371.jpg

35. NASA. (n.d.) Mars as art. Available at http://mars.nasa.gov/multimedia/marsasart/

36. Pumphrey, J.G. and Steinhardt, J. (1977) Crystallization of sickle hemoglobin from gently agitated solutions: an alternative to gelation. *Journal of Molecular Biology, 112*, 359–75.

37. Galkin, O. and Vekilov, P.G. (2004) Mechanisms of homogeneous nucleation of polymers of sickle cell anemia hemoglobin in deoxy state. *Journal of Molecular Biology, 36*, 43–59.

11 Statistical Mechanics, Diffusion, and Self-Assembly

We saw last chapter that it is surprisingly common to find examples of ordered structures that spontaneously self-assemble from disordered components. As we have described, this seems to go against intuition as well as the Second Law of Thermodynamics, both of which are founded on the observation that disorder tends to increase.

Thermodynamics, unfortunately, is often taught by rote, which disguises the elegant mechanisms by which nature achieves this order. Students of thermodynamics may recall the tedious and unilluminating impression of using Legendre transforms to derive the 28 so-called "Bridgman equations" that relate thermodynamic quantities. The unpleasantness of this impression is reinforced by the knowledge that an exceptional number of pioneers of the field suffered hard lives and tragic deaths. For example,

- Sadi Carnot (1796–1832), known for establishing limits for thermodynamic efficiency, worked and briefly fought in the French military from age 16 until retirement, after which he was admitted to an asylum where he perished in an early death from cholera.
- Rudolph Clausius (1822–1888), who formalized the Second Law of Thermodynamics, was wounded in the Franco-Prussian War, shortly thereafter suffered the death of his wife, and was left to raise his six children alone despite ongoing disability from his war injury.
- Josiah Gibbs (1839–1903), one of the most prolific founders of thermodynamics, suffered throughout his life from lung infections and ultimately died painfully of an acute intestinal obstruction.
- Ludwig Boltzmann (1844–1906) established the atomistic theory for thermodynamics, but was tormented by depression and ultimately hanged himself.
- Paul Ehrenfest (1880–1933) agonized over the birth of a son with a developmental disorder and shot himself, as did
- Percy Bridgman (1882–1961), who suffered from metastatic cancer and shot himself on what he wrote was the last day he had the strength to pull the trigger.

This isn't to say that thermodynamics isn't valuable: it governs the efficiency of mechanical engines and the outcome of chemical reactions. Nevertheless, seen in a pedagogical and historical light, no one could be criticized for turning away from thermodynamics—a pity, since the atomistic view of Boltzman reveals that the spontaneous formation of beautiful patterns can be understood from the simplest of considerations.[1] Boltzmann's view is now termed statistical mechanics, and this is what we focus on in this chapter.

Biomedical Fluid Dynamics: Flow and Form. Troy Shinbrot.

DOI: 10.1093/oso/9780198812586.001.0001

11.1 Statistical Mechanics: The Idea

Boltzmann proposed substituting the Byzantine calculations of traditional thermodynamics with something much simpler: the notion that matter chooses its state based only on Newton's laws of motion applied to molecules, rather than based on a dutiful obedience to arrays of differential equations. As an example, a standard relation familiar to students of thermodynamics is the following:

$$\Delta G = \Delta H - T \cdot \Delta S \qquad [11.1]$$

where G is the Gibbs free energy, H is the enthalpy, T is the temperature, S is the entropy, and the Δs are included in recognition of the fact that absolute energy has no meaning, and only its changes can be evaluated. This equation expresses changes in an energy, G (which cannot be directly measured) associated with the stability of a thermodynamic state, in terms of changes in another energy, H, and entropy, S (two more quantities that cannot be measured). Eq. [11.1] is important for determining equilibrium states, but is very difficult to understand and without serious effort sheds little light on what is going on. As Mark Twain put it in a different context, researchers "have already thrown much darkness on this subject, and it is probable that, if they continue, we shall soon know nothing at all about it."

Boltzmann's alternative was to observe that equations such as Eq. [11.1] are really only expressions of what a system's atoms and molecules are doing, and by examining their dynamics we can learn more—and understand it more clearly—than by poring over Eq. [11.1] and its kin. In Boltzmann's view, this equation tells us that the amount of energy that can be taken from a system's atoms is the difference between their total energy and their disorder. So the energy of a thousand atoms all bouncing up and down in phase on a platform can be turned into electricity, say by hooking the platform up to a generator, while very little of the energy of a disordered mass of randomly wandering atoms can be extracted.

We'll describe shortly how Boltzmann's view can be used to reveal how collections of objects self-assemble into ordered structures—as we saw in last chapter's description of rouleaux of blood cells. Before we do so, it is instructive to consider a couple of objections to Boltzmann's view.

A first objection was raised by the leading physicist, Ernst Mach (1838–1916). Mach maintained that Boltzmann's ideas were meaningless to begin with because at the time molecules could not be directly observed and in Mach's opinion should at most be viewed as mathematical artifices. Mach's colleague Wilhelm Ostwald (1853–1932) went further by asserting that,

> *"The proposition that all natural phenomena can ultimately be reduced to mechanical ones cannot even be taken as a useful working hypothesis: it is simply a mistake."*

Boltzmann was ultimately vindicated in his view, but as these objections indicate, his work was soundly rejected by leading authorities of the time, which exacerbated the depression leading to his suicide.

A second, and more scientifically founded, criticism was put forward by the German mathematician Ernst Zermelo (1871–1953). Zermelo was willing to entertain the notion of molecules as real entities, but noted that if we were to move all of the molecules in a room into one corner and release them, then we know from common sense and experience that the molecules would expand to fill the room, and would never return to the corner that they started from.

This is logical enough, however Zermelo pointed out that if the molecules are viewed as objects that simply move in response to Newton's laws of motion as Boltzmann contended, then a mathematical theorem requires that they must return arbitrarily closely to their initial positions and velocities an infinite number of times. This is termed the Poincaré recurrence theorem. Crucially, this is a theorem: it is true without regard to interpretation or equivocation, it does not depend on details of the molecules or the room involved, it simply requires that the molecules explore the entire room in an unbiased manner. Technically this is guaranteed for any bounded dynamical system that conserves probability (i.e. that does not concentrate or exclude moving particles). This condition is met both for particles that collide according to Newton's laws and for those that interact quantum mechanically (see, for example, Ott[2]).

So Zermelo asserted that Boltzmann's analysis of thermodynamics in terms of molecular, or atomistic, behaviors, contradicted well-established mathematical facts and must be rejected.

Boltzmann's response to Zermelo is revealing. Boltzmann accepted the validity of the Poincaré recurrence theorem but asked how long it would take for the molecules in the room to return to the corner that they started from. This is straightforward to calculate—for our purposes, let's make the problem simple by considering the case where the particles start in one octant of a room, so all of the particles initially have random velocities in $1/8^{\text{th}}$ of the volume available, as sketched in magenta in Fig. 11.1. Keeping the problem simple, we'll also assume that the particles don't interact with one another, in which case the *a priori* probability of finding one particle in the magenta octant is 1/8, the probability of two particles is $1/8^2$, three is $1/8^3$, and the probability of finding n particles in the octant is $(1/8)^n$. So if there were Avogadro's number of particles randomly wandering throughout the room, the probability would be $p = (1/8)^{6\times10^{23}}$. This is an impressively small number: to give an idea of its smallness, if we say that 1/8 is about 1/10, and Avogadro's number is about 10^{23}, then

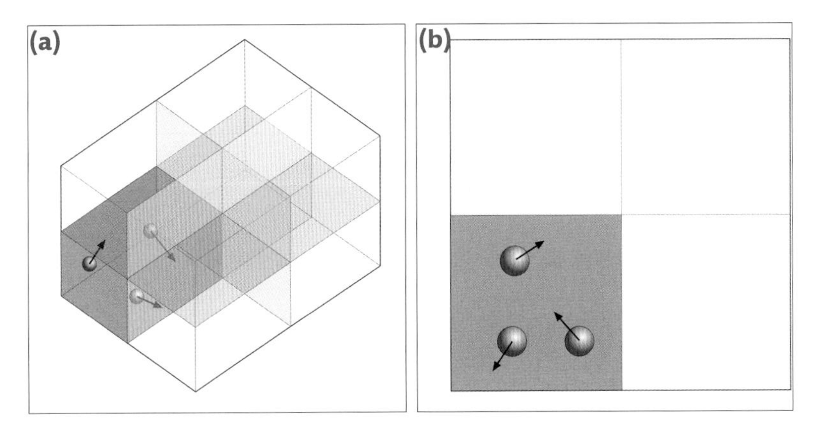

Fig. 11.1 (a) Depiction of particles starting in one octant (magenta) of a room with random velocities. (b) The same problem in two dimensions.

$p \sim 10^{-10^{23}}$, or a decimal point followed by 10^{23} zeros. Just writing this number down would occupy a stack of pages spanning 100 times the distance to the sun.

So if Avogadro's number of molecules were to bounce around in the room shown in Fig. 11.1, sure enough they would each periodically wander into the magenta octant, but very, very rarely would all of the molecules be found in the octant. How rarely? At standard temperature and pressure, the speed of a typical molecule is 400 m/s, so in 1 second in a room 1 meter on a side, we can work out that one molecule would find its way into the magenta octant about 250 times. How long would it take before all of the molecules wandered into the octant? About $250 \times 10^{10^{23}}$ seconds. The age of the universe is 4×10^{16} seconds. So this event would occur after $250 \times 10^{10^{23}} \div 4 \times 10^{16}$, or about $10^{10^{23}}$ times the age of the universe.

This implies that Poincaré recurrence will occur, but we will never witness it even after many, many ages of the universe have passed. This is in the simplest case, where particles do not interact with one another: if particles do interact, those inside the magenta octant will reject those outside, and although Poincaré recurrence must still occur, the probability of finding particles in the octant will be even smaller—as Exercise 11.1 will confirm.

The message of this discussion is that seemingly complex thermodynamic questions can be resolved through simple statistical calculations based on mechanical interactions (recall here our examples of mechanical ordering from last chapter). We will elaborate on these ideas, but first—in the continuing spirit of giving dessert first (recall Chapter 6)—let's build on what we have learned so far to draw a simple and surprisingly deep conclusion about what statistical counting tells us about self-assembly.

EXERCISE 11.1

Evaluate the recurrence probability for perfectly elastic colliding spheres by recording occurrences when the spheres all end up in either a three-dimensional octant (Fig. 11.1(a)) or a two-dimensional quadrant (Fig. 11.1(b). Necessary code for this problem is included in the Appendix, Section 8: Particle Dynamics. Show that recurrence times grow exponentially with the number of noninteracting spheres, and even faster than that for colliding ones. Explain why colliding spheres exhibit recurrence less often than noninteracting spheres, and based on this understanding propose a quantitative equation to describe the recurrence time of colliding spheres. Give some thought to, and then discuss briefly, why you might choose to solve this problem in two versus in three dimensions.

11.2 **Statistical Mechanics and Self-Assembly**

Last chapter, we saw that red blood cells (RBCs) self-assemble into organized rouleaux in the presence of much smaller organic proteins. Since Exercise 11.1 showed that it's simple enough to simulate identical colliding spheres, let's ask what happens when large and small colliding spheres are locked in a room. We can be imaginative and view the large spheres as representing blood cells and the small ones representing protein molecules, or we can be more literal and ask what happens to colloidal particles suspended in a carrier of smaller water molecules.[3] Both problems turn out to obey the same simple principle.

To understand the principle, consider what must happen when a smaller hard sphere interacts with a larger one. As shown in Fig. 11.2(a), the center of a small sphere can only fit within a fixed distance (its radius) from the larger sphere—so the centers of small spheres cannot fit within the gray excluded region of the large sphere. The red sphere shown is therefore allowed, but the black sphere is excluded. This geometrical fact is almost too simple to bear mentioning, but it has an important consequence. As shown in Fig. 11.2(b), both spheres and boundaries have excluded regions sketched in gray, and when these regions overlap—shown in Fig. 11.2(c)—the total excluded region must shrink. Meaning that there is more volume available for small particles to fit in the allowed cyan region of Fig. 11.2(c) than in Fig. 11.2(b).

What effect does this change in volume have? Suppose we put some number of small particles, completely at random, into the volumes shown in Figs. 11.2(b) and (c). We'll

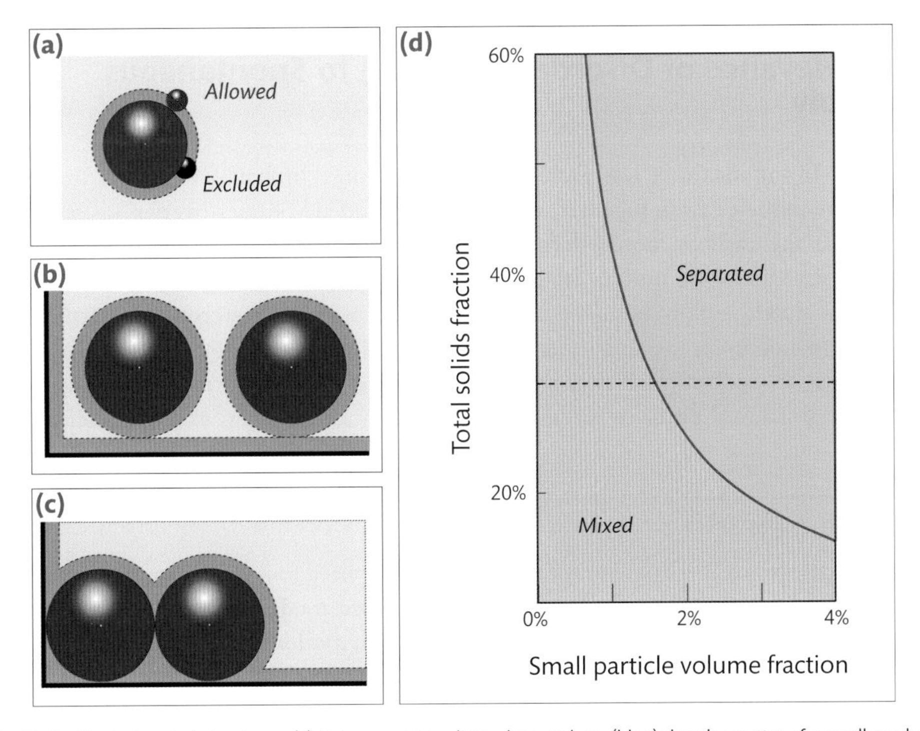

Fig. 11.2 Effect of excluded volume. (a) Volume surrounding a large sphere (blue) that the center of a smaller sphere is excluded from is shaded gray. So the red sphere can fit, but the black sphere cannot. (b) Excluded volumes apply to any hard surface, including boundaries as well as other spheres. (c) If large spheres are moved close to boundaries or close to one another, the total excluded volume is reduced. This means that more volume is available (cyan) to place small spheres. As discussed in the text, this in turn implies that in the presence of small spheres, the state in panel (c) is more likely than the state in panel (b). (d) Consequently, if the total volume occupied by large objects is held constant (dashed line), then by increasing the fraction occupied by small spheres, the most likely state will spontaneously change from mixed (as in (b)) to separated (as in (c)). Panel (d) uses rods rather than large spheres, but similar results are obtained for all shapes (see Fig. 11.3) (Panel (d) adapted from Adams et al.[4])

calculate the outcome numerically in a moment, but it should stand to reason that there are fewer different ways of randomly placing particles in a smaller volume than in a larger one. So if we can pack 10 small particles, say, into the available volume of Fig. 11.2(b), then there should be more ways of packing these same 10 particles into the larger available volume of Fig. 11.2(c). But *if there are more ways of doing a thing, then it must be more likely.*

Let's pause to appreciate this. This means that a state in which the large particles exclude less volume (Fig. 11.2(c)) *must be more likely* than a state in which they exclude more volume (Fig. 11.2(b)). That is, there are more ways of producing this state in a random universe. Crucially, this likelihood depends almost entirely on the number of *small* particles. We'll return to this point, but for now notice that this is exactly what was seen last chapter for rouleaux of RBCs (Fig. 10.6, Chapter 10): as the concentration of blood-borne *proteins* increases, the viscosity associated with rouleaux of RBCs grows. To calculate the likelihood of the formation of ordered structures such as rouleaux, we need a few tools from Statistics, which we present in a section following ("A brief Statistics review"). First, we qualitatively overview a few (of many!) structures that self-organize due to this "excluded volume" principle.

11.3 **Relevance of Disorder (Entropy) to Spontaneous Assembly**

According to the idea that less excluded volume is more probable, it follows that if we crowd large particles close together so that their excluded volumes overlap (as shown in Fig. 11.2(c)), there will be more volume available and so such a state will be more probable. This is an extremely odd notion, for it means that without analyzing any explicit force or pressure, it should simply be more probable to find big particles crowded together into whatever state occupies the smallest possible volume. Likewise it should be more probable to find particles concentrated near edges (which have their own excluded volumes)—or even more so to find particles in corners, where they can overlap excluded volumes on two sides.

ADVANCED EXERCISE 11.2

Can this really be so? It is easy enough to find out: all we need to do is to simulate particles of two sizes that bounce around at random. For this exercise, modify the code used in Exercise 11.1 to incorporate two particle sizes. Then add Brownian motion to the smaller particles: this can be done in any way that you like just so you cause the small particles to move slowly enough that the large particles have time to adjust their locations but fast enough for the small particles to ultimately adopt random positions.

This randomization is central to the mechanism of ordering through excluded volume: you will find that if there are too few small particles or if they jiggle about too little, the large particles will not migrate toward the edges or toward one another. By judicious manipulation of parameters, however, you will be able to produce results similar to the snapshot shown in Fig. 11.3(a). In that panel, we show such a simulation where as expected the large, blue, particles have rapidly moved to occupy the corners, and to crowd near one another.

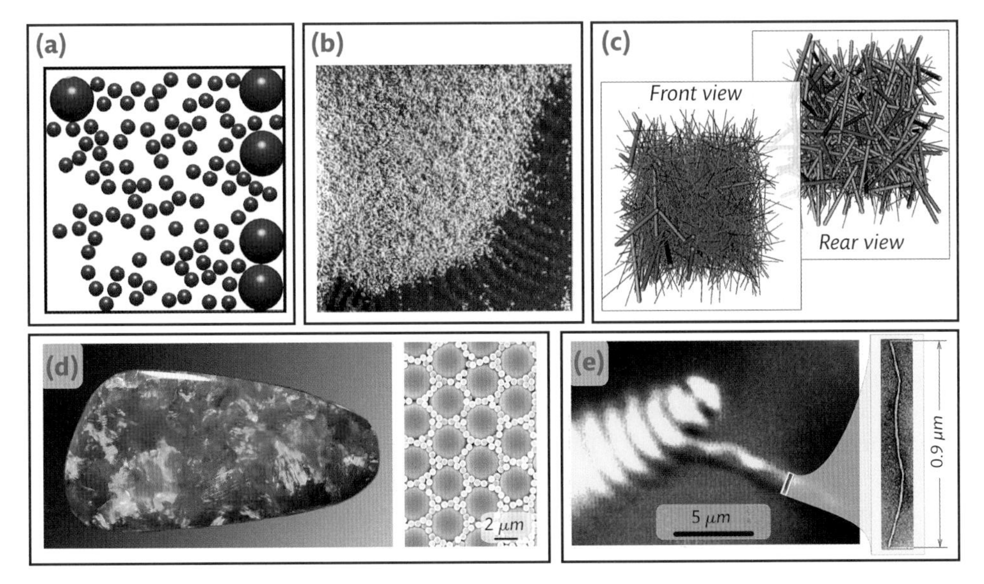

Fig. 11.3 (a) Simulation of large and small spheres confirms that large particles concentrate near edges and corners in the presence of small Brownian particles. Similarly (b) experiments show that in a bed of vibrated beads, the smaller (maroon) particles separate from the larger (brown) ones, and the same occurs in (c) simulations of rods of different sizes. (From van Roij et al,[5]. Courtesty of AIP.) (d, left) The same mechanism is also responsible for the formation of gemstones such as opal. (Dpulitzer/CC BY-SA 3.0,[6].) (d, right) the internal structure of a similar bidisperse self-assembled structure. (Courtesy of Harry Beeson,[7].) Biological structures are also produced by this mechanism. (e, left) False colored "tactoid" structure of self-assembled viral rods. (from Adams et al.,[4]. By permission of Springer Nature.) (e, right) enlarged view of a single filamentous bacteriophage virus that makes up the tactoid. (© The Rowland Institute at Harvard,[8].)

As simple as this idea is, one would think that concentrated structures of larger objects should be ubiquitous. Indeed, this is so. As the examples shown in Fig. 11.3 illustrate, only two things are required to form self-assembled structures: (1) excluded volume around the large objects (i.e. objects that are hard enough to keep smaller particles away), and crucially (2) a mechanism of randomizing the smaller particles. We return to this point, but note here the extraordinary fact that this implies: in these cases, *randomization* is required to produce order!

A first, very direct, example is shown in Fig. 11.3(b), where we display a vibrated bed of glass spheres of two sizes: the maroon and brown spheres have diameters 0.8 mm and 1.8 mm respectively. As the figure shows, the beads have spontaneously separated – an effect that only occurs above about 10% by mass of small spheres.[9] In this case, the particles are randomized by mechanical shaking and inter-particle collisions.[10] We saw previously (Chapter 10, Exercise 10.4) that macroscopic particles can spontaneously segregate—in one case due to inelastic collisions and in a second case due to competing dynamical effects. These effects are also at work in Fig. 11.3(b), but the emergence of separation with growth in small particle number is characteristic of the excluded volume mechanism. A second example is shown in Fig. 11.3(c), where rods, rather than spheres, of two sizes are intermixed in a computational model. Here randomization is performed using a Monte Carlo algorithm, in which particles are repeatedly rearranged to simulate thermal agitation.[5]

These two examples are artificial systems, but excluded volume effects are common in nature as well. Beyond rouleaux which we have already mentioned, a possibly surprising example is the formation of gemstones. In Fig. 11.3(d, left), we show an opal, whose iridescent colors are produced by so-called "Bragg diffraction" of light acting on self-assembled crystal facets. Within the opal is a mix of hexagonal, cubic, and other organized lattices,[11] which form over geological timescales through slow diffusion of silica spheres, and these ordered lattices are what produces the opal's colors. Here diffusion is the source of particle randomization (we will discuss diffusion and its effects later this chapter and next). Also in Fig. 11.3(d, right), we show a laboratory model in which an artificial opal structure is created from polystyrene spheres of diameters 0.5 and 3 μm that spontaneously self-assemble into a hexagonal lattice.[7]

Finally, in Fig. 11.3(e), we show the spontaneous concentration of elongated, rod-like, virus particles into a striped assembly.[4] The virus particles assemble into a smectic phase (see Chapter 9, Fig. 9.10), with rods aligned perpendicular to the direction of the stripes, as indicated by the snapshot of a single virus in the inset enlargement.[8] Here randomization of solvent molecules occurs through Brownian motion—again an example of diffusion.

Other examples are readily found:

- Block copolymers self-assemble into materials whose microstructures give them remarkable properties. Lego™ blocks become nearly indestructible using these materials.[12]
- So-called metamaterials can assemble[13,14] to produce perfect cloaking or "superlenses" that can image objects much smaller than the diffraction limit of light.[15,16]
- Flocculants, used to decant suspended solids from wine,[17] also make use of excluded volume effects in part.
- So do many biological systems, for example sickle-cell hemoglobin clumps in this way[18], leading to clots, pain, and infection.
- Also molecular chaperones prevent such clumping and more generally permit unimpeded protein folding and unfolding,[19] in crowded conditions, and
- patterned cellular structures such as stripes on the zebrafish form through related mechanisms (see Chapter 13).[20]

What unites these different phenomena is that disorder, in the form of randomization of small particle positions, is used to explore possible states to create order, in the form of large particle assemblies. Without a disordering mechanism, the ordered state would typically never be sampled. It is disordering that allows multiple states to be sampled, and in the presence of disordering the fact that ordered states, with smaller excluded volumes, are more probable causes them to emerge. For this reason, excluded volume effects are often termed "entropic ordering"—because entropy (disorder) alongside excluded volumes leads to ordered states.

Qualitatively then, excluded volume plus entropy causes disordered systems to self-assemble into ordered ones. We turn next to developing quantitative tools to understand this phenomenon.

11.4 A Brief Statistics Review

To get a firmer hold on the statistics part of statistical mechanics, let's briefly overview some basic ideas. On the downside, statistics can be stultifyingly tedious, but on the upside it is at its heart just a matter of simple counting. In studying the likelihood of an outcome, we just need to count how many ways it can be produced. So if one outcome—throwing a ball into a basket—is unlikely, then we expect that out of a hundred throws, almost all will miss the basket, whereas if another outcome—hitting a barn door with a rock—is likely, then we expect that out of a hundred throws, almost all will hit the door.

So let's count. Suppose we have five boxes into which we drop balls at random. As shown in Fig. 11.4(a), if the boxes are initially all empty, then the first ball ($k = 1$) can fall into any of *five* possible boxes, and the probability of landing in any one box is 1/5. When a second ball is

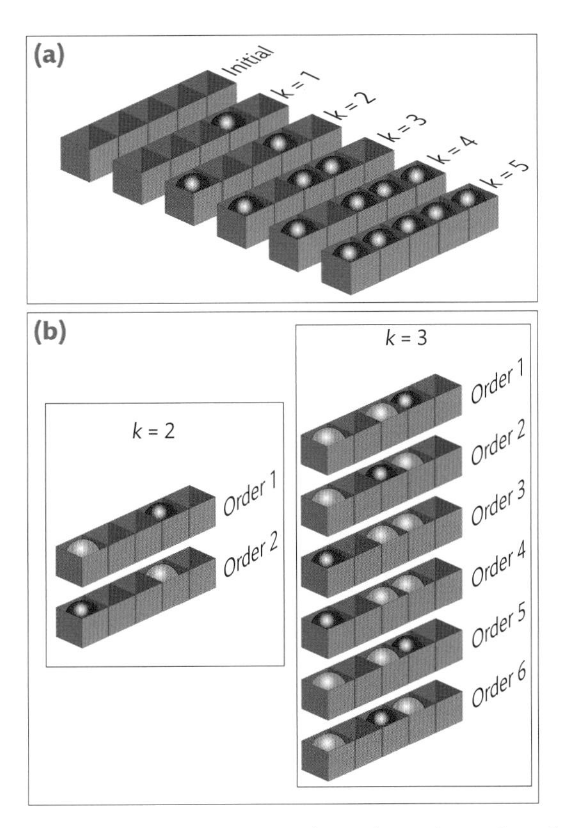

Fig. 11.4 (a) If a ball is dropped into a random box in an initial row of $N = 5$ boxes, the probability, p, of filling one particular box is 1/5–that is, $p = 1/N$. After 1 ball has been dropped, at $k = 1$, four boxes will remain, so $p = 1/(N - 1)$ to fill the next box at random. So the probability of both events is $P_{tot} = 1/(N \cdot (N - 1))$. Next, at $k = 2, p = 1/(N - 2)$, so the probability of all three events is $P_{tot} = 1/(N \cdot (N - 1) \cdot (N - 2))$, and so on. Ultimately filling k particular boxes has probability $P_{tot} = (N - k)!/N!$. (b) This analysis assumes an ordered sequence of filling boxes–yet the same state shown at $k = 2$ can be reached by dropping the yellow ball first and then the red ball or vice versa. Similarly there are six orderings of balls leading to the $k = 3$ state, and ultimately there are $k!$ possible orderings. Accounting for duplicate states generated from the different orderings, the probability becomes $P_{tot} = k! \cdot (N - k)!/N!$.

dropped ($k = 2$), only *four* boxes will be left, and the probability of the ball landing in any one of these boxes is $p = 1/4$. So the probability that the first box shown will be filled at $k = 1$ and then the second will be filled at at $k = 2$ is $P_{tot} = (1/5) \cdot (1/4)$. Similarly at $k = 3$, the probability of the *three* particular boxes shown being filled is $P_{tot} = 1/(5 \cdot 4 \cdot 3)$, at $k = 4$, $P_{tot} = 1/(5 \cdot 4 \cdot 3 \cdot 2)$, and at $k = 5$ there will only be one box left, so for consistency we write $P_{tot} = 1/(5 \cdot 4 \cdot 3 \cdot 2 \cdot 1)$. You can verify that these probabilities can be written compactly as $P_{tot} = (5 - k)!/5!$.

Good, so we're nearly done. If $k \leq 5$ are balls dropped into *five* boxes, then the probability of producing any particular sequence of filled boxes will be $P_{tot} = (5 - k)!/5!$, or more generally k balls in N boxes will give:

$$P_{tot} = \frac{(N-k)!}{N!}. \qquad [11.2]$$

This is the probability for balls that have been dropped in a particular order, or equivalently for balls with individual identities, to produce a particular pattern. So as shown in Fig. 11.4(b), if the balls are distinguishable from one another (either by order or by identity), then at $k = 2$ the yellow ball could be in either of the two occupied boxes. Similarly the figure shows that there are six possible orders of $k = 3$ colored balls.

This means that if the balls are identical, then all of the states shown in Fig. 11.4(b) will be the same, which makes many of these states indistinguishable and changes the probability (Eq. [11.2]). It is easily confirmed that there are $k!$ ways of ordering k objects, so if the order of objects matters, then fewer of any particular patterns of balls will be seen than if identical, unordered, objects are placed randomly. In particular, if the balls are identical, then the probability (Eq. [11.2]) of finding a particular pattern must be increased by $k!$. Consequently for unordered balls,

$$P_{tot} = \frac{k! \cdot (N-k)!}{N!}. \qquad [11.3]$$

EXERCISE 11.3

As an analogy, suppose I flip 100 coins and only one comes up heads. How many ways are there of making this happen? Suppose two come up heads? Three? A number k? How would this change if you used a six-sided die instead of a coin?

EXERCISE 11.4

We'd like to quantify how much more likely a crowded state, like Fig. 11.2(c), is than an uncrowded one, like Fig. 11.2(b). This is possible to determine in experiments or simulations (see, for example, Van Roij *et al.*[5] or Yodh *et al.*[21]), but doing so is more involved than need be.

For our purposes, consider the two configurations shown in Fig. 11.5. Here we have reduced the large blue spheres in Fig. 11.2 to square blocks. Excluded volumes are still shown in gray, and to keep things simple we will assume that smaller particles can only occupy the cyan squares (e.g. the two pink squares are allowed, but squares overlapping with gray areas are not). For this exercise, calculate which figure is more likely, and by how much, if a number, N, of small particles are added, where N ranges from 1 to 15. Plot the ratio as a likelihood: does it grow linearly, exponentially, or faster than exponentially?

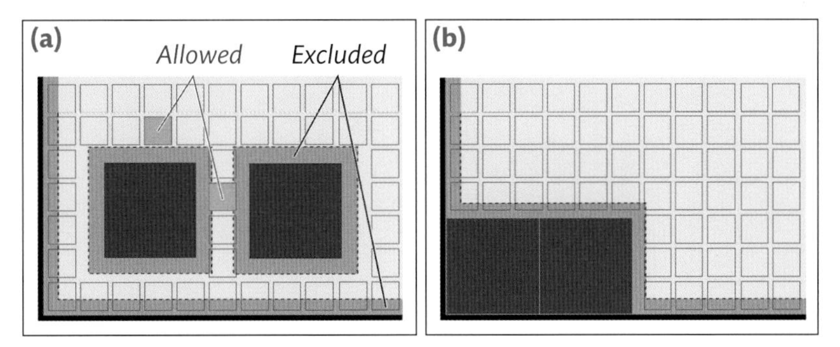

Fig. 11.5 Two configurations of blue blocks surrounded by gray excluded volume. (a) Uncrowded blocks have 28 allowed states for smaller blocks; to make calculations simple, allowed states are fixed on a grid (inside of red squares); two examples are highlighted. (b) When blue blocks are crowded, 45 states are allowed.

As the preceding exercises show, it is reasonably straightforward to calculate the number of possible states—and so the likelihood of obtaining a particular configuration in the presence of random variations. These calculations confirm first that it is more likely that crowded states will be found, and second that this likelihood grows dramatically with numbers of smaller particles. For this reason, larger objects typically spontaneously assemble into tightly packed structures—which tend to be highly ordered—in the presence of smaller agitated particles. So merely adding fibronectin to a suspension of RBCs causes the cells to aggregate into rouleaux.

The idea that disorder produces order in this way is not easy to absorb, but it is the case, and recognizing that entropy can produce self-assembled states opens up new insights that we will explore shortly. To take the next step, we must understand more clearly how diffusion of small particles operates. What we will find is that knowing how particles wander is valuable, both for what it reveals of how nature operates and for the ability that it provides for us to derive equations that govern both diffusion and pattern formation.

11.5 Statistical View of Diffusion

In Chapter 10 we described diffusion as being associated with randomizing collisions between particles. Let's build on that idea and explore what happens to a particle that undergoes a random "drunkard's walk." This is the wandering that occurs if a particle, or a drunkard, takes a step of constant distance but random direction every time step. At its simplest, the drunkard could start at a lamppost and take one step either to right or left, a distance $\Delta x_i = \pm 1$, along a sidewalk. In this case, as shown in Fig. 11.6(a), the steps would alternate randomly as the

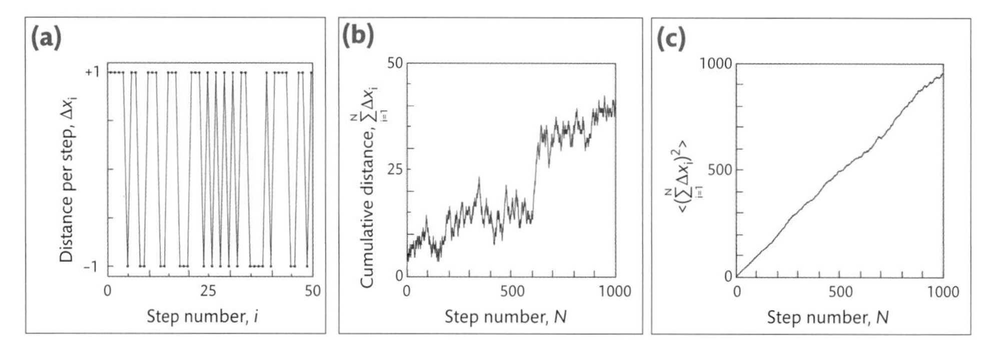

Fig. 11.6 (a) Successive random steps, Δx_i, of length 1, with equal likelihood of being in the positive or negative direction. (b) "Random walk:" cumulative distance traveled from origin. (c) Average of cumulative distance squared for 1000 random walks.

drunkard staggers away from the lamppost. The cumulative distance walked after N steps would be $\sum_{i=1}^{N} \Delta x_i$.

You might think that such a walk, entirely governed by randomness, could only yield more randomness: a typical plot of the cumulative distance traveled by such a drunkard is shown in Fig. 11.6(b). In fact, while each of these walks is indeed random, their average behavior obeys a very well defined law. If we set an army of drunkards walking away from a lamppost, we find that the *square* of their mean distance from the lamppost is very close to being linear, as shown in Fig. 11.6(c) for an average of 1000 simulated drunkards.

So this is interesting: although each drunkard wanders unpredictably, the "mean-squared displacement" of many drunkards is highly predictable. We'll see why this is after a brief exercise.

EXERCISE 11.5

Both molecules and biological cells travel in random walks. Simulating these walks illuminates several fine points to do with accurate modeling of chemical and biological behaviors. In this exercise, extend the results shown in Fig. 11.6 to two dimensions. Notice first that there are many ways of modeling a random walk: in Fig. 11.6(a), we illustrated a classic drunkard's walk, in which the drunkard takes a step of the same length each time. This needn't be the case though: depending on the speed of a molecule, its distance traveled in a unit time can vary; moreover, even at a constant speed a molecule will typically travel different distances between successive collisions.

In Fig. 11.7(a) we illustrate several possible random schemes. In the left panel, we show random numbers in which both Δx and Δy can take on any uniformly distributed value between 0 and 1. Notice that this model has preferred directions: the distance from the origin is greater along the diagonals than along horizontal or vertical directions. One alternative is shown in the central panel, where polar coordinates are used, and both Δr and $\Delta\vartheta/2\pi$ range from 0 to 1. In this case, due to geometry, although any Δr between 0 and 1 is equally probable, the probability of a point being near the origin is higher than further away. Another alternative is shown in the right panel, where points are uniformly distributed out to $\Delta r = 1$.

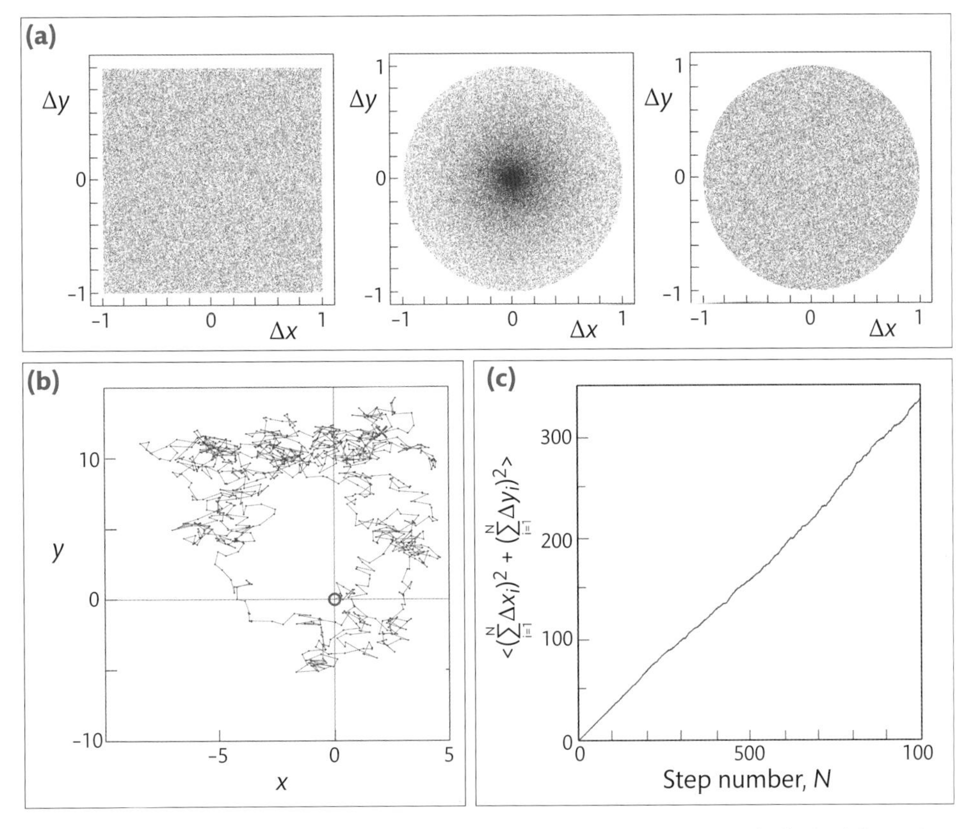

Fig. 11.7 (a) Three alternative sets of 100,000 random distances in two dimensions (2D). The left panel uses random distances uniformly distributed in x and y; the center panel uses random radii and angles; and the right panel uses uniformly distributed distances out to a maximum radius. (b) Random walk in 2D; red circle identifies origin, and red X identifies final point after 1000 steps. (c) Average of cumulative distance squared for 1000 random walks in 2D.

For this exercise, explain what algorithm you would use to produce each of these random distributions. Write the algorithm and plot the resulting points as in Fig. 11.7(a), and for each case simulate a random walk as shown in Fig. 11.7(b). Calculate mean-squared displacements for each case and determine whether they are linear—as shown in Fig. 11.7(c) for the algorithm used in Fig. 11.7(b).

ADVANCED EXERCISE 11.6

What distributions of random numbers would you use for the following problems? Program an algorithm for each case and show that it does what you expect. For some of these, you may need to invent a new way of walking randomly—in particular you may want to investigate "persistent random walks" and "integrated random walks."

(1) Diffusion of a molecule or small particle?[22]

(2) Random wandering of a bacterium on a surface? Note here that bacteria do wander randomly, but they also execute 'run and tumble' behaviors in which long-distance 'runs'

alternate with short-distance 'tumbles': this turns out to be an optimal search strategy when the location of a target (e.g. food) is unknown.[23]

(3) Chemotactic transport of a bacterium toward a fixed chemical source? At its simplest, growth here has both a component that is random and a component that is directed toward a target. For this exercise, model directed, or "persistent" motion, but be aware that cell motion responds to the *gradient* in a "chemoattractant's" concentration, which typically changes with distance from a target.

(4) Growth of a neurite tip? At the growing tip of a neurite (i.e. an incipient axon or dendrite) is a "growth cone" that causes the tip to travel predominantly forward, but permits it to deviate up to a maximum angle from the forward direction.

(5) Attraction of multiple wandering cells toward one another? This occurs for example in the slime mold *Dictyostelium discoideum* in response to environmental stresses, and leads to spectacular pattern formation.[24]

Finally on a practical note, it is worth knowing that just as all random walks are not equal, neither are all random number generators (RNGs). A good discussion can be found in Press *et al*[25], but in brief, typical RNGs consist of producing a new random number, R_{i+1}, at time $i+1$ from an old one, R_i, at time i, using the formula:

$$R_{j+1} = (A \cdot R_j + C) \bmod M \qquad [11.4]$$

where A, C, and M are constants.

This is relevant for several reasons. First, the numbers repeat after at most M applications of this formula, which can produce spuriously periodic solutions in simulations requiring large quantities of random numbers. Second, in fact recurrences can occur after much fewer than M applications—you may want to look up the notorious "RANDU" generator for an especially bad example of this. Third, modern computers with multiple processors often use the same RNG in each processor, so interactions can arise between computations that should be independent of one another. Resolutions to some of these issues are described in Press et al.[25], but the reader should be aware—especially in simulating self-assembled patterns—that a RNG can itself be the unintended source of ordered patterns.

We've seen that the mean-squared displacement of a random walk is simply linear with the number of steps—despite, or because of, the randomness of the process. Let's understand this. Again, this is only a matter of counting—and much of the counting has already been done. Let's say there are n_L steps to the left and n_R to the right out of a total of $N = n_L + n_R$ steps. We saw that the probability of producing k events (steps to the left, n_L, here) in N trials is given by Eq. [11.4]. Another way of saying this is that the number of ways of arriving at a particular location (k steps to the left and $N - k$ to the right) in N trials is the inverse of Eq. [11.4]:

$$NumberOfWays = \frac{N!}{k! \cdot (N-k)!}. \qquad [11.5]$$

This is useful for the random walk problem: if we call the probability of taking a step to the right p and the probability to the left q, then the probability, $P(n_R)$, of taking n_R and n_L of these steps is $p^{n_R}q^{n_L}$, multiplied by the number of ways of producing this outcome. We write n_L as $N - n_R$, so this probability becomes:

$$P(n_R) = NumberOfWays \cdot p^{n_R} \cdot q^{N-n_R}$$
$$= \frac{N!}{n_R! \cdot (N-n_R)!} p^{n_R} \cdot q^{N-n_R}. \quad [11.6]$$

For unbiased random walks, for example, a drunkard wandering at random without a preference for one direction over another, $p = q = \frac{1}{2}$, so:

$$P(n_R) = \frac{N!}{n_R! \cdot (N-n_R)!} \cdot \frac{1}{2^N}. \quad [11.7]$$

Using this probability, we can answer questions about how far drunkards walk from lampposts.

First, if we set 1000 drunkards out from the same lamppost, where will their average position be? If we call this position $\langle m \rangle$, then $\langle m \rangle = \langle n_R \rangle - \langle n_L \rangle = p \cdot N - q \cdot N = 0$. This means that on average, the position of the drunkards remains at the origin, where the lamppost is, for all time. This makes sense. Second then, how can the distance squared grow linearly with time? In Fig. 11.8, we redraw Fig. 11.6(b), tracking not just one drunkard, but a sampling of them. Evidently their average position $\langle m \rangle$ indeed remains close to the lamppost (black line), but also as time evolves, the distance that the crowd of drunks wanders from the lamppost grows.

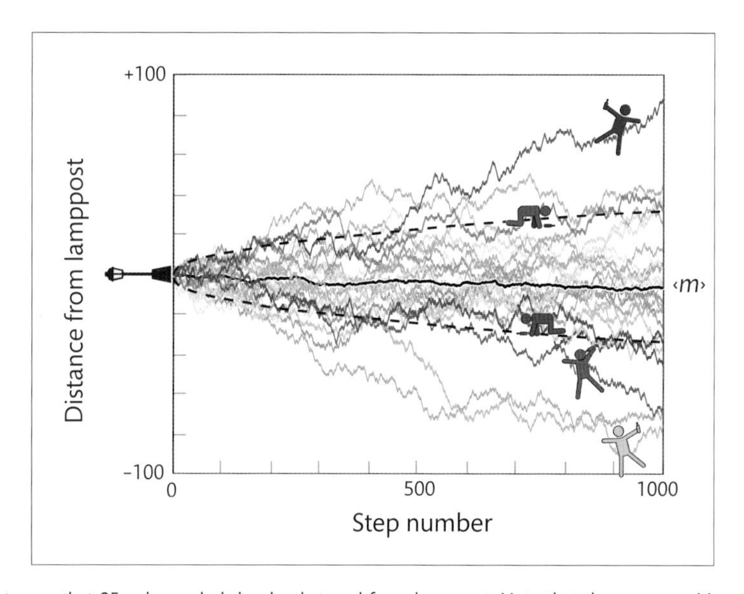

Fig. 11.8 Distances that 25 color-coded drunkards travel from lamppost. Note that the mean position, $\langle m \rangle$, shown in black remains close to the lamppost, yet the crowd expands with time: the average size of the crowd from Eq. [11.15] is included as broken lines.

So third, how can we characterize the growth of the crowd? One way is to ask what the average displacement from the lamppost is. That is, if the displacement of one drunkard from the lamppost is $\Delta m = m - \langle m \rangle$, then the displacement averaged over all of the drunkards is:

$$\langle \Delta m \rangle = \langle m - \langle m \rangle \rangle. \quad [11.8]$$

But we know that $m = n_R - n_L$, so:

$$\begin{aligned} \langle \Delta m \rangle &= \langle n_R - n_L - (\langle n_R \rangle - \langle n_L \rangle) \rangle \\ &= \langle n_R - (N - n_R) - (\langle n_R \rangle - \langle N - n_R \rangle) \rangle \\ &= 2\langle n_R - \langle n_R \rangle \rangle. \end{aligned} \quad [11.9]$$

This is fine, but the displacement could either be negative or positive, while the size of a crowd of drunkards is only positive. So traditionally, we square this expression to obtain a positive mean-squared displacement:

$$\langle \Delta m^2 \rangle = 4\langle n_R{}^2 - 2n_R\langle n_R \rangle + \langle n_R \rangle^2 \rangle. \quad [11.10]$$

We'll simplify this expression in Exercise 11.7.

EXERCISE 11.7

First, show that the last two terms in Eq. [11.10] equal $(-\langle n_R \rangle^2)$, so that Eq. [11.10] is just:

$$\langle \Delta m^2 \rangle = 4\langle n_R{}^2 - \langle n_R \rangle^2 \rangle. \quad [11.11]$$

This equals $4(\langle n_R{}^2 \rangle - \langle n_R \rangle^2)$, and $\langle n_R \rangle = Np$, so we need only solve for $\langle n_R{}^2 \rangle$ to be able to evaluate the mean-squared displacement. Solve for this using the fact that:

$$\langle n_R{}^2 \rangle = \sum_{n_R=1}^{N} P(n_R) n_R{}^2, \quad [11.12]$$

combined with Eq. [11.7]. Finally, simplify the result by making use of the trick that:

$$p \cdot n_R{}^2 = n_R \left(p \frac{d}{dp} \right) p^{n_R}. \quad [11.13]$$

Once Exercise 11.7 has been completed, you should be able to show that:

$$\begin{aligned} \langle \Delta m^2 \rangle &= 4\left(pN(1 + pN - p) - (Np)^2 \right) \\ &= 4Npq, \end{aligned} \quad [11.14]$$

and for $p = q = ½$ we finally obtain:

$$\langle \Delta m^2 \rangle = N. \quad [11.15]$$

This is why the mean-squared displacements in Figs. 11.6 and 11.7 grow linearly with number of steps. The mean position of the crowd remains fixed, and any one drunkard can in principle (with low probability) walk always in the same direction to reach distance N, but the mean size of a crowd of wandering drunkards grows as $\sqrt{N}$ (shown as broken lines in Fig. 11.8). We'll return to this fact next chapter: as a brief preview, the same $\sqrt{N}$ growth will be obtained using differential equations for diffusion as well as by statistical counting.

Before we close this chapter, let's consider some implications of what we've learned. We now know how drunkards deviate from a lamppost—or more usefully how randomly wandering cells or molecules diffuse from an initial point. This accounts for several otherwise perplexing phenomena.

11.6 Statistical Mechanics and Osmotic Pressure

As a first example, consider the experiment shown in Fig. 11.9(a), in which a sugar solution is initially separated from pure water by a membrane permeable to water, but not sugar. Such an experiment can be constructed by jamming a piece of apple into a the center of a plastic tube and filling the tube on both sides of the apple with water, one side containing food coloring and the other containing sugar. You can easily confirm that the sugar side of this experiment rises with time (and acquires food coloring) as depicted in the panel to the right of Fig. 11.9(a). The sugar solution has higher density than the pure water, so this isn't a hydrostatic phenomenon; also the diameters of the tubes can be quite large, so capillary forces play no role.

This effect is accurately described by a formula proposed by Jacobus van 't Hoff (1852–1911), the first recipient of the Nobel Prize in Chemistry. His formula states that the pressure, P, due to a material of concentration c is simply given by:

$$P = c \cdot R \cdot T, \qquad [11.16]$$

where R is the universal gas constant and T is the temperature. P here is termed the "osmotic pressure," and is derived in many texts by comparing chemical potentials across the membrane and asserting that water molecules move toward a lower chemical energy, which is obtained in higher sugar concentration solutions. How the materials involved know to do this and the mechanical process by which pressure is generated are both left somewhat abstract by this explanation, so let's instead focus on the motions of molecules as we have done previously in this chapter.

As shown in the expanded view at the bottom of Fig. 11.9(a), a semipermeable membrane is simply a material with pores smaller than the particles—here sugar—to be excluded. There are more water molecules on the water side of the membrane than on the sugar side, and so there must be more *collisions* by water molecules on the water side. Since each collision imparts momentum to the water in the pores shown, it follows that there is more momentum imparted on the water side. Force is change in momentum per unit time, so there is more force—and so more pressure—on the water side. As reflected by Eq. [11.16], this pressure naturally depends

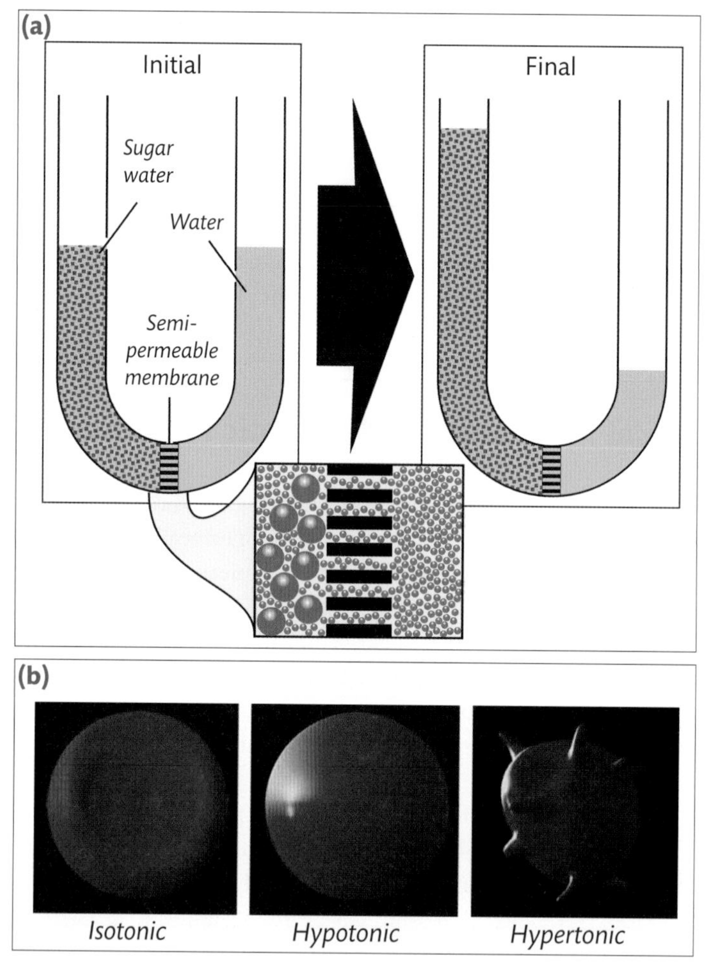

Fig. 11.9 (a) Schematic of experiment showing osmotic pressure in a U-tube half filled with sugar water and half with pure water. The two regions are separated by a semipermeable membrane (inset at bottom) that allows water to pass, but not sugar. As shown, water flows into the sugar solution to raise the pressure head on the left of the U-tube. A count of water molecules (blue) in the inset sketch yields 79 on the left, and 177 on the right, so there must be over twice as many collisions by water from the right than from the left – and hence there is more "osmotic pressure" on the right than on the left. (b) Red blood cells in a solution with salt concentrations outside the cells that are matched ("isotonic"), lower ("hypotonic") and higher ("hypertonic") than inside the cells. Imbalance of salt concentrations causes flow of water by Eq. [11.16] to inflate or deflate the cell.

on the difference in concentration, which governs the numbers of small molecule collisions across the membrane, and on the temperature, which governs the molecules' momenta. Notice that the large particles transmit force to the membrane but not to the fluid on its other side, so as with entropic ordering, entropic pressure depends ultimately on the smaller particles.

So we see that simple counting of water molecules on either side of a membrane accounts for the rise in the sugar solution column shown in Fig. 11.9(a). Membranes of course are ubiquitous in biology, and consequently examples of osmotic pressure resulting from this simple counting are plentiful. A few of these are listed below.

- We have focused on RBCs several times in this book. When a RBC is suspended in a salt solution, then if the solution has a lower salt concentration than the cell cytoplasm (a "hypotonic" state), water will flow into the cell, expanding and ultimately exploding it. On the other hand, a higher salt solution ("hypertonic") will cause the cell to shrivel. Snapshots of these states are shown in Fig. 11.9(b).
- Remarkably this same process can be used to trigger "artificial parthenogenesis," in which unfertilized sea urchin eggs, which normally require sperm to develop, reproduce asexually in response to a change in osmotic pressure.[26]
- Everyday examples also exhibit effects of osmotic pressure. Salt poured onto banana slugs causes them to shrivel as osmotic pressure pumps out their internal water. Similarly dry sugar poured onto apples being prepared for a pie produces a wet syrup by drawing liquid from the fruit. And honey or maple syrup draws water from the air, making the last dribbles from a bottle thin and tasteless.
- Plants pump water up their xylum by extending thin-walled root hairs into the soil: as water evaporates from leaves, the concentration of dissolved salts and sugars inside the roots grows, producing an inflow of water by what botanists term "root pressure."
- A corollary of the fact that osmotic pressure pushes dirty water up is that if this pressure is countered—by a mechanical pump for instance—clean water will flow out of dirty water. This is the mechanism used by reverse osmosis filters.

Example Donnan equilibrium

An interesting and physiologically important consequence of van 't Hoff's formula, Eq. [11.16], is the Donnan effect, named after a collaborator of van 't Hoff, Frederick Donnan (1870–1956). Donnan studied equilibrium concentrations of the dye, Congo red, across a membrane of parchment in the presence of various salts.[27] He found that salt ions set up competing forces across the parchment—one due to mechanical pressure, accounted for by Eq. [11.16], and a second due to electrical gradients, caused by charges on the ions. He found that when the two gradients are at equilibrium – termed a Donnan equilibrium – a net voltage can be maintained across the membrane. This voltage has broad effects, for example, causing transport of chloride into RBCs—which in turn regulates pH in the blood—and contributing to ionic imbalance across neuronal membranes—which leads to edema when the membranes are injured and fluid flows in to equalize the potential difference.

11.7 **Diffusivity**

As a second application of Eq. [11.15], note that another way of expressing this equation is to write that $\langle \Delta m^2 \rangle / N$ is constant, so if each of N random steps of distance ℓ takes a unit time, then the constant is:

$$D = \ell^2/t, \tag{11.17}$$

where we call D the "diffusivity," ℓ is a characteristic distance, and t is the total time. We will have much more to say about diffusivity next chapter; for now we summarize a few key features of diffusivity that are important to understanding its physical meaning. In overview, diffusivity has the same dimensions as viscosity, and is a measure of how much area (ℓ^2) a substance takes up over time as it diffuses. We intentionally write the distance as a curly ℓ to emphasize that this is the typical displacement occupied by a cloud of particles, and isn't the distance, L, traveled by a single particle (recall Fig. 11.8).

To clarify this distinction, at room temperature, the typical kinetic energy of a molecule is $\epsilon = kT$ where k is Boltzmann's constant and T is the temperature in degrees Kelvin. Since the kinetic energy is also $\epsilon = mv^2/2$ a molecule's typical speed must be $v = \sqrt{2kT/m}$. where m is the molecule's mass. So at room temperature, nitrogen molecules must typically move at about $v = 13$ m/s. After 1 s, an uninterrupted molecule of air would then typically *travel* a distance $L = 13$ m. By comparison, Eq. [11.17] and the value of diffusivity of nitrogen in air from a handbook, $D = 2 \times 10^{-4}\text{m}^2/\text{s}$, tell us that in that same second, a cloud of N_2 molecules would *diffuse* only $\ell = \sqrt{Dt} \simeq 4.5$ mm.

It is valuable to establish what the diffusivity of a substance depends on, for which we turn to a famous result from Einstein's doctoral thesis. There, Einstein showed that by balancing osmotic pressure acting on a single particle—which tends to accelerate the particle—with Stokes drag—which tends to slow it down—the diffusivity of a spherical particle can be derived. Einstein's approach is an early application of what is termed the fluctuation-dissipation theorem, and yields:

$$D = \frac{k}{6\pi}\frac{T}{\mu r}. \tag{11.18}$$

Here k again is Boltzmann's constant, T is the temperature, μ is the fluid viscosity, and r is the particle's radius. This tells us several things.

First, we can see that diffusivity goes as $1/r$. So, as shown in Fig. 11.10(a), larger particles diffuse more slowly than smaller ones. This plot shows diffusivity in air, with viscosity of 2×10^{-5} Pa s, so that a 1 μm particle diffuses a thousand times more slowly than a 1 nm molecule. This is straightforward enough, and from Eq. [11.18] we can see that the higher diffusivity is caused by the $T/\mu r$ term—that is, by the fact that temperature, which causes intermolecular collisions, has a larger effect on smaller than on larger particles. Turned on its head, by the way, this is a remarkable result of Einstein's derivation, for he showed that by measuring quantities like temperature, diffusivity, and viscosity, which are easily obtained on the macroscopic scale, we can accurately estimate the size of a molecule, which is far smaller than we can possibly see!

Second, combining Eqs. [11.17] and [11.18] and plotting displacement, ℓ, versus time, we see in Fig. 11.10(b) that longer displacements take greater times to reach, according to the square root law that we obtained earlier from Eq. [11.17]: $\ell = \sqrt{Dt}$. In Fig. 11.10(b), we plot the displacement of a 0.2 nm diameter molecule (about the size of O_2) in water. Evidently, as shown in the inset, it takes a million times longer for oxygen to diffuse 10 μm than 10 nm.

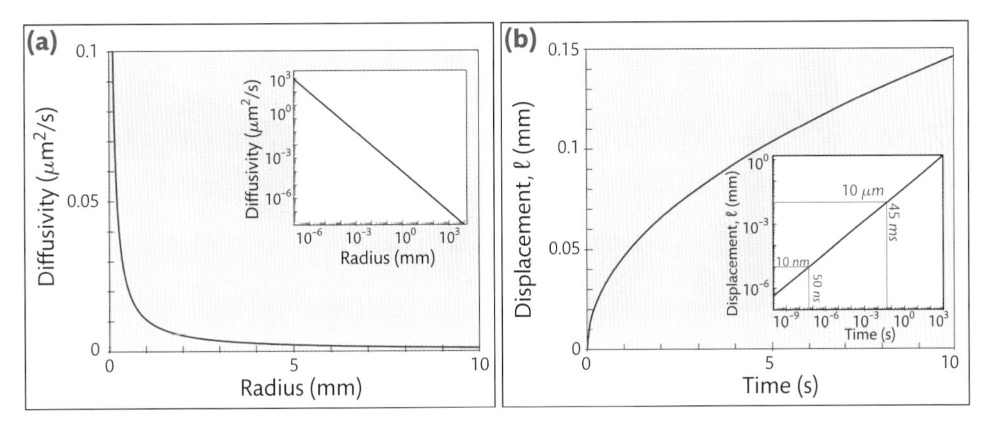

Fig. 11.10 (a) Diffusivity vs. radius. Main panel displays linear-linear plot of diffusivity in air, showing D ~ 1/r dependence. Inset displays $\log_{10}$-$\log_{10}$ plot. (b) Diffusion displacement, ℓ, vs. time. Main panel displays linear-linear plot, here of diffusion in water, showing $\ell \sim \sqrt{time}$ dependence. Inset displays $\log_{10}$-$\log_{10}$ plot: notice that as the displacement due to diffusion increases by a thousand, the time taken increases by a million.

EXERCISE 11.8

Estimate the diffusivity of O_2 in blood. We have seen in previous chapters that blood viscosity depends strongly on details of the experiment involved: for your purpose, what viscosity should you use? Human tissue cells are invariably within about 10 μm of the nearest capillary: use the diffusivity that you obtained to calculate how long it would take for O_2 to diffuse to a cell. How much longer would it take if the cell were 20 or 40 μm away? Is this time estimate reliable—what else might be relevant to accurately calculating O_2 transport times into tissues?

Next, turn to transport across neuronal synapses. How long does it take for a neurotransmitter to diffuse across a synaptic cleft? Given that the synaptic delay is about 500 μs, is your time estimate reasonable? What diffusivity did you choose, and why? You will need to estimate the radius, r, of a neurotransmitter: choose one and explain how you estimated r.

11.8 **Distribution of Energy**

A third example of notable phenomena that statistical mechanics can clarify deals with the energy distribution of liquid or gas particles.

To evaluate the distribution of energy of randomly colliding particles, once more we find that much of the counting has already been done. To see why, let's imagine that we already know how much energy a group of particles has—for example, consider the distribution of energy shown in Fig. 11.11(a). As always, we start with a simple case, here of just 28 particles. We have invented this distribution, but in general we want to know how many particles have energy in specified ranges. So we've shown 1 particle between 0 and 1 energy units, 3 between 1 and 2, and so forth. Here we use units of pico-Newton nanometers: air at room temperature happens to have about 4 pN-nm units of energy.

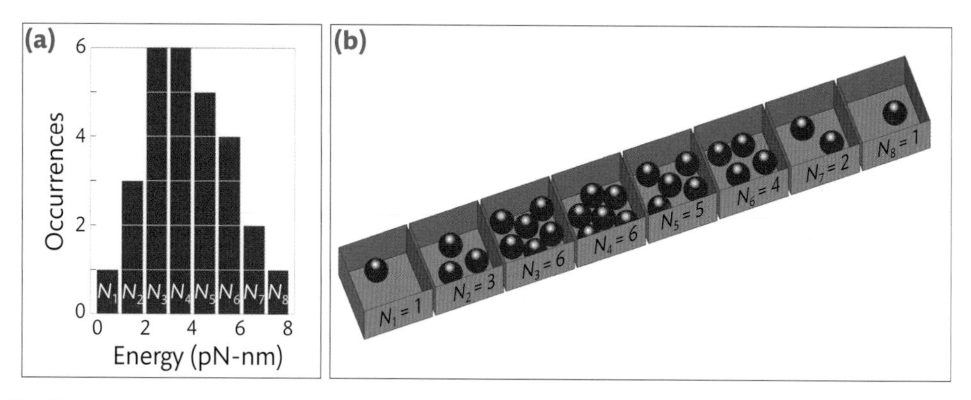

Fig. 11.11 (a) Simple example of distribution of energies, here of 28 particles. (b) Alternative representation of the same data: here we imagine eight boxes containing balls, rather than eight energy ranges. In this case, we already know how to count the probability of occurrence of any distribution of balls.

The number of ways of filling the first bin with N_1 out of N total particles is given by (Eq. [11.5]):

$$NumberOfWays_1 = \frac{N!}{N_1! \cdot (N - N_1)!}. \tag{11.19}$$

In the same way, the number of ways of filling the second bin with N_2 of the $N - N_1$ remaining particles is:

$$NumberOfWays_2 = \frac{(N - N_1)!}{N_2! \cdot \Big((N - N_1) - N_2\Big)!}. \tag{11.20}$$

So the number of ways of filling the first and second bins is:

$$\begin{aligned} NumberOfWays_{1,2} &= NumberOfWays_1 \cdot NumberOfWays_2 \\ &= \frac{N!}{N_1! N_2! (N - N_1 - N_2)!}. \end{aligned} \tag{11.21}$$

By extension, the number of ways of filling bins $1 \ldots n$ is:

$$NumberOfWays_{1..n} = \frac{N!}{N_1! N_2! \cdots N_n!}. \tag{11.22}$$

EXERCISE 11.9

We have previously established that the state chosen in a random system is the one with the highest probability—or the largest number of ways of being produced. So we only need to maximize Eq. [11.22] to determine that most probable state. We of course find maxima by differentiating, and differentiating Eq. [11.22] is made simple through Stirling's approximation:

$$\ln(N!) \simeq N \cdot \ln(N) - N. \tag{11.23}$$

This exercise has two parts. First, use the simplification in Eq. [11.23] to set the derivative of Eq. [11.22] equal to 0. Second, note that there are two constraints: (1) that $N = \sum_{i=1}^{n} N_i$ and (2) that $E = \sum_{i=1}^{n} \epsilon_i \cdot N_i$, where E is a constant total energy. Use Lagrange multipliers to solve this problem, and show that the maximum probability occurs when:

$$N_i = c \cdot e^{-\beta \epsilon_i}. \quad [11.24]$$

where c is a normalization constant.

This defines the famous Maxwell–Boltzmann statistics, which say that the most probable way to allocate energies of randomly moving particles into bins is exponential. By dimensional analysis, β must have units 1/energy, and for molecules, the characteristic energy should be proportional to temperature, T, so we write:

$$\beta = 1/kT, \quad [11.25]$$

where k once more is Boltzmann's constant.

Eq. [11.24] is named after both Maxwell and Boltzmann. The derivation in the preceding exercise follows directly from statistical counting, which is based on Boltzmann's atomistic view. As we have mentioned, historically this view was highly controversial and an alternative, continuum, derivation was proposed by Maxwell, who described the following logical argument.[28]

Let $f(x)$ be the distribution of N particles moving with velocity x, and similarly for orthogonal velocities y and z. Then the number of particles in a velocity range near (x, y, z) will be $N \cdot f(x)f(y)f(z)$. Fine. But Maxwell observed that the directions of x, y, and z are arbitrary, and so $f(x)f(y)f(z)$ can only depend on the magnitude, $x^2 + y^2 + z^2$, or:

$$f(x)f(y)f(z) = \phi(x^2 + y^2 + z^2), \quad [11.26]$$

for some function ϕ. And of course the function that turns products into sums is the exponential:

$$f(x) = C \cdot e^{A \cdot x^2} \textit{ and } \phi(r^2) = C^3 \cdot e^{A \cdot (x^2+y^2+z^2)}, \quad [11.27]$$

where C and A are constants. In his paper,[28] Maxwell notes that A must be negative, else the number of particles would diverge with velocity, and so for clarity he substitutes $-\beta$ for A. He then recalls that the energy, ϵ_i, of a moving molecule goes as its velocity-squared, $x^2 + y^2 + z^2$, to obtain Eq. [11.24].

EXERCISE 11.10

Critique both Boltzmann's and Maxwell's derivations. What hidden assumptions do they involve? What different relations could you obtain by following these same derivations? Do you feel that one or the other is more rigorous? Historically, there was bitter disagreement about the atomistic versus the continuum view, and as with many such scientific debates, every potential logical flaw was seen as evidence for the superiority of the opposing view. Does anything about either derivation support one view over the other? Every model uses simplifications of some sort: when should inevitable logical flaws be used to discredit scientific work?

Einstein wrote[29] the following of Ostwald and Mach's resistance to atomic theory:

> "*. . . even scholars of audacious spirit and fine instinct can be obstructed in the interpretation of facts by philosophical prejudices.*"

Can you identify current controversies in which philosophical prejudices or overemphasis on technical discrepancies are used to discredit a scientific result?

11.9 **Applications of Maxwell-Boltzmann Statistics**

Maxwell–Boltzmann statistics, Eq. [11.24], provide insights into a surprising variety of applications, some of which we summarize here.

11.9.1 **Energy Distribution**

It is important to clarify that Eq. [11.24] defines how many particles, N_i, will have a given energy ϵ_i. Eq. [11.24] does not define the distribution of energy itself: particles with lower energy contribute less to the total energy in a gas than those with higher energy, so the energy distribution obeys:

$$p(\epsilon_i) = \epsilon_i \cdot N_i = C_0 \cdot \epsilon_i \cdot e^{-\beta\epsilon_i}, \tag{11.28}$$

meaning that $p(\epsilon_i)$ is the probability of obtaining an energy near ϵ_i. Semantically, Eq. [11.28] is referred to as the Maxwell–Boltzmann energy distribution, as distinct from Maxwell–Boltzmann statistics, Eq. [11.24].

11.9.2 **Reaction Rates**

Eq. [11.28] is useful because in order for atoms to react with one another, they must collide with energy above an activation threshold, ϵ_{act}. We can write the fraction, f, of particles whose kinetic energy exceeds ϵ_{act} as:

$$f = \frac{\int_{\epsilon_{act}}^{\infty} C_0 \cdot \epsilon_i \cdot e^{-\beta\epsilon_i} d\epsilon_i}{\int_0^{\infty} C_0 \cdot \epsilon_i \cdot e^{-\beta\epsilon_i} d\epsilon_i} = e^{-\beta\epsilon_{act}}(1 + \beta\epsilon_{act}). \tag{11.29}$$

This tells us that for large activation energies, $\epsilon_{act} \gg kT$, reactions do not occur (because $e^{-\beta\epsilon_{act}}(1 + \beta\epsilon_{act}) \ll 1$), and for small activation energies, reactions occur with rates that go as

$$Reaction\ Rate = C_1 e^{-\beta\epsilon_{act}} \quad [11.30]$$

(because $\beta\epsilon_{act} \ll 1$). This is the well-known Arrhenius equation, named after Swedish chemist Svante Arrhenius (1859–1927), recipient of the 1903 Nobel Prize and, incidentally, the first to predict that CO_2 can act as a greenhouse gas to heat the planet.

11.9.3 **Equilibrium Sedimentation**

The Maxwell–Boltzmann distribution, Eq. [11.24], also defines how sedimentation of particles, cells, and molecules occurs. This field was pioneered by J.B. Perrin (1870–1942), who won the Nobel Prize in Physics for measurements of Brownian motion using the then novel "ultramicroscope," and for establishing the science of "equilibrium sedimentation:" the distribution of heights of varying density suspensions. Perrin's work dealt with sedimentation of colloidal suspensions under gravity (recall Fig. 6.5), but also has application to modern ultracentrifugation, used to separate proteins based on mass.

At its simplest, equilibrium sedimentation defines the process by which a particle's kinetic energy reaches equilibrium with its potential energy. We know this process by the adage that hot air rises: hot, low density, molecules adopt an equilibrium position higher—that is, with higher gravitational energy—than colder, more dense, molecules. By assuming that particles equilibrate so that their kinetic and potential energies are equal, Eq. [11.24] can be written as the "barometric formula," expressing the pressure as a function of height:

$$P = P_0 e^{\frac{-mgh}{kT}}. \quad [11.31]$$

Eq. [11.31] is often defined instead in terms of the molar mass, M, in which case the dependence becomes $e^{\frac{-Mgh}{RT}}$, where R is the universal gas constant. From this equation, we can surmise that the pressure, and hence the concentration, of molecules with larger mass M, decays more rapidly with height than for molecules with smaller M. Or put another way, there must be more heavy molecules lower down, and more light molecules higher up.

This is shown in Fig. 11.12(a), where we display the atmosphere from space, showing the highly concentrated troposphere, which appears as brown due to backlighting from the setting sun. In Fig. 11.12(b), we plot standard atmospheric data for the pressure in the lower atmosphere alongside a fit to Eq. [11.31].

It is perhaps worth remarking that this exponential pressure change cannot be not due to hydrostatic pressure caused by the weight of supported fluid. The weight of a column of liquid grows linearly with the vertical distance—and so pressure in the ocean increases linearly with depth to a very good approximation, whereas the exponential fit shown in Fig 11.12(b) is clear evidence that atmospheric pressure is associated with something more.

In principle, the barometric formula is an example of an idealized sort of sedimentation, insofar as gas molecules reach equilibrium between kinetic and potential energies, but most would not term this sedimentation. A different example is shown in Fig. 11.12(c), where we show clouds—a mist of fine water particles—beneath soaring skyscrapers. Evidently water,

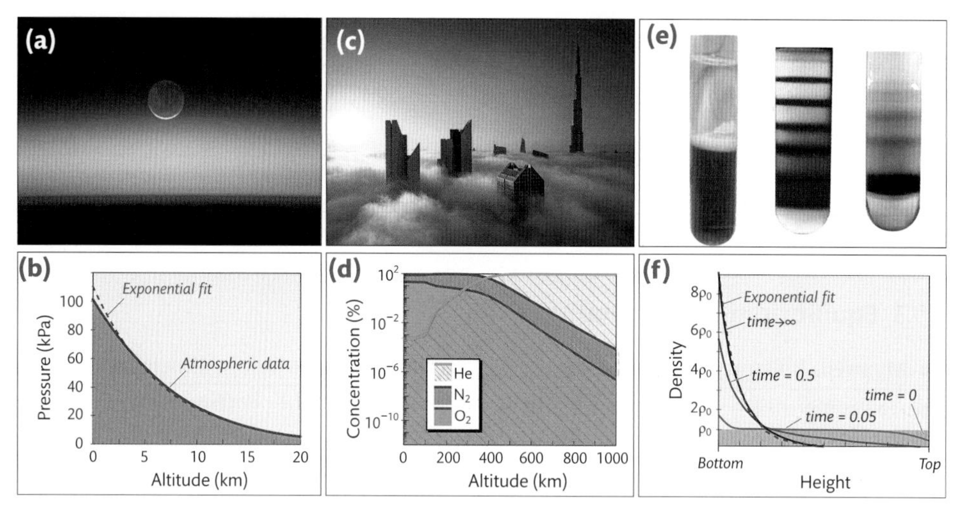

Fig. 11.12 Examples of equilibrium sedimentation. (a) Earth's atmosphere viewed from space; brown layer is the troposphere, containing three quarters of atmospheric mass (NASA/JSC Gateway to Astronaut Photography of Earth).[30] (b) Standard atmospheric pressure (blue) compared with fit to barometric formula from Eq. [11.31] (red broken curve).[31] (c) Density stratification of clouds in Dubai. (Marcelo de Castro Photography (http://www.marcelocastrophotography.com) and [32].) (d) Molecular stratification in Earth's atmosphere.[33] (e) Ultracentrifugation of (left) whole blood (from Amrein et al.,[34], by permission of Wolters Kluwer Health, Inc.); (center) parasite-infected red blood cells separated using density gradient (from Rüssmann et al.,[35], by permission of Springer Berlin Heidelberg); (right) single walled carbon nanotubes separated using density gradient. (from Ghosh et al.,[36], by permission of Springer Nature). (f) Evolution of density using Mason–Weaver equation in tall container: ρ_0 is initial suspension density; (data from Mason and Weaver,[37]) Exponential fit is shown as broken red curve.

when its temperature makes it heavier than air, can sediment in air. The same is so of different mass gas molecules, as Fig. 11.12(d) shows. In that figure, we plot the fractional concentrations of three components of air: at increasing altitude the concentration of the heaviest molecule (O_2, molecular mass 32) drops faster than the next heaviest (N_2, mass 28), and one of the lightest molecules (He, mass 4) grows in concentration with height. So as predicted by Eq. [11.31], heavier molecules sink and lighter ones rise.

Example Tragic consequences

Consequences of the barometric formula can be tragic. In 1986, carbon dioxide (CO_2) from ancient decaying vegetation bubbled up from Lake Nyos in Cameroon. The CO_2 eventually dissipated, but was trapped for many hours by the surrounding valley due to the fact that the mass (m in Eq. [11.31]) of CO_2 is nearly three times that of oxygen. As a consequence, nearly 1800 people and thousands of cattle were suffocated. CO_2 has also been released by Mammoth Lake in California, Lake Kivu in Congo, and elsewhere.

The same issue arises in modern laboratories that vent heavy gases in fume hoods—for example, enclosed courtyards beneath fume chimneys have been sealed at a university to prevent students from being exposed to trapped mercury and chlorine. Similarly, an extensive underground storage tank was constructed in Fassberg, Germany to capture releases of HF6, a heavy gas used to study turbulent energy cascades.

The examples of equilibrium sedimentation seen so far have fit well into the Maxwell–Boltzmann scheme, which after all was intended to model gas kinetics. More definitive examples showing that Maxwell-Boltzmann statistics govern sedimentation in its more literal sense can be seen by considering how particles, cells, or large molecules sink, as shown in Fig. 11.12(e). In that panel, we show centrifugation of whole blood (from Chapter 6), centrifugation of blood in a so-called density gradient, and centrifugation of carbon nanotubes, in a different type of density gradient. Density gradient centrifugation consists of preparing a substrate fluid that contains, as the name implies, a gradient of densities—for example, by suspending different sized coated silica particles in layers (center of Fig. 11.12(e)) or a suspension of sodium cholate with a strongly nonlinear density profile (right of Fig. 11.12(e)). In the latter case, the separated bands shown are formed from nanotubes of nearly identical densities produced only by nanotubes differing slightly in diameter.

More complex equations are available to describe sedimentation in density gradients as well as the companion case of sedimentation of multiple different materials, but in the simple case of a single sediment in a uniform density fluid, the change in time of the concentration c of sedimenting material can be well fit by the Mason–Weaver equation:

$$\frac{\partial c}{\partial t} = D\frac{\partial^2 c}{\partial z^2} + s \cdot g \cdot \frac{\partial c}{\partial z}, \qquad [11.32]$$

where t is time, z is vertical distance, D is the diffusivity (discussed more next chapter), s is an inverse drag coefficient, and g the applied (gravitational or centrifugal) acceleration. This equation says that evolution of the concentration is governed by a competition between diffusion, which tends to spread particles out, and gravity, which tends to draw denser particles downward subject to resistance by drag forces. This competition, again, was what Einstein introduced in his doctoral dissertation,[38] which he famously used to calculate the diffusivity, D. In Fig. 11.12(f), we plot the evolution of an initially uniform concentration of particulates subject to Eq. [11.32]. Overlaid as a red broken line, we show that the particulates asymptotically adopt a nearly exponential concentration profile, as expected from Maxwell–Boltzmann statistics.

EXERCISE 11.11

Consider Fig. 11.12(d) more carefully. Just using Eq. [11.31], show that this plot can qualitatively be reproduced. To do this, neglect other gases and make use of accepted values for molecular masses and total atmospheric concentrations of N_2, O_2, and He. Also, assume that the sum of pressures of N_2, O_2, and He is constant at all altitudes, and that concentration is proportional to pressure. You will find that the helium growth with altitude is easy to reproduce qualitatively, but that something differs between measurements of N_2 and O_2 concentrations in Fig. 11.12(d) and predictions from Eq. [11.31]. Why is this?

EXERCISE 11.12

In Chapter 7 (Exercise 7.1) we discussed swimming of bacteria, and in this chapter (Advanced exercise 11.6) we considered diffusion of both bacteria and particles. In this exercise, we'll compare both swimming and diffusion at bacterial length scales. Presumably bacteria swim to encounter particles of food. Let's see if that is so.

First, estimate the relative sizes of bacteria and the food they eat, and use that estimate to calculate the time scales over which bacteria and their food diffuse. Does diffusion of a bacterium affect its access to food? Second, estimate the time scale that a bacterium can swim its body length to find food and compare that with the time scale of diffusion of food over the same distance. To do this, calculate the ratio of the time for transport to occur on a bacterial length scale due to swimming with that time due to diffusion. Find the name of that dimensionless ratio. Does swimming affect a bacterium's access to food, or should it just sit still and let food diffuse to it? Finally, based on the answer to this question, explain why a bacterium swims.

REFERENCES

1. For a comprehensive overview of Boltzmann and related history, see Brush, S.G. (1976) *On the Kind of Motion We Call Heat*, Vol. 1. Amsterdam: North Holland
2. Ott, E. (1992) *Chaos in Dynamical Systems*, 2nd ed. Cambridge: Cambridge University Press.
3. Forsyth, P.A., Marelia S. and Mitchell, D.J. (1978) Ordering in colloidal systems. *Advances in Colloid & Interface Science, 9*, 37–60.
4. Adams, M., Dogic, Z., Keller, S.L. and Fraden, S. (1998) Entropically driven microphase transitions in mixtures of colloidal rods and spheres. *Nature, 393*, 349–52.
5. Van Roij, R., Mulder B. and Dijkstra, M. (1998) Phase behavior of binary mixtures of thick and thin hard rods. *Physica A, 261*, 374–90.
6. Wikipedia. (2018) Opal. Available at: https://en.wikipedia.org/wiki/Opal#/media/File:Coober_Pedy_Opal_Doublet.jpg
7. University of Cambridge. (2014) Self-assembled binary opal. Available at http://www.nanodtc.cam.ac.uk/News%20and%20Events/Phot%20Competition/08-self-assembled-binary-opal/view
8. The Rowland Institute at Harvard. (n.d.) *fd* Virus—a model system of rod-like colloids. Available at http://www.rowland.harvard.edu/rjf/dogic/fdvirus.php
9. Shinbrot, T. and Muzzio, F.J. (2001) Noise to order. *Nature, 410*, 251–8.
10. Shinbrot, T. (1997) Competition between randomizing impacts and inelastic collisions in granular pattern formation. *Nature, 389*, 574–6.
11. Shevchenko, E.V., Talapin, D.V., Murray, C.B. and O'Brien, S. (2006) Structural characterization of self-assembled multifunctional binary nanoparticle superlattices. *Journal of the American Chemical Society, 128*, 3820–37.

12. Hadjichristidis, N., Pispas, S. and Floudas, G. (2003) Block copolymers: synthetic strategies, physical properties, and applications. In: *Block Copolymers: Synthetic Strategies, Physical Properties, and Applications*. New York: John Wiley & Sons, 383–408.

13. Stebe, K.J., Lewandowski, E. and Ghosh, M. (2009) Oriented assembly of metamaterials. *Science, 325*, 159–60.

14. Cabane, B.; Li, J., Artzner, F., Botet, R., Labbez, C., Bareigts, G., Sztucki, M. and Goehring, L. (2016) Hiding in plain view: colloidal self-assembly from polydisperse populations. *Physical Review Letters, 116*, 208001.

15. Shaleev, V.M. (2007) Optical negative-index metamaterials. *Nature Photonics, 1*, 41–8; see also *Physical Review X*, 3, (2013), a special section on metamaterials.

16. Dupré, M., Lemoult, F., Fink, M. and Lerosey, G. (2016) Exploiting spatiotemporal degrees of freedom for far-field subwavelength focusing using time reversal in fractals. *Physical Review B, 93*, 180201.

17. Moreno, J. and Peinado, R. (2012) *Enological Chemistry*. San Diego, CA: Elsevier, 307–9.

18. Benesch, R.E., Edalji, R., Benesch, R. and Kwong, S. (1980) Solubilization of hemoglobin S by other hemoglobins. *Proceedings of the National Academy of Sciences of the USA, 77*, 5130–4.

19. Ellis, R.J. (2001) Macromolecular crowding: obvious but underappreciated. *Trends in Biochemical Science, 26*, 597–604.

20. Caicedo-Carvajal, C.J. and Shinbrot, T. (2008) In silico zebrafish pattern formation. *Developmental Biology, 315*, 397–403.

21. Yodh, A.G., Lin, K., Crocker, J.C., Dinsmore, A.D., Verma, R. and Kaplan, P.D. (2001) Entropically driven self-assembly of colloids in suspension. *Philosophical Transactions of the Royal Society A, 359*, 921–37.

22. Blum, J., Bruns, S., Rademacher, D., Voss, A., Willenberg, B. and Krause, M. Measurement of the translational and rotational Brownian motion of individual particles in a rarefied gas. *Physical Review Letters, 97*, 230601.

23. Bénichou, O., Coppey, M., Moreau, M., Suet, P.-H. and Voituriez, R. (2005) Optimal search strategies for hidden targets. *Physical Review Letters, 94*, 198101.

24. Ball, P. (1999) *The Self-Made Tapestry: Pattern Formation in Nature*. New York: Oxford University Press.

25. Press, W.H., Flannery, B.P., Teukolsky, S.A. and Vetterling, W.T. (1986) *Numerical Recipes: The Art of Scientific Computing*. Cambridge: Cambridge University Press.

26. Loeb, J. (1899) On the nature of the process of fertilization and the artificial production of normal larvae (Plutei) from the unfertilized eggs of the sea urchin. *American Journal of Physiology, 3*, 135–8.

27. Donnan, F.G. and Harris, A.B. (1911) The osmotic pressure and conductivity of aqueous solutions of Congo-red, and reversible membrane equilibria. *Journal of the Chemical Society, Transactions, 99*, 1554–77.

28. Maxwell, J.C. (1860) Illustrations of the dynamical theory of gases. *The London, Edinburgh, and Dublin Philosophical Magazine and Journal of Science, 19*, 19–32.

29. Schilpp, P.A. (Ed.) (2000) *Albert Einstein, Philosopher-Scientist: The Library of Living Philosophers, Volume VII*. Peru, IL: Open Court Publishing, p.49.

30. NASA. (2017) Earth's atmospheric layers. Available at http://www.nasa.gov/content/earths-atmospheric-layers/

31. Public Domain Aeronautical Software (PDAS). (2017) A sample atmosphere table (SI units). Available at http://www.pdas.com/atmosTable2SI.html

32. Nicholson, H. (2014) What a heavenly view! Stunning photographs appear to show Dubai's famous skyscrapers 'floating' in the fog above the city. Available at http://www.dailymail.co.uk/travel/travel_news/article-2755430/What-heavenly-view-Stunning-photographs-appear-Dubai-s-famous-skyscrapers-floating-fog-city.html

33. Virtual Ionosphere, Thermosphere, Mesosphere Observatory (VITMO). (no date) MSIS-E-90 atmosphere model. Available at http://omniweb.gsfc.nasa.gov/vitmo/msis_vitmo.html

34. Amrein, P. C., Kumar, J. R., Poulin, R. F., Umlas, J. and Weitzman, S. A. (1982) Comparison of filtration leukapheresis and centrifugation leukapheresis in treatment of lymphosarcoma cell leukemia. *Southern Medical Journal, 75* (8), 969–971.

35. Rüssmann, L., Jung, A., Heidrich, H.G. (1982). The use of percoll gradients, elutriator rotor elution, and mithramycin staining for the isolation and identification of intraerythrocytic stages of *Plasmodium berghei. Zeitschrift für Parasitenkunde. 66* (3): 273–280.

36. Ghosh, S. Bachilo, S.M. and Weisman, R.B. (2010) Advanced sorting of single-walled carbon nanotubes by nonlinear density-gradient ultracentrifugation. *Nature Nanotechnology, 5*, 443–50.

37. Mason, M. and Weaver, W. (1924) The settling of small particles in a fluid. *Physical Review, 23*, 412–26.

38. Stachel, J. (Ed.) (1989) A new determination of molecular dimensions. In *The Collected Papers of Albert Einstien. Volume 2*. Princeton, NJ: Princeton University Press, p. 63.

12 Diffusion

Last chapter we showed that by analyzing probabilities of random arrangements of particles, we can account for peculiar behaviors such as the paradoxical accumulation and self-assembly of cells, colloids and molecules. An essential element in this analysis was the conclusion that osmotic pressure results from a difference in collisional rates of particles across a membrane (Chapter 11, Fig. 11.9(a)). In this chapter, we will build on this idea to derive differential equations for concentrations of particles that govern both simple diffusion (from regions of high concentration to low) and self-assembly (i.e. paradoxical concentration of particles into ordered states). We will see in Chapter 13 that this leads to a simple and elegant description of so-called Turing patterns, used to describe everything from the growth of mutations to spots on a leopard.

12.1 Fick's Laws and the Diffusion Equation

As we have described, the starting point in this chapter is osmotic pressure, which we know depends on the number of collisions on either side of a membrane. We recall that van 't Hoff's formula (Eq. [11.14] in Chapter 11) tells us that the pressure on either side of a membrane is proportional to the concentration of particles, so in Fig. 12.1(a), we sketch a cartoon in which 12 particles are on one side of a membrane, and four are on the other. In this simple case, it should be clear that there will be $12 - 4 = 8$ more collisions per unit time from the left than from the right, and so there should be a net flux J_x, per unit time across the membrane of eight particles in the unit time needed for all particles to hit the membrane.

Of course the membrane drawn in Fig. 12.1(a) isn't doing anything, so more generally, if we define the flux of particles of concentration c to be $J_x \equiv \partial c/\partial t$, and if δc is the difference in concentration across a distance δx, then:

$$J_x = -D\frac{\delta c}{\delta x}, \qquad [12.1]$$

where D is a constant. This is known as Fick's first law, after German physiologist Adolf Fick (1829–1901). Fick took this equation to the next step by evaluating what happens not to a thin membrane-like slice, but to a box as shown in Fig. 12.1(b), if one flux $J_x|_x$ enters the left side of

Biomedical Fluid Dynamics: Flow and Form. Troy Shinbrot.

DOI: 10.1093/oso/9780198812586.001.0001

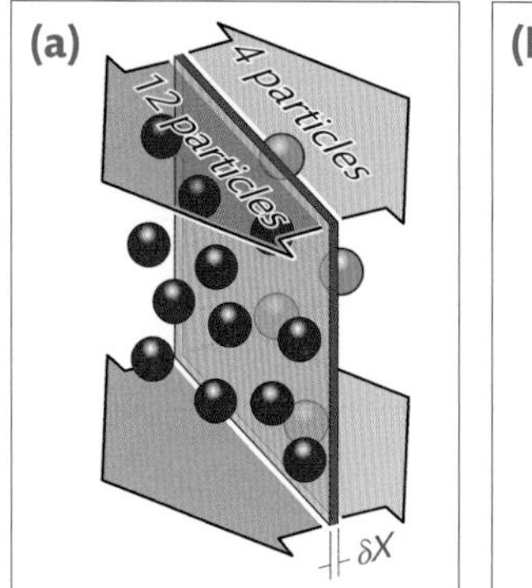

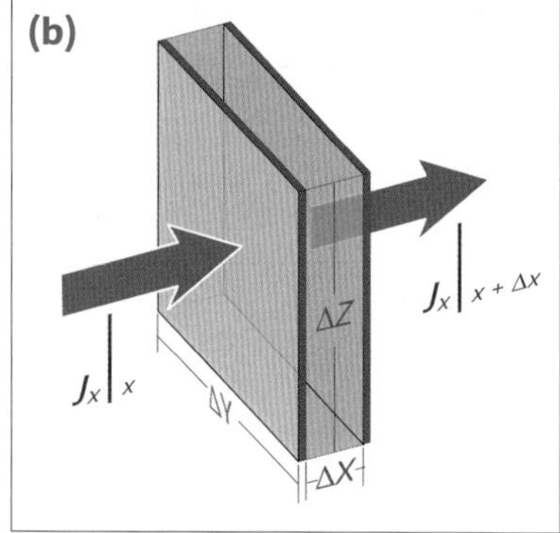

Fig. 12.1 (a) Effect of osmotic pressure: 12 particles (red) to left and four particles (green) to right cause more collisions on left than right of the surface shown. Consequently there is a net flux, J_x, from left to right, defined by Eq. [12.1]. (b) For a unit volume with different fluxes on left, $J_x|_x$, and right, $J_x|_{x+\Delta x}$, surfaces, a net flux, $J_x|_x - J_x|_{x+\Delta x}$, is produced, leading to a change in concentration within the volume given by Eq. [12.7].

the box and another flux, $J_x|_{x+\Delta x}$ exits the right. This is a problem that we solved in Chapter 1 (Fig. 1.3), where we showed that conserving mass in a unit volume requires:

$$\left\{\begin{matrix} rate\ of \\ increase \\ in\ mass \end{matrix}\right\} = \left\{\begin{matrix} rate\ of \\ mass\ in \end{matrix}\right\} - \left\{\begin{matrix} rate\ of \\ mass\ out \end{matrix}\right\}. \qquad [12.2]$$

Expressing mass in terms of concentration, c (mass per unit volume), Eq. [12.2] in the x-direction becomes:

$$\left\{volume \cdot \frac{\partial c}{\partial t}\right\} = \left\{area_x \cdot J_x|_x\right\} - \left\{area_x \cdot J_x|_{x+\Delta x}\right\}, \qquad [12.3]$$

where as in Chapter 1, $area_x = \Delta y \cdot \Delta z$ is the area perpendicular to the x-unit vector. So dividing by $volume = \Delta x \cdot \Delta y \cdot \Delta z$, we get:

$$\frac{\partial c}{\partial t} = \frac{J_x|_x - J_x|_{x+\Delta x}}{\Delta x}. \qquad [12.4]$$

Eq. [12.1] gives us the fluxes, so:

$$\frac{\partial c}{\partial t} = -D\frac{\frac{\delta c}{\delta x}\big|_x - \frac{\delta c}{\delta x}\big|_{x+\Delta x}}{\Delta x}, \qquad [12.5]$$

and as δx and Δx go to zero, this becomes:

$$\frac{\partial c}{\partial t} = D\frac{d^2 c}{dx^2}. \qquad [12.6]$$

As in Chapter 1, the minus sign disappears because of the convention of defining derivatives using differences evaluated from right to left. More generally, Eq. [12.6] can be written:

$$\frac{\partial c}{\partial t} = D \cdot \nabla^2 c. \qquad [12.7]$$

This is known as Fick's second law, or more commonly the diffusion equation (or the heat equation, if c is the temperature), and D is known as the diffusivity.

But wait, we already defined the diffusivity in Eq. [11.17] of Chapter 11, where we analyzed the random wandering of drunkards, or of drunken molecules. There we found that diffusivity of molecules can be expressed as the area that a cloud of molecules occupies per unit time: $D = \ell^2/t$. If we perform dimensional analysis on Eq. [12.7], we find that the relevant constant quantity, D, that governs concentration-mediated flow obeys the same relationship. In this case the result is obtained by conserving the quantity of a substance traveling into and out of an infinitesimal area, $area_x = \Delta y \cdot \Delta z$, per unit time, ∂t.

So apparently keeping track of trajectories of colliding molecules or accounting for conservation of concentrations of these molecules produces the same result. The advantage of Eq. [12.7] is that it provides an analytic formula for solving for the time evolution of a spatial allotment of concentration—for example, we will show that Eq. [12.7] can be used to evaluate the evolution of a bolus of drug injected into an artery, or the dissolution of a capsule of drug in the gut.

Let's begin by understanding qualitatively what Eq. [12.7] is telling us. Evidently the concentration in a unit volume changes with time in proportion to its *second* derivative in space. If the second derivative is zero, this means that the *first* derivative doesn't change from the left to the right sides of the volume. If the second derivative is positive, on the other hand, then more substance must be injected from the left than is removed from the right—and so the concentration, c, within the volume must increase. But we recall from calculus that a positive second derivative occurs when the curve $c(x)$ is concave up—and contrariwise a negative second derivative (causing c to decrease) occurs when $c(x)$ is concave down. This is shown in Fig. 12.2, where we plot in Fig 12.2(a) the behavior expected for a $c(x)$ curve with both concave down and concave up regions, and in Figs. 12.2(b)–(d) the consequent change in concentration of a blob of dye in a container of water.

From either of these views, Eq. [12.7] does evidently describe the qualitative behavior of diffusion that we are familiar with. In Fig. 12.2(a) we see that the decrease in c in a concave down region (pink, to the left), and the increase in a concave up region (blue, to the right) that are both required by Eq. [12.7] is the same as saying that material moves from higher concentration (left) to lower (right). Similarly in Figs. 12.2(b)–(d), we see that the decrease in concentration in a central concave-down region (lower plots) amounts to the spreading of material (upper plots) that we are all familiar with from daily experience.

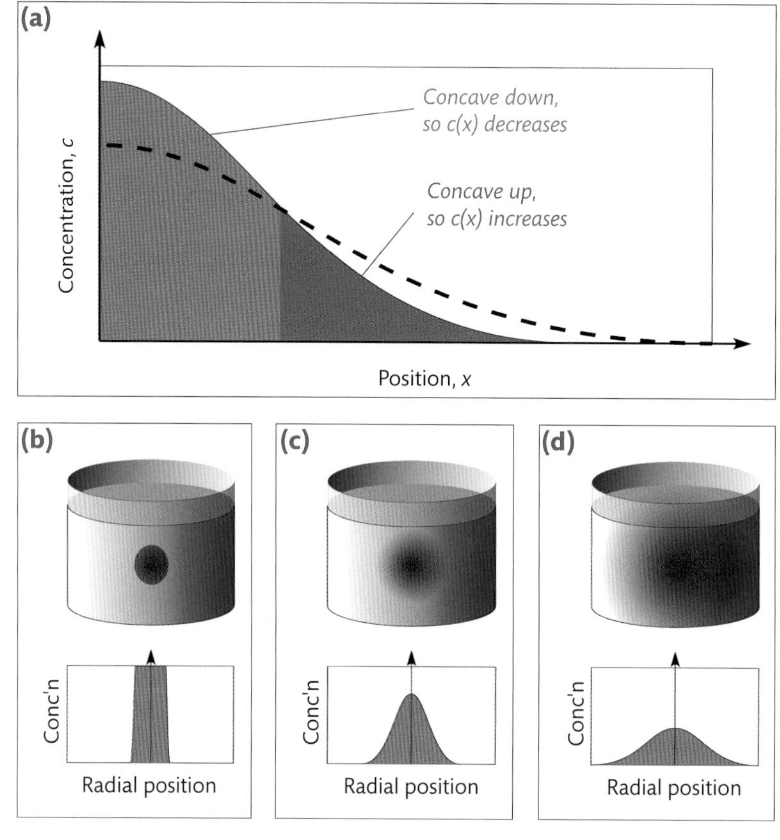

Fig. 12.2 Qualitative behavior of diffusion equation [12.7]. (a) Wherever the concentration, $c(x)$ is concave down, more material must leave a unit volume than enters, so $c(x)$ must decrease—and vice versa wherever $c(x)$ is concave up. This causes the red concentration curve to diminish and the blue to grow, toward the broken curve shown. (b)–(d) Consequently if a blob of dye is injected into a container of water, it must spread out as material leaves the high concentration, concave-down, regions toward lower concentration, concave-up, ones.

12.1.1 **Simple Steady Solution in One Dimension**

Let's turn to quantitative analysis by looking at some solutions to the diffusion equation [12.7], starting from a simple, one-dimensional (1D) problem. In one dimension, we already know that diffusion is governed by Eq. [12.6]. But we saw this equation before, in Chapter 9, where we solved for viscous flow between moving plates (Eqs. [9.56] and [9.57]). In that problem, the viscosity took the place of the diffusivity, but mathematically Eq. [12.6] is identical to Chapter 9's Eq. [9.56]. This gives further physical insight into diffusion: viscosity we recall was introduced in Chapter 1 as resulting from collisions between moving layers of particles (recall Chapter 1, Fig. 1.9). Viscosity is a measure of fluid collisions, which transform shear strain into heat, and without collisions, there would be no viscosity. This is the reason for the similarity between Newton's law of viscosity,

$$\frac{\partial \vec{V}}{\partial t} = \nu \nabla^2 \vec{V}, \qquad [12.8]$$

and Fick's second law, Eq. [12.7]. The context, of course, is different, and so Newton's law pertains to vector velocities rather than scalar concentrations, but the similarity between Eqs. [12.7] and [12.8] is inescapable.

Nevertheless, the context can matter, as Exercise 12.1 will illustrate. Let's use what we know from solving fluid flow equations to solve the diffusion equation in one dimension. As we did for fluid flow problems, we'll first solve for the steady case and then examine time-dependent solutions.

EXERCISE 12.1

We start with a steady 1D problem in which one end of a slab, shown in Fig. 12.3(a), is held at $c = c_0$, and the other end, a distance L away, is held at $c = c_L$. Solve the steady equation:

$$0 = D \cdot \frac{\partial^2 c}{\partial x^2}. \qquad [12.9]$$

for this problem, and determine $c(x)$ between *0* and L. Plot $c(x)$ from $x = -L$ to $x = 2L$. Then solve the same problem, where the region between 0 and L has two different diffusivities, as sketched in Fig. 12.3(b), where $D = D_1$ between $x = 0$ and $x = a$, and $D = D_2$ between a and L.

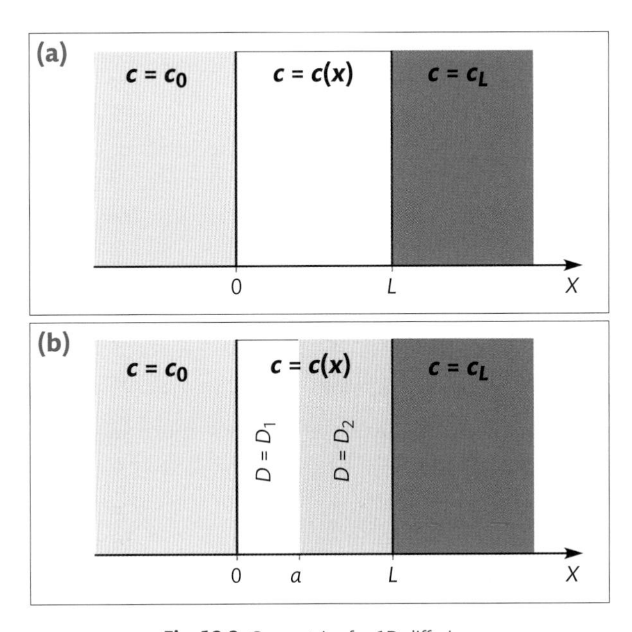

Fig. 12.3 Geometries for 1D diffusion.

This is a simple matching exercise as was done in Exercise 7.6 of Chapter 7, but you'll notice that the solution has a discontinuous derivative at $x = 0$ and $x = L$. Why is this? How does this compare with the viscosity problem, Eq. [12.8]? Does your solution have a discontinuous derivative at $x = a$? Why or why not?

12.1.2 **Self-Similar Solution**

Let's expand our scope and consider time-dependence—that is, how diffusion arrives at a steady solution in time. To do so, it is helpful to nondimensionalize the diffusion equation (Eq. [12.7]) by substituting $x = \chi/L$ and $t = \tau/(L^2D)$. Doing this, we obtain

$$\frac{\partial c}{\partial \tau} = \frac{d^2c}{d\chi^2}. \qquad [12.10]$$

This may bring to mind Landau's trick for self-similar entrance flow in a pipe (Equation [5.6] of Chapter 5). As in the earlier problem, Eq. [12.10] lacks any scale information, which tells us two things. First, that the solution applies to problems lacking a scale—as appears for example in diffusion across a sample with infinite length (see Fig. 12.4). Second, that in such a system—for example, an infinite or semi-infinite domain, $c(\chi, \tau)$ must look identical at all scales. But the same is true if we rewrite Eq. [12.10] in terms of a new variable, $\zeta = \chi^2/\tau$ like so:

$$4 \cdot c''(\zeta) + \left(\frac{2}{\zeta} + 1\right)c'(\zeta) = 0. \qquad [12.11]$$

As mentioned in Chapter 6, the standard method for solving a partial differential equation (PDE) is to reduce it to an ordinary differential equation (ODE), which we have done here by substitution. Although it isn't obvious, it's easily confirmed that Eq. [12.11] has solution:

$$c'(\zeta) = c_0\zeta^{-1/2}e^{-\zeta/4}. \qquad [12.12]$$

Almost done: we now have a solution for $c'(\zeta)$, while what seek is $c(\zeta)$—and recall once we have that, we can obtain what we really want, the concentration as a function of x and t, by substituting $\zeta \to \chi^2/\tau$ and $\chi \to xL, \tau \to tL^2D$. So we just have to integrate Eq. [12.12] so that:

$$c(\zeta) = c_0\int_0^{\zeta} \sigma^{-1/2}e^{-\sigma/4}d\sigma + c_1. \qquad [12.13]$$

At the risk of running out of Greek letters, let's do one more substitution to simplify this expression by setting $\sigma = 4\xi^2$, to get:

$$c(\zeta) = c_0\int_0^{\sqrt{\zeta}/2} e^{-\xi^2}d\xi + c_1, \qquad [12.14]$$

so that finally:

$$c(x,t) = c_0\int_0^{x/\sqrt{4Dt}} e^{-\xi^2}d\xi + c_1. \qquad [12.15]$$

EXERCISE 12.2

Derive Eq. [12.11] and confirm that Eq. [12.12] is a solution for constant c_0. Then substitute to obtain Eq. [12.14] and then Eq. [12.15].

Eq. [12.15] has no solution in terms of pre-existing functions, so as was the case when we defined the Bessel function in Chapter 4, we define the solution to be a new function, which we call the error function:

$$erf(X) = \frac{2}{\sqrt{\pi}} \int_0^X e^{-\xi^2} d\xi. \qquad [12.16]$$

As an aside, the reader may recognize that this is an integral of the Gaussian.[1] We'll return to this observation, but for now we remark that this is called the error function because it provides a measure of the "error," meaning the cumulative fraction of points between 0 and X in a normally distributed data set. The $2/\sqrt{\pi}$ term is merely a convention used to normalize the Gaussian—that is, to make the integral of the Gaussian from $-\infty$ to ∞ equal to one:

$$1 = \frac{1}{\sqrt{\pi}} \int_{-\infty}^{\infty} e^{-\xi^2} d\xi = \frac{2}{\sqrt{\pi}} \int_0^{\infty} e^{-\xi^2} d\xi. \qquad [12.17]$$

We'll make use of the normalization relation [12.17] shortly. Notice that our simplified Eq. [12.11] is second order, so there must be two solutions: the error function and an accompanying second function. That function is the "complementary error function," a measure of the fraction of points in a Gaussian data set that are <u>not</u> in *erf(X)*:

$$erfc(X) = 1 - erf(X). \qquad [12.18]$$

12.1.3 **Unsteady Solutions in One Dimension**

Eq. [12.15] allows us to easily evaluate how diffusion advances over time. As an example, suppose at time $t = 0$ we bring two long blocks of the same material together at $x = 0$, one with concentration $c = c_{left}$ to the left of the origin and a second with $c = c_{right}$ to the right. We can apply these initial conditions to find the solution for the initial step in concentration between the two blocks:

$$c(x,t) = \left(\frac{c_{right} - c_{left}}{2}\right) \cdot erf\left(\frac{x}{\sqrt{4Dt}}\right) + \frac{c_{left} + c_{right}}{2}. \qquad [12.19]$$

EXERCISE 12.3

Derive Eq. [12.19]; you'll need the normalization relation (Eq. [12.17]) to complete this. Then plot the result at several successive times. You should obtain plots like the ones in Fig. 12.4, where the thick black line shows the initial jump in concentration at time $t = 0$, and at later times the abrupt jump becomes smoother—as it must according to the reasoning shown qualitatively in Fig. 12.2.

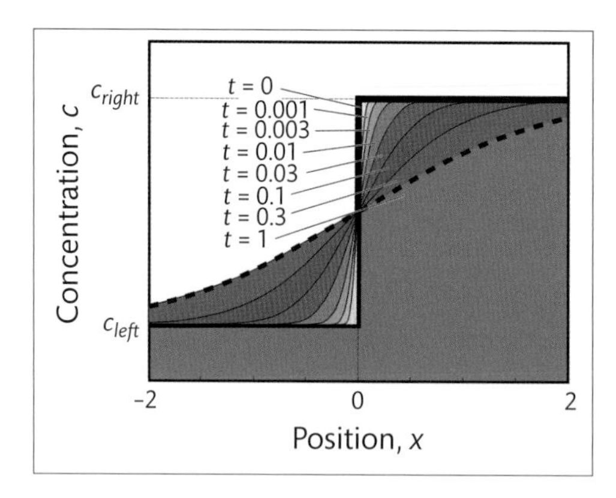

Fig. 12.4 Solution to diffusion equation at successive times, starting from an initial sharp step at $t = 0$ (thick black line), and evolving into a smooth slope at $t = 1$ (thick broken line). Time is expressed here in units of $(C_{right} - C_{left})^2/D$, using $(C_{right} - C_{left}) = 1$ and $D = 1$. Notice the times shown are at intervals that are exponentially spaced, in recognition of the fact that diffusion is initially rapid, and then slows.

Note that the diffusion equation is linear, and recall for a linear equation that if $f_1(x)$ and $f_2(x)$ are both solutions, then $f_1(x) + f_2(x)$ will also be a solution. This allows solutions to be superimposed on one another to compose solutions to more complicated problems.

As a simple example, we can superimpose two step functions, one facing left as in Exercise 12.3 and a second facing right, to produce a localized pulse. This is shown schematically in Fig. 12.5(a), and the sum of the solutions to these two functions obtained from Eq. [12.19] is shown in Fig. 12.5(b). As with Exercise 12.3, you can see that a sharp-edged initial pulse rapidly becomes smooth and "diffuse"—as shown qualitatively in Figs. 12.2(b)–(d), and as you would expect from a "diffusion" equation.

EXERCISE 12.4

For a followup exercise, use Eq. [12.19] to reproduce Fig. 12.5(b), choosing concentrations so that the amplitude of the initial pulse goes from 0 to 1 with a width of your choosing. Then use Eq. [12.19] to show why the solution evolves into a Gaussian for long times.

The self-similar solutions described so far have "free boundaries," so that the value of concentration at the endpoints is unconstrained. This is appropriate for a solution on infinite, or semi-infinite, domains, but it doesn't mean that there are no conditions on the solution: to solve any differential equation, we need conditions to evaluate constants. In the self-similar solutions shown, we defined the initial state everywhere (i.e. we chose an initial step or pulse with known positions and magnitudes). Thereafter, the value of the concentration at the

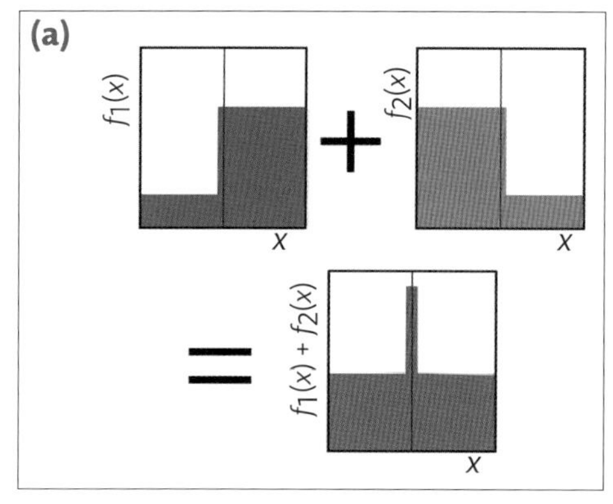

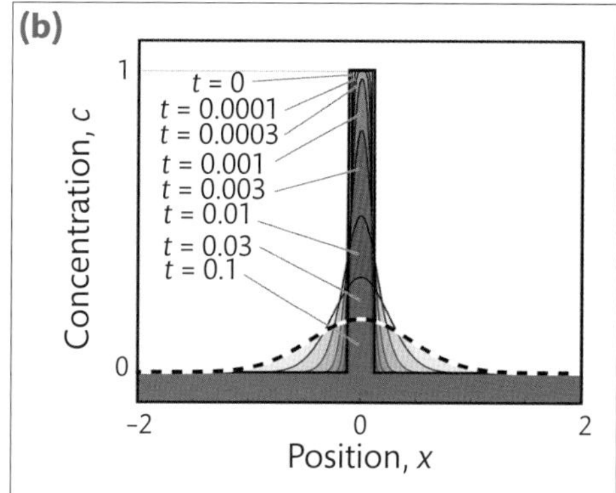

Fig. 12.5 (a) Adding a left-facing step (blue) to a right-facing step (red) produces a localized pulse (violet). Because of the principle of superposition for linear equations such as Eq. [12.7], the sum of solutions to the two step initial conditions must be the same as the solution to the pulse initial condition. (b) Consequently, we can easily compute how an initial pulse will diffuse. Time is expressed here in units of $Height^2/D$, using $Height = 1$ and $D = 1$. Notice that after a very short time, for example, $t = 0.0001$, the sharpest edges in an initial state have smoothed, whereas it takes much longer for the perimeter of the pulse to reach very far away.

boundaries was defined by the constraint that mass is conserved—Fig. 12.1 shows from the start of the derivation of the diffusion equation that the number of particles, or molecules, isn't changed as the particles diffuse. So the concentration integrated from $-\infty$ to $+\infty$ is constant in the problems discussed so far, and to achieve this, if the maximum sinks on any domain, as in Fig. 12.4, the boundary points must rise.

A common alternative is to use fixed boundaries—as occurs, for example, when boundaries are in contact with an infinite, rapidly diffusing, bath. So when we swallow a pill, our bodies provide effectively infinite sinks for the chemicals in the pill, and our circulatory systems typically spread the chemicals much more rapidly than diffusion within the pill. In this case, we treat the boundary of the pill as being in contact with a boundary at constant concentration, typically nearly zero. In principle, mass is still conserved, but in practice from the point of view of the pill, its boundary is effectively always at zero concentration, so the mass of chemical within the pill decreases to zero over time.

In fixed boundary problems, however, we cannot use a self-similar solution that lacks scale information—because the presence of boundaries at fixed distances provides a scale. We'll turn next to fixed boundary problems, but before we do so let's use the self-similar solution to provide a baby step toward more general problems. We do so by considering a 1D sample of material as before, but now let's imagine one end is immersed in a constant concentration bath. Eqs. [12.15]–[12.17] tell us that:

$$c(x,t) = \frac{c_0\sqrt{\pi}}{2}\operatorname{erf}(x/\sqrt{4Dt}) + c_1, \qquad [12.20]$$

and if the bath (to the left of the origin) is fixed at concentration c_{bath}, then we know that after a very long time $c(0,\infty) = c_{bath}$. Since $\operatorname{erf}(0) = 0$, c_1 must be c_{bath}. As for the sample (to the right of the origin), if its concentration is initially uniform at the value $c_{initial}$, then

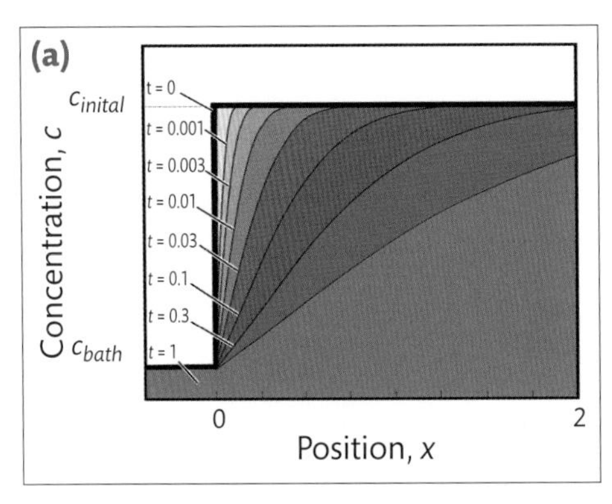

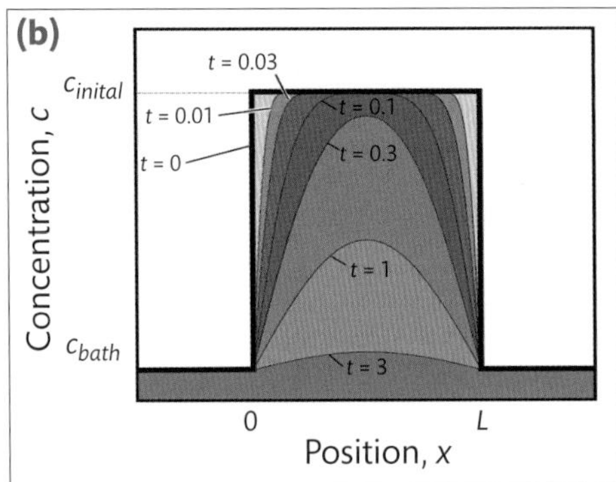

Fig. 12.6 (a) Diffusion from 1D sample with initially uniform concentration, $c_{initial}$, into rapidly diffusing bath with concentration, c_{bath}. Notice at $x = 0$ the sample has a fixed concentration: $c = c_{bath}$. (b) Diffusion from sample with concentrations fixed at $c = c_{bath}$ on two sides. Notice that unlike the case of Fig. 12.5, here the solution cannot be obtained by superposition: a superposition of left- and right-facing solutions shown in (a) would cause the concentration to change with time at the boundaries. Consequently, another method is needed, described in the next section.

$c(x, 0) = c_{initial}$. Since $\text{erf}(\infty) = 1$ (from Eq. [12.17]), Eq. [12.20] implies that $c_0 = 2c_{initial}/\sqrt{\pi}$. Having established the values of c_0 and c_1, it's easy to plot the outcome, shown in Fig. 12.6(a).

EXERCISE 12.5

Reproduce Fig. 12.6(a) as described above. Compare what you obtain with the result of Exercise 12.3. What can you conclude from this comparison? Suppose you wanted to produce a pill that dispensed a drug at a constant rate: by referring to Eq. [12.1], what can you conclude would be needed to generate this constant rate?

12.2 **General Solutions: Fourier Series**

The self-similar solutions that we have shown thus far are correct as far as they go, but as we mentioned, they only apply on infinite domains. Some limited headway can be made using superposition, but anything more complicated needs special treatment. Even the rather tame complication of making boundary concentrations fixed rather than free (as in Fig. 12.6(b)) requires more than superposition. Many texts are dedicated to numerical methods to solve complex problems,[2,3] and it's worthwhile to provide a flavor here for how such methods work. In this section we outline an idea used in one form or another by many of these methods. This idea uses Fourier series, mentioned briefly in Chapter 4.

To understand Fourier series, notice that instead of using the scaling argument following Eq. [12.10], we can use brute force, by separating the diffusion equation [12.6] as in Chapter 7, Eq. [7.43], here using $c(x,t) = X(x) \cdot T(t)$:

$$\frac{T'(t)}{T(t)} = D\frac{X''(x)}{X(x)}, \qquad [12.21]$$

to obtain a solution:

$$X(x) = X_n \sin\left(\frac{n\pi}{L}x\right)$$
$$T(t) = T_0 e^{-t/\tau}. \quad [12.22]$$

Here:

$$\tau = L^2/(n^2\pi^2 D), \quad [12.23]$$

where L is a length scale and n is an integer.

As we commented earlier, the diffusion equation is second order in x, and so there must be a second solution to accompany Eq. [12.22]; the other has a cosine in place of the sine: either will plainly solve [12.21]. We will show the complete solution including cosine terms, but for now analyzing cosine solutions will lead to the same lessons as analyzing sines, so we'll start by just focusing on the sine.

To make this concrete, let's recall Fig. 12.6(b) and imagine that we have a 1D sample of material between 0 and L in a bath of water that fixes the concentration of a chemical of interest at $c_0 = c_L = 0$ on the boundaries. Notice that by using sine solutions in Eq. [12.22], we guarantee that the boundary conditions are met for all time—that is, $c(x,t) \propto \sin(n\pi x/L)e^{-t/\tau} \equiv 0$ at $x = 0$ and L.

Well this is handy: the sine is always fixed at 0 at intervals of $L/n\pi$, so the problem of fixed boundary conditions is solved at a stroke. By including both sine and cosine terms, it can be shown we can match any choice of boundary values—but again we will keep things simple by neglecting cosine solutions for now.

Once more we make use of the fact that the diffusion equation is linear to apply superposition. That is, since a sine with any integer n solves the 1D diffusion equation, any *sum* of sines using different n values (i.e. different wavelengths, $2L/n$) must be a solution as well. So the following must solve the diffusion equation:

$$f(x) = \sum_{n=1}^{\infty} X_n \sin\left(\frac{n\pi}{L}x\right). \quad [12.24]$$

At this point, we remark without proof that there is a magical theorem, named after Fourier, that any sufficiently well behaved function, $f(x)$, can be written as a sum of sines (and cosines) as in Eq. [12.24]. There are technical conditions on the theorem, but for practical purposes the theorem tells us that we can write the solution to the diffusion equation like Eq. [12.24] for most situations: we only need to find the constant coefficients X_n that do the job.

And conveniently, Fourier's theorem comes with a straightforward way of calculating these coefficients: they are given by:

$$X_m = \frac{2}{L}\int_0^L f(x)\sin\left(\frac{m\pi}{L}x\right)dx. \quad [12.25]$$

It's worth taking a moment to understand why Eq. [12.25] provides the answer we need, for it sheds light on how the Fourier solution (Eq. [12.24]) works. Let's consider what we get when we substitute $f(x)$ from Eq. [12.24] into Eq. [12.25]:

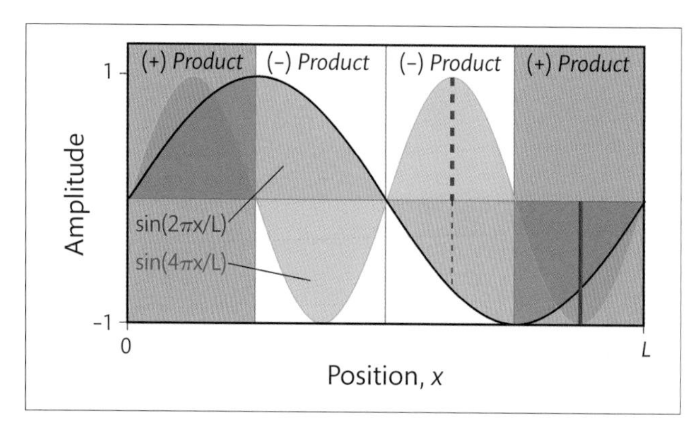

Fig. 12.7 Sin($2\pi x/L$) in magenta superimposed over sin($4\pi x/L$) in cyan. Notice that the product of the two functions is positive ("reinforce") exactly as often as it is negative ("interfere"), and for every point that the product attains a given positive value there is a point where the product reaches the same negative value. For example, corresponding to the red and blue broken lines of opposite sign are two red and blue solid lines of the same sign. Both pairs of lines have the same length, so their products have the same magnitude, but opposite sign.

$$X_m = \frac{2}{L}\sum_{n=1}^{\infty}\int_0^L X_n \sin\left(\frac{m\pi}{L}x\right)\sin\left(\frac{n\pi}{L}x\right)dx, \qquad [12.26]$$

where we have switched the order of the sum and the integral.

So what is the value of the integral shown? If $m = n$, it's just $\sin^2(n\pi x/L)$, but for any other m, $\sin(m\pi x/L)$ will have the opposite sign as $\sin(n\pi x/L)$ exactly as often as it has the same sign. This is illustrated in Fig. 12.7 for $m = 2$, $n = 4$, where we identify the regions in gray where $\sin(m\pi x/L)$ has the same sign as $\sin(n\pi x/L)$ (so the product of sines is positive) and in white the regions where the signs are opposite (so the product is negative). A little thought should convince you that because of the symmetry of the sine function, for every point where the value of the product in the integral is positive, there is a point with the same value but negative sign. Consequently, the integral exactly vanishes unless $m = n$.

But this really *is* magical: it means that no fancy or complicated mathematics is required to understand how Fourier series work: *just the fact that the sine function is symmetric* and spends as much time being positive as being negative tells us that the integral in Eq. [12.26] will pick out the specific components of $f(x)$ for which $m = n$, and all other wavelengths will cancel out of the calculation.

So what we've learned is that if $m = n$, the integral in Eq. [12.26] is $\sin^2(n\pi x/L)$, and if $m \neq n$, the integral vanishes. Also it's easy to show that:

$$\int_0^L \sin^2\left(\frac{n\pi}{L}x\right)dx = \frac{L}{2}, \qquad [12.27]$$

and since Eq. [12.26] has a pre-factor of $\frac{2}{L}$, the sum in that equation identically equals X_n.

To summarize, from Eq. [12.24] $f(x)$ is a sum of sines with amplitudes X_n and wavelengths $L/(n\pi)$, and we have just derived that Eq. [12.25] picks out the amplitude X_n for any chosen n.

There is a little more to Fourier's theorem to do with "completeness," meaning that the sums of sine functions (together with the cosines) can define essentially all functions $f(x)$, and don't maybe fail to define some types of functions. But at its heart, Fourier's theorem is just geometry. Wonderfully simple and wonderfully elegant geometry.

EXERCISE 12.6

The proof of course is in the pudding, so let's look at a few simple examples that illustrate how Fourier series work.

(1) As a first example, the step function shown in Fig. 12.4 can be reproduced using Fourier series. For this exercise, use Eq. [12.25] to solve for the Fourier coefficients where $f(x) = 1$ if $x > 0$, and $f(x) = 0$ otherwise. You'll find that $X_n = 0$ for n even, so plot the sums of odd terms up to $n = 1, 3, 5$, and 51 in Eq. [12.24], and show that they look like the first column of Fig. 12.8. As in that figure, you'll see that as n grows, the fit between the Fourier sum, Eq. [12.24], and the step function improves.

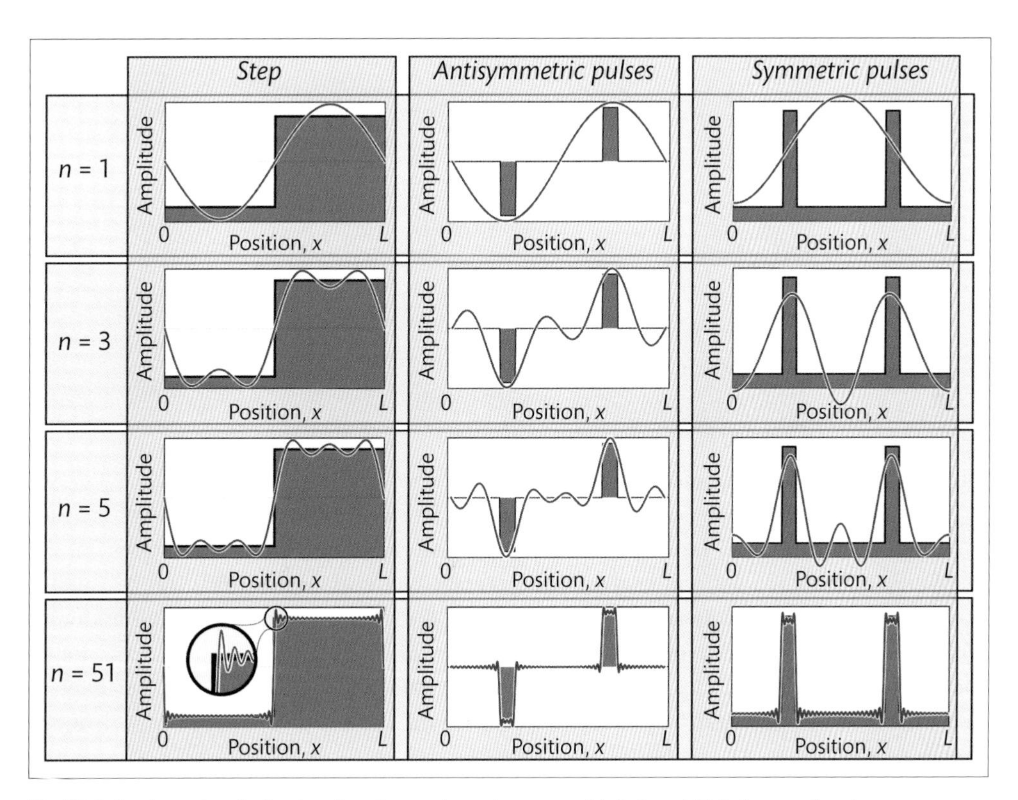

Fig. 12.8 Fourier series of a few functions for numbers of terms, n, of 1, 3, 5, and 51. Left: a step function; center: antisymmetric pulses, one downward and one upward, and right: symmetric pulses, both facing upward. Notice that the Fourier series solutions (red) fit the function (blue) increasingly well with higher number of terms, however short wavelength spikes appear at large n (highlighted in enlargement in lower left panel). As described in the text, the diffusion equation disperses these spikes rapidly, though they can cause spurious results in other equations—for example, in the wave equation used for medical imaging.

(2) Consider next two pulses, one facing down and the other facing up, as in the central column of Fig. 12.8. In Advanced exercise 12.8, we'll ask you to plot the evolution of these pulses: for this exercise, just plot the Fourier series for $n = 1, 3, 5$, and 51 as before.
By symmetry, at the midpoint between the downward and upward pulse, the concentration $c(x,t) = X(x) \cdot T(t)$ has to be zero at all times, right? But all initial shapes also have to spread out under the influence of the diffusion equation (Eq. [12.6]): we derived this result in this chapter using differential equations (e.g. Fig. 12.2) and in the last chapter by considering collisions between molecules (e.g. Fig. 11.10(b) of Chapter 11). Sketch how you think the amplitude, $c(x,t)$, of the concentration will evolve to satisfy both of these conditions.

(3) Last, consider two upward pulses, as shown in the right column of Fig. 12.8. In the first two examples, the functions to be fit were odd, and so sine functions, which are odd, could be used. In this case, the function is even—that is, it is the same for $x < L/2$ and for $x > L/2$. Consequently you will need to use cosine terms. Now in place of Eq. [12.24], we write the more complete solution to the diffusion equation:

$$f(x) = \sum_{n=1}^{\infty}\left[X_n \sin\left(\frac{n\pi}{L}x\right) + Y_n \cos\left(\frac{n\pi}{L}x\right)\right] + \frac{Y_0}{2}, \qquad [12.28]$$

where:

$$X_n = \frac{2}{L}\int_0^L f(x)\sin\left(\frac{n\pi}{L}x\right)dx$$
$$Y_n = \frac{2}{L}\int_0^L f(x)\cos\left(\frac{n\pi}{L}x\right)dx. \qquad [12.29]$$

Obtain the coefficients, X_n and Y_n, and plot $f(x)$ as before.

As in the second example, consider the midpoint, at $x = L/2$. The two pulses will increase the amplitude here, right? But also total concentration is conserved, so as time goes on, the amplitude at the midpoint will approach the average concentration—which can be small if the area inside the pulses is small. Does this mean that the concentration at the midpoint will rise and then fall? But doesn't the second part of Eq. [12.22], $T(t) = T_0 e^{-t/\tau}$, imply that the amplitude should decrease exponentially in time, and should not oscillate? Explain how these two observations are reconciled. We will discuss this further in Advanced exercise 12.8.

The examples of Exercise 12.6 illustrate that it is straightforward to reproduce functions using prescribed boundaries with sums of sines and cosines. Importantly, superposition implies that each term in the Fourier series travels independently—but notice that they travel at different *speeds*! This is made evident by recalling Eq. [12.23], $\tau = L^2/(n^2\pi^2 D)$, which tells us that as n increases, the timescale, τ, over which the solution drops by a factor of e shrinks rapidly—by the square of n. Eq. [12.23] is a form of a so-called "dispersion relation," meaning a formula that defines how rapidly oscillations disperse as wavelength changes. This relation tells us that $n = 51$ terms, at the bottom of Fig. 12.8, disperse $51^2 = 2601$ times faster than $n = 1$ terms, at the top of the figure. This has both a computational and a practical significance.

The term "dispersion relation" is from the physics and mathematics community, and, alas, is a problematic misnomer in biology and engineering, where "dispersion" has already been taken up as a synonym for diffusion. As with other semantic issues that have appeared in this book, forewarned is forearmed.

Computationally speaking, the dispersion relation is very useful, for it implies that many terms are needed in a Fourier expansion only if the short-term response is wanted: for times long compared with $L^2/(n_0{}^2\pi^2 D)$, terms with $n > n_0$ are of minor significance. It is worth remarking that other expansions have been invented using functions other than sines and cosines that capture as much of a function with as few terms as possible. This can be done in a number of ways, some of which are obvious, others less so. An obvious example is that we saw in Chapter 4 that Bessel functions arise naturally in cylindrical coordinates, so although in principle one could use sines and cosines to expand solutions within cylinders, Bessel expansions are much more sensible and require many fewer terms. Less obvious methods such as wavelets arise in data containing variations across many lengthscales: in the case of wavelets, self-similar functions are used in place of sines and cosines. Wavelets are used ECG and EKG analysis as well as in medical image processing,[4] and the image compression algorithm JPEG 2000 uses wavelets to more efficiently encode images than tradition JPEG.[5] More generally, the art of creating optimal functions to expand solutions to particular problems go by a number of titles, most commonly "proper orthogonal decomposition,"[6] or "Karhunen–Loève expansions."[7]

Practically, notice from the examples in Exercise 12.6 that as n grows, short wavelength spikes appear near sharp edges in the function being fit (enlarged in Fig. 12.8). The dispersion relation for the diffusion equation tells us that these short wavelength spikes diffuse much faster than longer wavelength terms, and so effects of sharp edges disappear very rapidly in time. Consequently if you were to inject a bolus of drug with a very sharp boundary into tissue, before long the concentration of the drug would be essentially indistinguishable from injection of a drug with a diffuse boundary. Likewise a fresh nicotine patch (with perfectly uniform concentration) and a slightly used patch (with lower concentration near the exposed surface) would provide essential identical dosing profiles.

The disappearance of high frequency features doesn't occur for all equations – for example short wavelength terms lead to so-called "ringing" artifacts in solutions to the wave equation: these are variants of the rapid spikes highlighted on the bottom left of Fig. 12.8 (more on this shortly). Fourier, and the related "Radon," transforms are used in medical imaging, and as a result ringing can introduce spurious images in MRI, ultrasound or CAT scans. Notwithstanding this caveat, most of the behavior of diffusive problems is captured using only a small number of Fourier terms, as shown in the exercise following.

EXERCISE 12.7

In the next exercises, you will apply the diffusion equation to evolve initial patterns in time.

First a warm-up: use Fourier series to solve for a 1D rod at temperature T_1 that has been instantaneously immersed into an infinite bath of temperature T_0: this is the problem shown in Fig. 12.6(b). For this problem, you know the solution: the rod should start with a sharp step from T_0 to T_1 at both ends, and should reach T_0 everywhere at long times. So before you start, consider Eqs. [12.28] and [12.29] to see if you can simplify matters by excluding solutions that don't match your boundary conditions. Then make sure the terms that remain have constant values at the edges ($x = 0$ and $x = L$) for all times.

Second, still warming up: to make sure we are clear about what is happening, note that as we mentioned in Exercise 12.6, the total concentration is conserved—this was built in to our derivation, beginning with Eq. [12.2] early in this chapter. So why does the concentration shown in Fig. 12.6(b) appear to decrease with time? Clearly there has been a cheat somewhere, and indeed when you reproduce this plot, you'll see that the concentration outside of $[0,L]$ is negative. So where do we get off omitting the concentration outside? How do we know the concentration inside the sample is correct?

The answer to these questions lies in the "uniqueness theorem" for linear differential equations. We have avoided mathematical proofs in this book, but the results of these proofs can be very important, as in this case. Here the uniqueness theorem says in short that there is only one solution to a linear differential equation (like the diffusion equation) that meets specified initial and boundary conditions on a domain of interest. So if by hook or by crook we can obtain a solution to the diffusion equation that meets our initial and boundary conditions, it must be the correct solution.

The solution shown in Fig. 12.6(b) does start with the initial pulse that we chose, it does meet the fixed boundary conditions, $x(0) = x(L) = 0$, and Eq. [12.28] does solve the diffusion equation between 0 and L. So according to the uniqueness theorem, it must be correct. But mass isn't conserved between *0* and L—why not, and doesn't this imply something is wrong? Where is the cheat, and is the plot shown in Fig. 12.6(b) correct, or not?

ADVANCED EXERCISE 12.8

In this exercise, you will evolve the initial patterns shown in Exercise 12.6, parts 2 and 3. This advanced exercise has two parts; use periodic boundary conditions for both.

(1) First, evaluate the accuracy of the Fourier series solution to the symmetric pulses shown in Fig. 12.9. Begin by using the superposition of steps described in Fig. 12.5: this will give you the exact solution for each part. Then propagate the individual Fourier terms obtained in Exercise 12.6 to evaluate how the accuracy of the Fourier series improves with numbers of terms. You will find, as expected, that very few terms (i.e. small n) are needed to provide long time accuracy, and that the shorter wavelength terms (i.e. large n) affect only short time results.

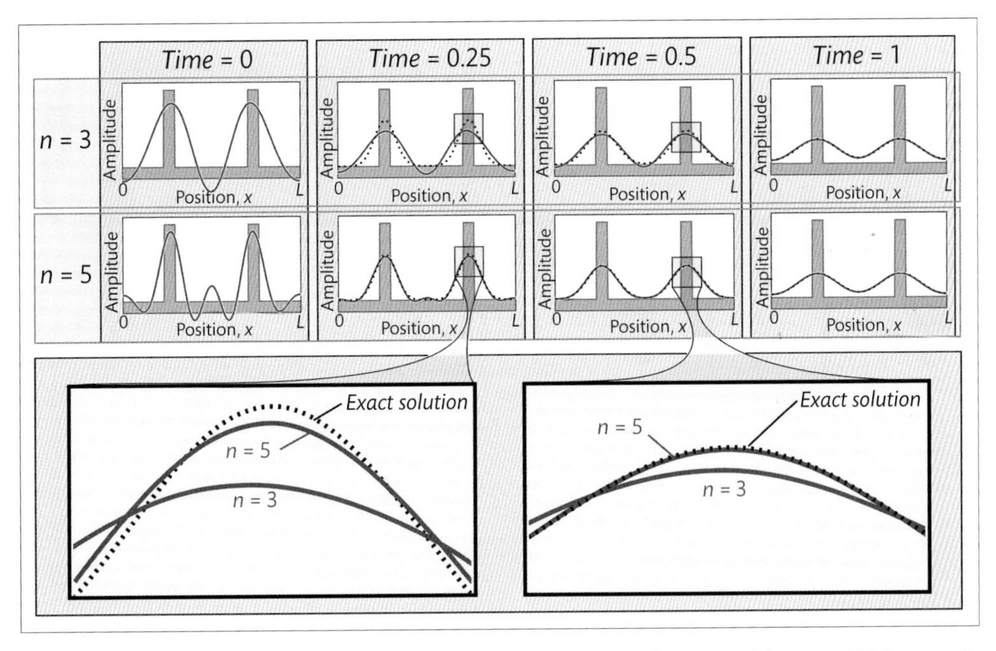

Fig. 12.9 Typical results showing comparison between exact time evolution (dotted lines) from two initial symmetric pulses (blue) with Fourier series using $n = 3$ and $n = 5$ terms. As shown in yellow enlargements, at *Time* = 0.25 for $n = 5$ the Fourier series produces a small but nonzero error, while for $n = 3$, the error is significant. At *Time* = 0.5 by comparison, the $n = 5$ error has become nearly imperceptible, while the $n = 3$ error persists until about *Time* = 1. The Fourier series is virtually indistinguishable from the exact solution at times > 0.1 for $n > 11$, or at times > 0.01 for $n > 29$. Time is expressed here in units of L^2/D, using $L = 2\pi$ and $D = 1$.

For example, in Fig. 12.9 we show typical results comparing the exact solution to the diffusion equation (dotted lines) as time progresses with Fourier series using $n = 3$ and $n = 5$. For this part of the exercise, plot the number of terms needed to maintain results within a small error of your choosing.

(2) Second, compare your expectations from Exercise 12.6 with computational simulations. For the antisymmetric case (part 2—center column of Fig. 12.8), does the solution agree with your prediction from Exercise 12.6? For the symmetric case (part 3—right column of Fig. 12.8), can you confirm that the solution rises and falls once as expected?

It is not hard also to show that multiple oscillations can be produced—for example, in Fig. 12.10 we show that four initial pulses can produce two peaks. Can you produce more peaks? Is there a limit to the number of peaks, and so could a truly oscillatory solution be produced? Does this contradict the analytic result, Eq. [12.22], that the time dependence is exponentially damped in time, that is, $T(t) = T_0 e^{-t/\tau}$, or is an oscillatory solution, $T(t) = T_0 e^{-it/\tau}$, possible?

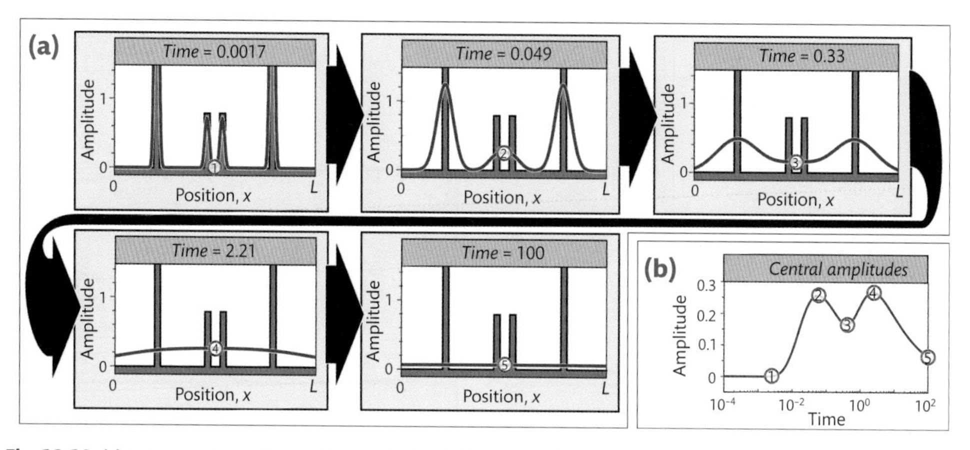

Fig. 12.10 (a) Time evolution of four pulses under the diffusion equation at successive times. Notice that after successive times the midpoint, circled, rises and falls repeatedly. Time units are as in Fig. 12.9. (b) Plotting the evolution of the circled points over time appears to produce an oscillatory solution.

For further reading, you may find the related topic of non-normal forms[8]—transient solutions to ODEs—to be of interest. Also, in Chapter 13 we will describe Turing oscillations: truly oscillatory solutions to coupled diffusion equations in the presence of reactions.

12.3 Summary and Extension to Three Dimensions

To conclude our overview of Fourier series, we have seen that the diffusion equation can be solved by superposition of known functions, including sines and cosines. Fourier series have exceptionally broad applicability in science and engineering, but for our purposes we emphasize just a few of their properties.

(1) **Fourier analysis is straightforward:** There is a straightforward formula, Eq. [12.29], for calculating the components of a Fourier series for (almost) any initial state.

(2) **Fourier series are efficient:** To evolve this initial state under the diffusion equation, only the first few terms matter for most timescales: more terms are useful only to establish shorter time behaviors. So rather than subdividing the range from $x = 0$ to $x = L$ into a large number of points and evaluating derivatives point-by-point, we need only evaluate a few Fourier terms, and the problem is solved.

    Note also that the dispersion relation, Eq. [12.23], tells us that effects of sharp edges (which require high order Fourier terms to resolve) disappear very rapidly.

(3) **Each Fourier term is an exact solution, so their sum is also:** We haven't stressed this point, but we emphasize that every Fourier term is an *exact* solution to the diffusion equation. So the evolution of a sum of Fourier terms is likewise exact: there are no numerical errors associated with evolving an initial state forward in time as would appear if we used an integration algorithm as discussed in the Appendix. We illustrate this shortly in an example in Fig. 12.11. Fourier series can produce errors, especially in

nonlinear problems that suffer from mode coupling (discussed in Chapter 4), but the fact that Fourier terms are exact solutions to many common differential equations accounts for popularity of their use.[3]

(4) **Fourier series work in two and three dimensions as well:** Finally, we have described Fourier series in one dimension, but nothing restricts the idea to being used in higher dimensions, and indeed Fourier series are widely used to solve three-dimensional (3D) problems. We touch on this point next.

12.3.1 Simple Spherically Symmetric Solution

We have so far only considered 1D solutions. Of course, the diffusion equation—and most practical problems—is 3D. As with fluid flow problems, complex geometries require numerical methods. As a step toward 3D solutions, and to provoke thought concerning issues that arise in numerical analysis, we consider diffusion from a sphere. This is a problem of practical importance in its own right, for it arises when a bolus of drug is injected into tissue, or when an idealized spherical drug tablet is dissolved in digestive fluids.

In spherical coordinates, the diffusion equation (Eq. [12.6]) reads:

$$\frac{\partial c}{\partial t} = D \cdot \nabla^2 c = D \cdot \left[\frac{1}{r^2}\frac{\partial}{\partial r}\left(r^2 \frac{\partial}{\partial r}\right) + \frac{1}{r^2 \sin^2 \vartheta}\frac{\partial^2}{\partial \phi^2} + \frac{1}{r^2 \sin \vartheta}\frac{\partial}{\partial \vartheta}\left(\sin\vartheta \frac{\partial}{\partial \vartheta}\right)\right] c. \qquad [12.30]$$

The blue terms represent variations in azimuthal and polar angles ϕ and ϑ, respectively, using the convention given in Fig. 6.8 of Chapter 6. If we neglect angular variations, we are left with:

$$\frac{\partial c}{\partial t} = D \cdot \left[\frac{1}{r^2}\frac{\partial}{\partial r}\left(r^2 \frac{\partial}{\partial r}\right)\right] c. \qquad [12.31]$$

We can solve this directly by using separation of variables, but the derivatives break into two parts, and it is easier to obtain the solution if we adopt the substitution $c(r,t) = \rho(r,t)/r$. This reduces Eq. [12.31] to:

$$\frac{\partial \rho}{\partial t} = D \cdot \frac{\partial^2 \rho}{\partial r^2}. \qquad [12.32]$$

This is highly convenient: we've seen this equation several times before, and we know its solution, so we can write straight away that:

$$c(r,t) = \frac{1}{r}\left\{\sum_{n=1}^{\infty}\left[X_n \sin\left(\frac{n\pi}{L}r\right) + Y_n \cos\left(\frac{n\pi}{L}r\right)\right] + \frac{Y_0}{2}\right\} e^{-t/\tau}, \qquad [12.33]$$

and as before, $\tau = L^2/(n^2\pi^2 D)$.

EXERCISE 12.9

Fill in the missing steps between Eqs. [12.31] and [12.33]. Then obtain the Fourier coefficients using Eq. [12.29]—for this exercise, assume that the concentration outside the sphere is always exactly zero (i.e. the drug is rapidly swept away for $r > R$), and that initially the concentration is constant inside the sphere. Plot the concentration as a function of radius at various times: you should obtain a result like Fig. 12.11(a).

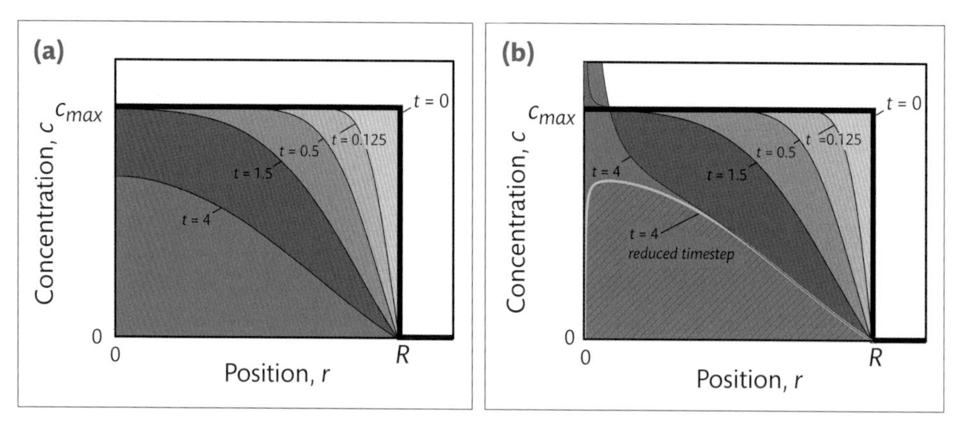

Fig. 12.11 (a) Fourier solution for diffusion of a chemical from a spherically symmetric sample into zero constant concentration bath. (b) Finite difference solution to the same problem. Notice that the solutions agree until after $t = 1$. By $t = 4$, finite difference solution either diverges or plunges to zero at $r = 0$, depending on time step. Time is expressed here in units of L^2/D, using $L = \pi$ and $D = ¼$.

ADVANCED EXERCISE 12.10

As a cautionary example, numerically integrate the diffusion equation, for example, by subdividing the domain from 0 to R into a large number of gridpoints and directly calculating d^2c/dr^2. This is termed a "finite difference" method, which is widely discussed on the web and in the literature and is not hard to implement (though you'll find that making it behave smoothly takes some work). Sample matlab finite difference code, in two-dimensional (2D) Cartesian coordinates, is included in the Appendix. Alternatively, use a pre-programmed solver such as Mathematica's NDSolve.

In either case, you will obtain time evolution plots like those shown in 12.11(b), and you will find that for long times it is exceedingly difficult to produce sensible behavior near $r = 0$. For example, for one choice of timestep, you will find that the concentration is well behaved for short times, but diverges near $r = 0$ at later times (shown at time $= 4$ in Fig. 12.11(b)). By reducing the timestep, you will be able to prevent the divergence (at the expense of computational speed), but you will find the concentration drops to zero at the origin (also shown in Fig. 12.11(b)). Clearly this is nonphysical: the concentration can neither diverge nor plunge to zero at the center of a diffusing sphere. Why does this occur, and why is this not seen for the Fourier solution, Fig. 12.11(a)?

The preceding exercise highlights the importance of being skeptical of computational simulations. It is absolutely standard across all disciplines and for all problems that simulations will produce nonphysical solutions under some conditions, and so it is essential to make contact with experimental or analytic validation wherever possible.

12.3.2 **Mechanisms Related to Diffusion**

Before we leave discussions of diffusion, it is important to note that in this book, we only deal with so-called passive diffusion, meaning effects of random motions of chemicals, cells, or particles. We have done so with the goal of providing necessary background for the next chapter, where we describe genuinely spectacular results that occur when we combine reaction with diffusion. Nevertheless, we make a few remarks here to provide guidance on where to look for related mechanisms.

A first remark is that in biological systems, diffusion is often accompanied by charge transfer—we mentioned this in a brief example in Chapter 11 when we described the Donnan effect: so-called "Nernst" voltages that arise when charged ions diffuse through membranes. This is a well-established topic that is important not only for biological membranes, but also for gel electrophoresis. For further reading, see Silver,[9] or the seminal works by Donnan and Harris[10] and Nernst.[11]

A second remark is there is a rich variety of what are termed "active" transport effects that involve an input of energy or the use of dedicated biological structures. These include so-called primary transport by energy-consuming pumps, secondary transport, or trafficking, such as endocytosis and transcytosis in which materials are encapsulated and transported into or within a cell, and carrier-mediated transport, in which chaperones or other dedicated structures carry materials either up or down an electrochemical gradient. Many of these active mechanisms are overviewed in Marieb and Hoehn.[12]

12.3.3 **Advection-Diffusion**

We conclude this chapter by discussing what happens when fluid flow, discussed in Chapters 1–10, is combined with particle diffusion, discussed in Chapters 11 and 12. This can arise in a couple of contexts.

First, material can flow and diffuse at the same time. This turns out to be simple to analyze, starting from the continuity equation, Eq. [12.4], which expressed in three dimensions is:

$$\frac{\partial c}{\partial t} = \nabla \cdot \vec{J}, \qquad [12.34]$$

where $\vec{J}$ is the flux through an infinitesimal volume. In Eqs. [12.5] and [12.6], we derived the diffusion equation from this by taking the flux to be due to random variations in concentration, c, from Eq. [12.1]. But concentrations can vary also due to material being advected (a synonym for convected, invented to be more general than the convection that is usually restricted to motion caused by differential heating). That is, if we know that material within a fluid flows with a known velocity, $\vec{v}$, obtained perhaps from stirring in a Couette device or from being pumped with Poiseuille velocity through a tube, then the flux due to the flow is just $\vec{J} = c \cdot \vec{v}$. Adding this to the flux due to diffusion, $\vec{J} = D \cdot \nabla c$, from Fick's first law, we obtain:

$$\frac{\partial c}{\partial t} = \nabla \cdot (D\nabla c + c\vec{v}). \qquad [12.35]$$

This is the advection–diffusion equation, explored in the following exercise.

EXERCISE 12.11

In Chapter 1, Exercise 1.1, we showed that transporting an initial circle of dye by an oscillatory flow produces chaotic mixing. Use the advection–diffusion equation to repeat this simulation in the presence of diffusion. The Matlab finite difference code in the Appendix will allow you to produce plots like those in Fig. 12.12 as a numerical solution to the influence of Eq. [12.35]. What do these results tell you about mixing in microfluidic devices? Specifically, suppose you have a tube of length L and diameter d: how would you determine whether the design is adequate to provide molecular mixing using diffusion alone? If diffusion isn't adequate, propose an alternative design that makes use of advection, again assuming microfluidic dimensions. Obtain a dimensionless number to help you decide: what terms must it have? Once you have obtained the number, find its name.

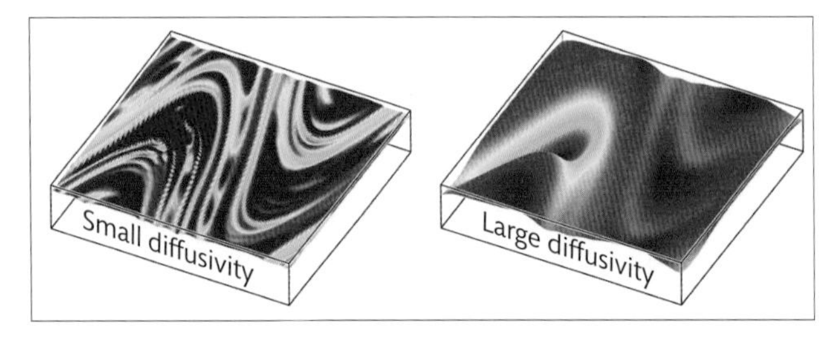

Fig. 12.12 Sine flow (first seen in Exercise 1.1, Chapter 1), combined with diffusion. Notice that as diffusivity grows, fine features vanish. Large features remain, but become smaller in amplitude as time progresses.

12.3.4 **Tissue Perfusion**

A second way of combining fluid flow with diffusion is sketched in Fig. 12.13, where we depict a "compartment model" for flow of blood over tissue. Compartment, or lumped parameter, models are widely used to evaluate spatially-averaged quantities in systems in which detailed spatial dependence isn't important or where it may be computationally cumbersome to

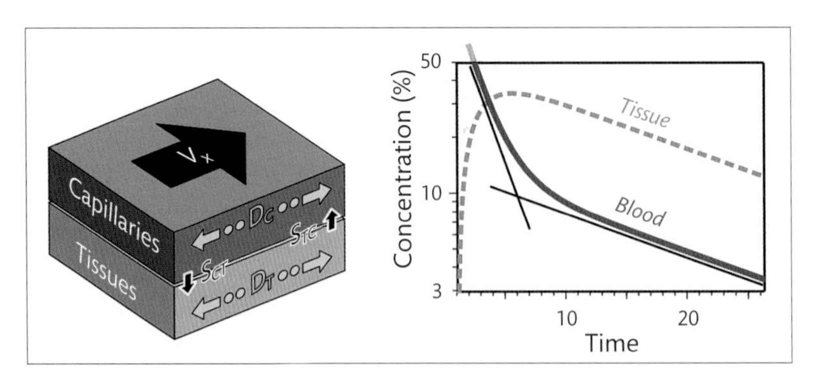

Fig. 12.13 (a) Schematic of compartment model of capillary flow over tissues, in which V_x is a flow speed in the capillaries, D_c and D_t are capillary and tissue diffusivities, and S_{ct} is a transfer rate from capillaries to tissues. Each compartment lumps together details such as spatial variations, and so this kind of approach is also termed a lumped parameter model. (b) Coupled equations (see Eqs. [12.41]) show that this produces a growth and then reduction of concentration within tissues, and two different exponential rates of decrease of concentration within the blood.

simulate the system in complete detail. An example is analysis of delivery of a drug into tissues, where microscopic details, such as what capillary is connected to which cell, are relatively unimportant. What is likely to matter is the cumulative dose response of the drug over the whole body, so it can be useful to neglect local differences and treat all capillaries as one compartment that distributes the drug to all tissues in a second compartment. Our job then is to collect the relevant physics by which the drug travels within and between compartments. As indicated by the arrows in Fig. 12.13, we can model flow (V_x) and diffusion in the capillary compartment (D_C), diffusion alone in the tissue compartment (D_T), and transfer from capillaries to tissues (S_{CT}) and back (S_{TC}).

A simple model for concentration, c_T, in the tissue compartment is:

$$\frac{\partial c_T}{\partial t} = \nabla \cdot (D_T \nabla c_T) + S_{CT}, \qquad [12.36]$$

where D_T is diffusivity in the tissue and S_{CT} (S for surface) is the rate of flow through the surface separating the blood from the tissue compartment. S_{CT} is commonly taken to be proportional to the difference between the capillary concentration, c_C, and the equilibrium concentration that the drug would reach in the tissues, αc_T, where α is a constant:

$$S_{CT} = k_{CT} \cdot A \cdot (c_C - \alpha \cdot c_T). \qquad [12.37]$$

Here k_{CT} is a constant rate of transfer from capillaries to tissues, and A is the interfacial area of contact between capillaries and tissues. Eq. [12.37] is reasonable insofar as it produces a realistic exponential approach to $c_C = \alpha \cdot c_T$, at which point flow between capillary and tissue stops. The constant, α, is optional, but is often included because, as described in last chapter, Donnan and related effects can produce equilibrium concentration imbalances across a membrane.

As for the capillary compartment, its model resembles the tissue equations (Eqs. [12.36] and [12.37], but also includes a flow term, has the opposite sign for S_{CT} to account for flow out of capillaries rather than into tissues, and has an optional term to model clearance of the drug by liver or kidneys. So for concentration in the capillaries, c_C, we write

$$\frac{\partial c_C}{\partial t} = \nabla \cdot (D_C \nabla c_C - c_C \vec{v}_x) - S_{TC} - R \cdot c_C, \qquad [12.38]$$

where D_C is the diffusivity in the blood, $\vec{v}_x$ is the flow speed, and R (for removal) is a clearance rate. Here S_{TC} has the same form as S_{CT}, but can in principle have a different coefficient, k_{TC}:

$$S_{TC} = k_{TC} \cdot A \cdot (c_C - \alpha \cdot c_T). \qquad [12.39]$$

Eqs. [12.36]–[12.39] can be combined into:

$$\begin{aligned} \frac{\partial c_T}{\partial t} &= \nabla \cdot (D_T \nabla c_T) - k_{CT} \cdot A \cdot \alpha \cdot c_T + k_{CT} \cdot A \cdot c_C \\ \frac{\partial c_C}{\partial t} &= \nabla \cdot (D_C \nabla c_C - c_C \vec{v}_x) + k_{TC} \cdot A \cdot \alpha c_T - (k_{TC} \cdot A + R) c_C. \end{aligned} \qquad [12.40]$$

Here the blue terms represent spatial variations, due to diffusion and advection. Let's study this problem in two parts, first neglecting these terms, and second including them.

Neglect spatial terms

If we neglect the blue diffusion and advection terms in Eqs. [12.40], we can express Eqs. [12.40] as:

$$\frac{\partial c_T}{\partial t} = -k_{CT} \cdot A \cdot \alpha \cdot c_T + k_{CT} \cdot A \cdot c_C$$
$$\frac{\partial c_C}{\partial t} = k_{TC} \cdot A \cdot \alpha c_T - (k_{TC} \cdot A + R)c_C. \qquad [12.41]$$

This approximate solution would apply when spatial variations and flow are small compared with rates of drug clearance and transport across capillary-tissue membranes. So, for example, we could justify using Eqs. [12.41] if a drug were intravenously administered so that its concentration became spatially nearly constant.

Eqs. [12.41] can be analyzed using standard techniques,[13] which we summarize here. To begin with, notice that if we attempt exponential solutions for c_T and c_C with the same rate, ω:

$$c_T = c_{To}e^{-\omega t}$$
$$c_C = c_{Co}e^{-\omega t}, \qquad [12.42]$$

then the exponential functions cancel out of Eqs. [12.41], leaving:

$$\omega \cdot c_T = -\hat{k}_{CT} \cdot c_T + \hat{k}_{CT} \cdot c_C$$
$$\omega \cdot c_C = \hat{k}_{TC} \cdot c_T - (\hat{k}_{TC} + R) \cdot c_C. \qquad [12.43]$$

Here to make the expressions less cumbersome, we have set $\alpha = 1$, and have absorbed the constant, A, into $\hat{k}_{CT}$ and $\hat{k}_{TC}$. The reader may recognize this as an eigenvalue solution; if not, we discuss the method further in Chapter 13.

Eqs. [12.43] are now algebraic equations, in keeping with the notion used throughout this book that the way to solve PDEs is to reduce them to ODEs, and the way to solve ODEs is to reduce them to algebraic equations. Long story short, we can solve these algebraic equations, obtaining

$$\omega = \frac{-b \pm \sqrt{b^2 - 4\hat{k}_{ct}R\alpha}}{2}, \qquad [12.44]$$

where we have used the abbreviation $b = \alpha\hat{k}_{ct} + \hat{k}_{tc} + R$. Notice that there are two possible values of ω, both of which are negative, so both blood and tissue concentrations decay to zero like $e^{-|\omega|t}$.

EXERCISE 12.12

Assume that initially the capillary concentration is a constant (make it simple by setting $c_c = 1$), and tissue concentration is $c_T = 0$. Use Runge–Kutta integration, discussed in Chapters 3 and 8 and described in the Appendix, to numerically solve Eq. [12.41]. Show that provided $R > 0$, the tissue and blood concentrations look like those shown in

Fig. 12.13(b). You will find that the tissue concentration grows and then diminishes: why is this? You will also find that for a substantial range of choices of k_{ct}, k_{tc}, A, and α, the tissue concentration will become higher than the capillary concentration. But the drug *originates* in the capillaries, so explain how the tissue concentration can grow beyond the capillary concentration?

If we plot the log of blood and tissue concentrations starting from $c_c = 1$, $c_T = 0$, we can see that the blood concentration has two linear regions (meaning regions of exponential decline on this semi-log plot), identified by black lines in Fig. 12.13(b). This is very typical of multi-compartment models, and occurs because the capillary compartment loses drug at one exponential rate, $k_{ct} - k_{tc}$, to the tissues, and at a second rate, R, to the kidneys or liver. The two rates of depletion produce two rates seen in measures of concentration in the blood. More broadly speaking, multi-component systems tend to exhibit multiple time responses.

How important are spatial terms?

Let's now restore the blue terms from Eqs. [12.40] to assess the significance of spatial variations. Notice that Eqs. [12.40] are of the form:

$$\frac{\partial c_T}{\partial t} = D_T \nabla^2 c_T + f(c_T, c_C)$$
$$\frac{\partial c_C}{\partial t} = D_C \nabla^2 c_C + g(c_T, c_C). \qquad [12.45]$$

These are "reaction-diffusion equations," where f and g are functions that here represent transport of a drug, but more generally could include reaction kinetics, interactions between cells, or other phenomena coupling c_T and c_C. In the next chapter we will explore some surprising ways that such equations can behave; for now, let's remain focused on the drug transport problem, where f and g are given in Eqs. [12.40].

Eqs. [12.45] can be simulated using finite differences, as was done in Exercise 12.11. This produces the results shown in Fig. 12.14. In the body of that figure, we show a color-coded plot of concentration after one pass of a drug bolus along a 2D capillary channel above a 2D tissue region, and on the lower right, we show the spatially averaged tissue and blood concentrations as they evolve over time. Evidently, the spatially averaged model, Fig. 12.13, and the more detailed model, Fig. 12.14, evolve in a qualitatively similar fashion. Differences can also be seen, both in that the detailed model reveals spatial localization of the drug bolus and in that the numerical fractions of drug in the two models differ quantitatively. In this sense, Fig. 12.13 is representative of compartment models: you can expect them to produce qualitatively appropriate results, but they cannot be relied upon for details.

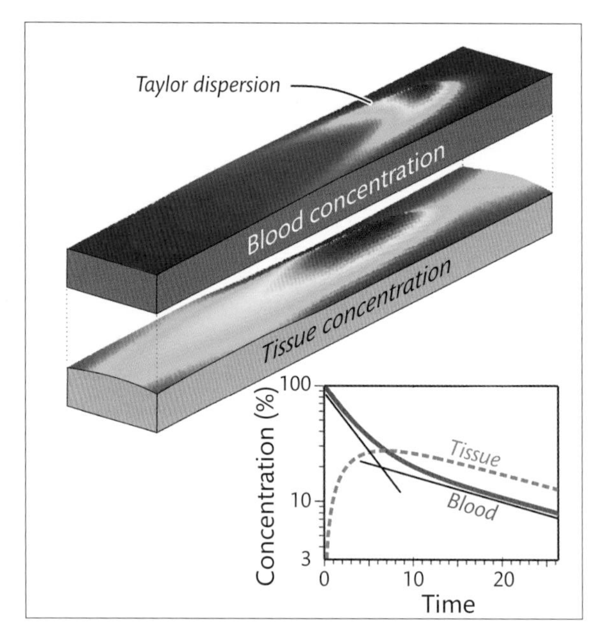

Fig. 12.14 Finite difference simulation of model including spatial variations, Eqs. [12.40], shows drug bolus in blood that is spread out over tissues. Red represents high drug concentration, and blue represents low. Notice that Taylor dispersion, introduced in Chapter 5, Fig. 5.3, is produced by the advection term introduced in Eq. [12.35]. Inset: The more detailed spatial model produces similar qualitative response, shown here, to the averaged compartment model shown in Fig. 12.13, although detailed fractions of drug concentration differ.

ADVANCED EXERCISE 12.13

Use a finite difference simulation, for example, in Matlab from the Appendix or Mathematica's NDSolve, to numerically solve Eqs. [12.40] on a long 2D channel, representing a capillary, over a 2D tissue of the same size. Make the channel periodic along its length, so that blood leaving one end re-enters into the other end. Use this model to display Taylor dispersion, shown in Fig. 12.14.

Suppose you want to design a drug that stays as long as possible in the tissues—what do you need to adjust to make this occur? Use your model to demonstrate that the tissue concentration decreases more slowly than before. Finally, your model is unlikely to look quite like the one shown in Fig. 12.14: it should produce a more uniform concentration in the tissue compartment, because each time the bolus travels over a region of tissue, it will add a layer of drug to that region. What change would you have to make to the model to generate a more localized concentration of blood in the tissue? (Hint: there are at least two ways to do this.)

Other compartment models

Now that you have been introduced to compartment models, you can conceive endless embellishments. For example, you could investigate how to best produce an extended release tablet—common now, but novel in the 1960s when they were first introduced.[14] A simple such model is sketched on the left of Fig. 12.15, where we envision a pill that dissolves slowly, releasing drug into the gut, which in turn transfers the drug to the blood, finally making the drug available to the tissues. This can be modeled using the following equations:

$$\begin{aligned}\frac{\partial p}{\partial t} &= -k_p \cdot p \\ \frac{\partial g}{\partial t} &= k_p \cdot p - k_g \cdot g \\ \frac{\partial b}{\partial t} &= k_g \cdot p - k_b \cdot b.\end{aligned} \qquad [12.46]$$

Here p, g, and b represent the drug concentration in the pill, gut, and blood compartments respectively, and transport between these compartments are defined by k_p, k_g. This leaves k_b, which is the rate at which drug leaves the blood for the tissue.

These equations can be solved analytically,[14] or numerically as in Exercise 12.12, either of which lead to plots for concentration in the blood, b, like those shown in Fig. 12.15(b). For fast dissolution of the pill (large k_p), one obtains the red "immediate release" curve for blood concentration, which releases drug rapidly and then decays exponentially. If dissolution is slowed, a blood concentration profile like the blue "controlled release" curve can be obtained. Notice in this illustrative example that the red box, identifying a twofold change in concentration, lasts 20 time units for the immediate release case, while the blue box, identifying the same twofold change, lasts more than twice as long: 46 time units.

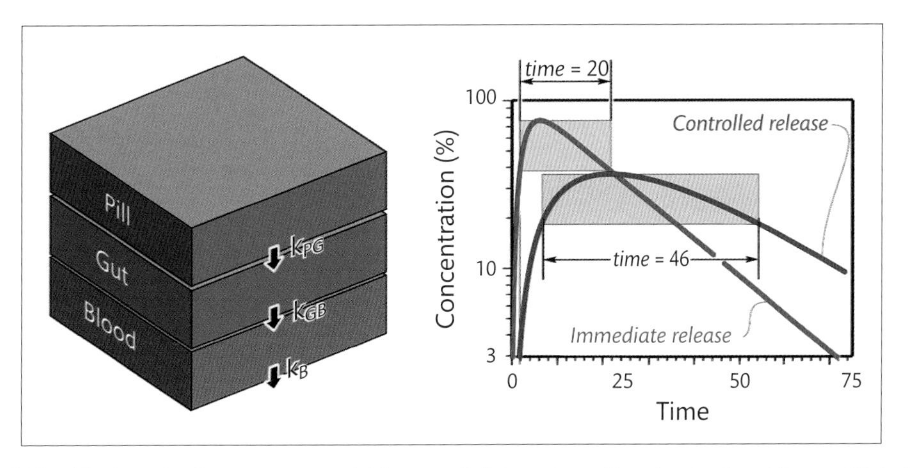

Fig. 12.15 (a) Schematic of compartment model for controlled drug release. Here drug is released from a pill into the gut over time at a rate k_{PG}, transported from the gut to the blood at a different rate, k_{GB}, and delivered from blood to tissues at a third rate, k_B. (b) This produces slower, controlled release, when the pill releases drug slowly (i.e. when k_{PG} is small) than in the immediate release case, when the pill releases the drug rapidly. Many-fold increases in therapeutic dosage times can be obtained using approaches such as this: in the blue and red boxes, we compare times over which the drug concentration in the blood changes by a factor of two.

EXERCISE 12.14

Much creativity has gone into designing approaches to produce desired drug release profiles over the years. At its simplest, the drug can be embedded in a matrix with slow diffusivity, or the drug's solubility can be reduced either by making chemical changes or by simply increasing the size of particles. More clever approaches have also been developed, such as making the drug buoyant in stomach fluids so that it can only slowly settle onto the stomach lining for absorption, or coating a tablet with a sacrificial layer that takes time to degrade—and that can itself deliver drug at an initial controlled rate. For this exercise, design a multi-compartment model with two delivery profiles—an initial profile that provides drug rapidly, along with a second profile that delivers a boost of drug some time later. Sketch the geometry of the compartments and simulate the ultimate performance, as in Fig. 12.15.

In this chapter we have explored several effects of diffusion. This story began in Chapter 11, when we saw that diffusion is the result of random wandering of particles, and where we learned that random wandering can produce self-assembled patterns. In Chapter 13, we will use what we now know about diffusion to analyze the mathematics behind self-assembly.

REFERENCES

A readable overview of statistics for the newcomer is:

1. Gonick, L. and Smith, W. (1993) *The Cartoon Guide to Statistics*. New York: Harper-Collins.
2. Press, W.H., Flannery, B.P., Teukolsky, S.A. and Vetterling, W.T. (1986) *Numerical Recipes: The Art of Scientific Computing*. Cambridge: Cambridge University Press.
3. Trefethen, L.N. (2000) *Spectral Methods in MATLAB*. Philadelphia, PA: SIAM.
4. Akay, M. (1997) Wavelet applications in medicine. *IEEE Spectrum, 34*(5), 50–6.
5. Usevitch, B.E. (2001) A tutorial on modern lossy wavelet image compression: foundations of JPEG 2000. *IEEE Signal Processing Magazine, 18*(5), 22–35.
6. Berkooz, G., Holmes, P. and Lumley, J.L. (1993) The proper orthogonal decomposition in the analysis of turbulent flows. *Annual Review of Fluid Mechanics, 25*, 539–75.
7. Graham, M.D. and Kevrekidis, I.G. (1996) Alternative approaches to Karhunen–Loéve decomposition for model reduction and data analysis. *Computers and Chemical Engineering, 20*, 495–606.
8. Trefethen, L.N. and Embree, M. (2005) *Spectra and Pseudospectra: The Behavior of Nonnormal Matrices and Operators*. Princeton, NJ: Princeton University Press.
9. Silver, B.L. (1985) *The Physical Chemistry of Membranes* New York: Solomon Press; see also Berne, R.M., Levy, M.N., Koeppen, B.M. and Stanton, B.A. (1998) *Physiology*, 4th ed. St. Louis, MO: Mosby, pp.21–9.

10. Donnan, F.G. and Harris, A.B. (1911) The osmotic pressure and conductivity of aqueous solutions of Congo-red, and reversible membrane equilibria. *Journal of the Chemical Society, Transactions, 99*, 1554–77.

11. Nernst, W. (1907) *Experimental and Theoretical Applications of Thermodynamics to Chemistry*. New York: Scribner.

12. Marieb, E.N. and Hoehn, K. (2007) *Human Anatomy and Physiology*, 7th ed. San Francisco, CA: Pearson, pp.70–81.

13. Strogatz, S.H. (2015) *Nonlinear Dynamics with Applications To Physics, Biology, Chemistry and Engineering*, 2nd ed. Philadelphia, PA: Westview Press, pp.125–45.

14. Wiegand, R.G. and Taylor, J.D. (1960) Kinetics of plasma drug levels after sustained release dosage. *Biochemical Pharmacology, 3*, 256–63.

13 Self-Assembly and Beyond

13.1 Introduction

In 1984, explorers in Antarctica uncovered a meteorite that was determined by isotopic concentration to have originated in the Valles Marineris, an ancient system of canyons on Mars. Electron microscopy of the surface of this meteorite revealed worm-like structures,[1] as shown in Fig. 13.1(a). By comparison, we show in Fig. 13.1(b) a sketch of a prehistoric segmented filamentous bacterium.[2] This comparison prompted intense interest in the possibility of life on Mars.

Several subsequent investigations,[3] however, revealed that similar worm-like structures can be produced by *nonbiological* assembly of mineral deposits. Indeed the very small size of the structures shown in Fig. 13.1(a)—only a few molecules across—make it unlikely that these could sustain biological metabolism. Disappointing as this may be for those of us who would like to see evidence of extraterrestrial life, from another perspective it might not be surprising that complex structures can form spontaneously.

To see this, we have to look no further than at water, one of the simplest of molecules. As we mentioned in Chapter 10, water produces an extraordinary variety of structures when it freezes:[4] a small sampling of these structures is shown in Fig. 13.1(c), all of which form on their own without any biological processes, external engineering or intervention. These structures are revealed through cryogenic electron microscopy—developed originally to limit damage to delicate structures in cells,[5] viruses,[6] and proteins,[7] but useful here because it preserves the structure of the frozen samples during imaging.

The similarity between nonbiological mineral aggregates and worm-like morphologies, alongside the self-assembly of multiple elaborate structures from molecules as simple as water, raises several essential questions. If there are mechanisms of self-assembly through which simple minerals can produce structures that resemble living organisms, do living organisms make use of similar mechanisms? Or must biological systems invent their own mechanisms for creating necessary structures that don't make use of self-assembly? And how, if at all, do biological and nonbiological mechanisms of morphogenesis differ?

Biomedical Fluid Dynamics: Flow and Form. Troy Shinbrot.

DOI: 10.1093/oso/9780198812586.001.0001

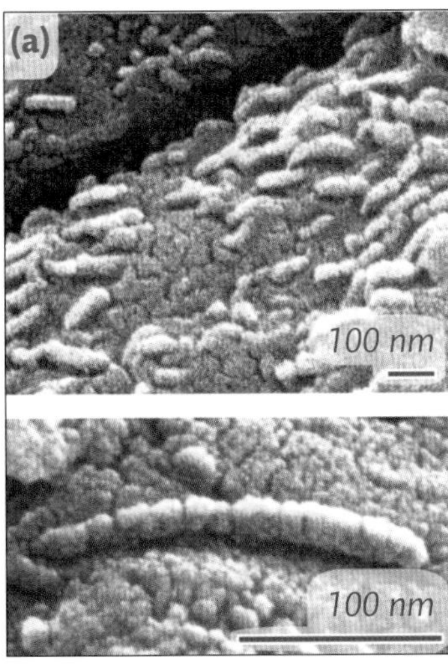

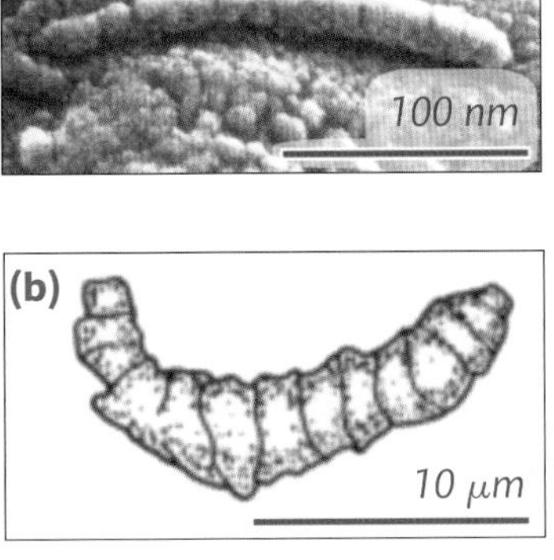

Fig. 13.1 (a) Worm-like structures found on surface of Martian meteorite, ALH84001. (From McKay et al.,[1].) (b) Sketch of *Primaevifilum amoenum* (a prehistoric filamentous bacterium) microfossil from Western Australia. (From Schopf,[2].)
(c) Ice crystals: all bars 100 μm long in this panel. (From Erbe et al.,[4]. By permission of John Wiley and Sons.) All images are false colored.

13.2 **Reaction-Diffusion Equations**

To answer these questions, we will spend some time in this chapter examining some fundamentals of of self-assembly, starting with reaction-diffusion equations such as we introduced last chapter, Eq. [12.45]:

$$\begin{aligned}\frac{\partial A}{dt} &= D_A \nabla^2 A + f(A,B)\\ \frac{\partial B}{dt} &= D_B \nabla^2 B + g(A,B),\end{aligned} \qquad [13.1]$$

where A and B are concentrations of different materials that diffuse at rates D_A and D_B, and the functions f and g define how A and B interact (or react).

We will use mathematical analysis and computer simulations to show that equations such as Eqs. [13.1] lead to the spontaneous formation of patterns, but first let's consider the concept. Eqs. [13.1] say that A and B diffuse through space, and do something when they interact. We'll see that this "something" could be a reaction, or a signal to reproduce, grow, die, activate, inhibit, or move, but for now we observe that the notion that complicated structures emerge from Eqs. [13.1] is not obvious. After all, the only things that were produced last chapter in studying tissue perfusion were smooth variations in concentration.

Indeed, the expectation from the last two chapters should be that when A and B diffuse, they should spread out, as in Fig. 12.2 of Chapter 12. We did see in Chapter 11 that entropy, which is at the heart of diffusion, can produce order out of disorder in a number of circumstances (Fig. 11.3 of Chapter 11), specifically by packing larger particles closer together. But Eqs. [13.1] do not enforce an increase in entropy, except in the vague sense that diffusion spreads things out, and the equations say nothing about particle sizes. Indeed, historically the assertion that chemical reactions can do anything other than approach a uniform asymptotic state was viewed as being plainly impossible.[8]

Historical background, east of the Iron Curtain

A first demonstration that chemical reactions do other than approach uniformity was provided by Boris Belousov (1893–1970), a Soviet biophysicist who sought an inorganic reaction model for the biological Krebs cycle, which allows energy from food products to be stored and released in the forms of ATP and ADP. Belousov observed cyclic, clock-like, reactions by combining a bromate oxidizer with an acid. We'll see that cyclic oscillations can indeed be produced during glycolysis (Eqs. [13.27]), but that is getting ahead of the story. Belousov merely wanted to study a simple analog of a common biological reaction, and observed surprisingly that this analog produced oscillations.

In terms of chemical kinetics, the reactions that Belousov worked with involved the slow consumption by the acid of a visualizing agent (e.g. free bromine), competing with fast autocatalysis (e.g. by potassium bromate). The emergence of chemical oscillations had been reported earlier in a little noted paper by W.C. Bray, who in 1921 reported exploratory chemical investigations combining slow and fast reactions of hydrogen peroxide and iodine.[9] Belousov, though, was unaware of that earlier work and was focused specifically on reproducing a biological cycle.

What Belousov saw was that the visible chemical slowly vanished, and then abruptly reappeared as the concentration of the autocatalytic product grew to a point where it produced large concentrations of the visualizing agent (as in Fig. 13.2(a)). At the time (in the 1950s), this was viewed as impossible (notwithstanding Bray's earlier demonstration), and Belousov despaired at trying to gain acceptance for his findings, only succeeding in publishing in an obscure collection of abstracts to do with radiation medicine.[10]

It should seem odd that Belousov had such difficulties, since his results were not theoretical or speculative, but were experimental and readily reproduced. Indeed, the reader can easily find recipes for clock reactions using common ingredients. Nevertheless, as we saw in Exercise 5.8 of Chapter 5, rejection is an unfortunately common response to new scientific findings. Ultimately, Belousov was vindicated when a graduate student, Anatol Zhabotinsky (1938–2008), reproduced the findings in the now famous "Belousov–Zhabotinsky" (BZ) reaction, and the pair was awarded the prestigious Lenin Prize—sadly only in 1980, a decade after Belousov's death.

The BZ reaction can be seen in both stirred tanks and on stationary gels. In a stirred tank, the BZ reaction produces repetitive clock-like cycles, as shown in Fig. 13.2(a), which as we will see correspond to solutions that appear when the diffusion terms are removed (on account of stirring) from Eqs. [13.1]. On the other hand, in a stationary gel where diffusion must be included, we will see that the fast and slow timescales result in spatial patterns, as shown in Fig. 13.2(b).

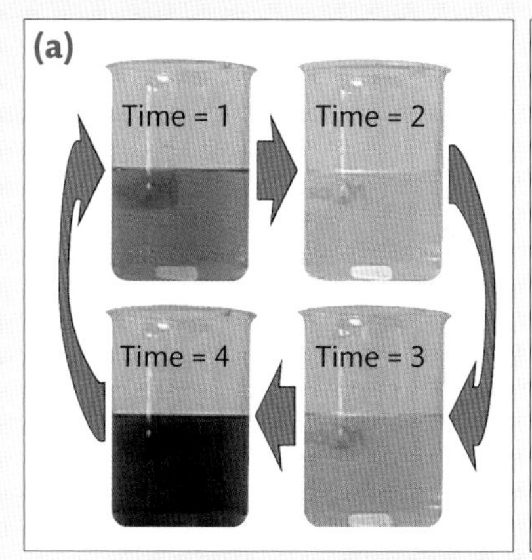

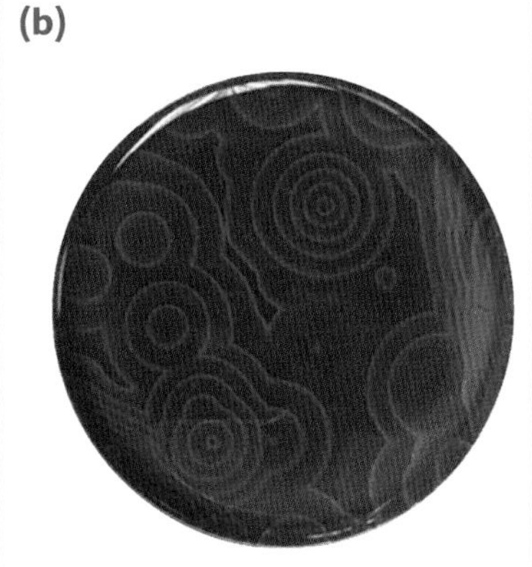

Fig. 13.2 Reaction-diffusion systems. (a) Oscillatory "clock" reaction, reported in 1973 by Briggs and Rauscher.[11] Here, the black state is produced by iodine reacting with starch; the iodine is consumed by malonic acid to produce the clear state shown.[12] Interaction between slow consumption of acid and fast autocatalysis leads to periodic oscillations. The clock reaction is made spatially uniform by stirring (stirrer visible at bottom of flask). (b) Spatial patterns in the Belousov–Zhabotinsky reaction are seen on a gel substrate, which prevents fluid flow. Again, a fast and a slow process interact, and colors indicate different reactant states. (From Tim Kench/YouTube.)

Historical background, west of the Iron Curtain

Meanwhile on the other side of the Iron Curtain in Britain, Alan Turing (1912–1954) wrote a seminal mathematical paper in which he sought to demonstrate that biological morphogenesis could be produced by diffusion of chemical reactants[13]—that is, through models like Eqs. [13.1].

Turing's treatment in Britain was no less tragic than Belousov's in the USSR, though for different reasons. Turing is known for many things beyond his morphogenesis study, including the first computer chess program, the Turing test (to establish whether a computer program can successfully impersonate a living person), the halting problem (showing that it is impossible to establish a general algorithm to determine whether a computer program will halt—and as a corollary whether a mathematical proof can be achieved), and decoding the Nazi's "enigma" code used in military communications. Thus beyond being a polymath and laying the foundations for cryptography, artificial intelligence and biological self-assembly, he as much as anyone else at the time brought the Nazi war in Europe to an end.

He was also a homosexual, which in the end was of greater importance to what is termed British Society than his scientific accomplishments and contributions to world peace. Following the war, Turing was convicted of "indecency" for having an affair with another man, and was sentenced to "chemical castration," leading to his suicide at the age of 41.

13.2.1 A simple first reaction-diffusion equation

We leave history behind for now and start our examination of reaction-diffusion equations with a simpler example than Eqs. [13.1] that will prove instructive when we set upon Eqs. [13.1] themselves. This example actually predates both Belousov and Turing: in the 1930s, statistician and geneticist Ronald Fisher (1890–1962) analyzed the presence of particular varieties of crabs along different regions of the British coastline[14] by reasoning that crabs with an advantageous mutation would reproduce and spread along the coast, as sketched in Fig. 13.3(a).

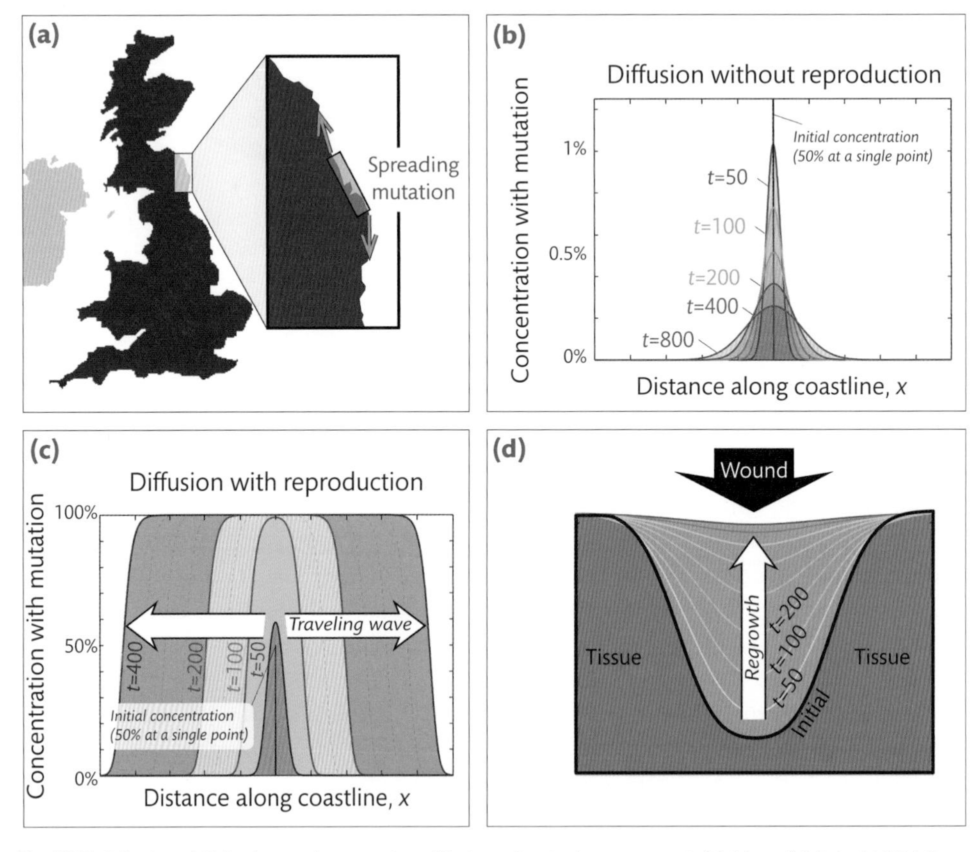

Fig. 13.3 Behavior of Fisher's equation: reaction-diffusion of a single component. (a) Map of Britain highlighting a region with an assumed advantageous mutation of crabs that can spread along the coastline (after Fisher[14], by permission of John Wiley and Sons). (b) Without a "reaction" term (here representing reproduction of crabs), diffusion would cause the mutated crab concentration to spread, but their total number would be conserved, and so the concentration could only decrease at any location. (c) With reproduction, a solitary wave emerges that travels with constant speed and shape—meaning that the advantageous mutation travels steadily along the coastline. Without diffusion, the initial state (the lone mutated crab) would reproduce, but would remain at its initial location. Thus both diffusion and reaction (here reproduction) terms are needed to produce steady spreading of an advantageous mutation. (d) The same model has been applied to describe the migration and growth of epithelial cells to generate new tissue over a wound.

Fisher defined the spread of crabs using an equation that combined diffusion and reaction as follows:

$$\frac{\partial u}{\partial t} = D_u \nabla^2 u + ru(1-u), \quad [13.2]$$

where u is the fraction of crabs with the advantageous mutation (i.e. a concentration) and D_u is a diffusivity (describing wandering of the crabs). Here what we refer to as a reaction is actually the reproduction of the crabs carrying the advantageous mutation. The reproductive term is called a logistic model (briefly mentioned in Chapter 5), which produces exponential growth at rate r for small populations, but penalizes growth as the population approaches 100%.

This model is simple enough that it can be (almost) completely analyzed: Fisher's intuition was that the advantageous mutation might travel along the coast like a wave with constant speed, $-v$, so he attempted a solution:

$$u(x,t) = U(x+vt) \quad [13.3]$$

With a little algebra, we can combine Eqs. [13.2] and [13.3] to produce:

$$D_u \frac{\partial^2 U}{\partial z^2} - v\frac{\partial U}{\partial z} + rU(1-U), \quad [13.4]$$

where $z = x + vt$. This has reduced the partial differential equation in Eq. [13.2] to an ordinary differential equation (ODE), which we can set about to solve as we have done for other equations elsewhere in this book: by reducing it further to an algebraic equation. We'll do this here by attempting a solution:

$$U = U_0 e^{\lambda z}, \quad [13.5]$$

which works if:

$$\lambda = \frac{v \pm \sqrt{v^2 - 4D_u r}}{2D_u}. \quad [13.6]$$

EXERCISE 13.1

This exercise has an analytic part and a computational part. First, show that Eq. [13.4] follows from Eqs. [13.2] and [13.3], and that Eq. [13.6] follows using Eq. [13.5]. Then prove that Eq. [13.6] implies that no growing solution is possible if either the diffusivity, D_u, or the reproductive rate, r, goes to zero. You will find that a questionable assumption is needed to obtain Eq. [13.6]: explain (a) what the justification for the assumption might be, and (b) whether you buy the assumption and why or why not.

Second, use the finite difference code in the Appendix (converted to 1D) to simulate the growth of an advantageous mutation: the result should look like Fig. 13.3(c). Confirm that the numerical solution to Eq. [13.2] is as expected: a wave that travels with constant shape and speed. Thus whether you accept the validity of the assumption made above or not, Fisher's equation does generate wavelike behavior.

The significance of this analysis is that both reaction (here the reproduction of crabs) and diffusion (their wandering) are needed to produce a viable growing solution. You showed this analytically in Exercise 13.1, and it can be seen from your simulations as well. In Fig. 13.3(b) we show simulation results demonstrating that without reproduction an initial colony of mutated crabs will spread, but in a way that conserves their total number. Thus as time advances, their extent will go to infinity but their concentration will go to zero. On the other hand, if the diffusion were zero, Eq. [13.2] would have no spatial dependence at all, so wherever there were mutated crabs, their concentration would grow, but they would have no mechanism to spread elsewhere along the coastline.

When both diffusion and reaction are included in Fisher's equation, a solitary wave emerges, which travels with constant shape and speed outward from the initial colony of crabs: this is shown in Fig. 13.3(c). The reader may recall that we discussed solitary waves as solutions to the Korteweg–de Vries (KdV) equation in Chapter 4. In that problem as well as this one, a diffusion-like term that tended to spread the solution (in that case the speed of a pressure pulse) was compensated by a nonlinear reaction-like term (produced by the elasticity of blood vessels) to produce a wave that traveled with constant shape and speed. Unlike the Fisher equation, the KdV equation conveniently has an exact wavelike solution that allowed us to show that the diffusion-like and reaction-like terms explicitly cancel out, but the lesson was the same: both mechanisms are necessary to generate coherent solutions.

This same process that Fisher considered for the spread of beneficial mutations has been applied to other problems, including flame propagation, defect migration in liquid crystals, and wound healing. In all of these problems, something—combustion, defects, or cells—migrates and grows. Significant differences in detail are involved in each of these examples; nevertheless to first order the simple Eq. [13.2] captures all of their qualitative features. Taking wound healing as a case in point, details including inflammation, formation of an extra cellular matrix substrate on which the cells migrate, fibrosis, and remodeling are important, yet Eq. [13.2] describes regrowth to fill in a wound.[15] This is illustrated in the finite difference solution shown in Fig. 13.3(d): here the same waves as before are present, but rather than traveling outward to propagate a mutation, they travel inward to fill the wound.

13.2.2 **Return to Reaction-Diffusion of Two Reactants**

OK, so combining diffusion and reaction produces a fixed shape traveling wave: what does this have to do with the patterns shown in Fig. 13.2 and described by Belousov experimentally and by Turing theoretically? To answer this question, let's consider what happens when a second reactant is included.

In Fig. 13.4, we show one of the simplest examples of diffusion and reaction, known as Liesegang bands. Originally discovered in 1855 by the notable chemist, Friedlieb Runge, the bands are named after Raphael Liesegang, who wrote about the patterns in 1896. Fig. 13.4(a) shows a gel containing dissolved potassium chromate onto which solid crystals of copper sulfate are placed. After about a week, bands—precipitated copper chromate—begin to emerge. Notice that there is no flow, so only diffusion brings the copper and chromate together. These patterns are believed to be the origin of stripes and rings in agates and other

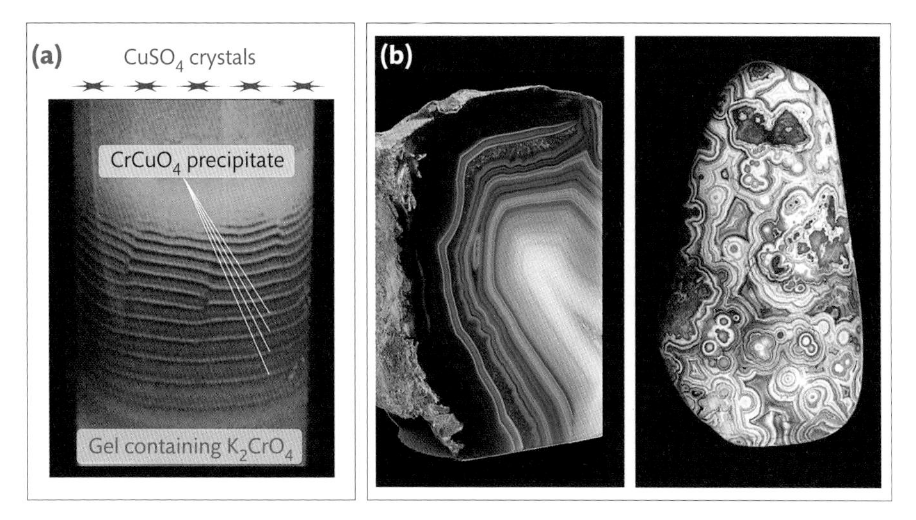

Fig. 13.4 Reaction-diffusion of two components. (a) Liesegang bands form through the reaction and diffusion of solid copper sulfate and dissolved potassium chromate through a gel to produce a precipitate of insoluble copper chromate in colored bands (Courtesy of Gene Stanley). (b) The same process is believed to produce colored bands in agate: these bands can either be simple or complex, depending on concentrations of inclusions within the stone. (left: Darren Pullman/Shutterstock.com; right: www.geologyin.com)

gemstones: both simple and complex examples of agate patterns are shown in Fig. 13.4(b). The reader may also want to compare the large scale patterns in agate shown with small scale patterns in opal (Chapter 11, Fig. 11.3(b)).

With this example as motivation, let's understand reaction and diffusion of multiple components by returning to Eqs. [13.1]. Consider a particular case: imagine that food (which we'll call F as a mnemonic), interacts with bacteria (which we'll call B). The picture we have in mind here is that bacteria consume food to produce more bacteria—Process 1 sketched in Fig. 13.5(a), reminiscent of the crabs in Fisher's equation. Differing from Fisher's problem, food is consumed, and we'll see that combining the dynamics of both bacteria and food leads to the emergence of patterned states. Written in terms of Eqs. [13.1], if two bacteria and one serving of food produces three bacteria, then we would write that:

$$\begin{aligned} \frac{\partial F}{dt} &= D_F \nabla^2 F - r_{consumption} \cdot F \cdot B^2 \\ \frac{\partial B}{dt} &= D_B \nabla^2 B + r_{reproduction} \cdot F \cdot B^2, \end{aligned} \qquad [13.7]$$

where $r_{consumption}$ and $r_{reproduction}$ are the rates of consumption of food and reproduction of bacteria (in time units required to make one new bacterium). Chemists will identify this as a second order reaction in B and a first order reaction in F. We'll obtain conditions (e.g. rates of reaction) that produce spatial and temporal patterns, but for now we just remark that this simple form of "reaction" between bacteria and food produces a plentiful variety of patterns. For simplicity, we'll save some writing by setting $r_{consumption} = r_{reproduction} = 1$, though by this stage the reader may recognize that rescaling the time variable and changing the units of diffusivity will achieve the same simplification.

EXERCISE 13.2

Show mathematically that Eqs. [13.7] lead to the steady depletion of food and growth of bacteria. There is a hard way to do this (solving the equations) and an easy way (stare at the equations until the answer occurs to you). I recommend the easy way—first, simplify the problem by assuming that the concentrations of bacteria and food are the same everywhere.

We can see either from Exercise 13.2 or through simple consideration of the cartoon in Fig. 13.5(a) that the model so far will lead to steady depletion of food and growth of bacteria.

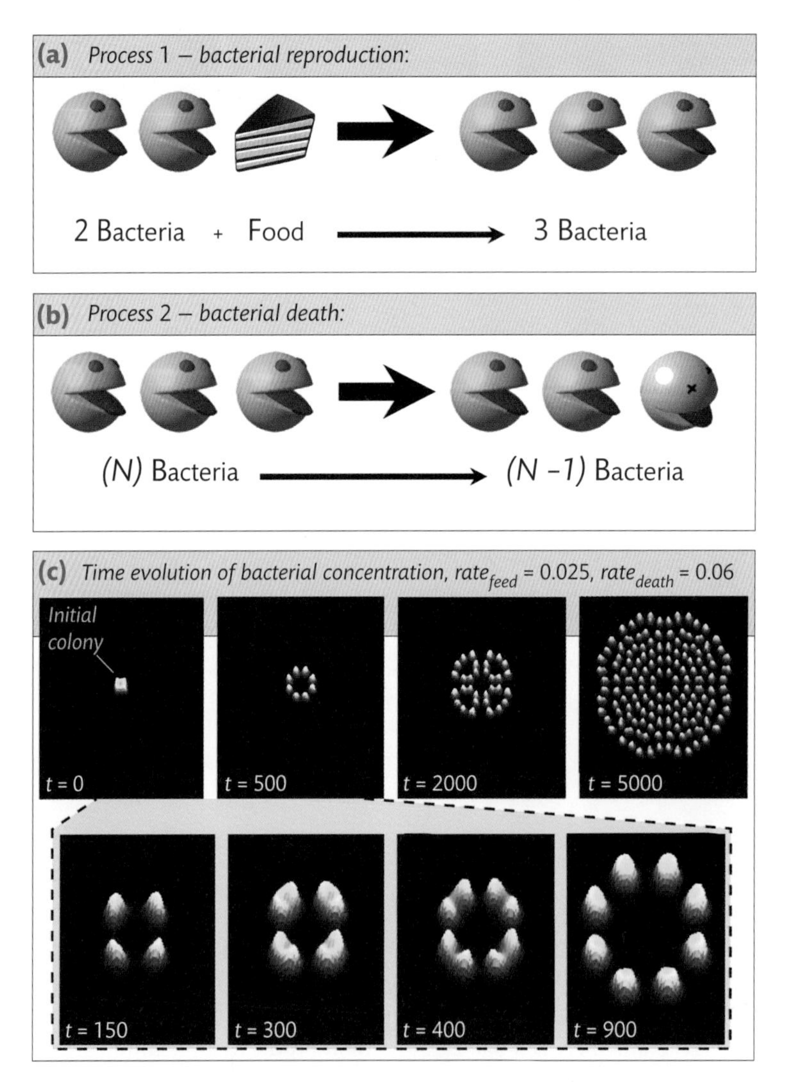

Fig. 13.5 (a) Cartoon of bacteria eating food and reproducing, leading to Eqs. [13.7]. (b) Cartoon of including death in the model, leading to Eqs. [13.8] and [13.9]. (c) Simulation of the growth of an initial colony of bacteria overlying an initial supply of food subject to Eqs. [13.9] using $D_B = 2D_F$ and rates as listed above panel. Black regions are all food with no bacteria, and orange regions are the opposite. Notice that bacterial regions reproduce, forming colonies with fixed size and spacing.

This is somewhat uninteresting, so let's add two more terms. First, consider the case where food is continuously supplied so that the system doesn't come to a stop when the food is all gone, and second let's permit bacteria to die, for example by getting old or from predation by another creature. Then we can write:

$$\frac{\partial F}{dt} = D_F \nabla^2 F - FB^2 + f(F, B)$$
$$\frac{\partial B}{dt} = D_B \nabla^2 B + FB^2 - g(F, B), \qquad [13.8]$$

where f is a function that increases the amount of food—and so must include a feed rate, $rate_{feed}$—and g is a function that defines cellular death—which again must include a rate, $rate_{death}$. We depict this as Process 2 in Fig. 13.5(b), completing the cartoon of the processes described in Eqs. [13.8]. Shortly we will describe how these functions relate to the patterns expressed, but for now we'll use a recipe that turns out to work well:

$$\frac{\partial F}{dt} = D_F \nabla^2 F - FB^2 + rate_{feed}(1 - F)$$
$$\frac{\partial B}{dt} = D_B \nabla^2 B + FB^2 - (rate_{feed} + rate_{death})B. \qquad [13.9]$$

Eqs. [13.9] are known as the Gray–Scott equations,[16] as described in Pearson.[17] Finite difference code that can be used to simulate their solution is provided in the Appendix (see also Exercise 13.3). Plainly enough, adding the $rate_{feed}$ term increases the quantity of food, F, and increasing the $rate_{death}$ term decreases the quantity of bacteria, B: you'll show this numerically in Exercise 13.3. More than this, consider what happens to a small colony of bacteria in the presence of food. This is shown in Fig. 13.5(c) for a particular choice of $rate_{feed}$ and $rate_{death}$.

As demonstrated in Exercise 13.2, the bacterial colony does grow and consume food, but as shown in Fig. 13.5(c), the bacteria also maintain fixed size and spacing of colonies. This has the appearance of cellular division, however we emphasize that this is pure serendipity: there are no discrete cells and there is no cellular machinery associated with mitosis. This is a case where the equations resemble biology unreasonably well: a situation presaged by Eugene Wigner (1902–1995) in his paper,[18] "The unreasonable effectiveness of mathematics in the natural sciences."

Moreover, the cell-like bacterial colonies shown in Fig. 13.5(c) are only one solution to Eqs. [13.9], and as we'll examine in Exercise 13.3 (see also Fig. 13.6(a)), most of the solutions aren't round or self-contained. The other solutions do share a property with Fig. 13.5(c) however: they tend to have a characteristic length scale. And at this point we already know enough to calculate this scale.

13.2.3 **Analysis 1: Dimensions and Scale**

To do this, let's return to the first tool that we introduced in this book: dimensional analysis. The concentrations of food and bacteria, F and B, in Eqs. [13.9] are dimensionless, so we'll ignore them. This leaves the dimensions:

$$\frac{1}{Time} \sim D_F \frac{1}{Length^2}$$
$$\frac{1}{Time} \sim D_B \frac{1}{Length^2}. \qquad [13.10]$$

If we subtract these equations one from another and rearrange terms, we get:

$$Length \sim \sqrt{(D_F - D_B) \cdot Time}. \qquad [13.11]$$

So the only thing in Eqs. [13.9] that has the dimension of length—and so the only thing that can produce a length scale in the ultimate patterns—depends on the difference between diffusion constants of food and bacteria. This makes sense, right? If food and bacteria diffuse at the same rate, Eqs. [13.9] will do their business in the same way to both F and B at every point in space, with nothing to establish more of either quantity here versus there. Only when bacteria or food overtake one another will there be more of one or the other at different locations—and indeed in Figs. 13.5(c) and (d), the bacteria diffuses twice as fast as the food. Incidentally, the reader may want to refer back to Chapter 7, where we discussed the advantages of bacteria diffusing faster than food.

EXERCISE 13.3

The length scale $\sqrt{(D_F - D_B) \cdot Time}$ obtained in Eq. [13.11] has numerous manifestations. In this exercise, embellish the finite difference solution to the diffusion equation given in the Appendix by including the terms shown in Eqs. [13.9]. You will have to include a second concentration variable (i.e. food, F, as well as bacteria, B), and you'll find that asymptotic patterns arise more rapidly if you add some randomization to the initial state (e.g. use several small colonies at random locations or add small random values to the initial colony's concentration). Use the resulting simulation to plot several distinct patterns by varying $rate_{feed}$ and $rate_{death}$. As indicated in Fig. 13.6(a):

(1) Show that as the death rate grows and the feed rate diminishes, the bacteria die out.

(2) Show that as the death rate diminishes and the feed rate grows, the bacteria grow until they reach a maximum "carrying capacity" allowed by the feed rate.

(3) Identify several distinct bacterial patterns between these extremes. A few examples are included in Fig. 13.6(a), but you can find many more—some steady and others time dependent.

(4) Measure the spacing between the states with steady patterns and show that they grow with $(D_F - D_B)$. You'll find that once the patterns are established, their spacing doesn't change with time: why not? Hint: consider the steady case, where the time derivatives vanish: what then sets the length scale?

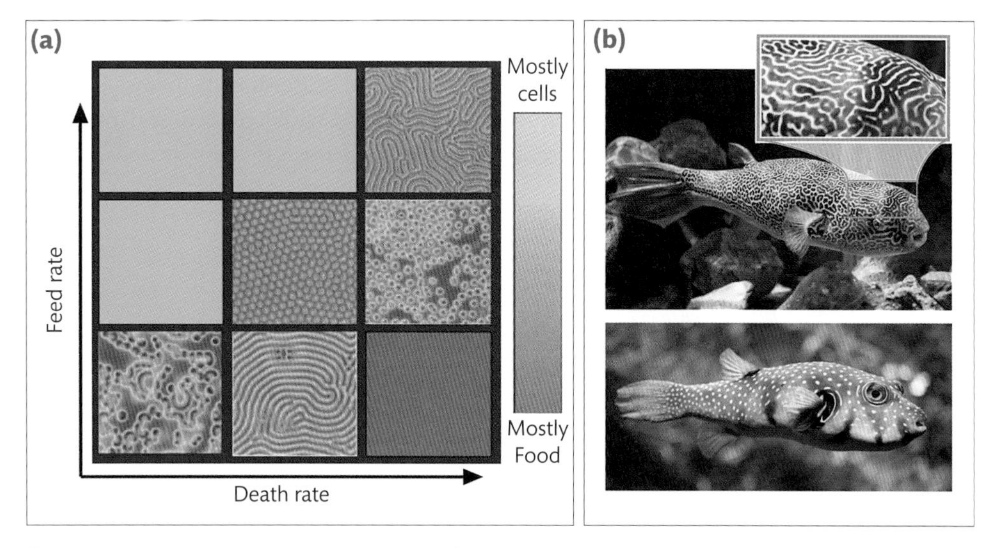

Fig. 13.6 Qualitative patterning behaviors from Eqs. [13.9]. (a) At high death rate and low feed rate, bacteria die out, while at high feed rate and low death rate, bacteria take over. Between these extremes, a variety of patterned states emerge: a small subset is shown here; see Pearson[17] for others. (b) Labyrinthine, striped, and spotted patterns on pufferfish. Notice from both panels that that although the patterns differ, the length scales of the patterns are similar (top: Dennis Jacobsen/Shutterstock.com; bottom: Vladimir Wrangel/Shutterstock.com).

ADVANCED EXERCISE 13.4

Consult the literature on bacterial growth patterns and identify both patterns that resemble those you can simulate using Eqs. [13.9] and patterns that you cannot. You will also find that there are patterns that you can simulate that do not resemble bacteria. Here you may compare your simulations with animal coat patterns—as shown, for example, in Fig. 13.6(b). Similar models to Eqs. [13.9] but using different interaction terms have been used to describe animal coats (see, for example, Ball[19]).

Your goal for this exercise is to identify as many solutions to Eqs. [13.9] with biological patterns as you can. If possible also propose (and if you are ambitious, try to implement) modifications to the model that could account for the some of the patterns that are not described by Eqs. [13.9].

Remark: the modeling of biological patterns is an open problem. The flip side of Wigner's observation that by the same token that some models are unreasonably effective in describing the natural sciences,[18] so the natural sciences can sometimes be unreasonably resistant to modeling. Many bacteria—as well as tissues in higher organisms (cf. Figs. 13.6(b) and 13.10)—produce highly intricate patterned structures that are very stereotyped and well controlled, but have proven challenging to describe mathematically. Some of these patterns may rely on inter- or intra-cellular signaling, but at the same time similarities to complex nonbiological reaction-diffusion systems call that proposition into question. Have a look, for example, at complex patterns in dendritic agate, or at a different diffusion-driven patterning phenomenon: phase separation through spinodal decomposition. Other topics involving both biological and nonbiological patterns are overviewed in Ball.[20]

13.2.4 **Analysis 2: Space Dependent Solutions**

Beyond dimensional analysis, we can derive onset criteria for patterning as follows. We will start by focusing on time-independent solutions to Eqs. [13.9]—we'll look at time dependence next, but it turns out that this will involve some minor mathematical complications, so let's ease our way into the problem.

Our starting place is Eqs. [13.9] with the time derivatives set to zero:

$$\begin{aligned} 0 &= D_F\nabla^2 F - FB^2 + rate_{feed}(1-F) \\ 0 &= D_B\nabla^2 B + FB^2 - (rate_{feed} + rate_{death})B. \end{aligned} \quad [13.12]$$

Notice that the red spatial derivatives will contribute whenever there is spatial variation. Which is to say that when these terms vanish, spatial patterns vanish.

So, *without even having to solve the problem*, we can tell when patterns will go away—or contrariwise when patterns might be expected to appear. This is easy: we just set the red terms to zero, and ask when it is possible to find a solution: these will be spatially uniform states. When it is *not* possible to find a solution, the states must either not be uniform (i.e. be patterned) or there will simply be no solutions.

This is now just algebra. The bottom equation of Eqs. [13.12] gives:

$$B = \frac{rate_{feed} + rate_{death}}{F}. \quad [13.13]$$

It's also possible that there are no bacteria ($B = 0$): that so-called "trivial" solution is wholly uninteresting and so we neglect it for now, but we'll return to it for completeness shortly. We can plug the nontrivial possibility, Eq. [13.13], into the top equation of Eqs. [13.12] to get:

$$F^2 - F + \frac{(rate_{feed} + rate_{death})^2}{rate_{feed}} = 0, \quad [13.14]$$

so

$$F = \frac{1}{2} \pm \sqrt{\frac{1}{4} - \frac{(rate_{feed} + rate_{death})^2}{rate_{feed}}}. \quad [13.15]$$

Problem solved: the uniform state is given by Eqs. [13.13] and [13.15], and Eq. [13.15] only exists so long as the square root is real, or:

$$rate_{feed} \geq 4(rate_{feed} + rate_{death})^2. \quad [13.16]$$

In Fig. 13.7(a), we plot this boundary between uniform and non-uniform solutions (solid black curve) overlying a color-coded phase plot that schematically identifies patterns seen in finite difference simulations of Eqs. [13.9]. The blue region to the left of the black curve represents states that are computationally found to be uniform, and the green regions near and to the right of the curve represent patterned states. The red region further to the right again represents uniform states, but in this case, these states have $B = 0$. The reason for this is simple:

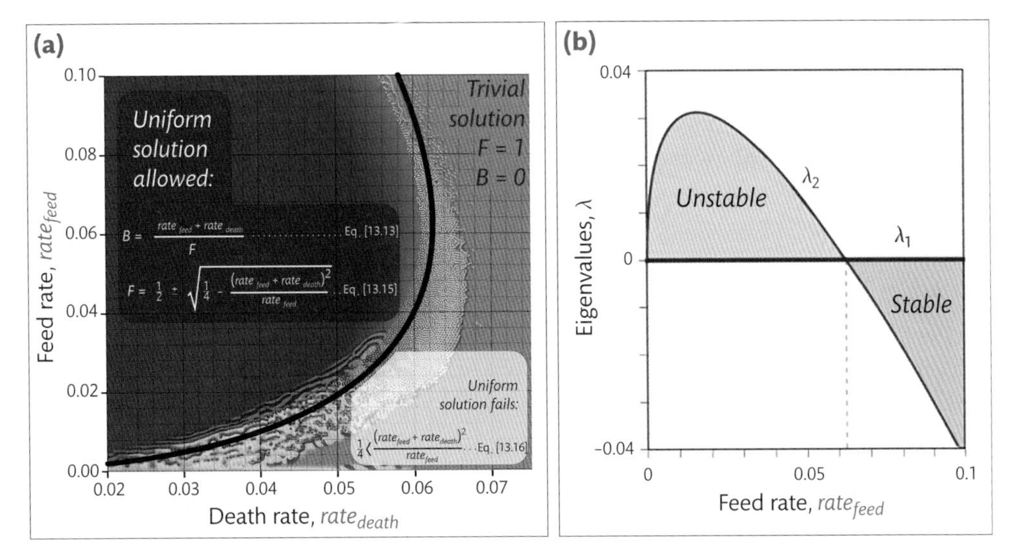

Fig. 13.7 (a) Parameter plot defining where spatially uniform solutions to the Gray–Scott equations (Eqs. [13.9]) are allowed. Black curve shows boundary between uniform and non-uniform solutions defined by Eq. [13.16]. Background shows computational solutions, where blue regions are uniform, red regions are trivial, and between are various non-uniform solutions. (Modified from: https://mrob.com/pub/comp/xmorphia/. Courtesy of Robert Munafo.) (b) Approximate eigenvalues for Gray–Scott model vs. feed rate, showing that solutions tend to be unstable in time for $rate_{feed} < 0.6$.

further to the right, the death rate is higher, the bacteria get killed off, and the trivial solution (which we neglected to this point) emerges. We turn next to analyzing when non-uniform steady solutions give way to unsteady ones.

13.2.5 **Analysis 3: Time Dependent Solutions in a Bounded System**

We've made some progress determining properties of steady solutions to Eqs. [13.9] without bothering to actually solve the equations, but this will only take us so far. To understand time dependent solutions, we need to ask when a stable patterned state becomes unstable and gives rise to time variations. This, it turns out, can happen in either of two ways. The first is a wave-like solution, similar to what we considered in Fisher's equation: we examine this using the Gray–Scott equations that we have already seen. Second, oscillations can result through the emergence of a limit-cycle, which we will describe using a different, but closely related, set of equations that we will introduce shortly.

13.3 **Wave Solution**

Recall that in order to solve Fisher's equation, we assumed a wave-like solution, where $z = x + vt$. This worked well enough, so let's try it again here. To keep things simple, we'll consider a one-dimensional wave—as did Fisher—in which case Eqs. [13.9] become:

$$\begin{aligned} vF' &= D_F F'' - FB^2 + rate_{feed}(1 - F) \\ vB' &= D_B B'' + FB^2 - (rate_{feed} + rate_{death})B, \end{aligned} \qquad [13.17]$$

where the primes denote differentiation with respect to z.

In principle, we could try to solve Eqs. [13.17] as they stand, but to keep things simple note that Fig. 13.7(a) shows that computational patterned solutions all lie close to the boundary between uniform and non-uniform solutions. We solved for that boundary by assuming the diffusion terms vanish, so let's do the same thing now with the goal of establishing when solutions near that boundary become unstable. In truth, this is only an approximation, because we know to the right of the boundary that there must be spatial variations—and so the diffusion terms cannot possibly vanish. Nevertheless we can use this approximation to establish not what happens to the right of the boundary curve (where diffusion cannot vanish), but where the first onset to instability lies along that curve (where diffusion can vanish, just).

In this case, the equations whose stability we want to analyze become:

$$\begin{aligned} F' &= \frac{1}{v}\left[-FB^2 + rate_{feed}(1 - F)\right] \\ B' &= \frac{1}{v}\left[FB^2 - (rate_{feed} + rate_{death})B\right]. \end{aligned} \qquad [13.18]$$

We've learned a thing or two about instability in this book: for example we've learned that we can establish where stability breaks down by asking when a small perturbation grows exponentially rather than shrinks. Mind, here we are differentiating with respect to z and not to time, but if a solution grows exponentially in time, it will also grow exponentially in $z = x + vt$. So let's see what happens if we try the exponential solution:

$$\begin{aligned} F &= F_0 + \epsilon \cdot e^{\lambda_F z} \\ B &= B_0 + \epsilon \cdot e^{\lambda_B z}. \end{aligned} \qquad [13.19]$$

Here F_0 and B_0 are solutions from Eqs. [13.13] and [13.15] on the boundary curve defined when Eq. [13.16] is an equality, and ϵ is an infinitesimal value.

As in Chapter 12, the reader may recognize this as an eigenvalue problem—but if not, no worries, we'll recap the high points of figuring out what the solution (Eqs. [13.19]) tells us. The idea behind eigenvalues is that the exponents λ_F and λ_B in a sense ought to be different from one another since the equations for F' and B' are different, but in fact there are two special choices, λ_1 and λ_2, that can typically be found such that $\lambda_F = \lambda_B$. Which is to say that for these choices, F and B grow or decay at the same exponential rate. These choices define behaviors that are in a sense more fundamental than other more general behaviors, and so are termed eigenstates after the German "eigen," meaning "own," or "inherent."

To find these eigen behaviors, we can plug the solution (Eq. [13.19]) into Eqs. [13.18] and solve for $\lambda = \lambda_F = \lambda_B$. Making use of the fact that ϵ is small so that we can neglect higher powers of ϵ, we obtain the general form commonly expressed as a matrix product:

$$0 = \begin{bmatrix} \frac{\partial F'}{\partial F} - \lambda & \frac{\partial F'}{\partial B} \\ \frac{\partial B'}{\partial F} & \frac{\partial B'}{\partial B} - \lambda \end{bmatrix} \begin{bmatrix} F \\ B \end{bmatrix}. \qquad [13.20]$$

If this isn't familiar to you, you can either work it through to convince yourself that it is correct, or you'll find many books and reference materials on eigenvalue analysis that describe this derivation (e.g. Strang[21]). In our case,

$$0 = \begin{bmatrix} -B^2 - rate_{feed} - \lambda & -2FB \\ B^2 & 2FB - (rate_{feed} + rate_{death}) - \lambda \end{bmatrix}. \quad [13.21]$$

This can be solved for λ (as we showed in Chapter 12, Eq. [12.44]), but the solution is a bit of a mess. Recall though that we want to solve the problem on the boundary curve, where Eqs. [13.13]–[13.16] tell us:

$$(rate_{feed} + rate_{death}) = \frac{\sqrt{rate_{feed}}}{2} \quad [13.22]$$

$$F = \frac{1}{2} \quad [13.23]$$

$$B = \sqrt{rate_{feed}}. \quad [13.24]$$

This is a help: so Eq. [13.21] becomes:

$$0 = \begin{bmatrix} -2rate_{feed} - \lambda & -\sqrt{rate_{feed}} \\ rate_{feed} & \frac{\sqrt{rate_{feed}}}{2} - \lambda \end{bmatrix}, \quad [13.25]$$

from which we can work out the eigenvalues:

$$\lambda_1 = 0$$
$$\lambda_2 = 2 \cdot rate_{feed} - \frac{\sqrt{rate_{feed}}}{2}. \quad [13.26]$$

These are plotted in Fig. 13.7(b). Apparently, near the boundary curve where patterned states first appear, solutions tend to become unstable for $rate_{feed} < 0.06$—and so time dependent solutions can be expected. Moreover, around $rate_{feed} = 0.02$ (cf. Fig. 13.7(b)), λ_2 is maximal, so, solutions are most unstable.

ADVANCED EXERCISE 13.5

Use your simulation from Exercise 13.3 to confirm that patterns are strongly time dependent around $rate_{feed} = 0.02$, and remain time dependent for $rate_{feed} \leq 0.06$. At higher $rate_{feed}$, you will find that the range in $rate_{death}$ that leads to patterns rather than the uniform or trivial solution is limited, and so it may be necessary to start at lower $rate_{feed}$ and slowly work your way up. To avoid restarting the program repeatedly, the Appendix provides a few ways of changing a program while it is running.

You will also find that time dependence persists (although it slows) above $rate_{feed} = 0.06$. Where specifically does the derivation that we have used to obtain that limit fail – that is, why aren't patterns stable above $rate_{feed} = 0.06$? Also, why don't we see time dependence to the left of the limiting curve shown in Fig. 13.7(a): shouldn't states there be unstable by the same argument that we have provided?

13.4 Limit Cycle Solution

We've seen that the Gray–Scott equation becomes unstable at a prescribed set of parameters. What happens to the solution at that point merits understanding, both for its own sake and because the behavior that results is broadly applicable. A different reaction-diffusion problem that both illustrates this breadth and is simpler to analyze than Gray-Scott is the Schnakenberg system, which arises as a model for glycolysis,[22] the fundamental biological process for breaking down sugar. In yeast and muscle, oscillations in the glycolysis are observed under conditions of stress (e.g. when baker's yeast is starved), and this has been modeled using the following equations:

$$\begin{aligned}\frac{\partial x}{dt} &= D_x\nabla^2 x + x^2 y - x + \alpha \\ \frac{\partial y}{dt} &= D_y\nabla^2 y - x^2 y + \beta.\end{aligned} \qquad [13.27]$$

Here x and y respectively define the concentrations of F6P (fructose 6-phosphate, an early product along the metabolic pathway) and ADP (the energy storage compound, adenosine diphosphate), and the parameters α and β provide feed rates for these quantities. These equations bear an obvious resemblance to the Gray–Scott system, Eqs. [13.9].

Let's start with the simplest thing by solving for the uniform, steady state by setting space and time derivatives to zero (as we did in the Gray–Scott case):

$$\begin{aligned}0 &= x^2 y - x + \alpha \\ 0 &= -x^2 y + \beta,\end{aligned} \qquad [13.28]$$

so:

$$\begin{aligned}x &= \alpha + \beta \\ y &= \frac{\beta}{(\alpha+\beta)^2}.\end{aligned} \qquad [13.29]$$

Next, let's ask when that state is stable (again as we did in Gray–Scott) by solving the eigenvalue equation. This is straightforward, as Exercise 13.6 shows.

EXERCISE 13.6

Solve the eigenvalue equation and show that for small α, the eigenvalues, become positive when $\beta < 1$—meaning the solution becomes unstable at this point. Plot the eigenvalues as a function of β for a small value of α, and show that their real part changes from positive to negative when $\beta = 1$.

So for small α, the steady uniform solution becomes unstable when $\beta < 1$. What happens then? There are only two possibilities, right? Either x and y grow without limit, or they don't. In any practical problem there is a limit to growth, and indeed in our problem a limit is imposed by the $-x$ term in Eqs. [13.28]: this term produces exponential damping that builds with x and so is sure to prevent boundless growth. This is especially easy to see by adding Eqs. [13.27] to obtain $\partial(x+y)/\partial t$, which has only a linear negative term and so is exponentially damped.

In any system on a plane (here x and y), a helpful and thoroughly logical theorem named after Henri Poincaré (mentioned in Chapter 5) and Swedish mathematician Ivar Bendixson (1861–1935) defines what can possibly happen. In short, the Poincaré–Bendixson theorem says that bounded solutions to ODEs in a plane (e.g. Eqs. [13.18]) must either end at a stable fixed point or on a limit cycle.

What does this mean? Consider Fig. 13.8(a). We depict there a steady state (termed a fixed point, because trajectories at that point are fixed in time) that is stable, surrounded by a boundary that no trajectories can leave. Imagine that, by changing parameters, the fixed point becomes unstable. In that case, trajectories will head outward toward the boundary, but of

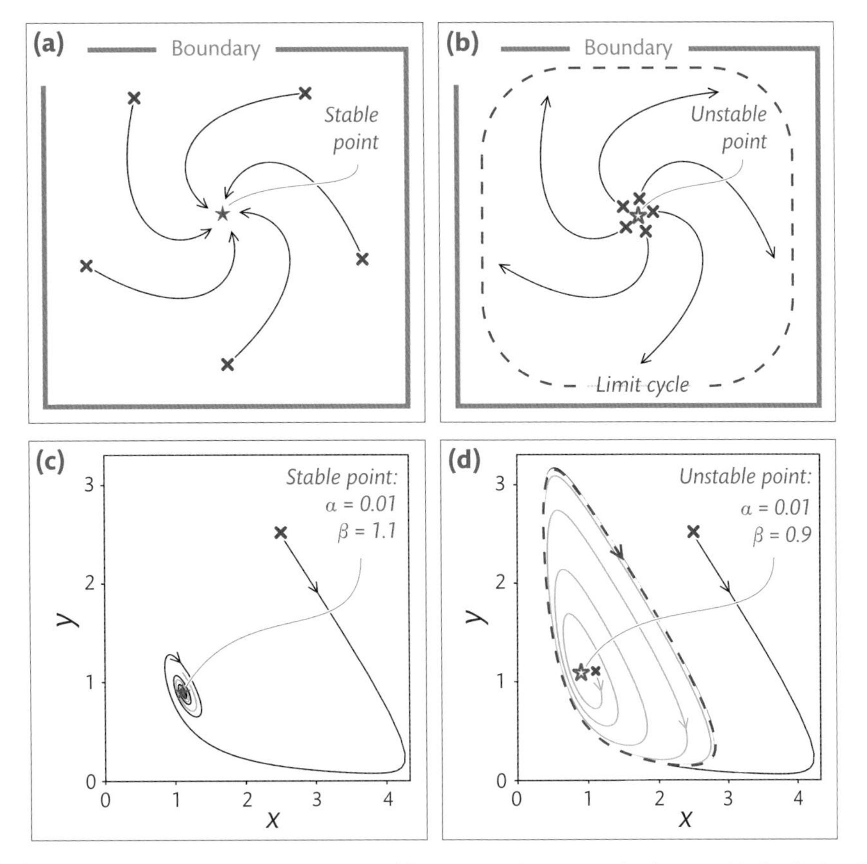

Fig. 13.8 Geometry of Poincaré-Bendixson theorem. (a) Depiction of trajectories leading to stable fixed point. Blue x's indicate starting points. (b) Depiction of trajectories leading away from unstable fixed point. Notice that when the domain is bounded (green rectangle), the outgoing trajectories cannot escape, and so must end in a limit cycle. (c) Typical trajectory leading to stable fixed point in a Schnakenberg system, Eqs. [13.27], without diffusion terms. (d) Typical trajectory in same system for parameters producing an unstable fixed point: note the inevitable emergence of a limit cycle. Green trajectory starting close to the fixed point is similarly attracted to the limit cycle, but from within.

course they cannot cross the boundary and so simple geometry dictates that any trajectory traveling outward from the fixed point will end up endlessly circling around, shy of the boundary, in a limiting cycle, as depicted in Fig. 13.8(b).

EXERCISE 13.7

As with any theorem, Poincaré–Bendixson's has caveats, commonly expressed in mathematically opaque language, but with simple geometric meaning. Describe (with sketches as needed) at least three things that a trajectory leaving a fixed point could do other than end at a limit cycle, and explain why these cannot occur. As part of your description, explain why trajectories cannot cross one another.

Next, reproduce the trajectories shown in Figs. 13.8(c) and (d) by numerically integrating Eqs. [13.27]. Show that the transition to a limit cycle coincides with the fixed point becoming unstable (e.g. by identifying parameters at which points very close to the fixed point change from being attracted to repelled, and showing that these trajectories end at a limit cycle). Note that this means that local changes, here nearby a fixed point, affect global dynamics, much further away (we mentioned this general property briefly in Chapter 8 in terms of topological index theorems). Show that this is a general result by examining a different model, for example, the Fitzhugh–Nagumo model for reactive media (e.g. in neurons, or in heart tissue) or the Gieirer–Meinhardt model for activation-inhibition (in intracellular regulation).

13.5 **Summing Up**

In earlier exercises in this chapter, we saw that spatial patterns can appear when a uniform solution to a reaction-diffusion equation fails. Exercise 13.7 shows that temporal oscillations appear (cf. the clock reaction of Fig. 13.2(a)) when a fixed point in a reaction-diffusion problem becomes unstable in a bounded system.

Mathematically speaking, this tells us that two essential ingredients are associated with pattern formation. The essential ingredient needed to generate a *spatial* pattern is the breakdown of the uniform state, and the essential ingredient needed to produce *temporal* oscillations is the breakdown of the steady state. Correspondingly, the wavelength of spatial patterns that result is prescribed by a difference in diffusivities between interacting ingredients, and the period of oscillations is prescribed by the size of the underlying limit cycle.

Mechanistically speaking, we again have seen that two essential ingredients are needed to produce interesting dynamics (solitary waves for the Fisher equation or patterns for Turing equations). These are a reaction—which allows distinct states to be generated—and diffusion—which allows the evolution of these states. There isn't a one to one correspondence between reactions and temporal cycles or diffusion and spatial patterns, but it is close.

More importantly, it's worth considering that our analysis of Turing patterns involves a twist on the entropic view presented in Chapter 11. In that chapter, we saw that ordered states emerge as a consequence of the requirement that entropy increases. In this chapter, we have seen that ordered states emerge as a consequence of diffusion—which we know from Chapter 11 is a description of entropic growth associated with molecular motion. Thus the earlier results showing that large and small particles separate to maximize entropy in fact have a more general analog in pattern formation: different chemicals or cells separate to form patterned states, again to maximize entropy.

Because the increase in entropy is universal and applies to all systems, similar patterns are found in very different problems, ranging from bacterial reproduction (using the Gray–Scott model) to electro-chemical reactions (Schnakenberg) to neuronal firing (Fitzhugh–Nagumo) to intracellular regulation (Gieirer–Meinhardt). The behaviors that result are complex in detail, but the mathematics and mechanisms associated with these behaviors are fundamentally identical in all of these very different systems.

We won't indulge in the philosophical aspects of this conclusion (readers so inclined may enjoy Erwin Schrödinger's short book, *What is Life?*), but we will emphasize that the connection between patterns and entropy explicitly addresses the questions raised at the start of this chapter. *Because* pattern formation is a consequence of the Second Law of Thermodynamics, its emergence in nonbiological as well as biological systems is not a surprise: it is to be expected. Likewise the similarity in appearance of nonbiological and biological patterns is to be expected: the same mechanisms and the same mathematics are at work in both classes of systems.

There is of course much more to in this story—for example, the selection of a pattern (stripes, spots, labyrinths, etc.) is governed by the fastest growing, i.e. most unstable, modes (as described in Chapter 2). Moreover the size of a limit cycle, and so the actual periodicity exhibited, is highly nontrivial to calculate. This is a wonderful field, full of intriguing problems, and the interested reader is referred to books by Murray[23] and Winfree[24] for accessible mathematical descriptions of pattern formation and biological rhythms. Additionally, some of the enormous variety of natural structures, originally examined on a case-wise basis, for example, in Thompson[25] and His[26], is attractively overviewed in Ball.[20]

13.6 Conclusion and Other Problems

We close this chapter—and this book—by re-emphasizing that the problems we have described represent only the small subset of problems that can be solved. Far more problems than these have not been solved. Let's therefore touch on a few of these.

13.6.1 Other Mechanisms and Caveats

We have told Alan Turing's story in this chapter, namely that diffusion combined with reaction provides a mechanism that can describe pattern formation in both inorganic and organic systems. This is a major and remarkable result, and again we emphasize the connection with Chapter 11, that without diffusion (i.e. increasing entropy), these patterns could not form. At the same time, this is only one of many stories about pattern formation. In Fig. 13.9, we illustrate this point by displaying three comparisons between physical and biological patterns that are associated with very different mechanisms.

In Fig. 13.9(a), we compare agglomerates of soap bubbles with retinal cells in the fruit fly. The evident similarity is not accidental: fly retinal cells agglomerate under the influence of unavoidable surface tension forces, and the same forces that act on bubbles also act on cell

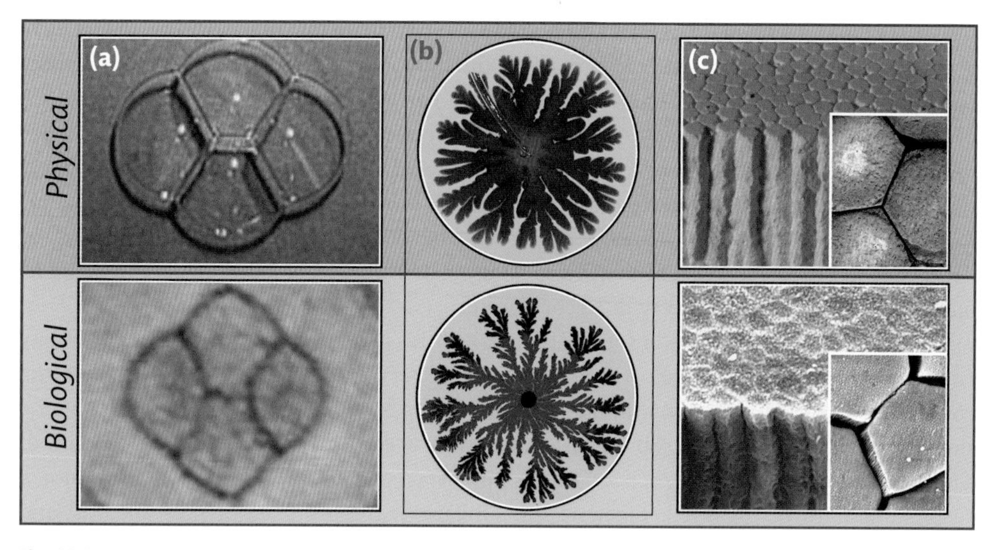

Fig. 13.9 Comparison between physical and biological structures. (a) Top: soap bubble, bottom: retinal cone cells in *Drosophila* (from Hayashi and Carthew,[27], by permission of Springer Nature); (b) Top: viscous fingering of colored water injected into glycerine,[28]; bottom: branching morphotype of *Paenibacillus dendritiformis* bacteria (from Ben-Jacob[29], by permission of the Royal Society). (c) Top: columnar evaporation in lab (from Goehring et al.,[30], courtesy of Lucas Goehring) (main display) and geology[31] (inset), bottom: epithelial cells in mammals[32] (main display) and frog toepads (from Federle et al.[33], by permission of the Royal Society) (inset). Panels (b) are false colored.

membranes.[27] Indeed, having had our thinking primed by the comparison between inorganic and organic patterning, it might in retrospect be surprising if biological morphogenesis did not have analogues in nonbiological systems. So surface tension is a first example of a mechanism that plays a significant and plausible role in cellular patterning that has little to do with the Turing model that we have studied.

OK, fine. But compare this conclusion with Fig. 13.9(b), where we show so-called Saffman–Taylor viscous fingering, which occurs as a low viscosity fluid (here water) is injected into a high viscosity one (here oil). This kind of fingering has implication in oil extraction, but is also seen in other systems, for example in aquifer and ocean mixing.

It has been argued that if fluids form these patterns when injected into a more viscous medium, cells might do likewise when growing into a more viscous substrate.[34] This is indeed what is seen, as found in bacteria shown at the bottom of Fig. 13.9(b), however in this case a reader could justifiably question the connection. For example, not all bacteria form these patterns, and bacterial reproduction at the leading edge of a colony is driven by different forces than fluid injected far from the edge. In the growth of bacteria, several mechanisms appear to be at work, some potentially describable by a Turing model, others related to classical fluid mechanisms, and still others due to entirely separate processes, for example swimming or crawling in response to chemical cues. Plainly this variety of processes—and the resulting variety of cellular structures—requires careful analysis, and it would be a mistake to believe either that a single picture—due to Turing, surface tension, viscosity or any other isolated cause—is all encompassing. So this is a first caveat to carry into understanding of further studies of pattern formation.

To illustrate a second caveat, in Fig. 13.9(c) we show another comparison. On the top, we show geological columns reproduced in the laboratory (main display) and seen in the "Giant's Causeway" in Northern Ireland (inset). On the bottom, we show columnar epithelial cells (main display) and cells from the toepad of a frog (inset). Here despite the similarity between three-dimensional as well as two-dimensional patterning, it is difficult to associate the growth of epithelial cells with the mechanism of formation of geological columns. The former, after all, is part of a developmental process steering cell shape and motion with chemical gradients and tightly controlled signalling mechanisms to produce a required form and function. The latter patterns, by contrast, are formed by the gradual evaporation of liquid upwards through a muddy suspension.

While future authors might yet find a commonality in the two mechanisms, it is safe for the time being to conclude that despite the successes of Turing's insight, visual similarities do not imply mechanistic ones. This is a departure from the initial premise of this book, namely that the sinking of the SS Fitzgerald holds lessons for biological flows; nevertheless the caveat remains that not everything that walks like a duck is a duck, and caution, as well as daring, is called for in modeling of complex systems.

13.6.2 **Open Areas**

Bearing these caveats in mind, there are several open areas for research in the patterning field that are both sweet and deep. A first of these is the *practical application* of patterning equations. As you will have discovered in Exercise 13.3 and Advanced exercise 13.4, there are features of biological patterning such as length scale that are moderately well described by reaction-diffusion models, but many other features remain poorly understood. For example, most bacterial growth patterns are not actually very well described by the Gray–Scott equations. On the other hand patterns on many animal coats—which are governed by very different mechanisms than those used in the Gray–Scott derivation—do appear to be qualitatively described by these equations. We saw this in the pufferfish in Fig. 13.6(b), which exhibit spotted, striped, and labyrinthine patterns similar to those generated by the Gray–Scott equations.

So the simple question of why equations for bacteria work only occasionally for bacteria but very often for animal coats is practically important, but seldom considered. Moreover, beyond relatively simple spatial patterning, organisms such as the cuttlefish[35] and octopus[36] can generate skin patterns resembling solutions to Eqs. [13.8] in a time dependent way that they appear to use for camouflage, mating, or to confuse prey.[37] Nature's richness in patterning far exceeds our ability to produce accurate models, and is sure to provide enticing material for research for decades to come.

A second, more narrow, area of ongoing research concerns the *effect of boundaries* on patterning. Here notice that the simulations that you used in Exercise 13.3 are periodic in both directions, and so lack realistic boundaries. Natural patterns, however, depend in complicated ways on boundaries—note, for example, the change from spots to stripes near the cuttlefish eyes in the bottom photograph of Fig. 13.6(b). Likewise, consider the cheetah in Fig. 13.10(a). Her body is spotted, but her tail is striped. The tail stripes are thought to be spots wrapped around a narrow tail,[23] but more must be involved: spots only go partway down her tail and her limbs have no stripes. Moreover, animal coats are enormously more complex and varied than this one cheetah can illustrate.[20] Nevertheless, the mathematical observation that a spot wrapped around a tail can look like a stripe has support in the observation that it is difficult to find examples of animals with spotted tails but striped bodies, so there is both need for better modeling and support for the understanding that existing modeling can provide. Significantly,

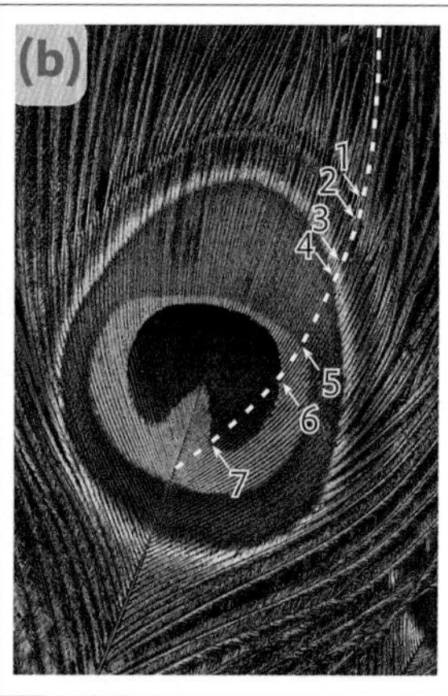

Fig. 13.10 Weakly understood patterning in nature. (a) The spotted body and striped tail of the cheetah are thought to be manifestations of the same essential patterning process, one on a broad surface and a second on a narrow cylinder. (b) The peacock feather must generate multiple orchestrated changes of color in each of its numerous barbs in order to produce an overall pattern.

the role of boundaries is crucial for development, where the location of an organ or a limb depends strongly on a coordinate system defined by boundaries,[38] and considerable work is ongoing to provide more realistic models of patterning. This topic is also important in problems ranging from cancer diagnosis to patterning in artificial tissue and organ design.

A final research area that illustrates the complexity of natural pattern formation concerns the *relationship between space and time in pattern formation*. This can be illustrated by considering the peacock: his feather, shown in Fig. 13.10(b), is a symphony of patterning. In order to produce the characteristic eye-shape, each barb on the feather must change color multiple times, in perfect concert with its neighbors, as it grows outward from the central shaft. For example, the broken line in Fig. 13.10(b) identifies a single barb whose color traces its history of growth. This barb started out brown (now its outer edge, but originally it began at the central shaft), then turned yellow, violet, green, tan, teal, blue, and teal again (now at the inner edge, the final part of the barb to emerge). This barb's neighbors must have followed different histories to produce their colorings, and all of these histories must be tightly choreographed to generate the overall feather pattern. And, moreover, this choreography must have been performed without feedback, since there is no central intelligence in the peacock feather monitoring or correcting its ultimate appearance.

How this signaling is regulated—and more generally how complex structures are generated in time as well as space during development—has been characterized (meaning measured) in numerous organisms, but is only weakly understood. We have mentioned artificial organs: it is worth remarking as we close that nature knows how to coordinate growth and migration of chemicals and cells to assemble elaborate patterns, and complex functioning structures, ranging from feathers to eyes to kidneys to livers. But we do not. Sometimes we can mimic nature's processes, but rarely can we understand them well enough to self-assemble a new structure or process from scratch in the way that we would build a bridge, design a car, or engineer an electronic circuit.

So although we have in these few hundred pages described methods for the modeling and understanding of flow and form, our ability to actually use these methods to do a fraction of what nature does is sorely lacking. This is a humbling recognition that should, I hope, lead us to listen more closely to what nature has to teach us.

REFERENCES

1. McKay, D.S., Gibson, E.K., Thomas-Keprta, K.L., Vali, H., Romanek, C.S., Clemett, S.J., Chillier, X.D.F., Maechling, C.R. and Zare, R.N. (1996) Search for past life on Mars: possible relic biogenic activity in Martian meteorite ALH84001. *Science, 273*, 924–30.
2. Schopf, J.W.(1993) Microfossils of the early Archean Apex Chert: new evidence of the antiquity of life. *Science, 260*, 640–6.
3. For an overview, see Treiman, A. (no date) Earlier scientific papers on ALH 84001 explained, with insightful and totally objective commentaries. Available at http://www.lpi.usra.edu/lpi/meteorites/alhnpapers_archive.html
4. Erbe, E.F., Rango, A., Foster, J., Josberger, E.G., Pooley, C. and Wergin, W.P. (2003) Collecting, shipping, storing, and imaging snow crystals and ice grains with low-temperature scanning electron microscopy. *Microscopy Research & Technique, 62*, 19–32.
5. Ferreira, S.A., Lopo, C. and Gentleman, E. (2016) Human stem cell. Available at http://www.nature.com/news/2016-in-pictures-the-best-science-images-of-the-year-1.21156
6. Adrian, M., Dubochet, J., Lepault, J. and McDowall, A.W. (1984) Cryo-electron microscopy of viruses. *Nature, 308*, 32–6.
7. Jiménez, J.L., Guijarro, J.I., Orlova, E. and Zurdo, J. (1999) Cryo-electron microscopy structure of an SH3 amyloid fibril and model of the molecular packing. *EMBO Journal, 18*, 815–21.
8. Winfree, A.T. (1984) The prehistory of the Belousov–Zhabotinsky oscillator. *Journal of Chemical Education, 61*, 661–3.
9. Bray, W.C., "A periodic reaction in homogeneous solution and its relation to catalysis," J. Am. Chem. Soc. 43 (1921) 1262-7.
10. Belousov, B.P. (1959) Periodically acting reaction and its mechanism. *Collection of Abstracts on Radiation Medicine (in Russian), 147*, 145.
11. Briggs, T.S. and Rauscher, W.C. (1973) An oscillating iodine clock. *Journal of Chemical Education, 50*, 496.
12. Kaiser Science. (2015) Briggs–Rauscher step by step. Available at https://kaiserscience.files.wordpress.com/2015/05/briggs-rauscher-step-by-step.png
13. Turing, A.M. (1952) The chemical basis of morphogenesis. *Bulletin of Mathematical Biology, 52*, 153–97.
14. Fisher, R.A. The wave of advance of advantageous genes. *Annals of Human Genetics, 7*, 355–69.
15. Sherratt, J.A. and Murray, J.D. (1990) Models of epidermal wound healing. *Proceedings of the Royal Society B, 241*, 29–36.
16. Gray, P. and Scott, S.K. (1984) Autocatalytic reactions in the isothermal continuous stirred tank reactor. *Chemical Engineering Science, 39*, 1087–97.

17. Pearson, J.E. (1993) Complex patterns in a simple system. *Science, 261*, 189–92.
18. Wigner, E. (1960) The unreasonable effectiveness of mathematics in the natural sciences. *Communications in Pure & Applied Mathematics, 13*, 1–14.
19. Ball, P. (2015) Forging patterns and making waves from biology to geology: a commentary on Turing (1952) 'The chemical basis of morphogenesis'. *Philosophical Transactions of the Royal Society of London B, 370*, 20140218.
20. Ball, P. (1999) *The Self-Made Tapestry: Pattern Formation in Nature*. New York: Oxford University Press.
21. Strang, G. (2016) *Introduction to Linear Algebra*. Wellesley, MA: Wellesley-Cambridge Press.
22. Schnakenberg, J. (1979) Simple chemical reaction systems with limit cycle behaviour. *Journal of Theoretical Biology, 81*, 389–400.
23. Murray, J.D. (1993) *Mathematical Biology*. Berlin: Springer.
24. Winfree, A.T. (2001) *The Geometry of Biological Time*. Berlin: Springer.
25. Thompson, D.W. (1942) *On Growth and Form*. Cambridge: Cambridge University Press.
26. His, W. (1874) Unsere Körperform und das physiologische Problem ihrer Entstehung (Our body form and the physiological problem of its emergence) Leipzig: FCW Vogel.
27. Hayashi, T. and Carthew, R.W. (2004) Surface mechanics mediate pattern formation in the developing retina. *Nature, 431*, 647–52.
28. Available at https://i.ytimg.com/vi/dJcomGs2YyA/maxresdefault.jpg
29. Ben-Jacob, E. (2003) Bacterial self-organization: co-enhancement of complexification and adaptability in a dynamic environment. *Philosophical Transactions of the Royal Society of London. Series A: Mathematical, Physical and Engineering Sciences, 361*, 1283–312.
30. Goehring, L. and Morris, S. W. (2005) Order and disorder in columnar joints. *Europhysics Letters, 69* (5), 739–745.
31. National Education Network. (2008) Basalt columns. Available at http://gallery.nen.gov.uk/asset82427-.html
32. Available at http://www1.udel.edu/biology/Wags/histopage/empage/eep/eep11.gif
33. Federle, W., Barnes, W.J.P., Baumgartner, W., Drechsler, P. and Smith, J.M. (2006) Wet but not slippery: boundary friction in tree frog adhesive toe pads. *Journal of the Royal Society Interface, 3*, 689–97.
34. Mather, W., Mondragon-Palomino, O., Danino, T., Hasty, J. and Tsimring, L.S. (2010) Streaming instability in growing cell populations. *Physical Review Letters, 104*, 208101.
35. Allen, J.J., Mäthger, L.M., Barbosa, A., Buresch, K.C., Sogin, E., Schwartz, J., Chubb, C. and Hanlon, R.T. (2010) Cuttlefish dynamic camouflage: responses to substrate choice and integration of multiple visual cues. *Proceedings of the Royal Society B, 277*, 1031–9.
36. Hanlon, R. (2007) Cephalopod dynamic camouflage. *Current Biology, 17*, R400–4.
37. Kardoudi, O. (2015) Cuttlefish hypnotize their prey performing these trippy light shows. Available at http://sploid.gizmodo.com/cuttlefish-hypnotize-their-pray-transforming-into-tripp-1689829464
38. Irvine, K.D. and Rauskolb, C. (2001) Boundaries in development: formation and function. *Annual Review of Cell and Developmental Biology, 17*, 189–214.

Appendix: Matlab Tools and Tricks

In this appendix, we present tools and tricks that may help you in Matlab programming. This isn't a primer on Matlab: there are many excellent examples of these in books and on the web. The goal here is to provide an understanding of how Matlab operates along with a stash of tools to perform common functions that will hopefully be useful. Some background will first be helpful for understanding how the tools and tricks—and how Matlab algorithms in general—work. In order, we describe in this appendix:

Background

In the early days of computer programming, there was Fortran—short for "formula translation"—a computer language developed at IBM. Fortran is an ancient and unforgiving *compiled* language that was superseded by many grammatically similar languages such as Basic, Pascal, C, etc. These languages are all "compiled," meaning that the user would have to write an entire, error-free, program, "compile" it, which means pass it through a program that turns the code into "machine code" that will operate on a specific computer, then "link" it, meaning attach other specific programs that the compiled code will need to run, "load" it, meaning put the program into memory so that the computer can use it, and finally "run" it, meaning wake the computer up and tell it to do its job.

The result of this wearisome multistep compile-link-load-run chore was—after a delay while the user's code made its way through a queue of other frustrated users—a printout, typically reading:

Biomedical Fluid Dynamics: Flow and Form. Troy Shinbrot.

DOI: 10.1093/oso/9780198812586.001.0001

```
            severe (8): Internal consistency check failure
or:         severe (62): Syntax error in format
or maybe:   error (76): IOT trap signal.
```

The user would then have to look up what these cryptic terms mean in a loose-leaf binder, figure out what the description—written by impatient and non-communicative computer nerds of the time—meant, guess *where* in the program the offending line could possibly be, rewrite the code, and then . . . yes, compile, link, load, and run again. For example, `error (76)` is "clarified" in the loose-leaf binder with:

```
FOR$IOS_SIGIOT. Core dump file created. Examine core dump for possible
cause of this IOT signal.
```

This procedure was followed throughout the 1950s, 1960s, 1970s, 1980s, and is even still carried out today. Meanwhile, little known to any but aficionados, in the early 1960s, a Canadian computer scientist named Kenneth Iverson (1920–2004) had the sense to recognize that the existing process made the user perform pointless repetitive tasks to accommodate the computer, rather than the other way around. Iverson therefore constructed APL: "A Programming Language"—the first *interpreted* computer language. This allowed the user to type instructions, have the computer spend its own time on the compile, link, load, and run steps, and give the user answers on the screen and in real time.

APL is a brilliant, elegant, and sophisticated language, still available with some effort, but it extensively uses mathematical abbreviations and is difficult to read, even by experts. Enter Matlab, introduced in the 1970s by New Mexico computer scientist, Cleve Moller (1939–), an interpreted language that uses near-English commands like "plot," to plot, "pause," to pause, and so on. Matlab is by no means the only language to do these things, but is widely used, especially in engineering fields, and is the linguistic choice for this book.

This history is intended to give the reader a fundamental message, namely that compiled languages are good, once debugged, for performing repetitive, identical, tasks without user intervention. Interpreted languages operate very differently, and are built around the expectation of user involvement.

Parallel computations

So what? Let's consider the snippet of Matlab code in Fig. A.1, labeled in red with summaries of relevant lines. The segment with yellow background has been written with an understanding of how an interpreted language operates; the segment with green background has been written as if Matlab were a compiled language. If you run this code in Matlab, you will find that the first segment takes a tenth of a second to run, while the second segment takes over 60 seconds to do exactly the same computations.

Why the difference? Note the lines in the blue box: these lines are read once, and operate on 1000 data points all at the same time. The reading and interpreting (as well as the plotting) takes much more time than the actual computations, so the reading is done once, and the computer is left to figure out the computations. It does this using the "rand" and "sin" operations, which are fast because they have themselves been efficiently compiled through the labor of computer scientists doing what they do best. So the slow, interpreted, stuff is done once; the fast, compiled, stuff is done many times.

```
figure (gcf)                          <bring current plotting window to the front
clf                                   <clear the window

%% parallel solution:                 < comment
subplot(1,2,1);                       < open a one by two matrix of plots, choose the first one
hold on;                              < make one plot print on top of the previous: don't clear between plots
axis([0,pi,0,1]);                     < set x-axis from 0 to π, y-axis from 0 to 1
tic;                                  < start clock: time between tic and toc will be recorded
x = rand(1000,1).*pi;                 < define x to be 1000 random numbers between 0 and π
x1 = sin(x);                          < define x1 to be the sin of x
plot(x,x1,'.');                       < plot, with x on the horizontal axis, x1 on the vertical, as single dots
drawnow;                              < draw it, for heaven sakes (rather than store the result, which Matlab will
                                        sometimes do.) This is known as "flushing the graphics buffer"

t1 = toc;                             < stop the clock and define t1 to be the time elapsed since tic.
t1 = ['Parallel time: ',...
   num2str(t1,'%6.3f'),...
   ' seconds'];                       < save the text "Parallel time: 2.3 seconds" (or however long it takes).
                                        note "..." means that the line is continued below.
title(t1);                            < print "Parallel time" (etc) at the top of the plot

%% serial solution:
subplot(1,2,2);                       < go to the one by two matrix of plots, choose the second one
hold on;
axis([0,pi,0,1]);
tic;
for i1 = 1:1000                       < run a loop 1000 times. By the way, it is a good idea not to use "i"
                                        as the increment name, because "i" is also sqrt(−1).
   x = rand(1,1).*pi;                 < define x to be one random number from 0 to π
   x1 = sin(x);
   plot(x,x1,'.');
   drawnow;
end;
t1 = toc;
t1 = ['serial time: ',num2str(t1,'%6.3f'),'seconds'];
title(t1) ;
```

Fig. A.1 Program comparing parallel with serial computation in Matlab. Note at the top of the plots that the parallel computation takes about 1/10 s, while the serial computation of the same thing takes 60 s.

The lines in the magenta box, on the other hand, are read and run 1000 times. In a compiled language, this would be fine, but in an interpreted language, it takes enormously longer to read and run a command many times, each on one number, than to run the same command once on many numbers. So here the slow, interpreted, stuff is done many times, and the fast, compiled, stuff is also done many times. Not a desirable way of doing things.

→**Moral:** don't run loops if you can avoid them.

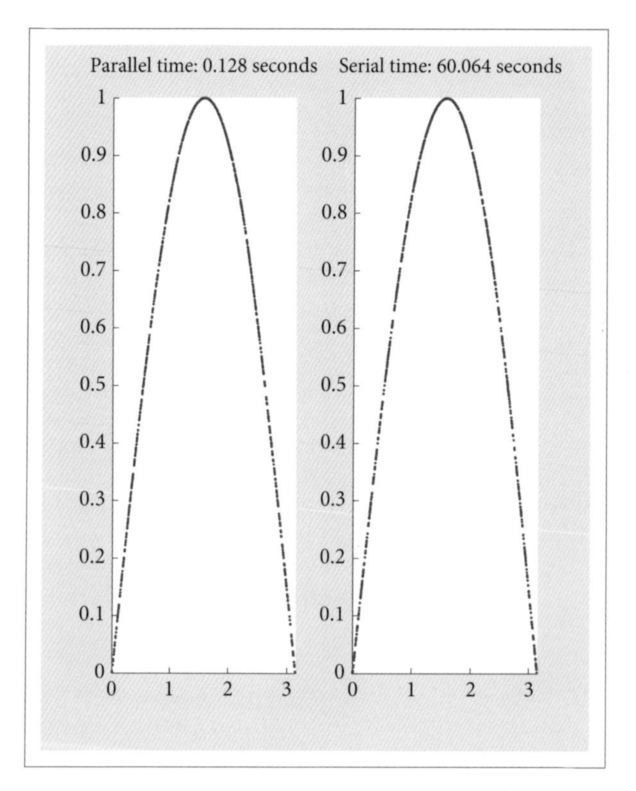

Fig. A.2 Output of program defined in Fig. A.1. Note at the top of the plots that the parallel computation takes about 1/10 s, while the serial computation of the same thing takes 60 s.

Allocating memory

A second attribute of interpreted languages like APL and Matlab is that they allocate memory for you. So if you want to know what π times 100 numbers is, you can just ask. Compiled languages, by comparison, need to know that you are going to use 100 numbers, each of which will be double precision. The program will then allocate that much space in memory, and you are good to go. Heaven help you, though, if you decide to calculate π times 101 numbers. In Fortran, the program will crash; in C, something much worse will happen: the compiler will happily occupy the space it needs at the end of wherever the memory was allocated for the 100 numbers. Maybe over top of where your word processing program is stored or maybe overwriting of some other file—an email you've been writing to your boss, for instance—or maybe where the operating system is trying to run.

Matlab doesn't do any of those things; on the other hand, the price is that it constantly needs to adjust its memory allocation. This is usually not a problem, but it's worth being aware of, as the program in Fig. A.3 illustrates.

A handy way of storing elements when you aren't sure how many you will use is to make use of the square brackets in the yellow segment of code. The format, `allnums = [allnums, newnum]` will put "`newnum`" at the end of "`allnums`." This is handy, but it requires Matlab to allocate more memory every time a new number is added. Usually this will cause no problem,

```
% not so good:
allnums = [];                 < open an empty variable to store list of numbers of
                                unknown length
for il = 1:100
  allnums = [allnums,il*pi];  < append each new number at the end of the last list of
                                numbers
end
```

```
% better:
allnums = zeros(1,100);       < open an empty variable to store list of numbers of length
                                100
for il = 1:100
  allnums(il) = il*pi;        < put each new number in a specified place
end
```

```
% best:
allnums = [1:100].*pi;        < who paid attention to the last section? Do it all at once
                                when possible
```

Fig. A.3 Memory allocation example. Note in the first case that memory has to be reallocated every time a new number is stored. In the second case, allocation is done three times: once to store the name "`allnums`," once to store 100 zeros, and once when Matlab discovers that the 100 elements are of size "`double`," and aren't integers.

but Matlab accomplishes this by allocating an empty block of space, moving the new contents of `allnums` to it, and then freeing up the old block of space. When allnums gets to be very big, or memory space gets to be short, this can become time intensive.

A better way of doing things is shown in the green highlighted segment, in which memory is explicitly allocated using the "`zeros`" command, which fills all of the necessary space with zeros, and better still is the option highlighted in violet: writing the relevant command and letting Matlab do whatever heavy lifting it needs—which you can be sure has already been optimized for efficiency—to get you the answer.

→**Moral:** make liberal use of Matlab's convenience, but when things bog down, be aware of what it is doing behind the scenes.

Local versus global

A final remark before we get to tools and tricks deals with local versus global variables. Most of the time, you won't need to know or care what these are, but sometimes it will be important to understand how variables are managed. As an example, suppose that you want to make the plots shown in Fig. A.2, not only to multiply 1000 random numbers by π at the start of the program, but to multiply any number of random numbers by any constant. One way of doing this would be to write Your Program shown at the top of Fig. A.4. You can then change `HowMany` and `TheMultiplier` at will to get your answer.

But one day, you run Your Program, and you find a strange result: maybe you see a gap in the plot that it produces, and you suspect that your random number generator has a problem (look up RANDU for an infamous example of this). Or maybe you have some other bug that you need to investigate further.

(a)

Your program

```
%set up figure:
figure(gcf)
clf
axis([0,pi,0,1]);

%define parameters:
HowMany = 1000;
TheMultiplier = pi;

%do function:
TheFunction(HowMany,TheMultiplier);
```

```
function TheFunction(HowMany1,TheMultiplier1)
%do calculations:
x = rand(1,HowMany1).*TheMultiplier1;
x1 = sin(x);

%plot results:
plot(x,x1,'.')
```

(b)

Solution 1: naughty

```
%set up figure:
figure(gcf)
clf
axis([0,pi,0,1]);

%define parameters:
HowMany = 1000;
TheMultiplier = pi;

%do function:
TheFunction(HowMany,TheMultiplier);
```

```
function TheFunction(HowMany1,TheMultiplier1)
global x x1
%do calculations:
x = rand(1,HowMany1).*TheMultiplier1;
x1 = sin(x);

%plot results:
plot(x,x1,'.')
```

Fig. A.4 Naughty and nice solutions to a programming problem. (a) a perfectly good program, that unfortunately doesn't allow access to the "local" variables, `x` and `x1`, from outside. (b) naughty solution that makes these variables global (see pink box). You can do this, but you may well regret it down the road. (c) nice solution that "passes" `x` and `x1` outside of `TheFunction` (see pink boxes).

(c)

Solution 2: nice

```
%set up figure:
figure(gcf)
clf
axis([0,pi,0,1]);

%define parameters:
HowMany = 1000;
TheMultiplier = pi;

%do function:
TheAnswers = TheFunction(HowMany,TheMultiplier);

    x3 = function TheFunction(HowMany1,TheMultiplier1)
    %do calculations:
    x = rand(1,HowMany1).*TheMultiplier1;
    x1 = sin(x);
    x3 = [x;x1];
    %plot results:
    plot(x,x1,'.')
```

Fig. A.4 Continued.

So you want to know what x and x1 are. No problem, says you, and you type x, and you get either the last value x of that you obtained from a previous calculation, or nothing at all, if you've erased that x. But what you *don't* get is the x that was used in the plot that you're looking at.

Why? Because anything that happens within a function (in this case calculating and plotting sin(x)) is "local," meaning it isn't accessible outside of the function. So after TheFunction has been run, its contents disappear. No problem, says you once again, and you confidently write the "naughty" solution shown in Fig. A.4 (b). This contains the line, "global x x1," which makes x and x1 accessible from outside of TheFunction. Problem solved.

Next week, you laboriously type in a huge array of numbers for an important project, and, naturally, you call the numbers x. And you run the naughty program to recall what that result looks like, and find... your huge array is gone, replaced by a bunch of random numbers. Pluck a duck: you *know* you spent hours typing the data in: where did it go? Answer: the naughty program reassigned *your* x to be *its* x. Because global made the x used inside TheFunction the same as x used outside. Or more commonly, you use x to refer to multiple different things at different parts of a very complex program, and one part of the program changes things that another part is relying on—a problem that is nearly impossible to either identify or locate in a long program.

For both of these reasons, the Nice Program in Fig. A.4 (c) is preferred: here, you formally pass the results that you want to `TheAnswers`. Notice that within `TheFunction`, this is called `x3`: this is not a problem, because `x3` is local, and is invisible outside of `TheFunction`: it can be called anything at all, including `TheAnswers`: without a `global` command, whatever it is called won't appear outside of the function, and won't interfere with any variable names that are used elsewhere. For good programming style, it is advisable to use unique names: this makes searching easier and prevents confusion, but strictly speaking, there is nothing preventing you from using the same names for different things inside and outside of functions – and using `global`, for that matter.

→**Moral:** don't run with scissors, and don't use globals. And if you do, don't say you learned it here.

Tools and Tricks

Here are several program snippets that do things that I've found useful over the years. Once upon a time, Matlab was truly interpreted, and it read each function every time it ran. As a consequence, you could make a change to a function while the program was running, and the change would instantly be acted on. This was immensely useful for simulations: for example if you needed to change plot axes, you could do that mid-computation, or if you wanted to simulate a change in a variable, you could make the change exactly as if the real physical system being simulated were changed.

Predictably, reading every function every time it was called took time, and Matlab's parent company, Mathworks, took to compiling functions to improve efficiency. We start with a description of two work-arounds that restore some of the functionality of earlier versions; then we describe some general tools that you may find useful.

(1) Change variables or parameters with a slider:

A first option to consider if you know in advance what you want to change is to use a graphical user interface (GUI). Matlab has an automated facility for creating GUIs, called GUIDE, but it takes a lot of fiddling, and once you are done, all that GUIDE actually does is to produce code that you need to fiddle with some more to get the GUI to actually do anything.

As an alternative to that process, I provide here sample code that manages a GUI slider to change a program parameter. This code does something useful: it fits a set of points, (x1,y1), with a "spline under tension," meaning a cubic spline whose smoothness can be adjusted.

Main program:

```
smthnss = 0.5; % default smoothness of spline fit
figure(gcf);
clf;
h1=uicontrol; % this defines a "pointer" to the GUI. To find all properties
        of the pointer, you can type "set(h1)." To set any property, you can
        type "set(h1,'propertyname',propertyvalue)"
set(h1,'style','slider','position',[10 10 50 100],...
    'value',smthnss,'min',0,'max',1,'callback','DoSplineSlider');
```

```
        % The style of the GUI is a slider
        % The position is 10 pixels from the left of the window, 10 pixels
              from the top, the slider is 50 pixels wide, and 100 pixels tall.
        % The value that the slider controls is the parameter "smthnss',
              which goes from a minimum of 0 to a maximum of 1.
        % When the GUI is used, the program "DoSplineSlider" is run.
DoSplineSlider
        % The program "DoSplineSlider" is run once to perform whatever
               functions are wanted (here, plotting the data (x1,y1) and
               performing a fit.
```

Program that must be put in the same folder as the main program, called "DoSplineSlider.m:"

```
% naughty programming, done out of laziness, to pass parameters between
       function and main program. Feel free to rewrite to pass these
       parameters: that would be better, but is time consuming.
global smthnss h1 x1 y1;

% The first thing that the function needs to do is to get the value that was
       chosen by the GUI:
smthnss = get(h1,'value');
% Then whatever is plotted in the current graphics window is erased:
cla
% And all points (x1,y1) are plotted, as blue plus signs with width 2
     pixels.
plot(x1,y1,'b+','LineWidth',2);
% We're going to plot a curve fit on top of the plus signs, so we don't want
       the screen erased when we do the curve fit:
hold on;
% This cryptic code does the actual curve fitting:
thefunc = csaps(x1,y1,smthnss);
% This cryptic code plots the curve fit as a red line:
fnplt(thefunc,'r');
% The rest just restores the graphing state, lets the user zoom with mouse
       clicks, and flushes the graphics buffer:
hold off;
zoom on;
drawnow
```

(2) Open and read a text file:

To run a program like a spline fit, it is of course likely that you'll need to read in some data, so this program does that. To read a file, put the snippet below at the start of the main program above. The code below has been tested on a Macintosh; it may work differently on other platforms.

```
% Every file has a path name (where it is stored) and a file name (what the
       file is called). Let's get them with a GUI:
[thefilename,thepathname] = uigetfile('','Choose Data File');
```

```
% Tell the computer where the file is:
path(path,thepathname);
% Tell the computer which file is of interest, and open it to be read:
fid = fopen(thefilename,'r');
F = fread(fid);
% Read the text in the file. If you have stored data as other than text,
    you'll need to use a different command:
s = char(F');
% convert the carriage returns in the file into row numbers, so that the
      data become a matrix of rows and columns:
t=str2num(strvcat(s));
% Assuming that the data consist of two columns, read the first and second
      column into x1 and y1, respectively:
x1 = t(:,1);
y1 = t(:,2);
```

(3) Change a program while it is running:

This is a much more general approach that allows the user to change anything whatsoever while a program runs. This can cause your program to abort if you mistype what you want done, but if you are careful, it is a very handy tool. The tool comes in three parts: a bit of code that reads what you want to do, a bit of code that contains your written instructions, and a bit of code that acts on what you have written.

Part 1a: Code that reads what you want to do: put this in your main program. At the start of your program, put these two lines:

```
% The first line defines how often you want to read your instructions. As
    I mentioned, Mathworks disabled reading files while running programs
    because it is inefficient, so you may want to read the instruction
    file more often (in which case put "ParseEvery = 1") or less often
    ("ParseEvery = 1000").
    The second line is because Matlab recognizes you are doing something
    fishy, and prints repeated warnings to the screen.
 ParseEvery = 10;
 warning('off')       % just to avoid printing rubbish during parsefile
```

Part 1b: Code that reads what you want to do. Assuming that you are simulating in a loop like: "For time = 1:100000," the following limits how often your instruction file is read. Put this in a loop in your main program.

```
if (time/ParseEvery) == round(time/ParseEvery)
    % The following runs a program called "parsefile.m" which reads and
          acts on your list of instructions in another program called
          "TheParsedFile.m"
     parsefile('TheParsedFile.m');
end % plot
```

Part 2: the function, which must be named "parsefile.m" and reside in the same folder as the main program, is:

```
function TheText = parsefile(filename)
        % Open "TheParsedFile.m" and read each line of instructions:
        TheText = [];
        fid1 = fopen(filename);
        TheLine = fgetl(fid1);
        TheText = [TheText,TheLine];
        while ischar(TheLine)
                TheLine = fgetl(fid1);
                TheText = [TheText,TheLine];
        end
        % Close the file:
        dummy = fclose(fid1);
        % Evaluate each line of instructions:
        evalin('base',TheText);
end
```

Part 3: the list of instructions, which must be named "TheParsedFile.m" and must reside in the same folder as the main program:

```
disp('Hello');
x=linspace(1:100);
clf; % sample instructions, which can be any commands at all.
       Note 1: proofread these lines carefully. Once you save the file, the
       lines will be acted on, and any typos or errors will cause the program
       to crash.
       Note 2: in this one file only, any instruction following the first
       comment will be ignored, so don't use comments until you have com-
       pleted all instructions that you want performed!!!
```

(4) Save figures to make a high quality movie:

Matlab comes with a facility to save movies, but the quality is erratic, depends on the operating system you are using, and what you see is often not what you get. As an alternative, you can save successive tiff or jpeg snapshots, and collect the results into a movie afterward, using any of a variety of graphics tools (e.g. GraphicConverter).

Sample code that plots and saves a wave:

```
for ii = linspace(0,2*pi,100)
     figure(gcf);
     clf;
     % The following is not at all necessary, but is useful to know about: if
            you are making large plots, the screen will flash between plots,
            because the computer erases the screen and replots every time -
            the following line tells the computer to do that off-screen, and
            swap the offscreen image with the onscreen one when it's done:
```

```
    set(gcf,'doublebuffer','on')
    % sample plot of a wave:
    [x,y]=meshgrid(linspace(0,2*pi,50),linspace(0,2*pi,50));
    surfl(x,y,-cos(y+ii).*cos(x./10));
    % gratuitous graphics commands to make the plot look nice:
    colormap copper
    shading interp
    light
    lighting gouraud
    drawnow;
   %name the first file "WavePic_0.tiff," the second "WavePic_0.06.tiff,"
         the third "WavePic_0.13.tiff," etc. (after the underscore is
         each value of ii, followed by ".tiff." The files will be put into
         the same folder as the program:
    filename = ['WavePic_',num2str(ii)];
    eval(['print -dtiff ',filename;]);
end
```

(5) Show an image file and find values of points on the screen:
From time to time, I need to extract data from an old paper or lab notebook or whatnot. To do this, I scan the original plot and then run the following code that displays the scan and allows the user to click on each visible data point, ultimately saving a scaled set of x and y data from the mouse clicks.

```
% open a .jpg file:
[thefilename,thepathname] = uigetfile('','Choose Image');
path(path,thepathname);
t1=imread(thefilename,'jpeg');

% show the .jpg:
figure; image(t1);

% to produce data values, the computer needs to know what the x- and y-
      ranges are; it then extrapolates each point between the minimum and
      maximum values provided:
t=title('Choose left-lower and right-upper limits');
set(t,'Color','r','FontSize',18);
t2=ginput(2);
lowxpix = t2(1,1);
upxpix  = t2(2,1);
lowypix = t2(1,2);
upypix  = t2(2,2);

% go to the command window to accept data range values:
commandwindow;
lowerxlim = input('type lower x limit: ');
upperxlim = input('type upper x limit: ');
```

```
lowerylim = input('type lower y limit: ');
upperylim = input('type upper y limit: ');

% go back to the plot screen to select data points with the mouse:
figure(gcf);
t=title('Choose all data points; then type CR');
set(t,'Color','r','FontSize',18);

% grab each mouse click:
pts = ginput;

% convert each mouse click to a scaled data point in x and y:
pts(:,1) = ((pts(:,1)-lowxpix)./(upxpix-lowxpix)).* ...
      (upperxlim-lowerxlim)+lowerxlim;
pts(:,2) = ((pts(:,2)-lowypix)./(upypix-lowypix)).* ...
      (upperylim-lowerylim)+lowerylim;
t=title('Data points are X and Y in main workspace');
set(t,'Color','r','FontSize',18);

disp('Data points are X and Y in main workspace');
X = pts(:,1);
Y = pts(:,2);
```

(6) Fit to line with uncertainties:

So now you have X and Y values: maybe you want to do a curve fit. A spline fit was shown in example 1; a polynomial fit can be done using the Matlab function polyfit, described in the Matlab help system. I hardly ever want to do an actual polynomial fit; usually I want to do a linear fit. Knowing the uncertainty in the fit is helpful also, so I describe here linear fits with uncertainties. Alternatives are lsqcurvefit or lsqnonlin.

```
% plot the data:
figure;
h2 = plot(X,Y,'ko');
set(h2,'MarkerSize',10);
hold on; zoom on;

% let Matlab's polyfit do its thing. pp are polynomial coeff's; ss are
      obscure scale structures that give error bars. Treat the following
      as magic; it'll be fine:
   [pp,ss] = polyfit(X,Y,1);
   [yy, dd] = polyval(pp, X, ss);
% yy are now the predicted values for the linear fit. Plot a red fit line:
   h3 = plot(X,yy,'r');
   set(h3,'LineWidth',2);

% now compute the uncertainties, assuming Poissonian errors from Taylor1,
      p.188:
```

```
commandwindow;
sigmasqrs = yy;
delta = sum(1./sigmasqrs)*sum((X.^2)./sigmasqrs) - sum(X./sigmasqrs).^2;
sigmab = sqrt(sum(1./yy)/delta);
disp(['slope = ', num2str(pp(1)), ' ± ', num2str(sigmab)]);
sigmaa = sqrt(sum((X.^2)./sigmasqrs)/delta);
disp(['intercept = ', num2str(pp(2)), ' ± ', num2str(sigmaa)]);
cc=corrcoef(X,yy);
disp(['Correlation Coefficient = ',num2str(cc(1,2))])
```

(7) Fourth order Runge–Kutta integration:

It is often desirable to integrate an ordinary differential equation—for example, in this book we have shown numerous examples of velocities that can be integrated to show fluid trajectories. One common numerical method for performing these integrations is the Runge–Kutta algorithm, named after German mathematicians, C.D.T. Runge (1856–1927) and G.W. Kutta (1867–1944). This is by no means the only, nor is it the most efficient, algorithm, but it is a solid and reliable method that has become the workhorse for most integration problems. For more details, and other algorithms, the reader should consult *Numerical Recipes*,[2] an outstanding and surprisingly readable book that candidly describes the pros and cons of a variety of numerical approaches.

The idea behind the Runge–Kutta algorithm is as follows. We describe here the second order algorithm; the fourth-order algorithm is an improvement that averages more derivatives at more points, but follows the same line of reasoning. Suppose we want to study the motion of a mass–spring system, as shown in Fig. A.5(a). Its trajectory for a particular initial condition is shown as a blue curve in Fig. A.5(b). If we start at any particular point, say the black circle in the enlarged inset (c) at time t_0, and calculate the instantaneous trajectory, it will take us to the green x after a short time, Δt. We'll call the green trajectory the "Euler trajectory," meaning we would use Euler integration, $x \rightarrow x + dx/dt\ \Delta t$, to get us from the circle to the green x.

The Euler trajectory departs from the actual trajectory (the blue curve) because Euler integration always follows a straight line, while real trajectories are almost always curved. Question: how can we correct for this difference? We consider the problem in general first, then we'll return to the particular problem of the mass–spring.

The general case of a curved trajectory, shown in blue in Fig. A.5(d) is misestimated by the Euler trajectory in green. We seek the "true" integrated trajectory sketched in red that intersects precisely with the actual trajectory. Our problem is that the slope at time t_0 will always take us *outside* of the curvature of the actual trajectory. How can we find the "true" slope that will lead us to intersect the blue curve at the red star? Here's a thought: as shown in Fig. A.5(e), if the curve were a perfect circle, the arc of a circle sketched would always have the slope of the midpoint of the arc, also sketched. So really all we have to do is to estimate what the slope is at the *midpoint* of a curve, rather than at the starting point.

This is a bit confusing, so let's refer back to our mass–spring problem. In Fig. A.5(f), we again plot a general trajectory in blue. In magenta, we plot the "true" slope, in magenta, evaluated at time $t_0 + \Delta t/2$: *half* the time it should take us to get to the target red star. We record the slope at that time, then go *back* to the original point, $(t_0, x(t_0))$, and we apply

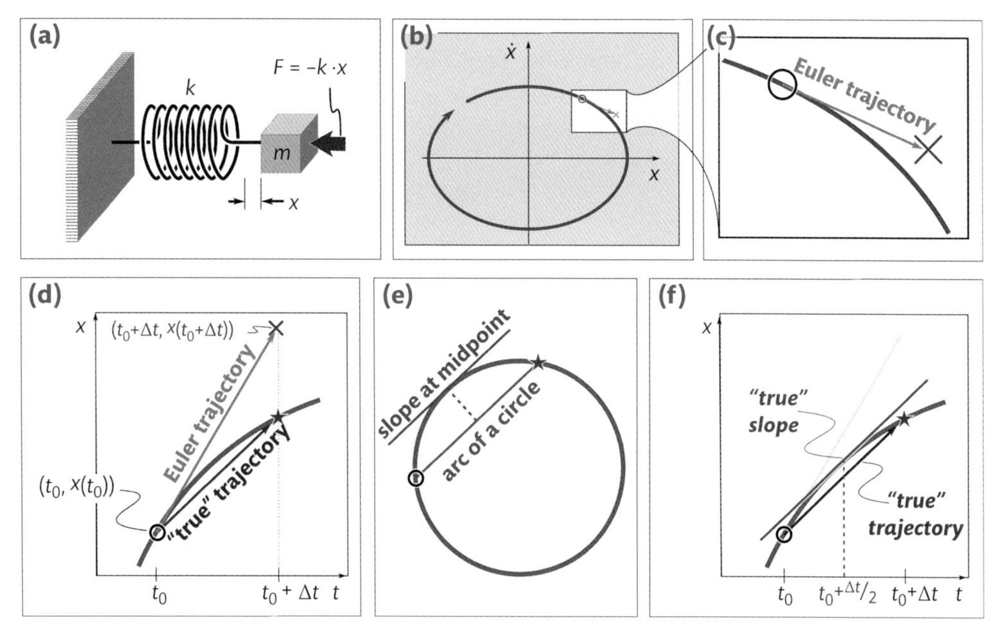

Fig. A.5 Schematic of numerical integration. (a) Mass-spring system that we'll use as an example. (b) Trajectory of oscillating mass (blue) as a function of $\dot{x} = dx/dt$ and x. (c) Enlargement of trajectories starting from the initial point in the circle: actual trajectory in blue; Euler integrated trajectory in green. (d) General problem, in which $x(t)$ follows blue trajectory, and the Euler integrated trajectory follows the green line. We seek the "true" integrated trajectory, following the red line to the red star. (e) Idea underlying second order Runge-Kutta integration is that if the actual trajectory were circular, the "true" slope would be the slope at the midpoint of an arc. (f) The second order Runge-Kutta trajectory estimates the "true" slope at the half-way time, $t_0 + \Delta t/2$.

Euler integration forward in time a *whole* time step, Δt, which takes us to the red star as desired.

It takes a little care to code this idea correctly, but this scheme works well provided the curvature is constant. If the curvature changes, something else has to be done, but the Runge–Kutta approach doesn't fail us there either: we just have to include corrections for the changes in curvature. This can be done as well, and is described in *Numerical Recipes*.[2] Luckily, both the second and fourth order systems have been coded many times before, and Matlab has a full suite of integration routines, all of which have the identical format. So let's turn to implementing Matlab's integration suite for any routine we might want to use.

For this purpose, we'll return to the problem we started with: a mass–spring system. The equation of motion is simple: we write it for spring constant k, viscous damping coefficient, β, and mass, m:

$$m \cdot \ddot{x} = -\beta \cdot \dot{x} - k \cdot x, \qquad [A.1]$$

where we use the abbreviations, $\dot{x} = dx/dt, \ddot{x} = d^2x/dt^2$. Damping isn't shown in Fig. A.5(a), but we include it here for generality.

There is a second derivative here, but the Runge–Kutta description only had a first derivative. What to do? Very simple: we take a lesson from algebra, and when we don't know something, we call it a name to disguise our ignorance, and then go about our business as if the problem had never arisen. Usually we call the unknown x, but here we'll use v:

$$v = \dot{x}, \tag{A.2}$$

so that Eq. [A.1] now becomes:

$$m \cdot \dot{v} = -\beta \cdot \dot{x} - k \cdot x. \tag{A.3}$$

In this way, we can always reduce an nth order differential equation to n first order differential equations, and each one of these can be solved using Runge–Kutta. There's a bit of a kerfuffle to deal with how the multiple equations couple together, but fortunately for us, that problem too has been solved, and Matlab's integration routines deal with the complications without pestering us with the details.

So with this all in mind, a code snippet that will solve Eqs. [A.2] and [A.3] is as follows.

```
%choose parameters of the mass-spring system. Mass=1 to make it a little
    simpler:
damping = 0.5;
spring = 4;

% set initial conditions. Here ThePos = [x,v], and these initial condi-
    tions start the mass of with 1 unit of displacement and no velocity.
ThePos = [1;0];

% how long are we integrating for? Notice that no time step is assigned: the
    Matlab system does that for us—comments on this to follow:
TimeTotal = 25;

% if you want, you can assign various integration options. Usually you will
    omit the following line, but it is included in case you should want to
    use options:

options = odeset('RelTol',1e-1,'Stats','on');

% finally you get to do the integration. Notice that spring and damping are
        being passed to the integrator: we'll see how they are used in a
        second. If you don't want to use options, erase the word "options,"
        below, but leave the two commas on either side of the word in place.
        The following line uses variable step 4th order Runge-Kutta inte-
        gration; variable step 2nd order Runge-Kutta is ode23, and numerous
        other alternatives are also described in the Matlab help system.
[t, ThePos] = ode45(@SHOvel, [0 TimeTotal],[ThePos],options,spring,
    damping);
```

```
% and you're done: you now have each of the times at which Matlab evaluates a
        point, and a long column matrix, where the first column is x-values
        at every evaluation time, and the second column is v-values at every
        evaluation time. So perhaps you will want to plot:
figure(gcf);
clf;
plot(ThePos(1,1),ThePos(1,2),'ro','MarkerSize',18,'LineWidth',6);
hold on;
plot(ThePos(:,1),ThePos(:,2),'c-*');

% ah: you probably noticed that we haven't defined the differential equa-
        tion - probably we should do that. The line starting with "[t, ThePos]
        = . . . " has something cryptic in it as well: this points the integra-
        tor to the function "SHOvel," which contains the derivatives:

% the following function has to be called "SHOvel," and has to reside in the
        same folder as the code above:
function vxyz = SHOvel(t,xyzvector,spring,damping)
% the next line is Eq. [A.2]. This says that the velocity is the second element
        in xyzvector that was passed to it ("[ThePos]" in the integration line:
v1 = xyzvector(2);

% the next line is Eq. [A.3]:
v2 = −(damping*xyzvector(2) + spring*xyzvector(1));

% and the derivatives produced by the differential equation, which Matlab
        requires be in column format as shown below, are:
vxyz = [v1;v2];
```

(8) Particle dynamics:

It is very common to have use for a simulation of individual particles. Examples include simulations of molecular dynamics, "agent based" simulations of biological cells, "discrete-element method" simulations of granular particles, simulations of colloidal assembly, and "smoothed particle hydrodynamics" in which small volumes of fluid are approximated as spheres. Details can vary: for example, molecules attract and repel through so-called Lennard-Jones potentials; cells adhere through cytoskeletal forces acting on membrane-bound adherins; grains can have complex shapes and frictional behaviors; and so on.

Common to simulating all of these systems is a central engine that allows spheres to interact with one another, and I provide a brief annotated code for this here. A couple of explanatory notes may be helpful to understand the algorithm.

First, the code uses "soft-particle" dynamics to evaluate forces on particles in contact. This is not the only way of doing things, see, for example, Pöschel and Schwager[3] for alternatives and details. For our purposes, the soft-particle approach is useful because it is straightforward to implement, and unlike some methods it can be used to simulate both moving and stationary particles. In soft-particle simulations, particles are treated as deformable spring-like spheres. So particles can deform slightly during contact, and the distance, δ, that two spheres would impinge into one another is directly evaluated. Using this distance, the restoration force needed to remove the impingement is easily calculated using Hooke's law: $F = -k \cdot \delta$.

Second, inelastic particle collisions cause a loss of energy expressed in terms of a restitution coefficient, κ. Explicitly, $\kappa = \dot{\delta}_{out}/\dot{\delta}_{in}$, where the speeds at which the spheres are approaching or receding are respectively $\dot{\delta}_{in}$ and $\dot{\delta}_{out}$. The most direct way of modeling this is to embellish Hooke's law by including viscous damping: $F = -k \cdot \delta - \beta \cdot \dot{\delta}$, where β is a damping constant. Here faster moving spheres experience stronger damping, in agreement with the restitution coefficient.

This is a perfectly acceptable approach, sometimes called a Voigt model. Predictably, however, if $\dot{\delta}$ is too large, the damping force can exceed the elastic restoring force, which leads to anomalous acceleration of high speed particles. An alternative, which turns out to agree with experiments for granular materials, is to use a different Hooke constant for approaching and receding particles.[4] So when particles approach one another, Hooke's law becomes $F = -k_{in} \cdot \delta$, and when they recede it is $F = -k_{out} \cdot \delta$. By setting $k_{out} < k_{in}$, kinetic energy is lost when particles deform and is only partially recovered when they restore their shape: it turns out that the restitution coefficient $\kappa = \sqrt{k_{out}/k_{in}}$. The code below uses this model, which improves computational stability and permits larger time steps than would otherwise be possible.

Third, obviously in order to use this approach we must determine whether particles are approaching or receding. To do this, we use a so-called "anonymous" function, `toward`. Details are available online, but in short to define the function $hokus(po, kus) = po^{kus}$, we would write $hokus = @(po, kus)po.\hat{}kus$. `toward` performs arithmetic on the positions of two particles, *x1* and *x2*, and their velocities, *v1* and *v2*, to decide if they are approaching or receding.

With these notes in mind, a code that simulates simple particle collisions follows.

```
%% Initialization:
numballs = 10;                      % number of spheres
Speed = 0.1;                        % maximum initial sphere speed
TwoD = 1;                           % whether to use 2D (if 1) or 3D (if 0)
dt = 1;                             % time increment
ballradius = 0.5;                   % radius of spheres
DomainSize = 10;                    % size of computational domain
mass = 1;                           % mass of each sphere
Kin = 1;                          % Hooke constant when spheres are approaching
Kout = 0.5;                         % Hooke constant when spheres are receding
X = (rand(1,numballs)-0.5).*2.*DomainSize;  % random initial x
Y = (rand(1,numballs)-0.5).*2.*DomainSize;  % random initial y
Z = (rand(1,numballs)-0.5).*2.*DomainSize;  % random initial z
XYZ=[X(:),Y(:),Z(:)];                     % positions of spheres
Vth = rand(numballs,1).*2.*pi;            % random angles of velocities
Vxyz = [cos(Vth),sin(Vth),Vth.*0].*Speed; % random amplitudes of
     velocities
figure;                                   % open plotting window
```

```
%% Anonymous function to establish if spheres are approaching or receding:
toward = @(x1,x2,v1,v2) sum(((x1+v1.*eps-x2+v2.*eps).^2),2) > ...
                                                  sum(((x1-x2).^2),2);
%% Run code:
while 1                        % run forever (ctrl-c to stop)
    XYZ = XYZ + Vxyz.*dt; % simple integration of sphere positions

    %% bounce spheres off of walls specularly:
    [ToRight,dum] = find((XYZ(:,1)>DomainSize) & (Vxyz(:,1)>0));
         Vxyz(ToRight,1)=-Vxyz(ToRight,1);
    [ToLeft,dum] = find((XYZ(:,1)<-DomainSize) & (Vxyz(:,1)<0));
         Vxyz(ToLeft,1)=-Vxyz(ToLeft,1);
    [ToRight,dum] = find((XYZ(:,2)>DomainSize) & (Vxyz(:,2)>0));
         Vxyz(ToRight,2)=-Vxyz(ToRight,2);
    [ToLeft,dum] = find((XYZ(:,2)<-DomainSize) & (Vxyz(:,2)<0));
         Vxyz(ToLeft,2)=-Vxyz(ToLeft,2);
    if TwoD
        XYZ(:,3) = 0;
    else
        [ToRight,dum] = find((XYZ(:,3)>DomainSize) & (Vxyz(:,3)>0));
               Vxyz(ToRight,3)=-Vxyz(ToRight,3);
        [ToLeft,dum] = find((XYZ(:,3)<-DomainSize) & (Vxyz(:,3)<0));
               Vxyz(ToLeft,3)=-Vxyz(ToLeft,3);
    end % if TwoD

    %% interactions between spheres:
    for c = 1:numballs
        distance = ( (XYZ(:,1)-XYZ(c,1)).^2+(XYZ(:,2)-XYZ(c,2)).^2+ ...
                  (XYZ(:,3)-XYZ(c,3)).^2 );
        overlaps = find(distance<((2*ballradius).^2) & distance~=0);
        distance = sqrt(distance);
        if ~isempty(overlaps)
            DD=2*ballradius-distance(overlaps); %NB: all radii identical
            Dx = ((XYZ(overlaps,1)-XYZ(c,1))./(distance(overlaps))).*DD;
            Dy = ((XYZ(overlaps,2)-XYZ(c,2))./(distance(overlaps))).*DD;
            Dz = ((XYZ(overlaps,3)-XYZ(c,3))./(distance(overlaps))).*DD;
            Do = [Dx Dy Dz];
            FF = Do./mass;
    end % if overlaps
    for jj = 1:3 % can be parallelized, but note that FF can be a matrix
                % if multiple spheres overlap
        T = toward(XYZ(overlaps,jj),XYZ(c,jj), ...
            Vxyz(overlaps,jj),Vxyz(c,jj));
        K1 = find(T==1);
        K2 = find(T==0);
        if ~isempty(K1)
            Vxyz(c,jj) = Vxyz(c,jj)-sum(Kin*(FF(K1,jj))).*dt;
        elseif ~isempty(K2)
            Vxyz(c,jj) = Vxyz(c,jj)-sum(Kout*(FF(K2,jj))).*dt;
```

```
            end % if k12
        end % for jj
    end % for c

    %% Plot output:
    plot3(XYZ(:,1),XYZ(:,2),XYZ(:,3),'.b','MarkerSize',40); %size approx.
    axis([-1 1 -1 1 -1 1].*(DomainSize+ballradius));
    view(2);
    drawnow;
end % for ii
```

(9) Finite differences:

A final piece of code that comes in handy from time to time is the following, which uses finite differences to simulate simple partial differential equations. In this case, the simulation is of the diffusion equation, with an added prescribed fluid advection velocity.

```
%% clear workspace & bring figure to front:
clear all;
figure(gcf);
colormap(jet)

%% set parameters:
width = 128;                    % number of gridpoints in x and y
height = width;
n = [height, 1:(height-1)];
e = [(2:width), 1];
s = [(2:height), 1];
w = [width, 1:(width-1)];
deltax = 5/width;               % distance between gridpoints
deltay= deltax;
PlotEvery = 25;                 % how often to plot
deltat = 0.00025;               % timestep
Dc = 0.01;                      % diffusivity
speed = 5;                      % optional: advection speed
timesteps = 50000;              % number of iterations

%% Initialize concentration:
C(1:height,1:width) = 0;
C(round(height/2)-10:round(height/2)+9, ...
round(width/2)-10:round(width/2)+9) = -1/2;

%% Define advection:
Velx = -1.*speed*sin(2*pi*[1:height]/height) ...+
1.25.*speed*cos(2*pi*[1:height]/height);
Velx = repmat(Velx, [width,1])';
Vely = speed*sin(2*pi*[1:width]/width);
Vely = repmat(Vely, [height,1]);
```

```
%% Evolve:
for tt = 1:timesteps
    % Vel*divergence of C:
    divCx = Velx.*(C(:,e) - C(:,w))./(2*deltax);
    divCy = Vely.*(C(n,:) - C(s,:))./(2*deltay);
    DivC = divCx + divCy;

    % Laplacian of C:
    DCX2 = (C(:,e)-2.*C +C(:,w))./deltax^2;
    DCY2 = (C(n,:)-2.*C +C(s,:))./deltay^2;

    Cdot = Dc.*(DCX2 + DCY2) + DivC;

    % Just Euler integration for simplicity:
    C = C + Cdot.*deltat;

    % Plot:
    if (tt/PlotEvery) == round(tt/PlotEvery)
      surf(C,'Edgecolor','none');
      view([15,64]); axis tight; axis off;
      drawnow
    end
end
```

REFERENCES

1. Taylor, J.R. (1997) *An Introduction to Error Analysis*, 2nd ed. Mill Valley, CA: University Science Books.
2. Press, W.H., Flannery, B.P., Teukolsky, S.A. and Vetterling, W.T. (1986) *Numerical Recipes: The Art of Scientific Computing*. Cambridge: Cambridge University Press.
3. Pöschel, T. and Schwager, T. (2005) *Computational Granular Dynamics: Models and Algorithms*. Berlin: Springer.
4. Braun, O.R. and Walton, R.L. (1986) Viscosity, granular-temperature, and stress calculations for shearing assemblies of inelastic, frictional disks. *Journal of Rheology, 30*, 949–80.

INDEX

A

B

C

Z